Lecture Notes in Electrical Engineering

Volume 68

Sio-Iong Ao · Burghard Rieger ·
Mahyar A. Amouzegar

Machine Learning and Systems Engineering

 Springer

Editors
Dr. Sio-Iong Ao
International Association of Engineers
Hung To Road 37-39
Hong Kong
Unit 1, 1/F
Hong Kong SAR
publication@iaeng.org

Prof. Dr. Burghard Rieger
Universität Trier
FB II Linguistische
Datenverarbeitung
Computerlinguistik
Universitätsring 15
54286 Trier
Germany

Prof. Mahyar A. Amouzegar
College of Engineering
California State University
Long Beach
CA 90840
USA

ISSN 1876-1100   e-ISSN 1876-1119
ISBN 978-94-007-3374-9   ISBN 978-90-481-9419-3 (eBook)
DOI 10.1007/978-90-481-9419-3
Springer Dordrecht Heidelberg London New York

Cover design: SPi Publisher Services

Printed on acid-free paper

Springer is part of Springer Science+Business Media (www.springer.com)

Preface

A large international conference on Advances in Machine Learning and Systems Engineering was held in UC Berkeley, California, USA, October 20–22, 2009, under the auspices of the World Congress on Engineering and Computer Science (WCECS 2009). The WCECS is organized by the International Association of Engineers (IAENG). IAENG is a non-profit international association for the engineers and the computer scientists, which was founded in 1968 and has been undergoing rapid expansions in recent years. The WCECS conferences have served as excellent venues for the engineering community to meet with each other and to exchange ideas. Moreover, WCECS continues to strike a balance between theoretical and application development. The conference committees have been formed with over two hundred members who are mainly research center heads, deans, department heads (chairs), professors, and research scientists from over thirty countries with the full committee list available at our congress web site (http://www.iaeng.org/WCECS2009/committee.html). The conference participants are truly international representing high level research and development from many countries. The responses for the congress have been excellent. In 2009, we received more than six hundred manuscripts, and after a thorough peer review process 54.69% of the papers were accepted.

This volume contains 46 revised and extended research articles written by prominent researchers participating in the conference. Topics covered include Expert system, Intelligent decision making, Knowledge-based systems, Knowledge extraction, Data analysis tools, Computational biology, Optimization algorithms, Experiment designs, Complex system identification, Computational modeling, and industrial applications. The book offers the state of the art of tremendous advances in machine learning and systems engineering and also serves as an excellent reference text for researchers and graduate students, working on machine learning and systems engineering.

Sio-Iong Ao
Burghard B. Rieger
Mahyar A. Amouzegar

Contents

Chapter 1
Multimodal Human Spacecraft Interaction in Remote Environments

A New Concept for Free Flyer Control

Enrico Stoll, Alvar Saenz-Otero, and Brent Tweddle

Abstract Most malfunctioning spacecraft require only a minor maintenance operation, but have to be retired due to the lack of so-called On-Orbit Servicing (OOS) opportunities. There is no maintenance and repair infrastructure for space systems. Occasionally, space shuttle based servicing missions are launched, but there are no routine procedures foreseen for the individual spacecraft.

The unmanned approach is to utilize the explorative possibilities of robots to dock a servicer spacecraft onto a malfunctioning target spacecraft and execute complex OOS operations, controlled from ground. Most OOS demonstration missions aim at equipping the servicing spacecraft with a high degree of autonomy. However, not all spacecraft can be serviced autonomously. Equipping the human operator on ground with the possibility of instantaneous interaction with the servicer satellite is a very beneficial capability that complements autonomous operations.

This work focuses on such teleoperated space systems with a strong emphasis on multimodal feedback, i.e. human spacecraft interaction is considered, which utilizes multiple human senses through which the operator can receive output from a technical device. This work proposes a new concept for free flyer control and shows the development of an according test environment.

1 Introduction

On-Orbit Servicing (OOS) has been an active research area in recent times. Two approaches have been studied: teleoperation by humans and autonomous systems. Autonomous systems use machine pattern recognition, object tracking, and

E. Stoll (✉)
Space Systems Laboratory, Massachusetts Institute of Technology, 77 Massachusetts Avenue, Cambridge, MA 02139-4307, USA
e-mail: estoll@MIT.edu

S.-I. Ao et al. (eds.), *Machine Learning and Systems Engineering*,
Lecture Notes in Electrical Engineering 68,
DOI 10.1007/978-90-481-9419-3_1, © Springer Science+Business Media B.V. 2010

acquisition algorithms, as for example DART [1] or Orbital Express [2]. The research is still in early stages and the algorithms have to be realized in complex systems.

In contrast, the human eye-brain combination is already very evolved and trainable. Procedures can be executed by the trained user from the ground. Unforeseen incidents can be solved with greater flexibility and robustness. Arbitrary spacecraft could be approached, i.e. spacecraft which were not explicitly designed for rendezvous and docking maneuvers. Analogously, inspections and fly-arounds can be controlled by the human operator. Based on the acquired information the human operator on ground can decide how to proceed and which servicing measures to take. Another element in the decision queue is the path planning approach for the target satellite to the capture object.

Multimodal telepresence, which combines autonomous operations with human oversight of the mission (with the ability to control the satellites), provides the benefits of autonomous free-flyers with the evolved human experience. In case autonomous operations cause the work area to exhibit an unknown and unforeseen state (e.g. when robotically exchanging or upgrading instruments) the human operator on ground can support the operations by either finishing the procedure or returning the system into a state which can be processed by autonomous procedures. The advantage of multimodal telepresence in this connection is the fact that the operator will not only see the remote site, but also feel it due to haptic displays. A haptic interface presents feedback to the human operator via the sense of touch by applying forces, vibrations or motion.

The applicability of the telepresence approach, with a human operator located in a ground station, controlling a spacecraft, is mostly limited to the Earth orbit. This is because the round trip delay increases with increasing distance from operator to the teleoperator. A decrease of the telepresence feeling is the consequence, which has a large impact on the task performance. Therefore, as the distance increases, the role of the autonomy must increase to maintain effective operations.

For an overall and significant evaluation of the benefits of multimodal telepresence a representative test environment is being developed at the MIT Space Systems Laboratory using the SPHERES satellites on ground and aboard the International Space Station (ISS).

2   The MIT SPHERES Program

The SPHERES laboratory for Distributed Satellite Systems [3] consists of a set of tools and hardware developed for use aboard the ISS and in ground based tests. Three micro-satellites, a custom metrology system (based on ultrasound time-of-flight measurements), communications hardware, consumables (tanks and batteries), and an astronaut interface are aboard the ISS. Figure 1 shows the three SPHERES satellites being operated aboard the ISS during the summer of 2008.

ISS017E017375

Fig. 1 SPHERES operations aboard the International Space Station (Picture: NASA)

The satellites operate autonomously, after the crew starts the test, within the US Destiny Laboratory.

The ground-based setup consists of an analog set of hardware: three micro-satellites, a metrology system with the same geometry as that on the ISS, a research oriented GUI, and replenishable consumables. A "guest scientist program" [4] provides documentation and programming interfaces which allow multiple researchers to use the facility.

2.1   General Information

The SPHERES satellites were designed to provide the best traceability to future formation flight missions by implementing all the features of a standard thruster-based *satellite bus*. The satellites have fully functional propulsion, guidance, communications, and power sub-systems. These enable the satellites to: maneuver in 6-DoF, communicate with each other and with the laptop control station, and identify their position with respect to each other and to the experiment reference frame. The computer architecture allows scientists to re-program the satellite with new algorithms. The laptop control station (an ISS supplied standard laptop) is used to collect and store data and to upload new algorithms. It uses the ISS network for all ground data communications (downlink and uplink). Figure 2 shows a picture of an assembled SPHERES satellite and identifies its main features. Physical properties of the satellites are listed in Table 1.

Fig. 2 SPHERES satellite

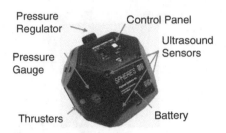

Pressure Regulator

Control Panel

Ultrasound Sensors

Pressure Gauge

Thrusters

Battery

Table 1 SPHERES satellite properties

Property	Value
Diameter	0.22 m
Mass (with tank and batteries)	4.3 kg
Max linear acceleration	0.17 m/s^2
Max angular acceleration	3.5 rad/s^2
Power consumption	13 W
Battery lifetime	2 h

SPHERES has been in operation aboard the ISS since May 2006. To date, 21 test sessions have taken place. The test sessions have included research on Formation Flight, Docking and Rendezvous, Fluid Slosh, Fault Detection, Isolation, and Recover (FDIR), and general distributed satellite systems autonomy.

2.2 Human-SPHERES Interaction

Most of the previous test sessions matured autonomous algorithms. However, future servicing missions and the assembly of complex space structures will not only depend on increased autonomy, but the ability of humans to provide high-level oversight and task scheduling will always be critical. SPHERES tests were conducted to develop and advance algorithms for adjustable autonomy and human system interaction. This research began with basic tests during Test Session 11, where the crew was asked to move a satellite to multiple corners in a pre-defined volume. The satellite autonomously prevented collisions with the walls of the ISS. The test demonstrated the ability of the crew to use the ISS laptop to control SPHERES. It provided baseline results for future tests. An ongoing sequence of ISS tests is being conducted in the framework of a program called "SPHERES Interact". The goal of the program is to conceive new algorithms that utilize both human interaction and machine autonomy to complete complex tasks in 6 degrees of freedom (DoF) environments. Tests during Test Session 19 and 20 included several scenarios where human interaction helps schedule tasks of a complex mission (e.g. servicing or assembly). The research area comprises human orientation, navigation, and recognition of motion patterns. Further, high level human abort commands and collision avoidance techniques for part of this ongoing research aboard the International Space Station.

2.3   SPHERES Goggles

The SPHERES Goggles is a hardware upgrade to the SPHERES satellites that adds cameras, lights, additional processing power and a high speed wireless communications system. Even though it was designed for autonomous operations, it can be used to support the operator with a visual feedback. The main objective of the SPHERES Goggles is to provide a flight-traceable platform for the development, testing and maturation of computer vision-based navigation algorithms for spacecraft proximity operations. Although this hardware was not intended to be launched to orbit, it was designed to be easily extensible to versions that can operate both inside and ultimately outside the ISS or any other spacecraft.

The Goggles, which are shown in Fig. 3, were designed to be able to image objects that are within few meters range and to possess the computational capability to process the captured images. They further provide a flexible software development environment and the ability to reconfigure the optics hardware.

The SPHERES Goggles were used in several parts of the telepresence environment setup at the MIT SSL ground facilities to support the human operator with a realistic video feedback which is representative for a camera system used on orbit. Apart from virtual reality animations of the remote environment it serves as the only source of visual data in the experiments.

Fig. 3 Front view of Goggles mounted on SPHERES satellite

3 Multimodal Telepresence

Servicing missions can be differentiated by whether or not a robotic manipulator is connected to a free flying base (the actual satellite). Different levels of autonomy can be applied to the control of either and the human operator receives the according feedback.

3.1 Areas of Application

Unlike robotic manipulators, where haptic feedback plays an important role for control as e.g. ETS-VII [5] or Rokviss [6], free flyers are commonly only steered using visual feedback. That means that even though free flying experiments can be steered with hand controllers, as for example Scamp [7] or the Mini AERCam [8], usually no haptic information is fed back to the human operator.

The implementation of haptic feedback into the control of free flyers enriches the telepresence feeling of the operator and helps the operator on ground to navigate. It paves the way for new concepts of telepresent spacecraft control. Collision avoidance maneuvers for example can be made perceptible for the human operator, by placing virtual walls around other spacecraft. Equipping these virtual walls with sufficient high stiffness means that the operator is not able to penetrate them by means of the haptic device, since it exerts to the operator a high resistance force.

Areas of fuel optimal paths can be displayed to the operator by implementing an ambient damping force, featuring a magnitude which is proportional to the deviation of the actual path from the fuel optimal trajectory and area, respectively. Docking maneuvers can be supported by virtual boundaries as a haptic guiding cone and damping forces which are increasing with decreasing distance to the target.

Summarizing the benefits it can be seen that the application of telepresence control will extend the amount of serviceable spacecraft failures by involving a well trained human operator. In this connection it is proposed that the task performance of the operator can be enhanced by feeding back high-fidelity information from the remote work environment. Here the haptic feedback plays an important role in human perception and will be tested in a representative test environment.

3.2 The Development of a Test Environment

The key element of the test environment is the Novint Falcon [9], which is a 3-DoF force feedback joystick. All degrees of freedom are of translational nature and servo motors are used to feed forces in 3-DoF back to the user. This system has high utility for space applications since it allows the human operator to control the space application in 3D dimensional space. The Falcon is implemented in a Matlab/ SIMULINK environment via the HaptikLibrary [10], which is a component based architecture for uniform access to haptic devices. It is used as the interface to Matlab and reads the positions and button states of the haptic device as well as feeds calculated forces back to it. By displacing the joystick handle, the human operator is able to interact with two instances of the remote environment - the virtual instance in SIMULINK and the hardware (SPHERES) instance on the SSL air table.

The joystick displacement is interpreted by the system, as either position, velocity or force commands. The received commands are communicated to a SIMULINK block, containing the satellite dynamics and a state estimator. The simulation returns the estimated state of the satellite in the virtual entity of the remote environment. This remote workspace is created using SIMULINK's Virtual Reality (VR) toolbox (cp. Fig. 4), allowing for satellite states and environmental properties to be displayed.

In addition to the Matlab environment, algorithms in C are used as the interface to the actual SPHERES hardware via the "SPHERES Core" API. Commands are transmitted via wireless communications to the SPHERES satellites. Torques and forces are calculated and directly commanded to the thrusters, which will cause a motion of the SPHERES satellite. The satellites measure their position and attitude and transmit the information in real-time to the laptop.

By transmitting the actual states to the VR, the operator obtains information of the estimated and the actual motion of the free flyer, which should be identical if the communication channel is not delayed and the virtual instance is a good

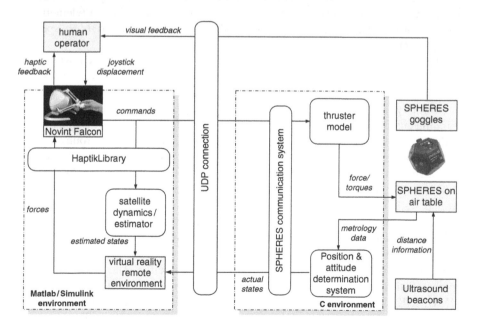

Fig. 4 General block diagram of the test environment

approximation of reality. In the presence of time delay the predictions should give the user a feeling for the behaviour of the system (cp. ETS-VII) and enhance the task performance. This is an important point if a human operator on ground will steer an application in space. That way the interactions between the autonomous space operations and a telepresence controlled free flyer can be tested.

4 Experimental Setup

If OOS missions are executed in low Earth orbit (LEO) only limited time windows are available for telecommands. The common approach for increasing those acquisition times is the usage of geostationary relay satellites. While those satellites do not have a profound impact on autonomous missions, they will influence the task performance of an operator on ground directly interacting with a satellite. Thus, this human spacecraft interaction was tested using a representative test scenario at SSL, which involved a geostationary relay satellite.

4.1   Control via ARTEMIS

Due to the orbit height of geostationary satellites, the relay of the signal increases the round trip delay between operator action and feedback to the operator to up to 7 s as in the case of ETS-VII. The delay between telecommand and telemetry is usually not very intuitively manageable for the human operator and thus a special area of interest if human spacecraft interaction is considered.

The effect on the human has already been shown for OOS missions in which the operator on ground steers robotic manipulators via geostationary relay satellites [11]. It has not been tested, yet, for multimodal human free flyer interaction. Accordingly, for initial tests a geostationary satellite was introduced in the commanding chain. The UDP connection (cp. Fig. 4) was utilized to send the commands of the Novint Falcon at SSL via a terrestrial internet connection to a ground station at the Institute of Astronautics of Technische Universitaet Muenchen in Germany. The telecommands were forwarded via the geostationary relay satellite ARTEMIS (Advanced Relay Technology Mission) of the European Space Agency (ESA) to a ground station of ESA in Redu, Belgium. The signal was mirrored in Redu and retransmitted analogously back to MIT, where again the UDP connection was used to feed the telecommand into the hardware on the air table and change the position of SPHERES in the test environment. That way the SPHERES satellites were controlled by the Novint Falcon via a geostationary satellite. The round trip delay characteristics were logged and subsequently implemented into the scenario as a SIMULINK block. That way the test scenarios could be evaluated in the absence of a satellite link but with round trip delays representative for commanding a spacecraft in orbit.

4.2   The Servicing Scenarios

To show the benefit of multimodal feedback to the operator, two scenarios were developed and tested. Both are based on a servicing operation, in which three satellites are involved. The *target satellite* is the satellite to be serviced. Therefore, the *servicer satellite* has to execute proximity operations, approach the target, and eventually dock with it. The *inspector satellite* is supporting the operator on ground with additional data of the remote environment. It carries a camera system and can yield information on the distance between the two other satellites.

4.2.1   The Human-Controlled Inspector Satellite

In this first scenario the control of the inspector satellite is handed over to the human operator, while the servicer and the target dock autonomously. The task of the operator is to ensure that the initial states of the other two satellites are appropriate for the docking maneuver. Thus, the operator commands the inspector satellite as depicted in Fig. 5 from its position in front of the other two satellites to a position behind the two satellites, which is indicated by a virtual checkered marker.

For efficiently accomplishing this circumnavigation, virtual obstacles were created to avoid collisions with the servicer, the target, and the borders of the experimental volume. As to be seen in Fig. 5 both of the satellites to dock feature a virtual collision avoidance sphere. Further, on the left and the right side of the volume, there are virtual (brick) walls introduced. The Novint Falcon generates forces in case the operator penetrates those objects. These resistance forces are fed back to the operator and thus prevent from colliding with the actual hardware on the SSL air table.

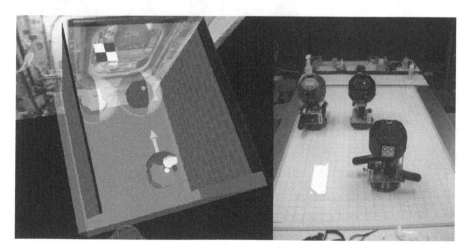

Fig. 5 Virtual and hardware instance of the inspection scenario

A further benefit of using the virtual reality techniques is that the environment can be augmented with additional data of the remote environment. For example, arrows can be used for indicating the current velocity and rotation rate (double arrow) of the inspector. Furthermore, there are two entities of the inspector satellite to be seen in the VR environment. The dark entity shows the commanded state, whereas the pale entity shows the actual state of the hardware in the remote environment. This is of great benefit for the human operator in the presence of time delays as they occur due to the use of relay satellites.

4.2.2 The Human-Controlled Servicer Satellite

Similar to the first scenario, the inspector, target, and servicer satellite are again involved in the second scenario. The servicer is supposed to dock with the target, whereas the inspector is transmitting additional data from the remote scene. In this scenario the target and the inspector (right upper corner in Fig. 6) are operating autonomously and the servicer satellite (lower right corner) is controlled by the human operator via the relay satellite.

Again, the virtual environment is enriched by collision avoidance objects (at the inspector and the borders of the volume). The task of the operator is to accomplish a successful docking maneuver. Therefore, the human operator is supposed to command the servicer at first to a position roughly aligned with centre of the docking cone, which can be seen in Fig. 6 and approx. 50 cm away from the target. In a second step the operator is commanding the servicer along the virtual cone until the berthing takes place.

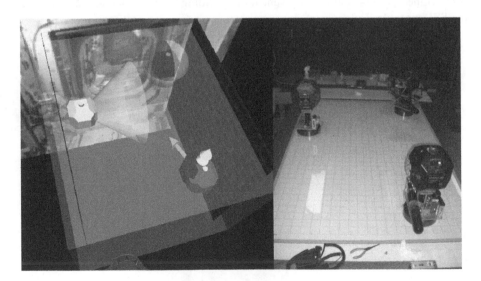

Fig. 6 Virtual and hardware instance of the docking scenario

The docking cone is a mean to simplify the proximity operations for the operator. Once the servicer has crossed the assistance horizon of the cone, a force field is applied to the Falcon, which drives the servicer into the docking cone. Inside the docking cone another force field drives the servicer towards the target. Here, the forces are proportional to the distance to the target. This helps the operator to concentrate on the precision of the docking point rather than to worry about relative velocities and collisions.

5   Results of the Experiments

The two scenarios were controlled via the German ground station [12] at the Institute of Astronautics and the ESA relay satellite. The human operator at MIT in Cambridge received instantaneous feedback from the haptic-visual workspace. To have a representative test conditions the operator had only visual feedback from the SPHERES Goggles and the Matlab Simulink virtual instance of the remote environment. Further, the haptic device yielded additional forces for an advanced human spacecraft interaction in the 3D environment.

5.1   Round Trip Delays due to the Relay Satellite

The occurring round trip delays were logged since they are a first indicator for the quality of the human task performance. Figure 7 shows an example graph of the delay characteristics over time. The round trip delays are plotted depending on the respective UDP packet number. They indicate that the delay in a real OOS mission can be, except for a couple of outliers, well below 1 s. The outliers occurred due to the use of a terrestrial internet connection and the lack of synchronization between

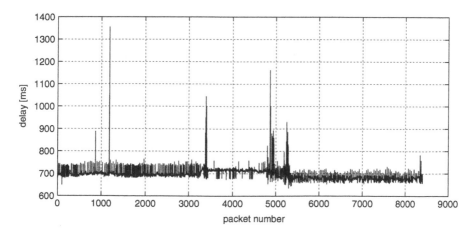

Fig. 7 Round trip delay characteristic of the free flyer control via ARTEMIS

the sampling rate of the hardware at MIT and the sampling rate of the satellite modem at LRT. Nonetheless, a mean of 695.5 ms with a sample standard deviation of 24.1 ms indicate an acceptable round trip delay [13] for telepresence operations.

5.2   Operator Force Feedback

Navigation in a 3D environment with a sparse number of reference points can be very complicated for a human operator. The motion with 6-DoF is not only very unintuitive since the motion in free space is no longer superimposed by gravity as it is on Earth the case. The equations of motions are further coupled in a way that an introduced torque about a main axis of inertia of the spacecraft will not necessarily cause the spacecraft to rotate about the respective axis but about all three axes.

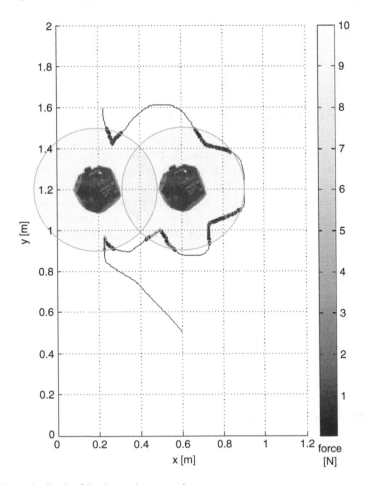

Fig. 8 Force feedback of the inspection scenario

Thus, the human operator has to be supported by technical means in order to solve complex problems in the remote environment. One of those means is, as shown in this work, to augment the feedback to the operator. Virtual reality can be used to show commanded/planned states versus actual states of spacecraft and can additionally visualize potential dangerous areas.

Since the 3D remote environment is usually projected onto 2D screens, it can be difficult for the operator to realize where exactly such an area, in which collisions could take place, is located. Consequently, a haptic device was used which utilizes another human sense and enriches the perception. Forces are fed back to the operator site, permitting the operator to enter the respective areas.

Figures 8 and 9 show example forces that were fed back to the operator depending on the position of the spacecraft in the remote environment. The path

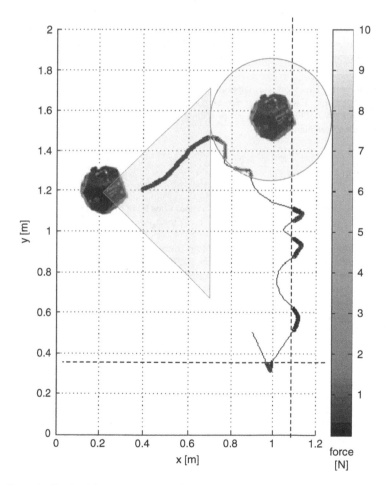

Fig. 9 Force feedback of the docking scenario

of the spacecraft is indicated by a solid line, whereas the force feedback is labeled by small circles in gray scale. If a collision avoidance sphere was penetrated as e.g. in Fig. 8 a restraining force was created proportional to the penetration depth and the velocity (spring-damper system) of the spacecraft. The same held true for virtual walls as can be seen in Fig. 9. This figure further shows the force feedback inside the docking cone. As can be seen, the haptic feedback prevented the human operator form colliding with the other spacecraft or the experimental boundaries. It gave the operator a feeling for critical areas and helped the operator to accomplish a very smooth docking/berthing approach.

6 Summary

This work presented the first tests on haptic feedback for free flyer systems. It proposes that multimodal feedback from servicer satellites enhances the human task performance. This feedback supports the operator with an intuitive concept for collision avoidance and relative navigation. That way, complex tasks in microgravity can be safely operated from ground.

Acknowledgements This work was supported in part by a post-doctoral fellowship program of the German Academic Exchange Service (DAAD). The authors would like to express their gratitude to the ESA ARTEMIS team for providing the opportunity to use the ARTEMIS relay satellite for their experiments.

References

1. T. Rumford, Demonstration of autonomous rendezvous technology (dart) project summary, in *Proceedings of Space Systems Technology and Operations Conference*, Orlando, USA, Apr 2003
2. T. Weismuller, M. Leinz, GN&C technology demonstrated by the orbital express autonomous rendezvous and capture sensor system. 29th AAS Guidance and Control Conference, Breckenridge, USA, 2006
3. A. Saenz-Otero, A. Chen et al., SPHERES: Development of an ISS laboratory for formation flight and docking research, *IEEE Aerospace Conference* (paper #81), Big Sky, Montana, USA, 9–16 Mar 2002
4. J. Enright, M.O. Hilstad et al., The SPHERES guest scientist program: collaborative science on the ISS, *2004 IEEE Aerospace Conference* (paper #1296), Big Sky, Montana, USA, 7–12 Mar 2004
5. T. Imaida, Y. Yokokohji, T. Doi, M. Oda, T. Yoshikawa, Ground-space bilateral teleoperation ex-periment using ETS-VII robot arm with direct kinesthetic coupling, in *Proceedings of IEEE International Conference on Robotics and Automation*, Seoul, Korea, 2001
6. K. Landzettel et al., ROKVISS verification of advanced light weight robotic joints and telepresence concepts for future space missions, in *Proceedings of 9th ESA Workshop on Advanced Space Technologies for Robotics and Automation* (ASTRA), Noordwijk, The Netherlands, Nov 2002
7. C. McGhan, R. Besser, R. Sanner, E. Atkins, Semi-autonomous inspection with a neutral buoyancy free-flyer, in *Proceedings of Guidance, Navigation, and Control Conference*, Keystone, USA, Aug 2006

8. S. Fredrickson, S. Duran, J. Mitchel, Mini AERCam inspection robot for human space missions, in *Proceedings of AIAA Space*, San Diego, USA, Sept 2004
9. Novint Technologies Inc. (February 2010). http://www.novint.com
10. M. de Pascale, D. Prattichizzo, The Haptik Library: a component based architecture for uniform access to haptic devices, IEEE Robotics Autom. Mag. **14**(4), 64–75 (2007)
11. E. Stoll, Ground verification of telepresence for on-orbit servicing. Dissertation, Lehrstuhl für Raumfahrttechnik, Technische Universität München, 2008, ISBN 978-3-89963-919-3
12. J. Letschnik, E. Stoll, U. Walter, Test environment for time delay measurements of space links via ARTEMIS, in *Proceedings of 4th ESA International Workshop on Tracking, Telemetry and Command Systems for Space Applications TTC 2007*, Darmstadt, Germany, 2007
13. E. Stoll et al., Ground verification of the feasibility of telepresent on-orbit servicing. J. Field Robotics **26**(3), 287–307 (2009)

Chapter 2
A Framework for Collaborative Aspects of Intelligent Service Robot

Joohee Suh and Chong-woo Woo

Abstract Intelligent service robot is becoming one of the most interesting issues in the recent Robot research. The service robot monitors its surroundings, and provides a service to meet a user's goal. The service often becomes too complex that one single robot may not handle efficiently. In other words, a group of robots may be needed to accomplish given task(s) by collaborating each other. We can define this activity as a robot grouping, and we need to study further to make better group(s) by considering their characteristics of the each robot. But, it is difficult and no formal methods to make such a specific group from the many heterogeneous robots that are different in their functions and structures. This paper describes an intelligent service robot framework that outlines a multi-layer structure, which is suitable to make a particular group of robots to solve given task by collaborating with other robots. Simulated experimentation for grouping from the generated several heterogeneous is done by utilizing *Entropy* algorithm. And the collaboration among the robots is done by the multi-level task planning mechanism.

1 Introduction

Ubiquitous computing [1] defines that various computing objects are connected through the network, so that the system automatically provides services in anytime in any place. An intelligent robot is an autonomous and dynamic object in this ubiquitous environment, and it becomes one of the most interesting issues in this area. It interacts with various surrounding computing devices, recognizes context,

J. Suh (✉)
Korea School of Computer Science, Kookmin University, 861-1 Jeongneung-Dong, Seongbuk-Gu, Seoul
e-mail: crazyDMP@gmail.com

S.-I. Ao et al. (eds.), *Machine Learning and Systems Engineering*,
Lecture Notes in Electrical Engineering 68,
DOI 10.1007/978-90-481-9419-3_2, © Springer Science+Business Media B.V. 2010

and provides appropriate services to human and the environment [2]. The definition of the context [3] is any entity that affects interactions among the computing objects in this environment. For instance, *user*, *physical location*, *robots* could be such entities. Therefore, information described for the characteristics of such entities are defined as a context, and recognizing a situation from the context is the main focus of the context awareness system [4]. Essentially, the robot is required to carry on two main issues in this study; understanding the context and carrying out the context.

First, understanding a context can be done in many different ways, but the general procedure includes that system perceives raw environmental information through the physical sensors. And then context modeling and reasoning steps follows right after preprocessing the raw data. The final result from this procedure is a context. The approaches on this procedure are vary, and it needs to be further studied in detail to make the system more efficient, and there are large amount of studies [2–4] are being reported recently. The second issue is that the robot has to carry out the generated context to meet the user's goal. The context can be further divided into some tasks, which often becomes too complex that one single robot may not handle in this environment. In other words, a group of robots may be needed to accomplish given task(s) by collaborating each other. We can define this activity as a robot grouping, and we need to study further to make better group(s) by considering their characteristics of the robot and the task.

In this paper, we are describing a development of the social robot framework for providing an intelligent service with the collaboration of the heterogeneous robots, in the context awareness environment. The framework is designed as multi-layers suitable for understanding context, grouping robots, and collaborating among the robots. In this study, we mainly focused and implemented on the grouping and collaborating parts of the system. The context understanding part of the system is designed, but will be discussed in the next study with comprehensive ontological knowledge representations along with context modeling and context reasoning mechanisms.

2   Related Works

2.1   Context-Awareness Systems

With the increase of mobile computing devices, the ubiquitous and pervasive computing is getting popular recently. One part of the pervasive system is the context awareness system, which is being studied explosively in various directions. Among the many research results on this issue, the most prominent systems are the CoBrA [5] (Context Broker Architecture), SOCAM [6] (Service Oriented Context-Aware Middleware), and Context-Toolkit [7]. The CoBrA and SOCAM used ontological model aiming for the benefits of easy sharing, reusing, and reasoning of context information. But, first, these systems were not flexible enough to extend to other domains when the system needs to expand with other devices or service with a new domain. And second, they were also limited in formalized and shared expression of the context, which is needed when the system interoperate or transplant with other systems. Therefore, the ontology becomes one of the most popular solutions to represent the data for the recent context awareness systems. Short reviews of the previous systems as follows.

Context-Toolkit: This early context awareness middleware system gains information from the connected devices. But, since it does not use ontology, it lacks of the standardized representation for the context, and also the interoperability between heterogeneous systems.

CoBrA: This system is developed based on the ontology, so that the standardized representation for the context is possible. But, since the use of ontology is limited only to a special domain, so called 'Intelligent Meeting Room', it does not guarantee any extensibility to the other diverse domains.

SOCAM: This system is based on Service-oriented structure, which is efficient middleware system for finding, acquiring, analyzing context information. But, since it depends on OWL (Web Ontology Language) for reasoning, its reasoning capability is limited to its own learning module and inference engine.

In our study, we adapted merits of the previously studied systems, and designed a framework that can overcome the above limitations, such as, the limited standardized representation or extensibility. For instance, we have adapted context-awareness layer by adapting the CONCON model [8] to provide extensibility of the ontological representation.

2.2   Robot Grouping and Collaboration

First of all, our research issue focused on the collaboration among the robots, which needs to form a group. Therefore, we first need to develop a method of grouping robots for a given task. Study on the robot grouping is just beginning, but some related researches are being reported as follows.

For instance, Rodic and Engelbrecht [9] studied initial investigation into feasibility of using 'social network' as a coordination tool for multi-robot teams. Under the assumption of multi-robot teams can accomplish certain task faster than a single robot, they proposed multi-robot coordination techniques. Inspired by the concept from the animal colony, Labella et al. [10], showed simple adaptation of an individual can lead task allocation. They developed several small and independent modules, called 's-bots', and the collaboration in this system is achieved by means of communication among them. They claimed that individuals that are mechanically better for retrieval are more likely to be selected. Another point of view for the collaboration is task allocation among the multi-robot colony. Mataric et al. [11] had an experimentation comparing between a simulated data with physical mobile robot experiment. The result showed that there is no single strategy that produces best performance in all cases. And other approaches are the multi-robot task allocation by planning algorithm [12–14].

All of these research efforts are being done in many different fields, and the selection of individual is rather random or simple, which may often result in inadequacy of performing a given task. Using 'entropy' of information theory [15] could be a good alternative compare to the other informal approaches. Goodrich argues that the behavioral entropy can predict human workload or measure of human performance in human robot interaction [16] (HRI) domain. Balch [17] demonstrated successfully in his experimental evaluation of multi-robot soccer and multi-robot foraging teams. In our study, we will use the 'entropy' metric for selecting an appropriate robot from the robot colonies by generating decision tree first, to minimize the complexities of adapting the entropy.

3   Design of the System

As in the Fig. 1, the overall architecture of our system is divided into three main layers, context-awareness layer, grouping layer, and collaboration layer. Also, we have two more sub-layers, Physical and Network layer, which will not be discussed here, since they are not the main issue here. The overall process of the three main layers works as follows, and entire structure of the system can be viewed in the next Fig. 2.

- The context-awareness layer generates a context from the raw information, and then does modeling and reasoning in order to aware of the context.
- The grouping layer creates decision classifier based on the entropy mechanism, and makes a necessary group.

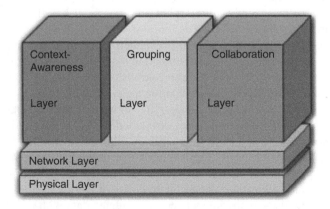

Fig. 1 General structure of social robot

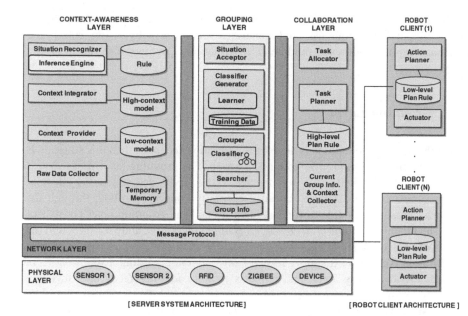

Fig. 2 Structure of the service robot in context awareness environment

- The collaboration layer does multi-level task planning; high-level task planning generates a set of tasks for the context, the low-level task planning generates a set of actions for the task.

3.1   Context-Awareness Layer

The context awareness layer receives raw data from surrounding computing devices including RFID, Zigbee, and so on. Then it transforms the raw data into a meaningful semantic data by going through some preprocessing steps, and finally it generates a situation. This work can be done from the next sub modules as follows.

Raw Data Collector: It simply receives raw data from the Physical layer passes it to the *context provider*.

Context Provider: It receives raw data from the *raw data collector* and transforms the data into *standardized context* as a preprocessing step according to low context model. The low context model means that the raw data is formalized but it is not semantic data.

Context Integrator: It receives standardized context from the *context provider* and generates inference level context through the high-level context modeling. The high level model supports converting the formalized context into a semantic context.

Situation Recognizer: The context awareness process goes through the above modules in sequence, and generates *situation(s)* by using rule-based inference engine in this sub module. This situation is delivered to the *grouping layer*.

3.2 Grouping Layer

When the situation generation is done by the situation acceptor, then this situation is delivered to the *grouping layer* to make a group. The grouping layer first receives information about which robot is connected to the server, and stores this information into *Group info* database. Because currently connected robots through the network can be the candidates for making a group. The grouping layer consists of three sub modules, *Situation Acceptor*, *Classifier Generator*, and *Grouper*, and the details of their work are as follows.

Situation Acceptor: It simply receives information regarding a *situation*, from the Context awareness layer, and requests to the *Classifier Generator* to begin grouping for this given situation.

Classifier Generator: It generates a *classifier* (i.e. decision tree) to make a group for a given specific situation. And also we need to have a set of predefined training data representing some characteristics of various kinds of robots. In this study, we generated the *classifier* based on ID3 decision tree algorithm.

Grouper: The Grouper has the next two sub-modules; the *searcher* requests instance information from the connected each robot through the network layer. In other words, after the request for the instance information is acquired, such as 'cleaning', the grouper makes a group by using the *classifier* that generated from the *Classifier Generator* module. The generated group information is stored in the *group info* repository, and will be used for collaboration in the collaboration layer later on.

3.2.1 Context Information for the Robot

For this experiment, we can set up several virtual situations, such as 'cleaning' situation, 'delivery' situation, 'conversation' situation, and so on. The Grouping layer receives one from the situations, and start making a group that is appropriate for the service.

The Fig. 3 is the user interface for entering robot attributes and a situation. From this interface, we can enter five attributes interactively, such as 'power', 'location', 'speed', 'possession', 'IL', and a situation, such as 'cleaning'. By using this interface, we can create robot instances as many as we want arbitrarily and they represent heterogeneous robots. And also we can set up a situation by just selecting from the window. This means that each robot has different characteristics and good for a certain work. For instance, we can set up a robot instance

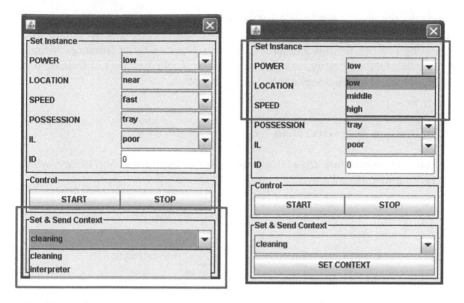

Fig. 3 User Interface for entering robot attributes and situation

Table 1 Training data to create a decision tree

Index	Power	Location	Speed	Possession	IL	Cleaning
1	Low	Near	Fast	Tray	Poor	Yes
2	Low	Far	Low	Gripper	Normal	Yes
3	Middle	Near	Fast	Interpreter	Smart	No
4	High	Near	Far	Tray	Normal	No
5	Low	Far	Normal	Tray	Smart	Yes
6	Low	Near	Normal	Interpreter	Normal	Yes
7	Middle	Far	Normal	Interpreter	Normal	No
8	High	Near	Slow	Gripper	Smart	Yes
9	Middle	Near	Far	Interpreter	Smart	No
10	High	Far	Slow	Gripper	Poor	Yes

good for a 'cleaning' job as follows. If the robot has low power, location is near, speed is low, possesses a tray, and so on, then we can consider this robot is good for a cleaning situation. Similarly, we can create several robot instances for our experimentation.

3.2.2 Training Data

Each robot's attributes are described as in the Table 1, which shows that each robot has different characteristics. For instance, 'power, and speed' means the current robot's basic characteristics, 'location' is the robot's location to perform the given context, 'possession' means a tool that robot can handle, and 'IL' means the robot's capability of language interpretation. Since we are not using sensed data from computing devices for this simulation, we can generate robot instances through the user interface arbitrarily, as many as possible. Table 1 is a set of training data that shows ten robot instances for '*cleaning*' situation.

3.2.3 Decision Tree Generation

The following equation is the entropy algorithm of the information theory, which will be used to generate a tree.

$$Entropy(S) = -_{P(+)} \log_2 P(+) -_{P(-)} \log_2 P(-) \tag{1}$$

$$Gain(S,A) = Entropy(S) - \sum_{v \in Values(A)} \frac{|S_v|}{S} Entropy(S)_v \tag{2}$$

We can compute the general entropy using the Eq. (1), and compute the gain entropy to select a single attribute using Eq. (2).

3.3 Collaboration Layer

It consists of following three sub-components on the server side, and action planner is on the client (robot) side. The distinctive features of them are as follows.

Group Info. and Context Collector: It collects information of selected situation and information for grouped robots from the *Grouping Layer*.

Task Planner: It generates a set of tasks using high-level planning rules (a global plan) on the server side. For instance, the generated tasks for "cleaning" situation can be the "sweeping" and "mopping".

Task Allocator: The task allocator sends the generated tasks to appropriate robots.

Action Planner: Generated tasks by the *task planner* are delivered to the client (robot), and further refined intro a set of actions by the action planner.

3.3.1 Multi-level Task Planning

In collaboration layer, the task is carried out by the multi-level task planning mechanism as follows (see Fig. 4).

- The task planner gets the situation(s), and generates a set of tasks based on the high-level planning rules.

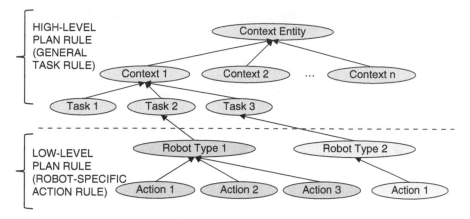

Fig. 4 Multi-level task planning mechanism

- Then the system allocates the tasks to the appropriate robot who can handle the specific task.
- When the task allocation is done, then each assigned robot activates the action planner to generate a set of actions.

4   Simulated Experimentation

The overall experimentation is divided into two parts, robot grouping and robot collaboration. Robot grouping is done by generating classifier using the Entropy metric, and the collaboration is done by task planning algorithm.

4.1   Robot Grouping

In this study, the robot grouping simulation experiment begins with generating virtual robot instances and a situation through the user interface as in the Fig. 3. We can set up characteristics of each robot by selecting five attributes and also can set up a virtual situation through the interface. When all the selection is done, then we can send the information to the server using the start/stop button in the interface.

Figure 5 shows the snapshot of the implemented simulation result for grouping. We designed the implementation result as six sub windows, and the function of each window is explained as follows.

- 'Context' window: It shows the selected virtual context, such as, 'cleaning'.
- 'Training Data' window: It shows ten training data for the selected situation.

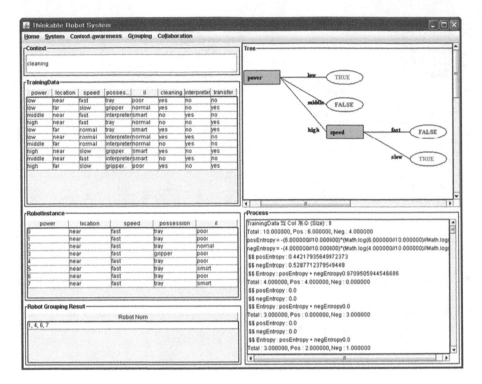

Fig. 5 Snapshot of simulated robot grouping

- 'Robot Instance' window: It shows activated instances.
- 'Robot Grouping Result' window: It shows grouped robots as instance numbers.
- 'Tree' windows: It shows the generated decision tree for the selected context.
- 'Process' window: It shows the entropy computation process for the generated decision tree.

The entire grouping process can be summarized as follows:

- Create arbitrary number of robot instances through the user interface (see Fig. 3), and this information is sent to server.
- Also, information of a 'virtual situation' is sent to server.
- Then, the server creates decision tree using the information of the robot instances, and decide a group of robots for the specific situation (e.g. cleaning).
- As a result of the final process, the grouped robot instances are shown in the bottom left window as robot instance numbers.

From the bottom left of the windows in the Fig. 5, we can see that the robot instance number 1, 4, 6 and 7 are grouped together to perform the task, 'cleaning'.

4.2   Robot Collaboration

Although there are many different approaches of collaborations in robot studies, we consider the collaboration as sharing a task among a group of robots. For instance, a generated task can be further refined into a set of subtask, which will be carried out by each individual robot. The system generates a set of tasks or subtasks by the task planning mechanism, which will be described in the below.

4.2.1   Task Planning Rules

The task planning rules for collaboration layer is divided into two levels, high level planning rules (general task rules), and the low-level planning rules (robot-specific action rules). The general task rule can generate set of tasks, and the robot specific action rule can generate a set of actions to perform the task. For example, if 'cleaning' is the selected context, then task planner generates a set of tasks for the 'cleaning' as 'sweeping', and 'mopping'. When the task 'sweeping' is assigned to a robot, the generated action plan is 'move', 'lift', 'sweep', and 'release'. The task planner works on the server side, and the action planner is located on the client side. The sample task planning rules are as in the Table 2.

4.2.2   Multi-level Planning

Our system divides the planning mechanism in two levels, a high-level planning and low-level planning. It is a kind of hierarchical planning mechanism that is efficient enough to make the planning mechanism as simple as possible. The high-level planner generates a set of subtasks to accomplish a given context, and saves them in a stack that will be taken one by one by the low-level planner. When the high-level planning is done, the low-level planner gets activated with each one of the subtasks. A single subtask becomes a goal to accomplish in the low-level planning process, and it generates a set of actions as a result.

Table 2 Task planning rules

High-level planning rules	Low-level planning rules
If (Context = 'Cleaning') then (NeedTask Sweeping, Mopping)	If (Task = 'Sweeping') then (NeedAction (Move, Lift, Sweeping, Release))
If (Context = 'Serving') then (NeedTask (Take_order, Food_service, Check_out)). . ..	If (Task = 'Take_order') then (NeedAction (Give_Menu, Help_order)). . ..

5 Conclusion

In this paper, we have described a development of intelligent robot framework for providing intelligent service in the context awareness environment. The significances of our research could be as follows.

First, we have designed intelligent service robot framework and successfully carried out simulated experimentation by generating several robot instances. The several heterogeneous robot instances can be generated by the user, through the user interface windows, as many as possible. Second, the robots are grouped by the decision classifier based on the entropy metric of the information theory. This approach will provide more reliable and systematic methods for the robot grouping. Third, in this study, we consider the collaboration as follows. Once a task is generated, then it gets refined into a set of subtasks, which will be assigned to each individual robot. If each robot accomplishes its own subtask, then the original task is being done as a result. We do not have any interaction among the robot at the moment. Fourth, the generation of task is being done by the multi-level task allocation planning mechanism. The high-level task planning is done on the server side, and detailed action planning is done on the client side (robot).

Our approach may provide some useful solutions for the intelligent service oriented applications, which require multiple robot instances. Our immediate next work is to complete the development of context-awareness layer using sensed raw data from the external physical computing devices. Therefore, when we complete the development of the entire framework, we are going to setup experimentations with several humanoid robots equipped with sensors and network communications. Also, more detailed strategy and also interaction for collaboration among each individual robot is needed.

Acknowledgement This work was supported by the Seoul R&BD program (10848) of Korea.

References

1. M. Weiser. The computer for the twenty-first century. Sci. Am. **265**(3), 94–104 (1991)
2. G.K. Mostefaoui, J. Pasquier-Rocha, P. Brezillon, Context-aware computing: a guide for the pervasive computing community. Proc. IEEE/ACS Int. Conf. Pervasive Serv. (ICPS'04) **19**(23), 39–48 (2004)
3. A.K. Dey, G.D. Abowd, Towards a better understanding of context and context-awareness. GVU Technical Report GIT-GVU-99-22, Georgia Institute of Technology, 1999
4. M. Baldauf, S. Dustdar, A Survey on context-aware systems. Int. J. Adhoc Ubiquitous Comput. **2**, 263–277 (2004)
5. H. Chen, An intelligent broker architecture for pervasive context-aware systems. Ph.D. thesis, University of Maryland, 2004
6. T. Gu, H.K. Pung, D.Q. Zhang, A Service-oriented middleware for building context-aware services. J. Netw. Comput. Appl. **28**, 1–18 (2005)
7. D. Salber, A. Dey, G. Abowd, The context toolkit: aiding the development of context-enabled applications, in *Proceedings of CHI'99*, Pittsburgh, PA, 1999, pp. 434–441

8. X.H. Wang, T. Gu, D.Q. Zhang, H.K. Pung, Ontology based context modeling and reasoning using owl, in *Proceedings of 2nd IEEE Annual Conference on Pervasive Computing and Communications Workshops, PERCOMW'04*, Orlando, Florida, USA, 2004, pp. 18–22

9. D. Rodic, A.P. Engelbrecht, Social network as a coordination techniques for multi-robot systems. International conference on system design and applications, Springer, Berlin, 2003, pp. 503–513

10. T.H. Labella, M. Dorigo, J-L. Deneubourg, Self-organized task allocation in a group of robots. Technical Report No. TR/IRIDIA/2004–6, Universite Libre de Bruxelles, 2004

11. M. Mataric, S. Sukhatme, E. Ostergaard, Multi-robot task allocation in Uncertain Environment. Auton. Robots **14**, 255–263 (2003)

12. M. Sterns, N. Windelinckx Combining planning with reinforcement learning for multi-robot task allocation. in *Proceedings of adaptive agents and MASII*, LNAI 3394, 2006, pp. 260–274

13. R. Alami, A. Clodic, V Montreuil, E. Akin, R. Chatila, Task planning for human-robot interaction, in *Proceedings of the 2005 joint conference on Smart Object and ambient intelligence*, 2005, pp. 81–85

14. B. Gerkey, and M. Mataric, Principled communication for dynamic multi-robot task allocation. Exp. Robotics **II**, 353–362 (2000)

15. C.E. Shannon, *The Mathematical Theory of Communication* (University of Illinois Press, Illinois, 1949)

16. M.A. Goodrich, E.R. Boer, J.W. Crandall, R.W. Ricks, M.L. Quigley, Behavioral entropy in human-robot interaction, Brigham University, Technical Report ADA446467, 2004

17. T. Balch, Hierarchic social entropy: an information theoretic measure of robot group diversity. Auton. Robots **8**, 09–237 (2000)

Chapter 3
Piecewise Bezier Curves Path Planning with Continuous Curvature Constraint for Autonomous Driving

Ji-Wung Choi, Renwick Curry, and Gabriel Elkaim

Abstract We present two practical path planning algorithms based on Bezier curves for autonomous vehicles operating under waypoints and corridor constraints. Bezier curves have useful properties for the trajectory generation problem. This paper describes how the algorithms apply these properties to generate the reference trajectory for vehicles to satisfy the path constraints. Both algorithms generate the piecewise-Bezier-curves path such that the curves segments are joined smoothly with C^2 constraint which leads to continuous curvature along the path. The degree of the curves are minimized to prevent them from being numerically unstable. Additionally, we discuss the constrained optimization problem that optimizes the resulting path for a user-defined cost function.

1 Introduction

Bezier Curves were invented in 1962 by the French engineer Pierre Bezier for designing automobile bodies. Today Bezier Curves are widely used in computer graphics and animation. The Bezier curves have useful properties for the path generation problem as described in Section 2 of this paper. Hence many path planning techniques for autonomous vehicles have been discussed based on Bezier Curves in the literature. Cornell University Team for 2005 DARPA Grand Challenge used a path planner based on Bezier curves of degree 3 in a sensing/action feedback loop to generate smooth paths that are consistent with vehicle dynamics [5]. Skrjanc proposed a new cooperative collision avoidance method for multiple robots with constraints and known start and goal velocities based on Bezier curves of degree 4 [6]. In this method, four control points out of five are placed such that

J.-W. Choi (✉)
Autonomous Systems Lab, University of California, Santa Cruz, CA 95064, USA
e-mail: jwchoi@soe.ucsc.edu

S.-I. Ao et al. (eds.), *Machine Learning and Systems Engineering*,
Lecture Notes in Electrical Engineering 68,
DOI 10.1007/978-90-481-9419-3_3, © Springer Science+Business Media B.V. 2010

desired positions and velocities of the start and the goal point are satisfied. The fifth point is obtained by minimizing penalty functions. Jolly described Bezier curve based approach for the path planning of a mobile robot in a multi-agent robot soccer system [3]. The resulting path is planned such that the initial state of the robot and the ball, and an obstacle avoidance constraints are satisfied. The velocity of the robot along the path is varied continuously to its maximum allowable levels by keeping its acceleration within the safe limits. When the robot is approaching a moving obstacle, it is decelerated and deviated to another Bezier path leading to the estimated target position.

Our previous works introduced two path planning algorithms based on Bezier curves for autonomous vehicles with waypoints and corridor constraints [1, 2]. Both algorithms join cubic Bezier curve segments smoothly to generate the reference trajectory for vehicles to satisfy the path constraints. Also, both algorithms are constrained in that the path must cross over a bisector line of corner area such that the tangent at the crossing point is normal to the bisector. Additionally, the constrained optimization problem that optimizes the resulting path for user-defined cost function was discussed. Although the simulations provided in that paper showed the generation of smooth routes, discontinuities of the yaw angular rate appeared at junction nodes between curve segments. This is because the curve segments are constrained to connect each other by only C^1 continuity, so the curvature of the path is discontinuous at the nodes. (Section 2 describes this more detail.)

To resolve this problem, we propose new path planning algorithms. The algorithms impose constraints such that curve segments are C^2 continuous in order to have curvature continuous for every point on the path. In addition, they give the reference path more freedom by eliminating redundant constraints used in previous works, such as the tangent being normal to the bisector and symmetry of curve segments on corner area. The degree of each Bezier curve segments are determined by the minimum number of control points to satisfy imposed constraints while cubic Bezier curves are used for every segments in previous works. The optimized resulting path is obtained by computing the constrained optimization problem that controls the tradeoff between shortest and smoothest path generation. Furthermore, the proposed algorithms satisfy the initial and goal orientation as well as position constraint while previous works only made the position constraint satisfied. The numerical simulation results demonstrate the improvement of trajectory generation in terms of smoother steering control and smaller cross track error.

2 Bezier Curve

A Bezier Curve of degree n can be represented as

$$\mathbf{P}(t) = \sum_{i=0}^{n} B_i^n(t)\mathbf{P}_i, \quad t \in [0, 1]$$

where $\mathbf{P}_i$ are control points and $B_i^n(t)$ is a Bernstein polynomial given by

$$B_i^n(t) = \binom{n}{i} (1-t)^{n-i} t^i, \quad i \in \{0, 1, \ldots, n\}$$

Bezier Curves have useful properties for path planning:

- They always start at $\mathbf{P}_0$ and stop at $\mathbf{P}_n$.
- They are always tangent to $\overline{\mathbf{P}_0\mathbf{P}_1}$ and $\overline{\mathbf{P}_n\mathbf{P}_{n-1}}$ at $\mathbf{P}_0$ and $\mathbf{P}_n$ respectively.
- They always lie within the convex hull consisting of their control points.

2.1 The de Casteljau Algorithm

The de Casteljau algorithm describes a recursive process to subdivide a Bezier curve $\mathbf{P}(t)$ into two segments. The subdivided segments are also Bezier curves. Let $\{\mathbf{P}_0^0, \mathbf{P}_1^0, \ldots, \mathbf{P}_n^0\}$ denote the control points of $\mathbf{P}(t)$. The control points of the segments can be computed by

$$\mathbf{P}_i^j = (1-\tau)\mathbf{P}_i^{j-1} + \tau \mathbf{P}_{i+1}^{j-1}, \tag{1}$$
$$\tau \in (0,1), \ j \in \{1, \ldots, n\}, \ i \in \{0, \ldots, n-j\}$$

Then, $\{\mathbf{P}_0^0, \mathbf{P}_0^1, \ldots, \mathbf{P}_0^n\}$ are the control points of one segment and $\{\mathbf{P}_0^n, \mathbf{P}_1^{n-1}, \ldots, \mathbf{P}_n^0\}$ are the another. (See an example in Fig. 1.) Note that, by applying the properties of Bezier Curves described in the previous subsection, both of the subdivided curves end at $\mathbf{P}_0^n$ and the one is tangent to $\overline{\mathbf{P}_0^{n-1}\mathbf{P}_0^n}$ and the other is to $\overline{\mathbf{P}_0^n\mathbf{P}_1^{n-1}}$ at the point. Since $\mathbf{P}_0^n$ is chosen on $\overline{\mathbf{P}_0^{n-1}\mathbf{P}_1^{n-1}}$ by using (1), the three points $\mathbf{P}_0^{n-1}$, $\mathbf{P}_0^n$, and $\mathbf{P}_1^{n-1}$ are collinear.

Remark 1. A Bezier curve $\mathbf{P}(t)$ constructed by control points $\{\mathbf{P}_0^0, \mathbf{P}_1^0, \ldots, \mathbf{P}_n^0\}$ always passes through the point $\mathbf{P}_0^n$ and is tangent to $\overline{\mathbf{P}_0^{n-1}\mathbf{P}_1^{n-1}}$ at $\mathbf{P}_0^n$.

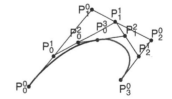

Fig. 1 Subdividing a cubic Bezier curve with $\tau = 0.4$ by the de Casteljau Algorithm

2.2   Derivatives, Continuity and Curvature

The derivatives of a Bezier curve can be determined by its control points. For a Bezier curve $\mathbf{P}(t) = \sum_{i=0}^{n} B_i^n(t)\mathbf{P}_i$, the first derivative is given by

$$\dot{\mathbf{P}}(t) = \sum_{i=0}^{n-1} B_i^{n-1}(t)\mathbf{D}_i \tag{2}$$

where $\mathbf{D}_i$, control points of $\dot{\mathbf{P}}(t)$ are

$$\mathbf{D}_i = n(\mathbf{P}_{i+1} - \mathbf{P}_i)$$

Geometrically, (2) provides us with a tangent vector. The higher order derivative can be obtained by using the relationship of (2), iteratively.

Two Bezier curves $\mathbf{P}(t)$ and $\mathbf{Q}(t)$ are said to be C^k at t_0 continuous if

$$\mathbf{P}(t_0) = \mathbf{Q}(t_0), \; \dot{\mathbf{P}}(t_0) = \dot{\mathbf{Q}}(t_0), \; \ldots, \; \mathbf{P}^{(k)}(t_0) = \mathbf{Q}^{(k)}(t_0) \tag{3}$$

The curvature of a Bezier curve $\mathbf{P}(t) = (x(t), y(t))$ at t is given by

$$\kappa(t) = \frac{|\dot{x}(t)\ddot{y}(t) - \dot{y}(t)\ddot{x}(t)|}{(\dot{x}^2(t) + \dot{y}^2(t))^{\frac{3}{2}}} \tag{4}$$

Lemma 1. *For the path constructed by two Bezier curve segments* $\mathbf{P}(t)|_{t\in[t_0,t_1]}$ *and* $\mathbf{Q}(t)|_{t\in[t_1,t_2]}$, *if* $\mathbf{P}(t)$ *and* $\mathbf{Q}(t)$ *are at least* C^2 *continuous at* t_1 *then the path has continuous curvature for every point on it.*

Proof. The curvature is expressed in terms of the first and the second derivative of a curve in (4). Since the Bezier curves are defined as polynomial functions of t, their k-th derivative for all $k = 1, 2, \ldots$ are continuous. Hence, they always have continuous curvature for all t. For two different Bezier curves $\mathbf{P}(t)$ and $\mathbf{Q}(t)$, it is sufficient that $\kappa(t_1)$, the curvature at the junction node is continuous if $\dot{\mathbf{P}}(t) = \dot{\mathbf{Q}}(t)$ and $\ddot{\mathbf{P}}(t) = \ddot{\mathbf{Q}}(t)$ are continuous at t_1.

3   Problem Statement

Consider the control problem of a ground vehicle with a mission defined by waypoints and corridor constraints in a two-dimensional free-space. Our goal is to develop and implement an algorithm for navigation that satisfies these constraints. That is divided into two parts: path planning and path following.

To describe each part, we need to introduce some notation. Each waypoint is denoted as $\mathbf{W}_i \in \mathbb{R}^2$ for $i \in \{1, 2, \ldots, N\}$, where $N \in \mathbb{R}$ is the total number of waypoints. Corridor width is denoted as w_j for $j \in \{1, \ldots, N-1\}$, j-th widths of each segment between two waypoints, $\mathbf{W}_j$ and $\mathbf{W}_{j+1}$. For the dynamics of the vehicle, the state and the control vector are denoted $\mathbf{q}(t) = (x_c(t), y_c(t), \psi(t))^T$ and $\mathbf{u}(t) = (v(t), \omega(t))^T$ respectively, where (x_c, y_c) represents the position of the center of gravity of the vehicle. The yaw angle ψ is defined to the angle from the X axis. v is the longitudinal velocity of the vehicle. $\omega = \dot{\psi}$ is the yaw angular rate. State space is denoted as $\mathcal{C} = \mathbb{R}^3$. We assume that the vehicle follows that

$$\dot{\mathbf{q}}(t) = \begin{pmatrix} \cos\psi(t) & 0 \\ \sin\psi(t) & 0 \\ 0 & 1 \end{pmatrix} \mathbf{u}(t)$$

With the notation defined above, the path planning problem is formulated as: Given an initial position and orientation and a goal position and orientation of the vehicle, generate a path λ specifying a continuous sequence of positions and orientations of the vehicle satisfying the path constraints [4]. In other words, we are to find a continuous map

$$\lambda : [0, 1] \rightarrow \mathcal{C}$$

with

$$\lambda(0) = \mathbf{q}_{init} \quad \text{and} \quad \lambda(1) = \mathbf{q}_{goal}$$

where $\mathbf{q}_{init} = (\mathbf{W}_1, \psi_0)$ and $\mathbf{q}_{goal} = (\mathbf{W}_N, \psi_f)$ are the initial and goal states of the path, respectively.

Given a planned path, we use the path following technique with feedback corrections as illustrated in Fig. 2. A position and orientation error is computed every 50 ms. A point $\mathbf{z}$ is computed with the current longitudinal velocity and

Fig. 2 The position error is measured from a point $\mathbf{z}$, projected in front of the vehicle, and onto the desired curve to point $\mathbf{p}$

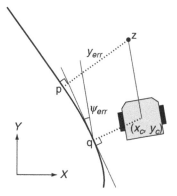

heading of the vehicle from the current position. $\mathbf{z}$ is projected onto the reference trajectory at point $\mathbf{p}$ such that $\mathbf{zp}$ is normal to the tangent at $\mathbf{p}$. The cross track error y_{err} is defined by the distance between $\mathbf{z}$ and $\mathbf{p}$. The steering control ω uses a PID controller with respect to cross track error y_{err}.

$$\delta\omega = k_p y_{err} + k_d \frac{dy_{err}}{dt} + k_i \int y_{err} dt$$

4   Path Planning Algorithm

In this section, two path planning methods based on Bezier curves are proposed. To describe the algorithms, let us denote $\hat{\mathbf{b}}_j$ as the unit vector codirectional with the outward bisector of $\angle \mathbf{W}_{j-1}\mathbf{W}_j\mathbf{W}_{j+1}$ for $j \in \{2, \ldots, N-1\}$ as illustrated in Fig. 3. The planned path must cross over the bisectors under the waypoint and the corridor constraints. The location of the crossing point is represented as $\mathbf{W}_j + d_j \cdot \hat{\mathbf{b}}_j$, where $d_j \in \mathbb{R}$ is a scalar value. The course is divided into segments G_i by bisectors. G_i indicates the permitted area for vehicles under corridor constraint w_i, from $\mathbf{W}_i$ to $\mathbf{W}_{i+1}$.

Bezier curves constructed by large numbers of control points are numerically unstable. For this reason, it is desirable to join low-degree Bezier curves together in a smooth way for path planning [6]. Thus both of the algorithms use a set of low-degree Bezier curves such that the neighboring curves are C^2 continuous at their end nodes. This will lead to continuous curvature on the resulting path by Lemma 1.

The Bezier curves used for the path plannings are denoted as ${}^i\mathbf{P}(t) = \sum_{k=0}^{n_i} B_k^{n_i}(t) \cdot {}^i\mathbf{P}_k$ for $i \in \{1, \ldots, M\}$, $t \in [0, 1]$, where M is the total number of the Bezier curves and n_i is the degree of ${}^i P$. The planned path λ is a concatenation of all ${}^i P$ such that

$$\lambda((i - 1 + t)/M) = {}^i P(t), \quad i \in \{1, \ldots, M\}, t \in [0, 1]$$

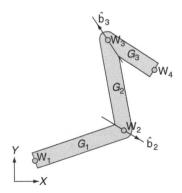

Fig. 3 An example of the course with four waypoints. *Gray area* is the permitted area for vehicles under a corridor constraint

4.1   Path Planning Placing Bezier Curves within Segments (BS)

In this path planning method (BS), the Bezier curve ${}^i\mathbf{P}$ for $i \in \{1, \ldots, N-1\}$ are used within each segment G_i. The adjacent curves, ${}^{j-1}\mathbf{P}$ and ${}^j\mathbf{P}$ are C^2 continuous at the crossing point, $\mathbf{W}_j + d_j \cdot \hat{\mathbf{b}}_j$. The control points of ${}^i\mathbf{P}$, ${}^i\mathbf{P}_k$ for $k = \{0, \ldots, n_i\}$ are determined to maintain these conditions.

- The beginning and the end point are $\mathbf{W}_1$ and $\mathbf{W}_N$.

$$
{}^1\mathbf{P}_0 = \mathbf{W}_1, \quad {}^{N-1}\mathbf{P}_{n_{N-1}} = \mathbf{W}_N \tag{5}
$$

- The beginning and end orientation of λ are ψ_0 and ψ_f.

$$
{}^1\mathbf{P}_1 = \mathbf{W}_1 + l_0(\cos\psi_0, \sin\psi_0), \quad l_0 \in \mathbb{R}^+ \tag{6a}
$$

$$
{}^{N-1}\mathbf{P}_{n_{N-1}-1} = \mathbf{W}_N - l_f(\cos\psi_f, \sin\psi_f), \quad l_f \in \mathbb{R}^+ \tag{6b}
$$

- ${}^{j-1}\mathbf{P}$ and ${}^j\mathbf{P}$, $\forall j \in \{2, \ldots, N-1\}$ are C^2 continuous at the crossing point.

$$
{}^{j-1}\mathbf{P}_{n_{j-1}} = {}^j\mathbf{P}_0 = \mathbf{W}_j + d_j \cdot \hat{\mathbf{b}}_j \tag{7a}
$$

$$
n_{j-1}({}^{j-1}\mathbf{P}_{n_{j-1}} - {}^{j-1}\mathbf{P}_{n_{j-1}-1}) = n_j({}^j\mathbf{P}_1 - {}^j\mathbf{P}_0) \tag{7b}
$$

$$
n_{j-1}(n_{j-1}-1)({}^{j-1}\mathbf{P}_{n_{j-1}} - 2 \cdot {}^{j-1}\mathbf{P}_{n_{j-1}-1} + {}^{j-1}\mathbf{P}_{n_{j-1}-2})
$$
$$
= n_j(n_j-1)({}^j\mathbf{P}_2 - 2\,{}^j\mathbf{P}_1 + {}^j\mathbf{P}_0) \tag{7c}
$$

- The crossing points are bounded within the corridor.

$$
|d_j| < \forall \frac{1}{2}\min(w_{j-1}, w_j) \tag{8}
$$

- ${}^i\mathbf{P}_k$, $k = \{1, \ldots, n_i - 1\}$ always lie within the area of G_i.

$$
{}^i\mathbf{P}_1 \in G_i, \quad \ldots, \quad {}^i\mathbf{P}_{n_i-1} \in G_i \tag{9}
$$

Equations (6a, b) are derived by using the tangent property of Bezier curves at their end points. Equations (7a–c) are obtained by applying (2) and (3). Equation (9) makes the resulting Bezier curve satisfy the corridor constraint by the convex hull property. At each crossing point, three control points of each adjacent Bezier curve are dedicated to the C^2 continuity constraint by (2), (4), and Lemma 1. So the minimum number of control points to satisfy the constraints independent of the others are six for ${}^i\mathbf{P}$, $i \in \{2, \ldots, N-2\}$. On other hand, ${}^1\mathbf{P}$ needs five: three to connect $^2\mathbf{P}$ plus two to connect $\mathbf{W}_1$ with slope of ψ_0. Likewise, $^{2N-3}\mathbf{P}$ needs five. Thus, n_i is determined as

$$n_i = \begin{cases} 4, & i \in \{1, N-1\} \\ 5, & i \in \{2, \ldots, N-2\} \end{cases}$$

Note that $^1\mathbf{P}_0$ and $^{N-1}\mathbf{P}_{n_{N-1}}$ are fixed in (5). $^1\mathbf{P}_1$ and $^{N-1}\mathbf{P}_{n_{N-1}-1}$ are fixed in (6a, b). $^{j-1}\mathbf{P}_{n_{j-1}}$ and $^j\mathbf{P}_0$ rely on d_j in (7a–c). $^{j-1}\mathbf{P}_{n_{j-1}-1}$ and $^{j-1}\mathbf{P}_{n_{j-1}-2}$ rely on $^j\mathbf{P}_1$ and $^{j-1}\mathbf{P}_2$. So the free variables are, $\forall j \in \{2, \ldots, N-1\}$, $\mathbf{P}_1 = \{^j\mathbf{P}_1\}$, $\mathbf{P}_2 = \{^j\mathbf{P}_2\}$, $\mathbf{d} = \{d_j\}$, and $\mathbf{L} = \{l_0, l_f\}$. The number of the variables or the degrees of freedom is $5N - 8$. The variables are computed by minimizing the constrained optimization problem:

$$\min_{\mathbf{P}_1, \mathbf{P}_2, \mathbf{d}, \mathbf{L}} J = \sum_{i=1}^{N-1} J_i \tag{10}$$

subject to (8) and (9), where J_i is the cost function of $^i\mathbf{P}(t)$, given by

$$J_i = \int_0^1 [(a_i(\dot{x}^2(t) + \dot{y}^2(t)) + b_i|^i\kappa(t)|^2]dt \tag{11}$$

where $a_i \in \mathbb{R}$ and $b_i \in \mathbb{R}$ are constants that control the tradeoff between arc lengths versus curvatures of the resulting path.

4.2 Path Planning Placing Bezier Curves on Corners (BC)

Another path planning method (BC) adds quadratic Bezier curves on the corner area around $\mathbf{W}_j$, $j \in \{2, \ldots, N-1\}$. The quadratic Bezier curves are denoted as $^j\mathbf{Q}(t) = \sum_{k=0}^2 B_k^2(t) \cdot {}^j\mathbf{Q}_k^0$ intersects the j-th bisector. The first and the last control point, $^j\mathbf{Q}_0^0$ and $^j\mathbf{Q}_2^0$ are constrained to lie within G_{j-1} and G_j, respectively. Within each segment G_i, another Bezier curve is used to connect the end points of $^j\mathbf{Q}$ with C^2 continuity and/or $\mathbf{W}_1$ with slope of ψ_0 and/or $\mathbf{W}_N$ with ψ_f. Hence, $^j\mathbf{Q}$ is the curve segments of even number index:

$$^{2(j-1)}\mathbf{P}(t) = {}^j\mathbf{Q}(t), \quad j \in \{2, \ldots, N-1\}$$

The degree of $^i\mathbf{P}$ is determined by the minimum number of control points to satisfy C^2 continuity constraint, independently:

$$n_i = \begin{cases} 4, & i \in \{1, 2N-3\} \\ 2, & i \in \{2, 4, \ldots, 2N-4\} \\ 5, & i \in \{3, 5, \ldots, 2N-5\} \end{cases}$$

Let t_c denote the Bezier curve parameter corresponding to the crossing point of $^j\mathbf{Q}(t)$ on the bisector, such that

$$^j\mathbf{Q}(t_c) = \mathbf{W}_j + d_j \cdot \hat{\mathbf{b}}_j. \tag{12}$$

Let θ_j denote the angle of the tangent vector at the crossing point from X-axis:

$$^j\dot{\mathbf{Q}}(t_c) = (|^j\dot{\mathbf{Q}}(t_c)| \cos \theta_j, \ |^j\dot{\mathbf{Q}}(t_c)| \sin \theta_j). \tag{13}$$

The notations are illustrated in Fig. 4. Due to the constraint of $^j\mathbf{Q}_0^0$ and $^j\mathbf{Q}_2^0$ within G_{j-1} and G_j, the feasible scope of θ_j is limited to the same direction as $\mathbf{W}_{j+1}$ is with respect to $\hat{\mathbf{b}}_j$. In other words, if $\mathbf{W}_{j+1}$ is to the right of $\hat{\mathbf{b}}_j$, then θ_j must point to the right of $\hat{\mathbf{b}}_j$, and vice versa.

Given $^j\mathbf{Q}_0^0$, $^j\mathbf{Q}_2^0$, d_j, and θ_j, the other control point $^j\mathbf{Q}_1^0$ is computed such that the crossing point is located at $\mathbf{W}_j + d_j \cdot \hat{\mathbf{b}}_j$ and the angle of the tangent vector at the crossing point is θ_j. Since each control point is two-dimensional, the degrees of freedom of $^j\mathbf{Q}(t)$ are six. Since d_j and θ_j are scaler, representing $^j\mathbf{Q}(t)$ in terms of $^j\mathbf{Q}_0^0$, $^j\mathbf{Q}_2^0$, d_j, and θ_j does not affect the degrees of freedom. However, it comes up with an advantage for corridor constraint. If we compute $^j\mathbf{Q}_1^0$ such as above, the points computed by applying the de Casteljau algorithm such that two subdivided curves are separated by the j-th bisector are represented as $^j\mathbf{Q}_0^0$ and $^j\mathbf{Q}_2^0$ as described in the following. The two curves are constructed by $\{^j\mathbf{Q}_0^0, \ ^j\mathbf{Q}_0^1, \ ^j\mathbf{Q}_0^2\}$ and $\{^j\mathbf{Q}_0^2, \ ^j\mathbf{Q}_1^1, \ ^j\mathbf{Q}_2^0\}$. We can test if the convex hull of $\{^j\mathbf{Q}_0^0, \ ^j\mathbf{Q}_0^1, \ ^j\mathbf{Q}_0^2\}$ lies within G_{j-1} and if that of $\{^j\mathbf{Q}_0^2, \ ^j\mathbf{Q}_1^1, \ ^j\mathbf{Q}_2^0\}$ lies within G_j in (26), instead of testing that of $\{^j\mathbf{Q}_0^0, \ ^j\mathbf{Q}_1^0, \ ^j\mathbf{Q}_2^0\}$. (Note that $^j\mathbf{Q}_1^0$ is not constrained to lie within corridor as shown in Fig. 4.) So, the convex hull property is tested for tighter conditions against the corridor constraint without increasing the degrees of freedom.

In order to compute $^j\mathbf{Q}_1^0$, the world coordinate frame T is transformed and rotated into the local frame ^jT where the origin is at the crossing point, $^j\mathbf{Q}(t_c)$ and X axis is codirectional with the tangent vector of the curve at the crossing point, $^j\dot{\mathbf{Q}}(t_c)$.

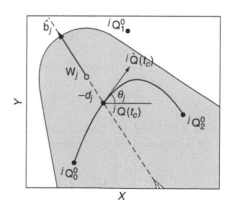

Fig. 4 The geometry of the $^j\mathbf{Q}(t)$ in corner area around $\mathbf{W}_j$

Fig. 5 The geometry of the control points of $^{j}\mathbf{Q}(t)$ with respect to ^{j}T

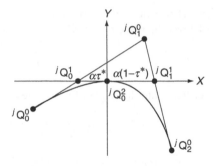

Let us consider the subdivision ratio, $\tau^{*} \in (0, 1)$ such that the location of $^{j}\mathbf{Q}_0^2$ computed by applying the de Casteljau algorithm with it is the crossing point. In other words, τ^{*} has $^{j}\mathbf{Q}_0^2$ be at the origin with respect to ^{j}T frame. Fig. 5 illustrates the control points of $^{j}\mathbf{Q}(t)$ with respect to ^{j}T frame. Note that $^{j}\mathbf{Q}_0^2$ is at the origin by the definition of ^{j}T and τ^{*}. $^{j}\mathbf{Q}_0^1$ and $^{j}\mathbf{Q}_1^1$ are on the X axis by the definition of ^{j}T and Remark 1. Let the coordinates of the control points be denoted as $\mathbf{Q}_i^0 = (x_i, y_i)$, $i \in \{0, 1, 2\}$, where all coordinates are with respect to ^{j}T.

Lemma 2. *Given d_j and θ_j, for $^{j}\mathbf{Q}(t)$ to intersect j-th bisector with the crossing point determined by the d_j and (12), and the tangent vector at the point determined by the θ_j and (13), it is necessary that $y_0 y_2 \geq 0$.*

Proof. Let $(x(t), y(t))$ denote the coordinate of $^{j}\mathbf{Q}(t)$ with respect to ^{j}T. By the definition of ^{j}T and Remark 1, $\mathbf{Q}(t)$ passes through the origin with tangent slope of zero with respect to ^{j}T. That is, $x(t_c) = 0$, $y(t_c) = 0$ and $\dot{y}(t_c) = 0$. Suppose that $y_0 = y(0) < 0$. Since $y(t)$ is a quadratic polynomial, $\dot{y}(t) > 0$ and $\ddot{y}(t) < 0$ for $t \in [0, t_c)$. Subsequently, $\dot{y}(t) < 0$ and $\ddot{y}(t) < 0$ for $t \in (t_c, 1]$. Thus, $y_2 = y(1) < 0$ and $y_0 y_2 > 0$. Similarly, if $y_0 > 0$ then $y_1 > 0$. If $y_0 = 0$ then $\dot{y}(t) = 0$ for $t \in [0, 1]$ and $y_2 = 0$. Thus, $y_0 y_2 = 0$. $\square$

We are to calculate $^{j}\mathbf{Q}_1^0$ depending on whether $y_0 y_2$ is nonzero. For simplicity, superscript j is dropped from now on. Without loss of generality, suppose that $y_0 < 0$ and $y_2 < 0$. $\mathbf{Q}_0^2$ is represented as

$$\mathbf{Q}_0^2 = (1 - \tau^{*})\mathbf{Q}_0^1 + \tau^{*}\mathbf{Q}_1^1$$

by applying (1). So the coordinates of $\mathbf{Q}_0^1$ and $\mathbf{Q}_1^1$ can be represented as

$$\mathbf{Q}_0^1 = (-\alpha\tau^{*}, 0), \quad \mathbf{Q}_1^1 = (\alpha(1 - \tau^{*}), 0), \quad \alpha > 0 \tag{14}$$

where $\alpha > 0$ is some constant. Applying (1) with $i = 0$ and $j = 1$ and arranging the result with respect to $\mathbf{Q}_1^0$ by using (14) gives

$$\mathbf{Q}_1^0 = \left(-\alpha - \frac{1 - \tau^{*}}{\tau^{*}}x_0, -\frac{1 - \tau^{*}}{\tau^{*}}y_0\right) \tag{15}$$

Similarly, applying (1) with $i = 1$ and $j = 1$ yields

$$\mathbf{Q}_1^0 = \left(\alpha - \frac{\tau^*}{1 - \tau^*} x_2, -\frac{\tau^*}{1 - \tau^*} y_2 \right) \tag{16}$$

where α and τ^* are obtained by equations (15) and (16):

$$\tau^* = \frac{1}{1 + \sqrt{y_2/y_0}}, \qquad \alpha = \frac{x_0 y_2 - y_0 x_2}{2 y_0 \sqrt{y_2/y_0}} \tag{17}$$

τ^* and α have the square root of y_2/y_0. So, if $y_0 y_2 < 0$ then τ^* and α are not determined, hence, $\mathbf{Q}(t)$ is infeasible. That is, (17) ends up with Lemma 2.

If $y_0 = y_2 = 0$ then all control points of $^j\mathbf{Q}$ are on X axis (see proof of Lemma 2). In the geometric relation of control points and the points computed by applying the de Casteljau algorithm as shown in Fig. 6, we obtain

$$\begin{aligned} x_0 &= -(\alpha + \beta)\tau \\ x_2 &= (\alpha + \gamma)(1 - \tau) \\ \alpha &= \beta(1 - \tau) + \gamma\tau \end{aligned} \tag{18}$$

where $\alpha > 0$, $\beta > 0$, $\gamma > 0$ are some constants. Using (18), $\mathbf{Q}_1^0 = (x_1, 0)$ is represented in terms of arbitrary $\tau \in (0, 1)$:

$$x_1 = -\frac{1}{2} \left(\frac{1 - \tau}{\tau} x_0 + \frac{\tau}{1 - \tau} x_2 \right) \tag{19}$$

The constraints imposed on the planned path are formulated as follows:

- The beginning and end position of λ are $\mathbf{W}_1$ and $\mathbf{W}_N$.

$$^1\mathbf{P}_0 = \mathbf{W}_1, \qquad ^{2N-3}\mathbf{P}_4 = \mathbf{W}_N \tag{20}$$

- The beginning and end orientation of λ are ψ_0 and ψ_f.

$$^1\mathbf{P}_1 = \mathbf{W}_1 + l_0(\cos\psi_0, \sin\psi_0), \quad l_0 \in \mathbb{R}^+ \tag{21a}$$

$$^{2N-3}\mathbf{P}_3 = \mathbf{W}_N - l_f(\cos\psi_f, \sin\psi_f), \quad l_f \in \mathbb{R}^+ \tag{21b}$$

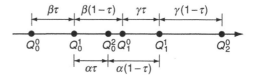

Fig. 6 The geometry of $^j\mathbf{Q}(t)$ with respect to ^jT when $y_0 = y_2 = 0$

- $^{i-1}\mathbf{P}$ and $^i\mathbf{P}$, $\forall i \in \{2, \ldots, 2N - 3\}$ are C^2 continuous at the junctions.

$$^{i-1}\mathbf{P}_{n_{i-1}} = {}^i\mathbf{P}_0 \tag{22a}$$

$$n_{i-1}({}^{i-1}\mathbf{P}_{n_{i-1}} - {}^{i-1}\mathbf{P}_{n_{i-1}-1}) = n_i({}^i\mathbf{P}_1 - {}^i\mathbf{P}_0) \tag{22b}$$

$$\begin{aligned} n_{i-1}(n_{i-1} - 1)({}^{i-1}\mathbf{P}_{n_{i-1}} - 2 \cdot {}^{i-1}\mathbf{P}_{n_{i-1}-1} + {}^{i-1}\mathbf{P}_{n_{i-1}-2}) \\ = n_i(n_i - 1)({}^i\mathbf{P}_2 - 2 \cdot {}^i\mathbf{P}_1 + {}^i\mathbf{P}_0) \end{aligned} \tag{22c}$$

- The crossing points are bounded within the corridor.

$$|d_j| < \frac{1}{2}\min(w_{j-1}, w_j), \quad \forall j \in \{2, \ldots, N - 1\} \tag{23}$$

- θ_j has the same direction as $\mathbf{W}_{j+1}$ is with respect to $\hat{\mathbf{b}}_j$.

$$\begin{aligned} \mathrm{mod}\,(\angle(\mathbf{W}_{j+1} - \mathbf{W}_j) - \angle\hat{\mathbf{b}}_j, 2\pi) > \pi \\ \Rightarrow \mathrm{mod}\,(\theta_j - \angle\hat{\mathbf{b}}_j, 2\pi) > \pi, \end{aligned} \tag{24a}$$

$$\begin{aligned} \mathrm{mod}\,(\angle(\mathbf{W}_{j+1} - \mathbf{W}_j) - \angle\hat{\mathbf{b}}_j, 2\pi) < \pi \\ \Rightarrow \mathrm{mod}\,(\theta_j - \angle\hat{\mathbf{b}}_j, 2\pi) < \pi \end{aligned} \tag{24b}$$

- $^j\mathbf{Q}_0^0$ and $^j\mathbf{Q}_2^0$ with respect to $^j T$ satisfies Lemma 2.

$$y_0 y_2 \geq 0 \tag{25}$$

where y_0 and y_2 are with respect to $^j T$.
- $^j\mathbf{Q}_0^0$ and $^j\mathbf{Q}_0^1$ lie within G_{j-1}. $^j\mathbf{Q}_2^0$ and $^j\mathbf{Q}_1^1$ lie within G_j.

$$^j\mathbf{Q}_0^0 \in G_{j-1}, {}^j\mathbf{Q}_0^1 \in G_{j-1}, {}^j\mathbf{Q}_2^0 \in G_j, {}^j\mathbf{Q}_1^1 \in G_j \tag{26}$$

- $\{^i\mathbf{P}_1, \ldots, {}^i\mathbf{P}_{ni-1}\}$ always lie within the area of G_i.

$$^i\mathbf{P}_1 \in G_i, \ldots, {}^i\mathbf{P}_{n_i-1} \in G_i, \quad i \in \{1, 3, \ldots, 2N - 3\} \tag{27}$$

The free variables are, for $j \in \{2, \ldots, N - 1\}$, $\mathbf{Q} = \{^j\mathbf{Q}_0, {}^j\mathbf{Q}_2\}$, $\mathbf{d} = \{d_j\}$, $\boldsymbol{\theta} = \{\theta_j\}$, and $\mathbf{L} = \{l_0, l_f\}$. The degrees of freedom is $6N - 10$. The variables are computed by minimizing the constrained optimization problem:

$$\min_{\mathbf{Q},\mathbf{d},\boldsymbol{\theta},\mathbf{L}} J = \sum_{i=1}^{N-1} J_i \tag{28}$$

subject to (23), (24,b), (25), (26), and (27), where J_i is the cost function of $^i\mathbf{P}(t)$, defined in (11). Notice that the convex hull property is tested for $^j\mathbf{Q}_0^1$ and $^j\mathbf{Q}_1^1$ of the divided curves instead of $^j\mathbf{Q}_1^0$ in (26). Thus, it comes up with tighter conditions for curves against the corridor constraint.

5 Simulation Results

Simulations were run for the course with four waypoints and identical corridor width of 8 as illustrated in Fig. 7. The initial and goal orientation are given by $\psi_0 = \psi_f = 0$. For path following, the constant longitudinal velocity $v(t) = 10 \; m/s$ is used. The magnitude of ω is bounded within $|\omega|_{max} = 25$ rpm. The PID gains are given by: $k_p = 2$, $k_d = 1$, and $k_i = 0.1$.

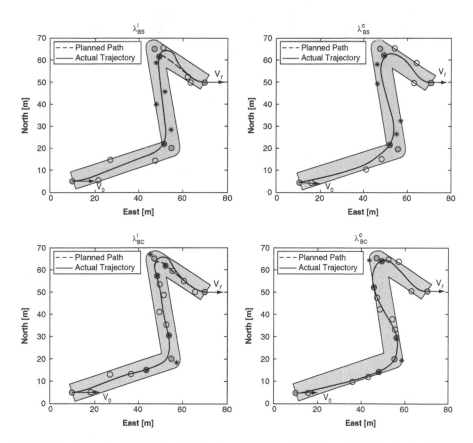

Fig. 7 The planned paths by BS method (*top*) and by BC method (*bottom*). Each pair of results are obtained by minimizing arc lengths (*left*) and by minimizing curvatures (*right*)

In Fig. 7, the reference paths (dashed curves) planned by applying the BS and BC method are placed in top and bottom row, respectively. The actual trajectories (solid curves) are generated by using the proposed tracking method. In the figures, stars indicate the location of control points of Bezier curve segments at an even number index, and empty circles indicate those of the others. The paths in the left column are planned by minimizing the summation of (11) with $a_i = 1$ and $b_i = 0$. Since only arc length is penalized, the cost function leads to paths with minimum arc length, which we denote λ_{BS}^l and λ_{BC}^l for the BS and BC method, respectively. On other hand, the paths in the right column are planned by minimizing the cost function with $a_i = 0$ and $b_i = 1$ so that resulting paths with larger radii of curvature are provided. We denote them λ_{BS}^c and λ_{BC}^c for the BS and BC method, respectively. λ_{BS}^c and λ_{BC}^c generate longer but smoother trajectory guidance than λ_{BS}^l and λ_{BC}^l. Taking a look at the tracking results on λ_{BS}^l and λ_{BC}^l, the vehicle overshoots in sharp turns around $\mathbf{W}_3$ resulting in a large position error (see Fig. 8) due to the limit on steering angle rate. The commanded steering angle rate on λ_{BC}^l, marked by '*' in Fig. 9 undergoes rapid changes and is constrained by the rate limit. However, on λ_{BS}^c and λ_{BC}^c, the vehicle tracks the part of the planned paths accurately thanks to larger radii of curvature. We can verify that more tangibly in cross track errors plot provided in Fig. 8. Also, the steering control signal on λ_{BC}^c, marked by 'o' in Fig. 9 is smoother than that on λ_{BC}^l.

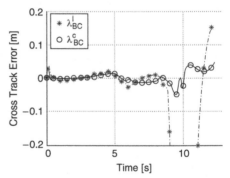

Fig. 8 The cross track errors by BC

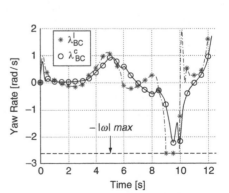

Fig. 9 The steering controls by BC

The main difference of the proposed algorithm from the previous ones of [1] is the degree of continuity at the junctions: C^2. Assuming that the vehicle tracks a reference path perfectly and v is continuous, if κ is continuous then ω is continuous because $\omega = \kappa v$. When v is constant as this simulation, since ω is proportional to κ, the continuity characteristic of ω tends to follow that of κ. Since the previous algorithms imposed only C^1 continuity on junction nodes of the path, κ is discontinuous at the nodes. Hence the path underwent the discontinuity of angular rate, that is, large angular acceleration, which leads to large torque on the vehicle. On other hand, the proposed algorithm has κ continuous along the resulting paths. If the path has curvature small enough so that the vehicle is able to track it accurately given a maximum steering angle rate, then the steering control signal will be smooth, as that of λ_{BC}^c in Fig. 9.

6 Conclusions

This paper presents two path planning algorithms based on Bezier curves for autonomous vehicles with waypoints and corridor constraints. Bezier curves provide an efficient way to generate the optimized path and satisfy the constraints at the same time. The simulation results also show that the trajectory of the vehicle follows the planned path within the constraints.

References

1. J. Choi, R.E. Curry, G.H. Elkaim, in *Path Planning Based on Bezier Curve for Autonomous Ground Vehicles* (Chapter 19, IEEE Computer Society, 2009), pp. 158–166
2. J. Choi, G.H. Elkaim, Bezier curves for trajectory guidance. *Proceedings of the World Congress on Engineering and Computer Science, WCECS 2008*, San Francisco, CA (2008)
3. K.G. Jolly, K.R. Sreerama, R. Vijayakumar, A bezier curve based path planning in a multi-agent robot soccer system without violating the acceleration limits. Robot. Autonom. Syst. **57**, 23–33 (2009)
4. J.C. Latombe, in *Robot Motion Planning* (Kluwer, Norwell, MA, 1991)
5. I. Miller, S. Lupashin, N. Zych, P. Moran, B. Schimpf, A. Nathan, E. Garcia, in *Cornell University's 2005 DARPA Grand Challenge Entry*, vol. 36, chapter 12 (Springer, Heidelberg, 2007), pp. 363–405
6. I. Skrjanc, G. Klancar, Cooperative collision avoidance between multiple robots based on bezier curves. *Proceedings of the Information Technology Interfaces, 2007 (ITI 2007)*, pp. 451–456

Chapter 4
Combined Heuristic Approach to Resource-Constrained Project Scheduling Problem

Miloš Šeda, Radomil Matoušek, Pavel Ošmera, Čeněk Šandera, and Roman Weisser

Abstract This chapter deals with the resource-constrained project scheduling problem that belongs to NP-hard optimisation problems. There are many different heuristic strategies how to shift activities in time when resource requirements exceed their available amounts. We propose a transformation of the problem to a sequence of simpler instances of (multi)knapsack problems that do not use traditionally predefined activity priorities and enable to maximise limited resources in all time intervals given by start or end of an activity and therefore to reduce the total time.

1 Introduction

A classical task of project management is to create the network graph of a project and, on the basis of knowledge or an estimation of time of activities, to determine critical activities that would have influenced a project delay. Each activity draws on certain resources, e.g. financial resources, natural resources (energy, water, material, etc.), labour, managerial skills. The solution of this task is well known in the case when we do not take limited resources into consideration.

The scheduling problem is a frequent task in the control of various systems such as manufacturing processes [1], project management [2], and service system control (reservation systems, timetabling).

M. Šeda (✉)
Institute of Automation and Computer Science, Faculty of Mechanical Engineering, Brno University of Technology, Technická 2, Brno 616 69, Czech Republic
e-mail: seda@fme.vutbr.cz

S.-I. Ao et al. (eds.), *Machine Learning and Systems Engineering*,
Lecture Notes in Electrical Engineering 68,
DOI 10.1007/978-90-481-9419-3_4, © Springer Science+Business Media B.V. 2010

However, in real situations, available capacities of resources are constrained to certain limits. Project planning under limited resources [1] is difficult because of

- Interdependence among activities due to sharing the same resources
- Dependence of resource consumption on the manner in which the activity is subdivided
- Dependence on the limits of availability of the various resources

The character of all the three dependencies is basically non-linear. In general, scheduling problems are *NP-hard*, consequently no polynomial-time algorithms are known to guarantee an optimal solution.

Classical methods of network analysis offer some approaches which depend on whether activities that have a lack of resource(s) may be suspended or not and whether activities will be put off until the moment when it is possible to perform them. A question is: Which concurrent activities should be suspended or put off because of a lack of resource(s)?

An excellent survey [3] with 203 references presents classification, models and methods for solving the *resource-constrained project scheduling problem* (RCPSP) and, in [4], it is critically reviewed and completed. There are single-mode and multi-mode variants of the problem, resources can be non-renewable or renewable, activities of project can be interrupted or not and so on.

With respect to NP-hardness of the RCPSP, mainly heuristic methods are used for its solving. Papers frequently present applications of stochastic heuristics such as simulated annealing [5, 6], genetic algorithms [6–9], tabu-search [10, 11], ant systems [12] and swarm optimization method [13].

In this paper we assume that, when resources are not sufficient, a suitable method for shifting one or more activities has been selected and we touch another problem. In real situations, the durations of activities can only be estimated and are subject to change. If we included this new information in our considerations, this could result in quite a different optimal schedule. Since, obviously, the dates of activities in the past cannot be changed, it is necessary for the calculation of the entire project with new data to modify the dates of only those activities that have not yet been finished.

2   Basic Notions

In this paragraph we introduce the notation used in this paper and the basic concepts of CPM. We consider a network graph G with n *topologically ordered* vertices [14], which means that, for each edge (i, j), i appears before j ($i < j$) and the starting vertex has number $n_0 = 1$ and ending vertex number n. This ordering can be gained as follows:

1. Start from the origin and assign n_0 to it.
2. Leave all edges outgoing from n_0 and assign numbers $n_0 + 1, \ldots, n_0 + k_1$ to k_1 vertices that have no input edge.

3. Leave all edges outgoing from the vertices numbered in the previous step and assign numbers $n_0 + k_1 + 1, \ldots, n_0 + k_1 + k_2$ to k_2 vertices that have no input edge.
4. Continue this way until all vertices are numbered.

Assume that edges represent activities of a project and vertices correspond to beginnings or ends of activities. Denote $E(G)$ the set of edges of a graph G, and $V(G)$ its set of vertices. If $e = (i, j)$ is an edge in $E(G)$, then we denote its duration by t_{ij} or $t(i, j)$, or t_e in short. Similarly, requirements of activity (i, j) for a resource r will be denoted by r_{ij} etc.

The following notions refer to start and end vertices of the network graph: $T_i^{(0)}$ represents the earliest possible start time of vertex i, $T_j^{(1)}$ the latest allowable finish time of vertex j and $T_S(i, j) = T_j^{(1)} - T_i^{(0)} - t_{i\,j}$ is the *total slack of activity* (i, j) or *total (activity) float* (the amount of time by which the start of a given activity can be delayed without delaying the completion of the project).

Finally, assume that $V_i(e)$ denotes the starting vertex of an edge $e = (i, j)$ and $V_j(e)$ is its ending vertex.

Further notation will be introduced when needed.

3   Algorithm

The algorithm is based on time shifting of activities when their total requirements are higher than the resource limit. This is implemented by prolonging their duration but distinguishing, for each activity, its starting duration and current duration which equals the length of shift and starting duration. The greatest advantage of this access is that whenever we need to compute new earliest possible start times and latest allowable finish times for activities after shifts or update the actual time duration of some activities, we can compute the whole project using a simple CPM method and, in spite of this, the dates of finished activities remain unchanged in the result of the new calculation. In other words, any change in the present has no effect on results in the past.

Let us denote

t_{ij}^s starting duration of activity (i, j)

t_{ij}^c current duration of activity (i, j)

$\delta_{ij} = t_{ij}^c - t_{ij}^s$ interval when activity has no requirements ("sleeps")

Now we will formulate an algorithm. The symbol:$=$ stands for the assignment operator.

1. *Initialization*

Using CPM method, we determine for each edge the earliest possible start time and the latest allowable finish time. Let us assign

$$\tau_1 := 0 \tag{1}$$

$$\delta_{ij} := 0, \; t_{ij}^c := t_{ij}^s \text{ for every } (i,j) \in E(G) \tag{2}$$

2. *Test for finishing*

If $\tau_1 = T_n^{(0)}$ then algorithm finishes else we continue to step 3.

3. *Determination of interval* $[\tau_1, \tau_2]$

The left bound is given and the right bound we determine from the following formula

$$\tau_2 = \min_{(i,j)\in E(G)} \left(\left\{ T_i^{(0)} \mid T_i^{(0)} > \tau_1 \right\} \cup \left\{ T_i^{(0)} + t_{ij}^c \right\} \right) \tag{3}$$

4. *Determination of activities requiring resource in* $[\tau_1, \tau_2]$

Let

$$A = \left\{ (i,j) \in E(G) \mid [\tau_1, \tau_2] \subseteq [T_i^{(0)} + \delta_{ij}, T_i^{(0)} + t_{ij}^c] \right\} \tag{4}$$

Let us determine the total requirements q of activities from A.

$$q = \sum_{(i,j)\in A} r_{ij} \tag{5}$$

If $q >$ resource limit, then we continue to step 5 else to step 7.

5. *Ordering activities from A by their priorities*

Let $A = \{e_1, ..., e_k\}$. We order all m activities in A into a non-ascending sequence B as follows. If b_k, b_l are any elements from B, then the order relation is defined as follows:

$$
\begin{aligned}
b_k \succ b_l &\Leftrightarrow T_{V_i(b_k)}^{(0)} + \delta_{b_k} < \tau_1 \\
&\text{else if } T_S(b_k) < T_S(b_l) \\
&\text{then if } T_S(b_k) = T_S(b_l) \\
&\text{else if } r_{b_k} > r_{b_l}
\end{aligned}
\tag{6}
$$

The first condition on the right-hand side means that b_k has begun in the previous interval.

6. *Right-shifting*

Because of step 4, there exists $j < m$ such that

$$\sum_{i=1}^{j} r_{b_i} \leq \text{ limit and } \sum_{i=1}^{j+1} r_{b_i} > \text{ limit} \tag{7}$$

We shift activity b_{j+1} so that new values of its parameters are obtained from the old values by the following assignments:

$$t^c_{b_{j+1}} := t^c_{b_{j+1}} + \tau_2 - \left(T^{(0)}_{PV(b_{j+1})} + \delta_{b_{j+1}} \right)$$
$$\delta_{b_{j+1}} := t^c_{b_{j+1}} - t^s_{b_{j+1}} \tag{8}$$

for the other activities in B (according to their priorities), we either add their resource requirements and place them into the schedule or shift them if limit has been exceeded.

Finally, we apply the CPM method again.

7. *Next interval*

$$\tau_1 := \tau_2 \tag{9}$$

and return to step 2.

4   Generalisation for Multiproject Schedule

The algorithm described in the previous section can be very simply adapted for multiproject scheduling with limited resources. The steps of this generalised algorithm must take into account that resources are shared by all the projects. If the number of projects is N, then, e.g., Step 3 can be modified as follows

$$\tau_2 = \min_{(i,j) \in E(G) k \in [1,N]} \left(\left\{ T^{(0)}_i | T^{(0)}_i > \tau_1 \right\} \cup \left\{ T^{(0)}_i + t^c_{ij} \right\} \right) \tag{10}$$

The other steps may be adapted in a similar way.

5   KNapsack-Based Heuristic

If we consider a schedule as a sequence of time intervals where neighbouring intervals based on the time an activity starts or finishes, the scheduling problem may be reduced to a set of the optimal choices of activities in intervals by the order relation (6).

In the network analysis literature, we may find many other strategies describing how the ranking of competing activities is accomplished, for instance:

- *Greatest remaining resource demand*, which schedules as first those activities with the greatest amount of work remaining to be completed

- *Least total float*, which schedules as first those activities possessing the least total float
- *Shortest imminent activity*, which schedules as first those activities requiring the least time to complete
- *Greatest resource demand*, which schedules as first those activities requiring the greatest quantity of resources from the outset

Using defined priorities as prices, the task investigated in one interval corresponds to the well-known *knapsack problem*. If the number of the activities sharing a source in an interval is low, e.g. up to 20, and this is satisfied in most of real situations, then this problem can be solved exactly by a branch and bound method or by dynamic programming.

Assuming only one resource with a limited capacity, we can deal with the 0–1 *knapsack problem* (0–1 KP), which is defined as follows: A set of n items is available to be packed into a knapsack with capacity of C units. Item i has value v_i and uses up to w_i of capacity. We try to maximise the total value of packed items subject to capacity constraint.

$$\max\left\{\sum_{i=1}^{n} v_i x_i \mid \sum_{i=1}^{n} w_i x_i \leq C,\ x_i \in \{0, 1\},\ i = 1, ..., n\right\} \tag{11}$$

Binary decision variables x_i specify whether or not item i is included in the knapsack.

The simplest way is to generate all possible choices of items and to determine the optimal solution among them. This strategy is of course not effective because its time complexity is $O(2^n)$.

Finding the solution may be faster using the *branch and bound method* [15], which restricts the growth of the search tree. Avoiding much enumeration depends on the precise upper bounds (the lower the upper bounds, the faster the finding of the solution is). Let items be numbered so that

$$\frac{v_1}{w_1} \geq \frac{v_2}{w_2} \geq \cdots \geq \frac{v_n}{w_n} \tag{12}$$

We place items in the knapsack by this non-increasing sequence. Let $x_1, x_2, \ldots, x_p$ be fixed values of 0 or 1 and

$$M_k = \left\{\mathbf{x} \mid \mathbf{x} \in M,\ x_j = \xi_j,\ \xi_j \in \{0, 1\},\ j = 1, \ldots, p\right\} \tag{13}$$

where M is a set of feasible solutions. if

$$(\exists q)\ (p < q \leq n\} : \sum_{j=p+1}^{q-1} w_j \leq C - \sum_{j=1}^{p} w_j \xi_j < \sum_{j=p+1}^{q} w_j \tag{14}$$

then the upper bound for M_k can be determined as follows:

$$U_B(M_k) = \sum_{j=1}^{p} v_j \xi_j + \sum_{j=p+1}^{q-1} v_j + \frac{v_q}{w_q} \left(C - \sum_{j=1}^{p} w_j \xi_j + \sum_{j=p+1}^{q-1} w_j \right) \tag{15}$$

In more complex and more frequent situations when we have more than one limited resource, we transform the resource-constrained scheduling into a sequence of *multi-knapsack problem* (MKP) solutions. MKP is defined as follows:

$$\text{maximise} \quad \sum_{i=1}^{n} v_i x_i$$

$$\text{subject to} \quad \sum_{i=1}^{n} w_{ri} x_i \leq C_r, \; r = 1, \ldots, m, \tag{16}$$

$$x_i \in \{0,1\}, \; i = 1, \ldots, n$$

For the solution of the task with multiple constraints, we must generalise the approaches mentioned above.

The combinatorial approach can be applied without any changes, but using the branch and bound method, we must redefine the upper bound. For its evaluation we use the following formula

$$U_B(M_k) = \min \left\{ U_B^1(M_k), \ldots, U_B^m(M_k) \right\} \tag{17}$$

where auxiliary bounds $U_B^i(M_k)$, $i = 1, \ldots, m$ correspond to the given constraints and are determined as in the 0–1 knapsack problem. Before evaluation of these auxiliary bounds, the other variables must be sorted again by decreasing values v_j/w_{ij}. Evidently, the run time will increase substantially.

6   Stochastic Heuristic Methods

The branch and bound method discussed above is deterministic. Now we will pay attention to stochastic heuristic methods. These methods are used in situations where the exact methods would fail or calculations would require a great amount of time.

Heuristic [16] is a technique which seeks goal (i.e. near optimal) solutions at a reasonable computational cost without being able to guarantee either feasibility or optimality, or even in many cases, to state how close to optimality a particular feasible solution is. The most popular heuristics – genetic algorithms, simulated annealing, tabu search and neural networks are reviewed in [16]. Examples of their possible use are described in [13].

Let us briefly deal with the genetic algorithms now. The skeleton for GA is shown as follows:

Generate an initial population.
Evaluate fitness of individuals in the population.
Repeat select parents from the population.
Recombine parents to produce children.
Evaluate fitness of the children.
Replace some or all of the population by the children.
Until a satisfactory solution has been found.

In the following paragraphs, we briefly summarize GA settings to our scheduling problem.

Individuals in the population (*chromosomes*) are represented as binary strings of length n, where a value of 0 or 1 at the i-th bit (*gene*) implies that $x_i = 0$ or 1 in the solution respectively.

The *population size N* is usually chosen between n and $2n$.

An *initial population* consists of N feasible solutions and it is obtained by generating random strings of 0s and 1s in the following way: First, all bits in all strings are set to 0, and then, for each of the strings, randomly selected bits are set to 1 until the solutions (represented by strings) are feasible.

The *fitness function* corresponds to the objective function to be maximised:

$$f(x) = \sum_{i=1}^{n} v_i x_i \tag{18}$$

Pairs of chromosomes (parents) are selected for recombination by the *binary tournament selection* method, which selects a parent by randomly choosing two individuals from the population and selecting the most fit one.

The *recombination* is provided by the *uniform crossover* operator. That means each gene in the child solution is created by copying the corresponding gene from one or the other parent, chosen according to a binary random number generator. If a random number is 0, the gene is copied from the first parent; if it is a 1, the gene is copied from the second parent. After crossover, the *mutation* operation is applied to each child. It works by inverting each bit in the solution with a small probability. We use a mutation rate of $5/n$ as a lower bound on the optimal mutation rate. It is equivalent to mutating five randomly chosen bits per string.

If we perform crossover or mutation operations as described above, then the generated children can violate certain capacity constraints. We can assign *penalties* to these children that prevent infeasible individuals from entering the population. A more constructive approach uses a *repair operator* that modifies the structure of an infeasible individual, so that the solution becomes feasible. Its pseudo-Pascal code is shown below for a more general case of multiple constraints. Once a new feasible child solution has been generated, the child will replace a randomly chosen solution. We use a *steady-state* replacement technique based on eliminating the individual with the lowest fitness value. Since the optimal solution values for most problems are not known, the termination of a GA is usually controlled by specifying a maximum number of generations t_{max}. Usually we choose $t_{max} \leq 5,000$.

7 Experimentation

The approach discussed in the previous paragraphs has been implemented in Borland Delphi. Its interface is similar to Microsoft Project. Although Microsoft Project often is not able to solve a constrained scheduling problem even for very simple projects (it reports some over-allocation and underallocation may be unavoidable), we have never come across this situation when using our programme. Besides the GA approach, the programme implements another popular heuristic method – *simulated annealing* – for a comparison, and a user may choose between these two possibilities. It enables a verification of the quality of computational results, as the exact results for large projects are not known.

We have tested the proposed approach in many situations. Table 1 shows results for one real project with 183 activities, 54 vertices in the network graph and 12 limited resources. This project deals with the innovation action of a Benson boiler in a local power station in the surroundings of Brno. This table contains durations of schedule for this project designed in 30 tests by the genetic algorithm (GA) and by the simulated annealing (SA). Parameters for both methods were set in a way such that the computational time for design of the project schedule should be approximately 2.4 s. The branch and bound (BB) method found a schedule with a duration of 291 days in a several times longer CPU time. The results of BB achieved are favourable because the parallelism of activities was low. The dynamic programming approach could not be used because of insufficient memory.

While the results appear comparable, statistically, the results gained by GA are better than the SA results. The well-known *Kruskal-Wallis test* for balanced one-way design was used to show that samples were taken from the same population. It has yielded the following results: Average rank for SA = 39.75, for GA = 21.25, the value of test statistic = 17.5472 and computed significance level = 2.8×10^{-5}.

Table 1 Durations for schedule (days) designed by GA and SA

GA	GA	GA	SA	SA	SA
291	294	293	294	293	295
292	292	291	292	294	296
293	291	292	293	296	292
293	293	293	294	292	297
292	292	291	297	294	294
294	291	293	292	291	294
292	291	292	293	298	295
292	292	292	294	294	291
291	293	291	294	292	295
293	291	293	295	293	293

Thus the hypothesis of sampling from the same population is rejected at all traditionally used significance levels (0.05; 0.01; 0.001).

8 Conclusion

In this chapter, a new technique for computing of the resource-constrained project scheduling was proposed. The strategy of activity-shifting was replaced by prolonging their duration and dividing them into active and sleeping parts. It makes it possible to apply a simple CPM algorithm. The proposed algorithm was designed in a mathematical form and verified for a single version of RCPSP. We also sketched how to adapt the proposed algorithm for multiproject scheduling with limited resources.

However, the deterministic approaches are not effective or may not be used in complex projects with multiple constraints because of the exponentially growing time in branch and bound method calculations. A genetic algorithm approach is proposed and compared with simulated annealing.

Further investigation will include fuzzy versions of the problem.

References

1. J. Blazewicz, K.H. Ecker, G. Schmidt, J. Weglarz, *Scheduling Computer and Manufacturing Processes* (Springer-Verlag, Berlin, 1996)
2. S.E. Elmaghraby, *Activity Networks: Project Planning and Control by Network Models* (Wiley, New York, 1977)
3. P. Brucker, A. Drexl, R. Möhring, K. Neumann, E. Pesch, Resource-constrained project scheduling: notation, classification, models, and methods. Eur. J. Oper. Res. **112**: 3–41 (1999)
4. W. Herroelen, E. Demeulemeester, B.D. Reyck, A note on the paper "resource-constrained project scheduling: notation, classification, models, and methods" by Brucker et al. Eur. J. Oper. Res. **128**, 679–688 (2001)
5. K. Bouleimen, H. Lecocq, A new efficient simulated annealing algorithm for the resource-constrained project scheduling problem and its multiple mode version. Eur. J. Oper. Res. **149**, 268–281 (2003)
6. M. Mika, G. Waligóra, J. Weglarz, Simulated annealing and tabu search for multi-mode resource-constrained project scheduling with positive discounted cash flows and different payment models. Eur. J. Oper. Res. **164**, 639–668 (2005)
7. A. Azaron, C. Perkgoz, M. Sakawa, A genetic algorithm for the time-cost trade-off in PERT networks. Appl. Math. Comput. **168**, 1317–1339 (2005)
8. K.W. Kim, M. Gen, G. Yamazaki, Hybrid genetic algorithm with fuzzy logic for resource-constrained project scheduling. Appl. Soft Comput. **2/3F**, 174–188 (2003)
9. K.W. Kim, Y.S. Yun, J.M. Yoon, M. Gen, G. Yamazaki, Hybrid genetic algorithm with adaptive abilities for resource-constrained multiple project scheduling. Comput. Indus. **56**, 143–160 (2005)
10. C. Artigues, P. Michelon, S. Reusser, Insertion techniques for static and dynamic resource-constrained project scheduling. Eur. J. Oper. Res. **149**, 249–267 (2003)

11. V. Valls, S. Quintanilla, F. Ballestin, Resource-constrained project scheduling: a critical activity reordering heuristic. Eur. J. Oper. Res. **149**, 282–301 (2003)
12 L.-Y. Tseng, S.-C. Chen, A hybrid metaheuristic for the resource-constrained project scheduling problem. Eur. J. Oper. Res. **175**(2), 707–721 (2006)
13. H. Zhang, X. Li, H. Li, F. Huang, Particle swarm optimization-based schemes for resource-constrained project scheduling. Autom. Constr. **14**, 393–404 (2005)
14. T.H. Cormen, C.E. Leiserson, R.L. Rivest, C. Stein, *Introduction to Algorithms* (MIT Press, Cambridge, MA, 2001)
15. J. Klapka, J. Dvořák, P. Popela, *Methods of Operational Research* (in Czech) (VUTIUM, Brno, 2001)
16. C.R. Reeves, *Modern Heuristic Techniques for Combinatorial Problems* (Blackwell Scientific Publications, Oxford, 1993)

Chapter 5
A Development of Data-Logger for Indoor Environment

Anuj Kumar, I.P. Singh, and S.K. Sud

Abstract This chapter describes a development of data logger for indoor environment. Present work concentrates to environmental parameter (temperature and humidity) and more polluted contaminants (concentration level of CO and CO_2). In this work four channels have been used for data logger and other four channels is open to external sensor module. The data collected will be stored in the EEPROM and output can be taken in note-pad in tabular corresponding to month/date/year using graphical user interface.

1 Introduction

Environment monitoring system is a complete data logging system. It automatically measures and records temperature, humidity and other parameters and provides warnings when readings go out of range [1]. Indoor environment monitoring is required to protect the building occupant's health by providing thermally comfortable and toxicant free environment [2]. We need a system that monitors as well as records the indoor environment data easily. In this chapter, we are discussing about the developed indoor environment monitoring system that monitors and records indoor temperature, humidity, CO and CO_2.

Available technologies for sensing these environmental parameters are developed by Onseat HOBO, Spectrum Technologies (Watch Dog weather conditions), TandD, Telaire (Wireless Monitoring Systems), Testo, Log Tag, Measurement Computing Corporation, Monarch Instruments, MSR Electronics GmbH, P3 International, Quality Thermistor, S3 Crop, Sensaphone, Sansatronics, Lascar, ICP, Graphtech, Extech Instruments, Dickson, Dent Instruments, Davis, ACR System Inc, 3M International,

A. Kumar (✉)
Instrument Design Development Centre, Indian Institute of Technology Delhi, Electronics Lab, New Delhi, India 110016
e-mail: anujkumariitd@gmail.com

S.-I. Ao et al. (eds.), *Machine Learning and Systems Engineering*,
Lecture Notes in Electrical Engineering 68,
DOI 10.1007/978-90-481-9419-3_5, © Springer Science+Business Media B.V. 2010

and Acumen [3]. The drawback of these available systems is that, both air quality and thermal comfort can not be measured simultaneously. So there is a need to develop an economical system (prototype data logger) which can help in collecting data to analyze the necessary environmental parameters simultaneously.

A prototype data logger has been developed to monitor the environmental parameters. The developed data logger consist (i) sensors module (ii) LCD (iii) Real Time Clock (RTC) (iv) EEPROM and (v) PC serial communication. This data logger is operated through PC using graphical user interface (GUI) in visual basic.

2 Sensors Module

A sensor is a device that measures a physical quantity and converts it into an equivalent analog or digital signal which can be read by an observer or by an instrument [4]. We have used temperature, relative humidity, CO, and CO_2 sensors in the developed system.

A gas sensor detects particular gas molecules and produces an electrical signal whose magnitude is proportional to the concentration of the gas [5]. Till date, no gas sensor exists that is 100% selective to only a single gas. A good sensor is sensitive to the measured quantity but less sensitive to other quantities. Available gas sensors are based on five basic principles. These can be electrochemical, infrared, catalytic bead, photo ionization and solid-state [6, 7]. We have selected these sensors because they produce a strong signal for the selected variable especially at high gas concentrations with adequate sensitivity. They have a fast response time, high stability, long life, low cost, low dependency on humidity, low power consumption, and compact size [5]. Four sensors along with their signal conditioning circuit are used to sense the desired parameter such as temperature, humidity, CO, and CO_2. Signal conditioning circuit for that sensor needs to be connected externally. In software we can select any of the analog channels. The interface of temperature, humidity, CO, and CO_2 sensors with microcontroller PIC 18F4458 is described as follows.

2.1 Temperature Sensor

National semiconductor's LM 35 IC has been used for sensing the temperature. It is an integrated circuit sensor that can be used to measure temperature with an electrical output proportional to the temperature (in °C). The temperature can be measured more accurately with it than thermistor. The operating and amplification circuit is shown in Fig. 1. The output voltage of IC LM 35 is converted to temperature in °C is given by the following expression [7]

$$Temp.(^{\circ}C) = (V_{out} \times 100)/7^{\circ}C$$

The measuring temperature range of the instrument is between 15°C and 70°C.

Fig. 1 Operating and amplification circuitof temperature sensor

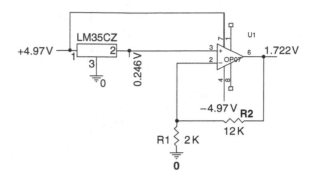

Fig. 2 Operating circuit of humidity sensor

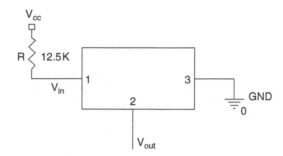

2.2   Humidity Sensor

The sensor circuit develops a linear voltage vs. RH (relative humidity) output, which is ratio metric to the supply voltage. This means when the supply voltage varies, the sensor output voltage follows the same proportion. It can operate over a range of 4–5.8 V supply. At 5 V supply voltage (at room temperature), corresponding to relative humidity variation from 0% to 100% (noncondensing), the output voltage varies from 0.8 to 3.9 V. The humidity sensor functions with a resolution of up to 0.5% of relative humidity (RH), with a typical current draw of only 200 μA, the HIH4000 series is ideally suited for low drain, battery operated systems.

The operating circuit is shown in Fig. 2. The change in the RH of the surroundings causes an equivalent change in the voltage output. The output is an analog voltage proportional to the supply voltage. Consequently, converting it to relative humidity (RH) requires both the supply and the sensor output voltages (At 25°C) and is given by the following expression [7].

$$RH = \left(\left(V_{out}/V_{\sup ply} \right) - 0.16 \right)/0.0062$$

The output of the humidity sensor is 2.548 V i.e. the relative humidity is 56% at 25°C.

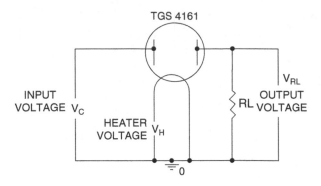

Fig. 3 Circuit for CO and CO_2 sensor [7]

2.3 CO and CO_2 Sensor

The operating circuit of CO and CO_2 sensors is shown in Fig. 3. The relationship between output voltage and gas concentration is given by the following expression

$$c = \left[\left(\frac{(V_C R_L / V_{OUT}) - R_L}{R_0} - 1 \right) \frac{1}{K} \right]^2$$

where, V_{OUT} = output voltage; V_C = input voltage, R_0 = electrical resistance of sensor at zero ppm, K = a constant for particular, R_L = sensor load resistance [5, 7].

3   LCD Interface to the Microcontroller

In this work, we are using on-chip analog to digital converter which is on the microcontroller. This analog to digital converter is having the 12 bit resolution with programmable acquisition time. It is sensing the analog signal from the sensor at the variable sampling rate (1 s to 1 h). The sensed value is converted to its digital equivalent. This digital value is displayed on the LCD (liquid crystal display) and is interfaced to the microcontroller [8, 9–12].

4   Real Time Clock Interface to the Microcontroller

The IC DS1307 operates as a slave device on the I^2C bus. Access is obtained by implementing a START condition and providing a device identification code followed by a register address. Subsequent registers can be accessed sequentially until a STOP condition is executed. When V_{CC} falls below 1.25 V_{BAT}, the device terminates an access in progress and resets the device address counter. Inputs to the device will not be recognized at this time to prevent erroneous data from being written to the device from an out of tolerance system. When V_{CC} falls below V_{BAT}, the device switches into a low-current battery backup mode. Upon power up, the device switches from battery to V_{CC} when V_{CC} is greater than $V_{BAT} +0.2$ V and recognizes inputs when V_{CC} is greater than 1.25 V_{BAT}. We are using IC DS1307 as real time clock which have features such as real-time clock counts in seconds, minutes, hours, day of month, day of week, month, and year with leap year compensation valid up to 2100, 56 Byte, Nonvolatile (NV) RAM for data storage, I^2C serial interface, programmable square wave output signal, automatic power fail detect and switch circuitry (consumes less than 500 nA in battery backup mode oscillator running), and temperature range $-40°C$ to 85°C [9, 10]. We are using I^2C to interface RTC and EEPROM to the micro-controller. The I^2C bus is the most popular of three serial EEPROM protocols. The I^2C chips include address pins as an easy way to have multiple chips on a single bus while only using two connections to the microcontroller [9].

5   EEPROM Interface to the Microcontroller

The EEPROM will store the digital value which is coming from analog to digital converter. We will require 52.73 MB of EEPROM if we are sampling all analog channels at the rate of 1 sample/s. We are using the EEPROM AT24C256 (ATMEL). This will store the sample data at different instants [10–14].

6   PC Interface Using RS-232 Serial Communication

PIC 18F4458 using MAX-232 is interfaced with PC. IC (MAX-232) used to convert TTL logic level to RS-232 logic level. RS-232 is the serial communication protocol that does not require the clock along with data lines. Two data lines are there one is T_X and another is R_X for serial communication. MAX-432 has two receivers (converts RS-232 logic level to TTL logic) and two drivers. Separate power supply has been provided because minimum power supply needed is 5 V and MAX-232 consumes a lot of current for operation. External capacitors are required for internal voltage pump to convert TTL logic level to RS-232 level. For battery operated application MAX-232 can be used as level converter instead of MAX-232. It is low supply low power consumption logic converter IC for RS-232 [9, 10, 13].

7   Graphical User Interface

The GUI is one of the important parts for this device as it displays the data from microcontroller for data monitoring and analysis. The design template has to be user friendly for best usage. For this chapter, the main objective is to display data received in graphical form. As transducer detects and translate an analog signal, the data will go through a conversion at the ADC. This digital data will be stored in EEPROM chip with the help of Visual Basic 6.0 software. Since the data is using serial RS232 communication, an initialization needs to be done having baud rate, data bits, parity, stop bit, and the COM port at PC. The baud rate is the number of signal changes per second or transition speed between Mark (negative) and Space (positive) which ranges from 110 to 19,200, data bits is the length of data in bit which has one Least Significant Bit and one Most Significant Bit, the parity bit is an optional bit mainly for bit error checking. It can be odd, even, none Mark, and Space. Stop bit is used to frame up the data bits and usually combined with the start bit. These bits are always represented by a negative voltage and can be 1, 1.5 and 2 stop bits, and COM port is the selection of the available COM port at PC. The commonly used setting to establish a serial RS232 communication is 9600 baud rate, none parity, 8 data bits, 1 stop bit, and COM port 1. This can be done by using the GUI monitoring system where it automatically saves the data received in a notepad. The data saved is the date and time at which the data collected and the data value it self. Figures 4 and 5, represents the graphical user interface and logged data in file respectively [12, 13].

8   Schematic of the Data Logger

Figure 6 shows, the full schematic diagram of the data logger for indoor environment. This data logger has four embedded sensor module and other four channels are open to be used for the measurement of other environmental parameters.

9   Software Design of Data Logger

This section includes the discussion on software design for all the modules interfaced with PIC 18F4458. It also explains the functions of software designed for data logger [11].

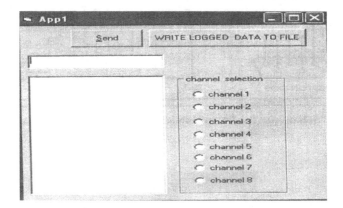

Fig. 4   GUI for the data logger

| SET_1 - Notepad |
| File Edit Format View Help |

```
Date Time        Temperature (*F) (1)        Temperature (*C) (1)      RH (%) CO (ppm) CO2 (ppm)
06/01/09 06:00:00.C     89.48    31.93     53        3.2       418
06/01/09 06:10:00.C     89.48    31.93     53.3      3.2       419
06/01/09 06:20:00.C     89.48    31.93     53.5      3.2       418
06/01/09 06:30:00.C     89.48    31.93     53.8      3.2       418
06/01/09 06:40:00.C     89.48    31.93     53.8      3.1       419
06/01/09 06:50:00.C     89.48    31.93     53.8      3.1       420
06/01/09 07:00:00.C     89.48    31.93     53.8      3.1       420
06/01/09 07:10:00.C     89.48    31.93     53.8      3.3       420
06/01/09 07:20:00.C     89.48    31.93     53.8      3.3       420
06/01/09 07:30:00.C     89.48    31.93     54        3.4       421
06/01/09 07:40:00.C     89.48    31.93     53.8      3.4       421
06/01/09 07:50:00.C     89.48    31.93     54.3      3.4       421
06/01/09 08:00:00.C     89.48    31.93     54.3      3.4       421
06/01/09 08:10:00.C     89.48    31.93     54.5      3.2       422
06/01/09 08:20:00.C     89.48    31.93     54.5      3.2       422
06/01/09 08:30:00.C     89.48    31.93     54.5      3.2       422
06/01/09 08:40:00.C     89.48    31.93     54.3      3.4       423
06/01/09 08:50:00.C     89.48    31.93     54.3      3.4       423
06/01/09 09:00:00.C     89.48    31.93     53.8      3.5       423
06/01/09 09:10:00.C     89.48    31.93     54.8      3.5       424
06/01/09 09:20:00.C     88.74    31.52     54.9      3.5       424
06/01/09 09:30:00.C     88.74    31.52     53.1      3.5       425
06/01/09 09:40:00.C     88.74    31.52     52        3.4       425
06/01/09 09:50:00.C     88.74    31.52     51.3      3.4       428
06/01/09 10:00:00.C     88.74    31.52     50.3      3.4       429
06/01/09 10:10:00.C     89.48    31.93     49.4      3.4       428
06/01/09 10:20:00.C     89.48    31.93     48.7      3.2       429
06/01/09 10:30:00.C     89.48    31.93     49        3.2       430
06/01/09 10:40:00.C     89.48    31.93     48.5      3.2       430
06/01/09 10:50:00.C     89.48    31.93     47.9      3.3       430
06/01/09 11:00:00.C     89.48    31.93     47.6      3.4       431
06/01/09 11:10:00.C     89.48    31.93     46.9      3.4       431
```

Fig. 5   Representations of the logged data in file

9.1   Programming Steps for I^2C Interface

I^2C interface is bi-directional. This is implemented by an "Acknowledge" or "ACK" system allows data to be sent in one direction to one item on the I^2C bus, than, that item will "ACK" to indicate the data received. Normally, the master device controls the clock line, SCL. This line dictates the timing of all transfers on the I^2C bus. Other devices can manipulate this line, but they can only force the line low. This action means that item on the bus cannot deal with more data in to any device.

9.1.1   Writing to an I^2C Chip

The function of writing to the EEPROM is shown here as "Control IN", which represents putting the EEPROM in an "input" mode. Since we are only sending data to the EEPROM (as shown in Fig. 7), we use "Control IN" byte and later used "Control OUT". Next, the EEPROM acknowledges this byte. This is shown by the "A" after the byte. It is put on the next line to indicate that this is transmitted by the EEPROM. The Address Byte contains the address of the location of the EEPROM where we want to write data. Since the address is valid, the data is acknowledged by the EEPROM. Finally, we send the data we want to write. The data is then acknowledged by the EEPROM. When that finishes, we send a stop condition to complete the transfer. Remember the "STOP" is represented as the "T" block on the end. Once the EEPROM gets the "STOP" condition it will begin writing to its memory.

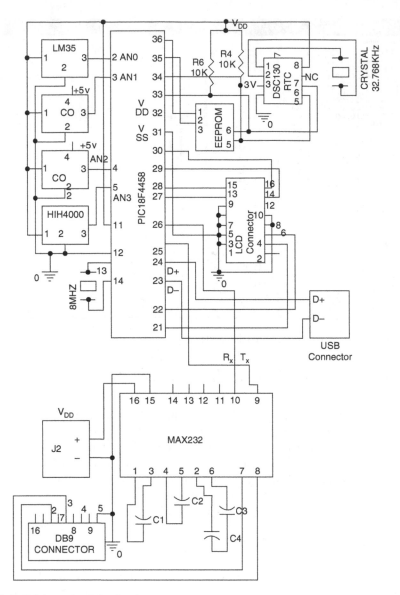

Fig. 6 Full Schematic of the data logger

9.1.2 Reading from an I²C Chip

The transfer will use the "Control IN" byte to load the address into the EEPROM (as shown in Fig. 8). This sends data to the EEPROM which is why we use the control in byte. Once the address is loaded, we want to retrieve the data. So, we send a "Control OUT" byte to indicate to the EEPROM that we want data FROM it.

Fig. 7 Writing the data in
I²C chip (Controlling
Window)

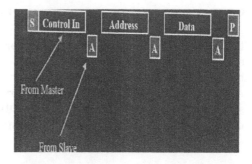

Fig. 8 Reading the data from
an I²C chip

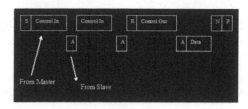

The EEPROM will acknowledge this and then send the data we requested. When we are done getting data, we send a "NACK" to tell the EEPROM that we do not want more data. If we were to send an ACK at this point, we could get the next byte of data from the EEPROM. Since we only want to read one byte, we send a "NACK".

9.2   Programming Steps for LCD Interface

Set RS = 0 to send command; Send 0b0010 to data lines three times with a delay of 2 ms; to send a byte on four data lines, send higher nibble first and give a RE pulse of 100 µs at RE; send a set of instruction one after another with a delay of 2 ms between each command to configure various setting as given in instruction set of LCD datasheet [9, 13]; send instruction set again.

Set RS = 1; Send higher nibble at four data lines. Send 100 µs RE pulse; Send lower nibble at data lines. Send RE pulse; Keep track of number of character already displayed on display panel using LCD_count. Go to line 2 or line 1 according to that.

9.3   Programming Steps for Sensor Data Collection

There are four sensor module connected such as temperature, humidity, CO, and CO_2. Data is collected by the ADC inbuilt in PIC. ADC provides 12 bit of data after the conversion is completed.

9.3.1   Temperature Sensor Data Collection

Data collection from the temperature sensor needs following actions to be carried out (a) Selecting the analog channel AN_0, sampling frequency, and alignment of bits for ADRESH and ADRESL, (b) Vref and power on the ADC module by setting $ADCON_0$, $ADCON_1$ and $ADCON_2$ registers, (c) starting analog to digital conversion by setting ADGO bit high (wait till ADIF flag will not indicate the completion of conversion), and (d) copy of results from ADRESH and ADRESL to variables.

9.3.2   Humidity Sensor Data Collection

Select the AN_3 and set other features of ADC as temperature sensor; after completion of conversion copy the result in variable.

9.3.3   CO and CO_2 Sensor Data Collection

Data collection from the CO sensor needs following actions to be carried out (a) Selecting the analog channel AN_1, sampling frequency, and alignment of bits for ADRESH and ADRESL, (b) Vref and power on the ADC module by setting $ADCON_0$, $ADCON_1$ and $ADCON_2$ registers, (c) starting analog to digital conversion by setting ADGO bit high (wait till ADIF flag will not indicate the completion of conversion), and (d) copy of results from ADRESH and ADRESL to variables.

Now repeat the same process to collect the CO_2 data on the channel number AN_2.

10   Results and Discussion

Sensors module, EEPROM, RTC, and LCD have been successfully interfaced to the microcontroller. EEPROM is successfully storing the logged data with time and date tag. The sensors data is being displayed on LCD module. A simple GUI has been designed to store a logged data to a text file, so that it can be analyzed further.

11   Conclusions

We have developed a low cost, 12 bit resolution data logger and successfully measured temperature, humidity, and concentration of CO and CO_2 gases. The GUI designed gives a lucratively look to the functioning of data logger. Initial results of the data logger are encouraging and we are working on to improve the GUI model as well as the accuracy of data logger.

References

1. G.L. Tang, *Lecture notes on Health and the built environment: Indoor air quality* (University of Calgary, Alberta, Canada)
2. J.D. Richard, G.S. Brager, Thermal comfort in naturally ventilated buildings: revisions to ASHRAE standard 55. Energy Buildings **34**, 549–561 (2002)
3. Microdaq (March 3, 2009); http://www.microdaq.com/data-logger/
4. D.D. Lee, D.S. Lee, Environment gas sensors. IEEE Sensors J. **1**(3), 214–215 (2001)
5. N. Kularatna, B.H. Sudantha, An environmental air pollution monitoring system based on the IEEE 1451 standard for low cost requirements. IEEE Sensors J. **8**(4), 415–422 (2008)
6. Sensor industry development and trends, Sensors express (Nov 2002)
7. RS India (April 4, 2009); http://www.rsonlineindia.com
8. R. Luharuka, R.X. Gao, A microcontroller-based data logger for physiological sensing. IEEE Proc. on Instrument and Measurement Technology Conference, Anchorage, AK, USA, 21–23 May, 175–180 (2002)
9. Data acquisition logging circuits (20 Mar 2009); http://www.hobbyprojects.com/A/acquistions_data_circuits.html
10. Data sheet of real time clock DS 1307 (8 May 2008). http://www.maxim-ic.com/products/rtc/real-time-clocks.cfm
11. Microchip (10 Jan 2009). http://www.microchip.com
12. Introduction to data acquisition (10 May 2008). http://zone.ni.com/devzone/concepted.nsf/webmain/
13. G. Mason, A handheld data acquisition system for use in an undergraduate data acquisition course. Data Acquisit. **45**, 338–393 (2002)
14. D. Malone, Build a versatile data logger. Popular Electronics, 35–42, July 1994

Chapter 6
Multiobjective Evolutionary Optimization and Machine Learning: Application to Renewable Energy Predictions

Kashif Gill, Abedalrazq Khalil, Yasir Kaheil, and Dennis Moon

Abstract The inherent variability in climate processes results in significant impacts on renewable energy production. While a number of advancements have been made over the years, the accurate energy production estimates and the corresponding long-term variability at the full wind farm remains a big challenge. At the same time, long-term energy estimates and the variability are important for financial assessment of the wind farm projects. In this chapter, a machine learning approach to model wind energy output from the wind farm is presented. A multiobjective evolutionary optimization (MOEO) method has been applied for the optimization of an Artificial Intelligence learning methodology the "Support Vector Machines" (SVM). The optimum parameter search is conducted in an intelligent manner by narrowing the desired regions of interest that avoids getting struck in local optima. The National Center for Environmental Prediction (NCEP)'s global reanalysis gridded dataset has been employed in this study. The gridded dataset for this particular application consists of four points each consisting of five variables. A 40-years, 6-hourly energy prediction time series is built using the 40-years of reanalysis data (1968-present) after training against short-term observed farm data. This is useful in understanding the long-term energy production at the farm site. The results of MOEO-SVM for the prediction of wind energy are reported along with the multiobjective trade-off curves.

1 Introduction

It is well-known that the weather characteristics in a given time frame are determined by a relatively small number of discrete weather systems, each of which may exhibit very different influences and patterns of development. These fundamental

K. Gill (✉)
WindLogics, Inc., 1021 Bandana Blvd., E. # 111, St Paul, MN 55108, USA
e-mail: kgill@windlogics.com

S.-I. Ao et al. (eds.), *Machine Learning and Systems Engineering*,
Lecture Notes in Electrical Engineering 68,
DOI 10.1007/978-90-481-9419-3_6, © Springer Science+Business Media B.V. 2010

causes of variations in on-site wind speeds are inherent in the atmosphere and must be understood, to the extent possible, for accurate wind resource assessment. In order to make useful predictions of wind speed/energy, it is therefore important to develop statistically sound relationships between those wind regimes and the atmospheric conditions. This is the core of the current methodology described in this chapter. Machine learning tools have gained immense popularity in the geosciences community due to their success against physically-based modeling approaches. In brief, machine learning tools are used to determine the relationship between inputs and output in an empirical framework. These models do not employ traditional form of equations common in physically-based models or as in regressions, instead have flexible and adaptive model structures that can abstract relationships from data.

The chapter describes a method for training Support Vector Machine (SVM) in applications for wind energy predictions at a wind-farm level. The SVM is a powerful learning algorithm developed by Vapnik and is known for its robust formulation in solving predictive learning problems employing finite data [1]. The method is well-suited for the operational predictions and forecasting of wind power, which is an important variable for power utility companies. The proposed methodology employs a Multiobjective Evolutionary Optimization approach for training the SVM. The goal of an optimization method is to efficiently converge to a global optimum, in the case of a single objective function, and to define a trade-off surface in the case of multiobjective problems. Overall, global optimization (GO) methods have two main categories: deterministic and probabilistic. Deterministic methods use well-defined mathematical search algorithms (e.g., linear programming, gradient search techniques) to reach the global optimum, and sometimes use penalties to escape from a local optimum. Probabilistic methods employ probabilistic inference to reach the global optimum. Evolutionary optimization algorithms generally fall in probabilistic category of optimization methods.

2   Material and Methods

The details on support vector machine, the multiobjective evolutionary optimization, and the SVM trainings are presented in the current section.

2.1   Support Vector Machines

Support Vector Machines (SVM) has emerged as an alternative data-driven tool in many traditionally Artificial Neural Network (ANN) dominated fields. SVM was developed by Vapnik in the early 1990s mainly for applications in classification. Vapnik later extended his work by developing SVM for regression. The SVM can "learn" the underlying knowledge from a finite training data and develops generalization capability which makes it a robust estimation method.

The mathematical basis for SVM is derived from statistical learning theory. SVM consists of a quadratic programming problem that can be efficiently solved and for which a global extremum is guaranteed [2]. It is sparse (robust) algorithm when compared against ANN, but what makes it sparse is the use of Structural Risk Minimization (SRM) instead of Empirical Risk Minimization (ERM) as is the case in ANN. In SRM, instead of minimizing the total error, one is only minimizing an upper bound on the error. The mathematical details on SVM can be found in [3–5]. The function $f(x)$ that relates inputs to output has the following form:

$$f(\mathbf{x}) = \sum_{i=1}^{N} w_i \phi(\mathbf{x}_i) + b \tag{1}$$

Where N (usually $N \ll L$) is the number of support vectors which are selected from the entire dataset of 'L' samples. The support vectors are subsets of the training data points that support the "*decision surface*" or "*hyper-plane*" that best fits the data.

The function $f(x)$ is approximated using the SRM induction principle by minimizing the following objective function:

$$\text{Minimize } R[f] = C \frac{1}{K} \sum_{i=1}^{N} |y_i - f(\mathbf{x}_i)|_\varepsilon + \|\mathbf{w}\|^2 \tag{2}$$

The first term $(|y_i - f(\mathbf{x}_i)|_\varepsilon)$ in the above expression is what is called ε-insensitive loss function. The samples outside the ε-tube are penalized by a penalty termξ. The above formulation makes the solution sparse in the sense that the errors less than ε are ignored. The sparse solution is the one that gives the same error with minimum number of coefficients describing $f(x)$. The second term in the objective function is the regularization term added to avoid the consequences of the ill-posedness of the inverse problem [6].

The Eq. (2) is solved in dual form employing Lagrange multipliers as [3]:

$$f(\mathbf{x}) = \sum_{i=1}^{N} (\alpha_i^* - \alpha_i) K(\mathbf{x}, \mathbf{x}_i) + b \tag{3}$$

The α_i and α_i^* are the Lagrange multipliers and have to greater than zero for the support vectors $i = 1, \ldots, N$, and $K(\mathbf{x}_i, \mathbf{x})$ is a kernel function.

There are three main parameters to be determined as part of SVM trainings; the tradeoff 'C', the degree of error tolerance 'ε', and parameter related to the kernel, the kernel width, 'γ'. In practice, these may be determined either using a trial-and-error procedure or an automatic optimization method [7]. The optimization scheme can be single objective or multiobjective depending upon the nature of the problem. The initial formulation for training SVM employed a single objective approach to optimization. It was noticed that single objective methods to train model can result in optimal parameter sets for the particular single-objective value; but fail to provide reasonable estimates for the other objective. It is therefore preferred to employ multiobjective search procedures. In the current implementation, Multiobjective Particle Swarm Optimization (MOPSO) algorithm is used to determine the SVM parameters.

2.2 Multiobjective Evolutionary Optimization

The goal in an optimization algorithm is to find the minimum (or maximum) of a real valued function $f : S \rightarrow \Re$ i.e., finding optimum parameter set $x^* \in S$ such that:

$$f(x^*) \leq f(x), \quad \forall x \in S \tag{4}$$

Where $S \subset \Re^D$ is the feasible range for x (the parameter set, having D-dimensions) and $\Re^D$ represents the D-dimensional space of real number.

The current multiobjective methodology employs Swarm Intelligence based evolutionary computing multiobjective strategy called Multiobjective Particle Swarm Optimization (MOPSO) [8]. The PSO method has been developed for single objective optimization by R. C. Eberhart and J. Kennedy [9]. It has been later extended to solve multiobjective problems by various researchers including the method by [8].

The method originates from the swarm paradigm, called Particle Swarm Optimization (PSO), and is expected to provide the so-called global or near-global optimum. PSO is characterized by an adaptive algorithm based on a social-psychological metaphor [9] involving individuals who are interacting with one another in a social world. This sociocognitive view can be effectively applied to computationally intelligent systems [10]. The governing factor in PSO is that the individuals, or "particles," keep track of their best positions in the search space thus far obtained, and also the best positions obtained by their neighboring particles. The best position of an individual particle is called "local best," and the best of the positions obtained by all the particles is called the "global best." Hence the global best is what all the particles tend to follow. The algorithmic details on PSO can be found in [8, 9, 11, 12]. The approach in [8] presents a multiobjective framework for SVM optimization using MOPSO.

A multiobjective approach differs from a single objective method in that the objective function to be minimized (or maximized) is now a vector containing more than one objective function. The task, therefore, of the optimization method is to map out a trade-off surface (otherwise known as *Pareto* front), unlike finding a single scalar-valued optimum in case of single objective problems. The multiobjective approach to the PSO algorithm is implemented by using the concept of *Pareto* ranks and defining the *Pareto* front in the objective function space. Mathematically, a *Pareto* optimal front is defined as follows: A decision vector $\vec{x}_1 \in S$ is called *Pareto* optimal if there does not exist another $\vec{x}_2 \in S$ that dominates it. Let $P \subseteq \Re^m$ be a set of vectors. The *Pareto* optimal front $P^* \subseteq P$ contains all vectors $\vec{x}_1 \in P$, which are not dominated by any vector $\vec{x}_2 \in P$:

$$P^* = \{\vec{x}_1 \in P| \ \not\exists \vec{x}_2 \in P : \vec{x}_2 \prec \vec{x}_1\} \tag{5}$$

The idea of *Pareto* ranking is to rank the population in the objective space and separate the points with rank 1 in a set *P** from the remaining points. This establishes a *Pareto* front defined by a set *P**. All the points in the set *P** are the "behavioral" points (or non-dominated solutions), and the remaining points in set *P* become the "non-behavioral" points (or inferior solution or dominated solutions). The reason behind using the *Pareto* optimality concept is that there are solutions for which the performance of one objective function cannot be improved without sacrificing the performance of at least one other.

In the MOPSO algorithm, as devised in [8], the particles will follow the nearest neighboring member of the *Pareto* front based on the proximity in the objective function (solution) space. At the same time, the particles in the front will follow the best individual in the front, which is the median of the *Pareto* front. The term follow means assignments done for each particle in the population set to decide the direction and offset (velocity) in the subsequent iteration. These assignments are done based on the proximity in the objective function or solution space. The best individual is defined in a relative sense and may change from iteration to iteration depending upon the value of objective function.

An example test problem is shown in Fig. 1. The test function presents a maximization problem (as in [13]):

$$\text{Maximize } F = (f_1(x, y), f_2(x, y)) \tag{6}$$

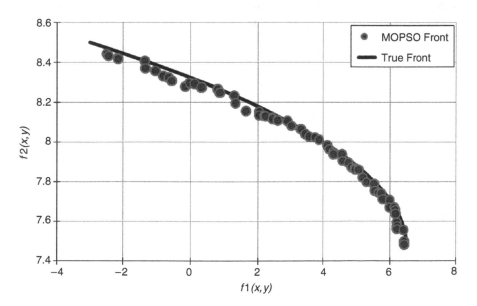

Fig. 1 Test function results

where

$$f_1(x,y) = -x^2 + y, \quad f_2(x,y) = \frac{x}{2} + y + 1$$

subject to

$$0 \geq \frac{x}{6} + y - 6.5, \quad 0 \geq \frac{x}{2} + y - 7.5, \quad 0 \geq 5x + y - 30, \quad x, y \geq 0$$

in the range $0 \leq x, y \leq 7$. The true front and the front from the MOPSO algorithm are shown in Fig. 1 after 5,000 function evaluations. It can be noticed that MOPSO was able to reproduce the true front for this test case.

2.3   SVM-MOPSO Trainings

The unique formulation of MOPSO helps it to avoid getting struck in local optima, when making a search in the multi-dimensional parameter domain. In the current research, the MOPSO is used to parameterize the three parameters of SVM namely; the trade-off or cost parameter 'C', the epsilon 'ε', and the kernel width 'γ'. The MOPSO method uses a population of parameter sets to compete against each other through a number of iterations in order to improve values of specified multi-objective criteria (objective functions) e.g., root mean square error (RMSE), bias, histogram error (BinRMSE), correlation, etc. The optimum parameter search is conducted in an intelligent manner by narrowing the desired regions of interest and avoids getting struck in local optima.

In the earlier efforts, a single objective optimization methodology has been employed for optimization of three SVM parameters. The approach was tested on a number of sites and results were encouraging. However, it has been noticed that using a single objective optimization method can result in sub-optimal predictions when looking at multiple objectives. The single objective formulation (using PSO) employed coefficient of determination (COD), as the only objective function, but it was noticed that the resulting distributions were highly distorted when compared to the observed distributions. The coefficient of determination (COD) is linearly related to RMSE and can range between $-\infty$ and 1; the value of 1 being a perfect fit. It is shown in Fig. 2 where a trade-off curve is presented between BinRMSE vs. COD. It can be noticed that COD value increases with the increase in BinRMSE value. The corresponding histograms are also shown in Fig. 3 for each of the extreme ends (maximum COD and minimum BinRMSE) and the "compromise" solution from the curve. The histograms shown in Fig. 3 make it clear that the best COD (or RMSE) is the one with highest BinRMSE and indeed misses the extreme ends of the distribution. Thus no matter how tempting it is to achieve best COD

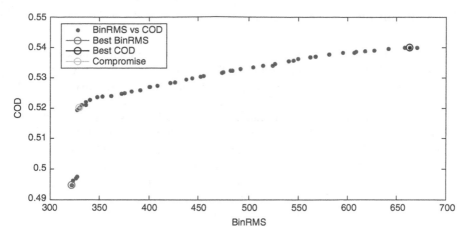

Fig. 2 Trade-off curve between BinRMSE and COD

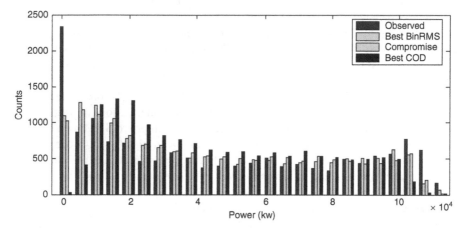

Fig. 3 Histogram comparison between Observed, BinRMS best, COD best, and the best compromise solution

value it does not cover the extreme ends of the distribution. On the other hand the best BinRMSE comes at the cost of lowest COD (or highest RMSE) and is not desired either. Thus it is required to have a multiobjective scheme that simultaneously minimize these objectives and provide a trade-off surface and therefore a compromise solution can be chosen between the two objectives. Figure 3 also shows the histogram for the "compromise" solution which provides a decent histogram when compared to observed data.

3 Application

The current procedures primarily employ SVM for building regression models for assessing and forecasting wind resources. The primary inputs to the SVM come from the National Center for Environmental Prediction (NCEP)'s reanalysis gridded data [14] centered on the wind farm location. The target is the measurements of wind farm aggregate power. The current training uses a k-fold cross validation scheme referred to as "Round-Robin strategy". The idea within the "Round-Robin" is to divide the available training data into two sets; use one for training and hold the other for testing the model. In this particular "Round-Robin strategy" data is divided into months. The training is done on all the months except one, and the testing is done on the hold-out month. The previous operational methods employ manual calibration for the SVM parameters in assessment projects and a simple grid-based parameter search in forecasting applications.

The goal in using MOPSO is to explore the regions of interest with respect to the specific multiobjective criteria in an efficient way. Another attractive feature of MOPSO is that it results in a so-called *Pareto* parameter space, which accounts for parameter uncertainty between the two objective functions. Thus the result is an ensemble of parameter sets cluttered around the so called global optimum with respect to the multiobjective space. This ensemble of parameter sets also gives tradeoffs on different objective criteria. The MOPSO-SVM method is tested on the data from an operational assessment site in North America. The results are compared with observed data using a number of evaluation criteria on the validation sets.

As stated above, the data from four NCEP's grid points each consisting of five variables (total 20 variables) is used. It has been noticed that various normalization and pre-processing techniques on the data may help to improve SVM's prediction capabilities. In the current study, input data has also been tried by pre-processing it using Principal Component Analysis (PCA). In that case, the PC's explaining 95% of the variance are included as inputs to SVM. The comparison is made with SVM that does not use PCA as pre-processing step.

MOPSO require a population consisting of parameter sets to be evolved through a number of iterations competing against each other to obtain an optimum (minimum in this case) value for the BinRMSE and RMSE. In the current formulation, a 50 member population is evolved for 100 iterations within MOPSO for wind power predictions at the wind farm.

4 Results and Discussion

The wind farm site is located in Canada and has 29 months of energy data available. The MOPSO is used to train SVM over the available 17 months of training data (at 6-hourly time resolution) using a "Round-Robin" cross-validation strategy. This gives an opportunity to train on 16 months and test the results on a 'hold-out' 1 month test set. By repeating the process for all the 17 months, gives a full 17 months of test data to compare against the observed. Since there is 29 months of data available for this site, a full 1 year of data is used in validation (completely unseen data). The results that follow are the predictions on the validation set. The results are shown on the normalized (between −1 and 1) dataset. The MOPSO-SVM results are shown with and without PCA pre-processing.

The trade-off curve for MOPSO SVM optimization for the two objectives is shown in Fig. 4. The trade-off between BinRMS vs. RMSE is shown for the SVM using original input compared against SVM using Principal Components (PCs) as inputs. It can be noticed that there is little difference between the two approaches. The PCA-SVM produced a better objective function result for BinRMS, where as simple SVM provided a better objective function result for RMSE.

Figure 5 shows the monthly mean wind power for the 12 months compared against the observed data. The results are shown for SVM prediction with and without the pre-processing using PCA. As stated above, there is a very little difference between the two approaches and a good fit has been found. It can be noticed that predictions are in reasonable agreement with the observed data. The results in Fig. 6 show histogram of observed vs. the predicted wind power data at the 6-hourly time resolution (the prediction time step). The results are shown for SVM prediction with and without the pre-processing using PCA. It can be noticed that the distributions are well-maintained using MOPSO methodology and a reasonable agreement between observed and predicted power is evident from Fig. 6. A number of goodness-of-fit measures are evaluated in Table 1, which are monthly root mean square error (RMSE), monthly coefficient of determination (COD), instantaneous RMSE, instantaneous COD, and BinRMSE (histogram bin RMSE). The results in Table 1 are presented for SVM prediction with and without the pre-processing using PCA. Both monthly and instantaneous wind power are of significant interest, and thus are

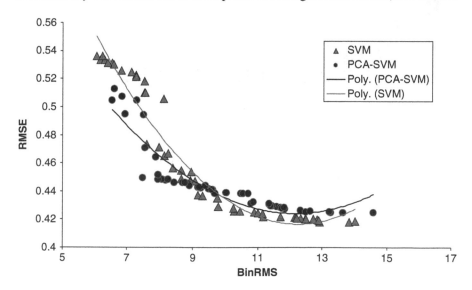

Fig. 4 Trade-off curve between the two objectives BinRMS vs. RMSE for SVM and PCA-SVM

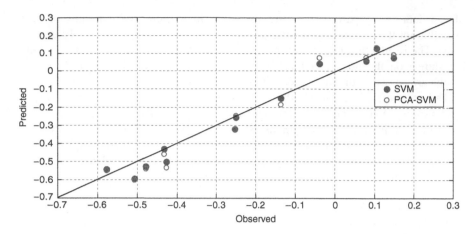

Fig. 5 Scatter plot for the mean monthly wind energy data for SVM and PCA-SVM

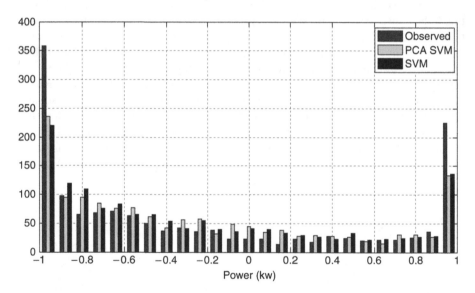

Fig. 6 Histogram for SVM and PCA-SVM along with observed

included in the current analysis. It can be noticed that not only monthly but also instantaneous power predictions are in close agreement with the observed.

5   Conclusions

Machine learning methods have gained in popularity in solving prediction problems in the area of geosciences. Due to their ability to model complex processes, the machine learning algorithms are preferred over computationally rigorous process-based

Table 1 Wind energy goodness-of-fit

Goodness measure	SVM	PCA-SVM
Monthly RMSE	0.053	0.064
Monthly COD	0.954	0.934
Instantaneous RMSE	0.381	0.381
Instantaneous COD	0.734	0.735
BinRMSE	37.758	36.491

models which are usually over-parameterized and require immense amounts of data to calibrate. Support Vector Machines are well-suited for the problems exhibiting high degrees of spatial and temporal variability, issues of nonlinearity, conflicting scales, and hierarchical uncertainty.

In the current chapter, a multiobjective evolutionary computing method MOPSO is used to optimize the three parameters of SVM for wind energy predictions. The approach has been tested on data from a wind farm using NCEP's re-analysis grid data. The prediction strategy employs SVM which is parameterized for the two objective functions. The approach is also tested by pre-processing the input data using PCA. A number of graphical and tabular results in the form of goodness-of-fit measures are presented for wind energy predictions. The results also show a trade-off curve for the two objectives employed in the MOPSO. The trade-off curve is helpful in identifying the appropriate parameter set for SVM in order to achieve the desired accuracy for the two objective problem. The SVM predictions at the farm level produced excellent agreement with the observed data for the validation set. Overall, the results have been encouraging and it is recommended to use MOPSO-SVM approach for other operational projects in the area of renewable energy predictions and forecasting. While further modifications and advancements are underway, the current procedure is sound enough to be applied in operational settings.

References

1. V. Cherkassky, F. Mulier, *Learning from Data - Concepts, Theory, and Methods* (Wiley, New York, 1998), p. 441
2. B. Schölkopf, K. Sung, J.C. Chris, C. Burges, F. Girosi, T. Poggio, V. Vapnik, Comparing support vector machines with Gaussian kernels to radial basis function classifiers. IEEE Trans. Signal Process. 45(11), 2758–2765 (1997)
3. V. Vapnik, *The Nature of Statistical Learning Theory* (Springer, New York, 1995), p. 188
4. V. Vapnik, *Statistical Learning Theory* (Wiley, Hoboken, NJ, 1998), p. 736
5. N. Cristianini, J. Shaw-Taylor (eds), *An Introduction to Support Vector Machines and Other Kernel-Based Learning Methods* (Cambridge University Press, Cambridge, 2000), p. 189
6. A. Tikhonov, V. Arsenin, *Solution of Ill-posed Problems* (W.H. Winston & Sons, Washington, DC, 1977), p. 258
7. M.K. Gill, T. Asefa, Y. Kaheil, M. McKee, Effect of missing data on performance of learning algorithms for hydrologic predictions: Implications to an imputation technique. Water Resour. Res. 43, W07416 (2007). doi:10.1029/2006WR005298.

8. M.K. Gill, Y.H. Kaheil, A. Khalil, M. McKee, L. Bastidas, Multiobjective particle swarm optimization for parameter estimation in hydrology. Water Resour. Res. 42, W07417 (2006). doi:10.1029/2005WR004528 (2006)
9. R.C. Eberhart, J. Kennedy, A new optimizer using particle swarm theory, in *Proceedings of the Sixth International Symposium on Micro Machine and Human Science*, 1995, MHS'95. doi:10.1109/MHS.1995.494215 (IEEE Press, Piscataway, NJ (1995), pp. 39–43
10. J. Kennedy, R.C. Eberhart, Particle swarm optimization, in *Proceedings of IEEE International Conference on Neural Networks*, IV, vol. 4. doi:10.1109/ICNN.1995.488968 (IEEE Press, Piscataway, NJ (1995), pp. 1942–1948
11. J. Kennedy, R.C. Eberhart, Y. Shi, *Swarm Intelligence* (Morgan Kaufmann, San Francisco, CA, 2001)
12. R. C. Eberhart, R. W. Dobbins, P. Simpson, *Computational Intelligence PC Tools* (Elsevier, New York, 1996)
13. H. Kita, Y. Yabumoto, N. Mori, Y. Nishikawa, Multi-objective optimization by means of the thermodynamical genetic algorithm, in *Parallel Problem Solving From Nature—PPSN IV*, eds. by H.-M. Voigt, W. Ebeling, I. Rechenberg, H.-P. Schwefel (Springer-Verlag Berlin, Germany, 1996), Sept. Lecture Notes in Computer Science, pp. 504–512
14. Kalnay et al. The NCEP/NCAR 40-year reanalysis project. Bull. Am. Meteor. Soc. 77, 437–470 (1996)

Chapter 7
Hybriding Intelligent Host-Based and Network-Based Stepping Stone Detections

Mohd Nizam Omar and Rahmat Budiarto

Abstract This paper discusses the idea of hybriding intelligent host-based and network-based stepping stone detections (SSD) in order to increase detection accuracy. Experiments to measure the True Positive Rate (TPR) and False Positive Rate (FPR) for both Intelligent-Network SSD (I-NSSD) and Intelligent-Host SSD (I-HSSD) are conducted. In order to overcome the weaknesses observed from each approach, a Hybrid Intelligent SSD (HI-SSD) is proposed. The advantages of applying both approaches are preserved. The experiment results show that HI-SSD not only increases the TPR but at the same time also decreases the FPR. High TPR means that accuracy of the SSD approach increases and this is the main objective of the creation of HI-SSD.

1 Introduction

When the Internet was created, security was not a priority. The TCP/IP protocol security mechanism was thought to be sufficient at the beginning. However, as the Internet usage increases, its security mechanism has become more and more problematic [1]. In fact, as the internet is used widely, the number of attacks also continues to increase. Therefore, attacks or intrusions have always occurred from time to time. There are many techniques that can be used by an attacker or intruder to execute network attacks or network intrusions. A persistent attacker usually employs stepping stones as a way to prevent from being detected [2]. By using Stepping Stone Detection (SSD), the attacker can be detected.

However, due to the complicated patterns of the stepping stones used by the attackers, detection of these stepping stones becomes a challenging task. More

M.N. Omar (✉)
College of Arts and Sciences, Universiti Utara Malaysia, 06010 UUM Sintok, Kedah, Malaysia
e-mail: niezam@uum.edu.my

S.-I. Ao et al. (eds.), *Machine Learning and Systems Engineering*,
Lecture Notes in Electrical Engineering 68,
DOI 10.1007/978-90-481-9419-3_7, © Springer Science+Business Media B.V. 2010

intelligent techniques are required in order to detect accurately the existence of stepping stones by analyzing the traffic pattern from one node to another. Beginning with a research conducted by Standiford-Chen and Herberlein [3] to the latest research by Wu and Huang [4], problems such as accuracy [5] and Active Perturbation Attacks (APA) [6] still remain the top issues among the researchers. An Intelligent-Host-Based SSD (I-HSSD) [7] and an Intelligent Network-based SSD (I-NSSD) [8] have been introduced in our previous research. Nevertheless, they have yet to provide a high rate of accuracy in detecting the stepping stones. We propose a Hybrid Intelligent SSD (HI-SSD) so as to tackle this accuracy problem. The proposed HI-SSD will be compared with other approaches: the I-HSSD and I-NSSD to examine the hybrid approach that has been applied in this research. Experiments will be conducted using a well-planned dataset, and the performance in terms of True Positive Rate (TPR) and False Positive Rate (FPR) [9] will become our main benchmark.

The rest of this article is structured as follows. Section 2 gives research terms used in this research. Section 3 discusses related works and Section 4 describes the proposed approach. In Section 5, we discuss further the experiment and then, a discussion of the results is in Section 6. Finally, we summarize the overall research and present possible future works in Section 7.

2   Research Terms

Before we start in more detail on the experiment, there are several research terms or terminologies used in this work which need to be clarified. In Fig. 1, there are five hosts involved in an SSD environment. Host A is a source of attack and Host E is a victim.

From Fig. 1, Stepping Stones (SS) are hosts A, B, C, D and E. SS = {A, B, C, D, E} where hosts A, B, C, D and E contain the same packet that flows through each host. A Connection Chain (CC), on the other hand, is the connection between hosts A, B, C, D and E. Therefore CC is a, b, c and d. CC = {a, b, c, d}. In Stepping Stone Detection (SSD) research using the Network-based approach (N-SSD), either SS = {A, B, C, D, E} or CC = {a, b, c, d}, can be used to denote the existence of stepping stones.

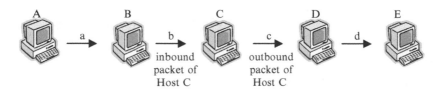

Fig. 1 Detecting a stepping stone chain

Host-based SSD (HSSD), in contrast, works by determining whether or not inbound and outbound connections contain the same packet flow. By referring to Fig. 1, if Host C is referred to, the inbound connection is $CC_c = b$ and the outbound connection is $CC_c = c$. In this case, $CC_2 = CC_3$ or $b = c$.

The challenge to SSD is to find the right solution by determining SS or CC. Overall, SSD research can be divided into statistical- and AI-based SSD, with most of the AI approach comprising the most recent research on SSD. As explained before, SSD research begins with the introduction of statistical-based research such as ON/OFF [10], Thumbprint [3], Deviation [11], and so forth.

3   Related Works

Research by Thames et al. [12] applied the hybrid concept by combining Bayesian Learning Network (BLN) and Self-Organizing Map (SOM) for classifying network-based and host-based data collected within LAN for network security purposes. The experiment was conducted by using four types of analyses (i) BLN with network and host-based data, (ii) BLN with network data, (iii) hybrid BLN-SOM analysis with host and network-based data and (iv) hybrid BLN-SOM analysis with net-work-based data. The four different types of analyses were required to compare one result to another.

Meanwhile, Bashah et al. [13] proposed a system that combines anomaly, misuse and host-based detection for Intrusion Detection System (IDS). This research only proposed an architecture that combines fuzzy logic and the SOM approach without any implementations/experiments.

Inspired by the above two research works, we hybrid the I-HSSD and I-NSSD approaches and compare the hybrid approach (HI-SSD) with the non-hybrid approaches (I-HSSD and I-NSSD) in terms of accuracy.

4   Proposed Approach: Hybrid Intelligence Stepping Stone Detection (HI-SSD)

The main components of HI-SSD are intelligence and hybrid. The intelligence component comes from the use of the Self-Organization Map (SOM) approach and the hybrid component is created from the combination of Host-based SSD and Network-based SSD. Both components of intelligence and the hybrid are discussed in detail our previous research [7, 8, 14]. Figure 2 shows the HI-SSD architecture.

Figure 2 shows the overall HI-SSD architecture that involves I-HSSD and I-NSSD. In this architecture an intrusion detection system (IDS) is used as a trigger to detect any network intrusion. When an intrusion occurs, I-NSSD starts to capture the network packet in a defined range. At the same time, each host also captures the

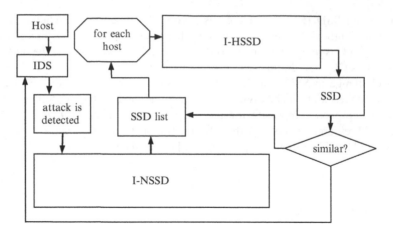

Fig. 2 HI-SSD architecture

network packet as well. When I-NSSD finishes its process to detect stepping stones, information about related hosts involved in the chain of stepping stones is produced. Each host listed in I-NSSD as a stepping stone node then executes a self-examination to check whether it is being used as a stepping stone or not. Results on both I-NSSD's list and I-HSSD's list then are compared. Similarity between these lists shows the real stepping stone host.

For testing purposes, only the functions of HI-SSD, I-NSSD and I-HSSD are involved in the experiment. The development of a fully-functional HI-SSD will become our future work.

The HI-SSD will contain a stepping stone list each from I-NSSD and I-HSSD, while the I-NSSD and I-HSSD will contain a stepping stone list for every network and host respectively. Comparisons will be measured on the TPR and the FPR on each component.

5 Experiment

For the dataset arrangement, Telnet Scripting Tool v.1.0 [15] is used. This is to guarantee a uniform pattern of telnet operations during the execution of the experiment. Here, Telnet represents the interactive connection most frequently used by SSD-related research. Moreover, there is no other dataset that is suitable in this research. Jianhua and Shou-Hsuan [16] used their own dataset. Staniford-Chen and Herberlein [3] also agreed with that.

The experiment is run in a controlled environment so as to avoid any interference with outside networks. Wireshark [17], on the other hand, is used to capture network packets that flow in each host. After the Telnet Scripting tool has ended its run, information pertaining to the packets is converted into text-based form. This is done in order to proceed to the consequent processes to acquire the appropriate information needed later. In this research, only time information is needed. This time information from the experiment is transferred into m-type file to be used later with Matlab 6.1 [18] software. In Matlab 6.1, the time information is used as the input to create, train, and lastly, to plot the SOM graph. The result from the visualization is taken as the result of this research and will be discussed in the next section.

Although the result for HI-SSD is obtained from I-NSSD and I-HSSD, this work will show the benefit of stepping stone detection when information obtained from a combination of the two approaches is compared to two separate sets of information when using I-NSSD or I-HSSD alone.

Figure 3 shows the layout of the overall experiment. From the figure, it explicitly shows that four hosts are involved in the stepping stone.

From these four stepping stone hosts, there are three connection chains (C_1, C_2, and C_3) involved. That means that only four stepping stones or three connections should be detected by the SSD approach. Increased number of detections causes lower TPR and decreased number of detections causes higher FPR. Table 1 shows the relationships for each host and its connection chains.

Based on the location of each host in Fig. 1, Table 1 shows the number of possible connection chains for each host and its list. The number of connection chains has been made based on the assumption that Host 1 and Host 4 are the starting and ending points respectively. Host 2 and Host 3 contain two different types of connection chains, one and two. However, connection chains that contain

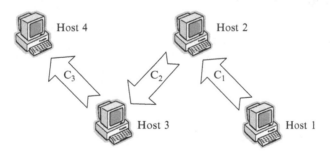

Fig. 3 Experiment layout

Table 1 Host and its relationship

Host	No. of connection chain	Connection chain list
Host 1	3	c_1, c_2, c_3
Host 2	1	c_1
	2	c_2, c_3
Host 3	2	c_1, c_2
	1	c_3
Host 4	3	c_1, c_2, c_3

just one connection can be eliminated because one connection does not mean that it is a stepping stone connection. In fact, research by RTT-based SSD [19–21] agreed that only connection chains with more than three connections can be considered as stepping stone connections. Based on our previous research [8], the existence of two connections onwards is enough for a chain to be identified as a possible stepping stone chain.

To calculate the effectiveness of the tested approach, TPR and FPR are used.

$$FPR = \frac{number\ of\ false\ positive}{number\ of\ possible\ negative\ instances} \tag{1}$$

False Positive Rate (FPR) refers to the fraction of negative instances that are falsely reported by the algorithm as being positive. In this situation, the algorithm has detected the connection chains which exist even though it is not true.

$$TPR = \frac{number\ of\ true\ positive}{number\ of\ possible\ true\ instances} \tag{2}$$

True Positive Rate (TPR) refers to the fraction of true instances detected by the algorithm versus all possible true instances. The discussion on the result and its analysis will be presented.

6   Result and Analysis

Result on each type of SSD will be discussed in each of the following sub-section. As described before, we have chosen to use SOM as the intelligent approach. As a result, the SOM graph will represent each compared solution.

6.1   *Intelligence Network Stepping Stone Detection (I-NSSD)*

In I-NSSD, the results are obtained from the execution of SOM approach through the arrival time for the overall captured data in the network. In this case, only one graph has been produced. It is different from I-HSSD that needs one graph for each host involved. Figure 4 shows the result.

In contrast to the I-HSSD approach that depends on the number of possible straight lines which can be created to determine the number of connection chains, the I-NSSD approach is based on the number of possible groups generated. In Fig. 4, four possible groups of SOM nodes can be counted (labeled as a, b, c and d). Thus, there are four connection chains involved in the experiment. However, the true number of connection chains involved in this research is only three. Therefore,

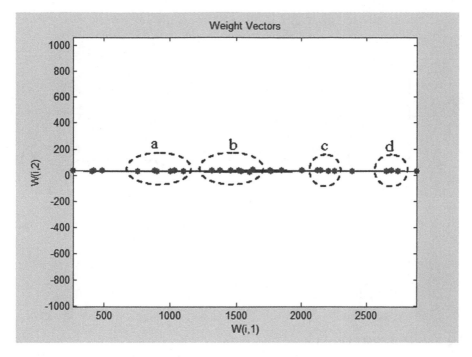

Fig. 4 Node of SOM in I-NSSD

there exist false positive reports in the I-HSSD experiment. By using formula (1), FPR for I-NSSD is 33.3%. For TPR, I-NSSD shows 100% achievement.

6.2 Intelligence Host-Based Stepping Stone Detection (I-HSSD)

As described previously, I-HSSD involves every host that has been listed in I-NSSD. In this case, arrival time for each host is run by using the SOM approach. Each result is an output from the execution.

Figure 5 shows that three possible directions can be traced (e, f and g). That means there are three possible connections which exist in the originating host, Host 1. Based on the value from Table 1, Host 1 obtains 100% TPR and 0% FPR.

Figure 6 on the other hand shows that there are two connection chains (h and i) that could possibly exist in Host 2 as the monitored host. Based on the number of connection chains from Table 1, Host 2 got 100% TPR and 0% FPR.

In Fig. 7, similar to Fig. 6, there are two connection chains (j and k) that could possibly exist in this graph. Based on the number of connection chains from Table 1, Host 3 also got 100% TPR and 0% FPR.

Figure 8 shows the last node of SOM that needs to be observed. From the graph, there are two possible obtainable directions. That means that two connection chains

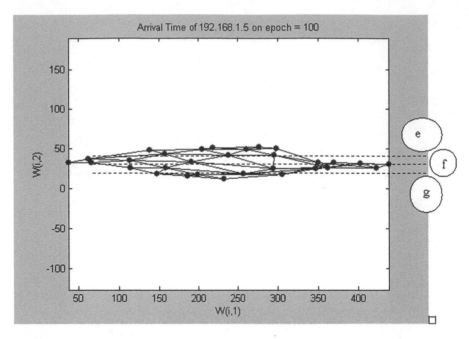

Fig. 5 Node of SOM on Host 1

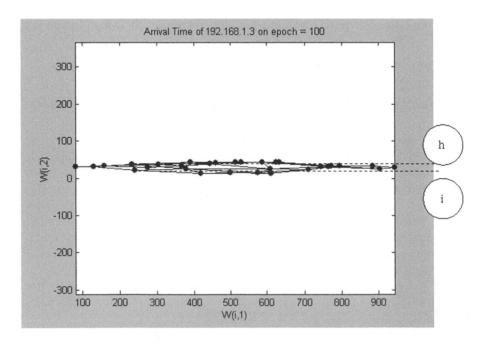

Fig. 6 Node of SOM on Host 2

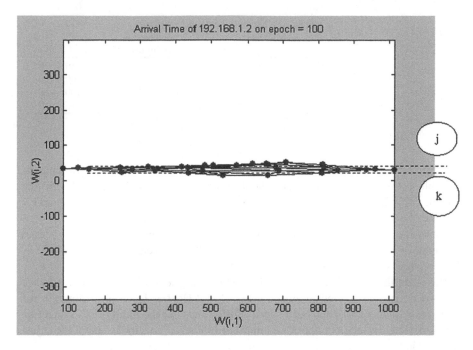

Fig. 7 Node of SOM on Host 3

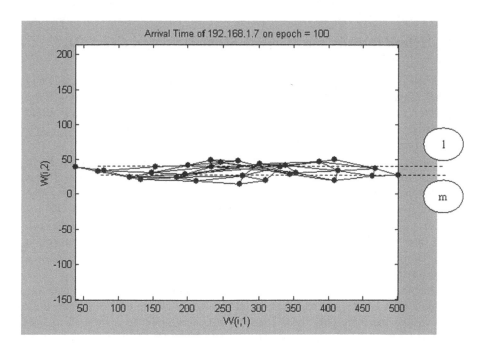

Fig. 8 Node of SOM on Host 4

Table 2 I-HSSD experiment result

Host	No. of connection chain	No. of connection chain detected	TPR (%)	FPR (%)
1	3	3	100	0
2	2	2	100	0
3	2	2	100	0
4	3	2	66.7	0

(l and m) have been detected. However, based on the number of connection chains in Table 1, three connection chains should have been detected in Host 4. For that reason, I-HSSD in Host 4 miss-detects another connection. In this case, the TPR becomes 66.7%. The FPR on the other hand also gives 0%.

From the result on each host, it shows that only the result of Host 4 does not achieve 100% TPR. This can be considered as a weakness of the I-HSSD. Table 2 shows the result of the I-HSSD experiment.

Based on the overall result of the I-HSSD experiment, connection chains have been successfully detected in Host 1, Host 2 and Host 3. In Host 4, although the connection chains have also been successfully detected, the number of connection chains detected is less than the actual number involved. This makes the TPR at just 66.7%. On the other hand, the FPR is the same for all hosts.

6.3   Hybrid Intelligence Stepping Stone Detection (HI-SSD)

As described earlier, HI-SSD is constructed from the combination of I-HSSD and I-NSSD. Therefore, the information from I-HSSD and I-NSSD is still used in the HI-SSD. As shown in Fig. 2, the information obtained from I-HSSD and I-NSSD are rearranged so as to generate a more accurate stepping stone detection result compared to the use of I-HSSD or I-NSSD alone.

For this purpose, the first information acquired from I-NSSD is observed. From the result, four connection chains are detected. At this level, false detection still not has been discovered. All information needed is distributed to the related hosts. The host which is related to the list then runs the I-HSSD function by itself. This is to check whether or not the host is used as a stepping stone. In this case, Host 1, Host 2, and Host 3 have successfully detected the number of existing connection chains. However, Host 4 has just detected two connection chains compared to three connection chains that should have been detected.

Although miss-detection has occurred here, the number of maximum possible hosts from Host 1 helps a lot to balance the Host 4 result. From the I-HSSD result, it is shown that there are only three connection chains involved compared to four proposed by I-NSSD. Combining I-HSSD and I-NSSD avoids false and miss-detections. By using both I-HSSD and I-NSSD, it is shown that 100% TPR and

Table 3 Experiment result

Type of stepping stone detection approach	TPR (%)	FPR (%)
I-NSSD	100	33.3
I-HSSD	91.67	0
HI-SSD	100	0

0% FPR has been achieved. For a clear picture on the overall HI-SSD, the related algorithm is given as follows.

Begin
  I-NSSD:
  capture network packet
  collect arrival time
  execute SOM
  count the group of node, n
  list involved host, l
  sent n & l to I-HSSD
  I-HSSD:
  activate I-HSSD on selected host based on l
  for 1 to n
  execute SOM
  count the possible straight line,
  identify the existence of connection chain for the host, e
  end for
  End

Result of the experiment according to the percentage of TPR and FPR obtained from I-NSSD, I-HSSD and HI-SSD respectively is tabulated in Table 3.

Table 3 shows the experiment result of I-NSSD, I-HSSD and HI-SSD. As described previously, HI-SSD is a combination of I-NSSD and I-HSSD. Therefore, the TPR and FPR for HI-SSD is the combination of I-NSSD's TPR and FPR and I-HSSD's TPR and FPR. In the other words, the average TPR and FPR of I-NSSD and I-HSSD become the result of HI-SSD. In this case, 100% TPR and 0% FPR for HI-SSD is better than the use of I-NSSD or I-HSSD alone. Although I-NSSD shows 100% TPR, but the 33.3% FPR can adversely affect the overall function of stepping stone detection. In the I-HSSD, 100% FPR makes this approach free of false detection. However, 91.67% TPR is not enough to render this approach to be fully-functional. As a result, combining both I-NSSD and I-HSSD to form the HI-SSD will balance the TPR and the FPR at the same time.

7 Conclusion and Future Work

The goal of this research is to prove the effectiveness of the proposed HI-SSD which is actually a hybrid or combination of I-NSSD and I-HSSD. From the experiment, it is shown that the HI-SSD is more accurate compared to both

I-NSSD and I-HSSD. Therefore, it is proven that HI-SSD is more accurate compared to I-NSSD or I-HSSD.

In the future, we will improve our approach to be more robust towards active perturbation attacks such as delay, packet drop and chaffing. Testing on the active perturbation problem would not only be executed onto our proposed solution, but it would also involve our closest research that used the Data Mining approach. By doing this, we not only can measure the proposed approach's capabilities but at the same time can also compare it with other similar approaches.

Only one dataset is used in this experiment. With the intention of verifying the capability of the proposed HI-SSD, we need different types of dataset. Our observation on the datasets used in SSD-based research has shown that there are two kinds of datasets which can be used; datasets generated by ourselves and public datasets. By using various kinds of datasets instead of just one type of dataset, the true capabilities of the approach can be examined more. As such, testing the proposed HI-SSD on top of different datasets is one of our future works.

Another future plan is to compare the proposed HI-SSD approach with the statistical-based approaches. Therefore, to ascertain that the proposed approach is better than statistical-based approaches, it is good to involve the different approaches together in the experiment.

Lastly, for a true experience of a fully-functional stepping stone detection, the overall approach should be translated into a full system. If a complete system is created, comparisons with other approaches could be made more easily.

References

1. C. Chamber, J. Dolske, J. Iyer, *"TCP/IP Security"*, *Department of Computer and Information Science* (Ohio State University, Ohio 43210). Available at: http://www.linuxsecurity.com/resource_files/documentation/tcpip-security.html
2. K.D. Mitnick, W.L. Simon, The art of intrusion: the real stories behind the exploits of hackers, intruders & deceivers" (Wiley, 10475 Crosspoint Boulevard, Indianapolis, IN 46256, 2005)
3. S. Staniford-Chen, L.T. Herberlein, Holding intruders accountable on the Internet, in *Proceedings of the 1995 IEEE Symposium on Security and Privacy*, Oakland, CA, 1995, pp. 39–49
4. H. Wu, S.S. Huang, Neural network-based detection of stepping-stone intrusion. Exp. Sys. Appl. **37**(2), 1431–1437 (March 2010)
5. A. Blum, D. Song, S. Benkataraman, *Detection of Interactive Stepping Stone: Algorithm0020and Confidence Bounds*, vol 3224/2004. Lecture Notes in Computer Science (Springer Berlin/Heidelberg, 1 Oct 2004), pp. 258–277
6. X. Wang, *Tracing Intruder BStepping Stone*. Ph.D. Thesis, North Carolina State University, Raleigh, 2004.
7. M.N. Omar, R. Budiarto, Intelligent host-based stepping stone detection approach, in *Proceeding of The World Congress on Engineering and Computer Science 2009*, vol II, San Francisco, USA, 20–22 Oct 2009, pp. 815–820
8. M.N. Omar, R. Budiarto, Intelligent network-based stepping stone detection approach. Int. J. Appl. Sci. Eng. Technol. **53–135**, 834–841 (2009)
9. A. Almulhem, I. Traore, Detecting connection-chains: a data mining approach. Int. J. Netw. Security. **11**(1), 62–74 (Jan 2010)

10. Y. Zhang, V. Paxson, Detecting stepping stones, in *Proceedings of the 9th USENIX Security Symposium*, Denver, CO, 2000, pp. 67–81.

11. K. Yoda, H. Etoh, Finding connection chain for tracing intruders, in *Proceedings of the 6th European Symposium on Research in Computer Security (LNCS 1985)*, Toulouse, France, 2000, pp. 31–42

12. J.L. Themes, R. Abler, A. Saad, Hybrid intelligent system for network security. ACM southeast regional conference 2006, 2006.

13. N. Bashah, I.B. Shanmugam, A.M. Ahmed, Hybrid intelligent intrusion detection system, Proc. World Acad. Sci. Eng. Technol. **6** (2005)

14. M.N. Omar, L. Serigar, R. Budiarto, Hybrid stepping stone detection method. in *Proceeding of 1st International Conference on Distrubuted Framework and Application (DFmA 2008)*, Universiti Sains Malaysia, Penang, 21–22 Oct 2008, pp. 134–138

15. Wareseeker (2008) [Online] Available: http://wareseeker.com/freeware/telnet-scripting-tool-1.0/19344/TST10.zip 8 Feb 2008

16. Y. Jianhua, S.H. Shou-Hsuan, A real-time algorithm to detect long connection chains of interactive terminal session", in *Proceedings of the 3rd International Conference on Information Security*, China, 2004, pp. 198–203

17. Wireshark (2009). [Online] Available: http://www.wireshark.org 8 Feb 2009

18. H. Duane, L. Bruce, *Mastering MATLAB A Comprehensive Tutorial and Reference* (Prentice-Hall, New Jersey, 1996)

19. K.H. Yung, Detecting long connection chains of interactive terminal sessions, in *Proceedings of the International Symposium on Recent Advance in Intrusion Detection (RAID 2002)*, Zurich, Switzerland, 2002, pp. 1–16

20. J. Yang, S.S. Huang, A real-time algorithm to detect long connection chains of interactive terminal sessions, in *Proceedings of the 3rd International Conference on Information Security (INFOSECU 2004)*, Shanghai, China, 2004, pp. 198–203.

21. J. Yang, S.S. Huang, Matching TCP packets and its application to the detection of long connection chains on the internet, in *Proceedings of the 19th International Conference on Advance Information Networking and Applications (AINA 2005)*, Tamkang University, Taiwan, 2005, pp. 1005–1010

Chapter 8
Open Source Software Use in City Government

Is Full Immersion Possible?

David J. Ward and Eric Y. Tao

Abstract The adoption of open source software (OSS) by government has been a topic of interest in recent years. National, regional, and local government are using OSS in increasing numbers, yet the adoption rate is still very low. This study considers if it is possible from an organizational perspective for small to medium-sized cities to provide services and conduct business using only OSS. We examine characteristics of municipal government that may influence the adoption of OSS for the delivery of services and to conduct city business. Three characteristics are considered to develop an understanding of city behavior with respect to OSS: capability, discipline, and cultural affinity. Each of these general characteristics contributes to the successful adoption and deployment of OSS by cities. Our goal was to determine the organizational characteristics that promote the adoption of OSS. We conducted a survey to support this study resulting in 3,316 responses representing 1,286 cities in the Unites States and Canada. We found most cities do not have the requisite characteristics to successfully adopt OSS on a comprehensive scale and most cities not currently using OSS have no future plans for OSS.

1 Introduction

All city governments seek to deliver services in the most efficient manner possible. Whether in direct support of service delivery or support of conducting the business of government, Information Technology (IT) has become and integral component of operations at all levels of government.

In the past 5 years there has been a trend by some national, regional, and local governments toward use of open source software (OSS) and open standards as a

D.J. Ward (✉)
CSU Monterey Bay, 100 Campus Center, Seaside, CA 93955, USA
e-mail: dward@csumb.edu

S.-I. Ao et al. (eds.), *Machine Learning and Systems Engineering*,
Lecture Notes in Electrical Engineering 68,
DOI 10.1007/978-90-481-9419-3_8, © Springer Science+Business Media B.V. 2010

first choice rather than a curiosity. Recently, The Netherlands has mandated that all national government agencies will use open standards formatted documents by April 2009 [1]. The U.S. Navy has clarified the category and status of OSS to promote a wider use of OSS to provide seamless access to critical information [2]. The U.S. Congress is recognizing the potential value of OSS in the National Defense Authorization Act for fiscal year 2009. OSS is identified as an objective in the procurement strategy for common ground stations and payloads for manned and unmanned aerial vehicles, and in the development of technology-neutral information technology guidelines for the Department of Defense and Department of Veteran Affairs.

Small to medium size cities, populations less than 500,000 [3], may have serious limitations in funding IT efforts. Escalating costs of service delivery coupled with reduce revenue will force governments to seek novel ways to reduce operating costs. With limited revenue their budgets seem to force cities to under fund IT infrastructure in favor of applying resources to increasing labor requirements to deliver services. Careful and deliberate selection of IT solutions can reduce the labor required for service delivery freeing that labor to be harvested for other purposes.

Considerable research has been conducted on the topic of e-government and the development of models to explain e-government maturity. There have also been ample studies of the trends in adoption of OSS by government at various levels. However, little research has been done in the characteristics of regional and local government that would promote adoption and successful deployment of OSS.

The support of city leadership and management as well as the IT staff are required for successful adoption of any technology, not just OSS. Vision, strategy and government support are important for success of IT projects, while insufficient funding and poor infrastructure are major factors for failure [4].

This research focused on the perspectives of city leadership, management, and IT staff with respect to OSS and its adoption by city government. While OSS was the focus, the approach used in this study can be applied to investigating the adoption of most technologies or practices.

2   Related Research

The literature review revealed little research with similar characteristics of this study. While much work can be found on OSS in a wide variety of topics, the body of research covering municipal OSS adoption is relatively limited. Of the literature found relating to OSS adoption by government, a significant portion of that work examines the adoption of OSS by developing countries.

A fair number of studies have been conducted to examine the extent to which government entities are using OSS. The studies tend to collect, categorize, and report the current state of open source adoption [5, 6]. A study of Finnish municipalities [7] had a methodology similar to the methodology used in this study.

While the Finnish study considered only IT managers as survey subjects, our survey considered city leaders, managers, and IT staff for subjects, as our focus is examining organizational behavioral influences on OSS adoption rather than technical and budgetary influences.

In the current economic climate government at all levels are facing funding crises as costs of operations increase and revenue decreases. The question of what can be done to reduce IT operating costs is now a very important one. There are more benefits to using OSS than just reduced acquisition costs [8]. Restrictive licensing, vendor lock-in, and high switching costs can be eliminated, which, in the long term, also may reduce costs.

The level of knowledge of a user with respect to OSS and Closed Source Software (CSS) will influence their decision to use OSS.

> There are two topologies of consumers: a) "informed" users, i.e. those who know about the existence of both CSS and OSS and make their adoption decision by comparing the utility given by each alternative, and b) "uniformed" users, i.e. those who ignore the existence of OSS and therefore when making their adoption decision consider only the closed source software [9].

We can apply this observation to municipal organizations. The informed municipal organization, which knows about the existence of open-source alternatives to commercial products, may make adoption decisions based on the value provided by each. The uninformed organization either ignores the existence of open-source alternatives or is unaware of OSS alternatives to commercial products. The uninformed organization may have misperceptions of OSS. These misperceptions may include OSS usability, deployment, and support. A common misperception is that an organization must have a programmer on staff in order to deploy and maintain OSS. While in the distant past within the open source era, this may have been true, the current maturity of most open source applications may require only a competent IT technician.

The adoption of OSS by municipal government is a technology transformation problem. Technology transformation is not about the technology, but the organization adopting the technology, in this case city governments. Implementation of technologies to support government efforts in its self does not guarantee success [10]. Organizations must change to embrace the new technologies in order to use them effectively [11].

e-Government cannot be achieved by just simply implementing available software [12]. Fundamental differences between the public and private sectors influence the rate at which ICT is employed [13]. The real opportunity is to use IT to help create fundamental improvement in the efficiency, convenience and quality of service [14]. Adopting new or different technology requires an organization to change, with varying degree, at all levels. An organization with a culture that embraces change will be more successful at adopting new technologies than an organization with a strong status quo bias.

3   Research Goals

Our approached in this study was to focus on organizational characteristics required to promote OSS adoption instead of OSS as technology needing to be implemented. One fundamental question ultimately coalesced; is it possible for small to medium cities to use only OSS to deliver services and conduct city business? This is not a question of technical feasibility or product availability, but one of organizational characteristics. If we assume OSS products exist in the application domains city government requires for IT operations, then successful adoption and deployment becomes an organizational behavior issue.

Three general domains of interest evolved from the core question; capability, discipline, and cultural affinity. Having the capability to deploy and manage information technology was deemed very important for a city to successfully deploy OSS technologies. A city that has a mature management structure in place that demonstrates forethought in planning and a deliberate approach to budgeting and acquisition will have the level of discipline to deploy OSS on a comprehensive scale. Cultural affinity we define as the predisposition toward or against the use of OSS. The awareness, support, and understanding of OSS by key city personnel (leadership, management, and IT Staff) will influence a city's affinity toward the use of OSS. From these domains we developed additional questions to guide this research effort: Do cities have the necessary organizational characteristics to adopt and use only OSS to deliver services and conduct city business? What are the basic IT capabilities of cities? Do these capabilities support the adoption and deployment of open source technologies? Do cities plan and budget for IT in a deliberate manner that would support the adoption and deployment of open source technologies? Does the organizational culture of cities promote the adoption of open source technologies?

4   Methodology

The following sections describe the methodology used for this study. We begin by describing in general the process, followed by the survey design, survey execution, and subject selection.

A survey was conducted to collect data from municipal IT managers, IT staff, city leadership, city management, and city employees. The survey was administered online using SurveyMonkey.com. The collection period was initially scheduled for 30 days from June 1, 2008 through June 30, 2008.

The survey required soliciting responses from subjects to provide insight into the characteristics of their cities with respect to IT capability, organizational discipline, and cultural affinity to OSS. Presenting direct questions would not produce useful data, as subjects may not have the requisite knowledge in the subject areas. One goal in the design of the survey was to reduce the number of aborted attempts by subjects. An aborted attempt is the failure to complete the survey once started.

We identified six subject classifications; IT Manager, IT Staff, City Leadership, City Management, City Employee, Other. These classifications were used to tailor the set of questions presented to the subject to keep the set within the knowledge domain of the subject. All survey questions were framed in the context of the subject and their perception of their city. This was an important aspect of the design as the survey questions solicit information that indirectly relates to the research questions.

The survey was divided into four sections. The first three sections address in three interest domains; capability, disciple, and cultural affinity. The fourth section solicited demographic information.

The first section of the survey collected data regarding city IT capability addressing the capability dimension characteristic of cities. The second section solicited responses related to the city's IT strategy addressing the discipline dimension characteristic of cities. The third section solicited responses related to the subject's perspectives and opinions about IT and OSS and the subject's impression of the city leadership, management, and IT staff's perspectives of IT and OSS. These questions were intended to reveal the city's cultural affinity to the adoption of OSS.

5   Survey Execution

For the announcement strategy we used three channels to contact potential subjects; magazines related to city management, municipal associations, and direct e-mail.

Announcing the survey through a magazine was deemed to have potential for generating significant level of exposure for those subjects who are more likely to read city government related magazines. Several magazines were contacted for assistance to announce the survey. Two magazines responded, the Next American City Magazine and American City and County Magazine. The Next American City magazine provided a half page ad space to announce this survey. American City and County Magazine announced the survey in an article posted on the front page of its website. Although the potential exposure was thought to be high, the magazine announcement channel only produced 20 responses of which only one response was valid for analysis.

Municipal associations were thought to be the best vehicle for reaching the largest number of subjects. The rationale behind this was the assumption that individuals affiliated with municipal associations might be more inclined to respond to a survey announcement received from their association. The expectation was that the greatest number of responses would result from municipal associations. Individuals affiliated with municipal associations may also have greater interest in supporting this research as they may see a potential benefit for their city.

A total of 116 municipal associations were contacted to assist with announcing the survey to their members, 28 associations approved the request for assistance and forwarded the announcement to their members.

The municipal associations were identified via a search of the Internet. Most of the associations found were regional providing representation within a county, multi-county, state, or multi-state area. Each municipal association was sent an initial survey announcement with a reminder sent within 7 days of the initial announcement.

We received 207 responses from subjects indicating they learned of the survey through a municipal association.

To reach the greatest number of potential subjects a direct e-mail approach was also used. Direct e-mail has proven to be a very effective and economical means of reaching the greatest number of survey subjects [15].

Individuals were contacted via email addresses harvested from municipal websites. A commercial email-harvesting program, Email Spider (www.gsa-online.de), was used to collect the municipal email addresses.

The collection process harvested over 80,000 email addresses from the municipal websites. Invalid email addresses, those not associated with a city government domain name (i.e. yahoo.com, aol.com), were excluded to produce a set of 60,000 addresses to which the announcements were sent.

Survey announcements were emailed to the potential subjects over a 2-week period. Reminder emails were sent within 7 days of the initial announcements. Survey responses can be increased with a reminder email that includes more than just a link to the survey[15]. With this in mind the reminder email included the original announcement text with an introductory paragraph explaining the email was a reminder.

The collection period defined in the research design was for 1 month running from June 1, 2008 through June 30, 2008. Toward the end of the collection period, between 18 and 25 June, the response activity maintained a significant level (an average of 143 per day) prompting this researcher to extend the collection period 15 days ending on July 15, 2008. Additionally, many automated email responses indicated the recipients were on vacation during the month of June. In the days following distribution of the initial survey announcement email and reminder emails increased response activity was observed. We anticipated response activity would increase during the first part of July as potential subjects returned from vacation. During the 15 day extended collection period 1443 responses were collected, 43.5% of the total responses.

6 Survey Results

The total estimated exposures to the survey announcement were in excess of 60,000. An exposure for the purpose of this study is defined as the delivery of a survey announcement to a potential subject. Several factors prevent an accurate tally of total survey announcement exposures. We did not have access to the membership numbers for the municipal associations that forwarded the survey announcement or access to the web site page hit counts for the article on the magazine website.

Of the 60,000 e-mails sent directly to city leaders, managers, and employees, 53,900 may have reached the addressee. 6,100 of the original 60,000 email announcements were returned as undeliverable, a non-existent address, or reported by an email server as potential "spam".

A total of 3,316 individuals responded to the survey announcement resulting in a response rate of 6%. There are 1,286 distinct cities represented in this response set.

A sample set was created to include cases from cities with populations less than 300,000 and an indicated primary duty of IT Manager, IT staff, City Leader, or City Manager. While the survey data included responses from cities with populations greater than 300,000, those responses were too few in number to permit valid analysis (Fig. 1).

Reponses from city employees (not city leaders, managers, or IT staff) and individuals not affiliated with a city were excluded from the sample set used for analysis. The responses from city employees had very limited or no value as these subjects had little knowledge of city IT capability, strategy, or the views of the city leadership, management, and IT staff regarding OSS. These criteria produced a sample set of 1,404 cases representing 1,206 cities.

Cities use a wide variety of desktop operating systems. Within the IT staff sample sub-set, 15 different operating systems were identified. Virtually all cities (99.7%) deploy one or more versions of Microsoft Windows on desktop computers. Twenty percent of the respondents indicate Linux is used on desktop computers. The survey instrument did not collect the degree to which Linux is deployed on the desktop in the respondents' city.

An interesting observation was 13% of the IT staff indicating that Mac OS X was deployed on desktop computers in their city. Since Mac OS X can only be installed on Apple Inc. hardware [16], we can conclude these cities are using Apple computers to support service delivery or to conduct city business (Fig. 2).

Server side operating systems showed similar deployment rates with Microsoft Windows most widely used at 96.5% and Linux following in second at 40%. A small number of cities continue to use Novell Netware and DecVMS/OpenVMS indicating some cities may be maintaining legacy software to support service delivery or city operations (Fig. 3).

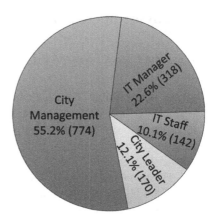

Fig. 1 Response by primary duty

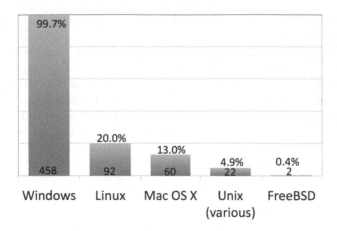

Fig. 2 Desktop OS deployment

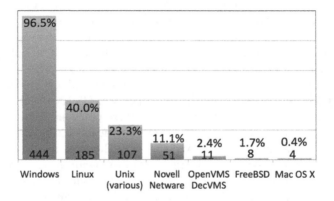

Fig. 3 Server OS deployment

7 Analysis: Interesting Findings

7.1 Few Cities Have All Characteristics

Analysis of the survey data indicates few cities have all the characteristics that would enable successful comprehensive adoption and deployment of OSS. Of the 1,206 distinct cities in the sample set, just ten cities satisfied all characteristics within the three dimensions.

Ten cities, listed in Table 1, satisfied the following criteria; has an IT Department, IT support is handled in-house, currently uses OSS, has well defined IT strategy, has IT line item in budget, IT is sufficiently funded, total cost of ownership acquisition strategy, uses budget for software acquisition.

Table 1 Cities satifying all selection criteria

City	State	Population
Balwin	Missouri	30,000
Northglen	Colorado	31,000
Houma	Louisiana	32,400
Ipswitch	Massachusetts	12,000
Largo	Florida	73,000
Layton	Utah	64,300
Redding	California	80,800
Santa Monica	California	87,200
Tomball	Texas	10,200
Ulysses	Kansas	5,600

Largo, Florida is of particular interest. Largo, has embraced the use of OSS. It has deployed Linux as the operating system for its 400 desktop clients saving the city an estimated $1 million per year in hardware, software, licensing, maintenance, and staff costs [17].

7.2   Possible Aversion to OSS If Not Currently Using OSS

Of the 460 Municipal IT managers and staff in the sample set, 56% indicated their city was not currently using OSS while 39% indicated their city was using OSS. Considering the widespread use of OSS in the commercial sector, the relatively high percentage of cities in this survey not currently using OSS required further investigation (Fig. 4).

Of the Cities currently using OSS, 76% are planning to use OSS in the future and 10% have no plans to use OSS in the future. The high percentage of cities planning to use OSS in the future that currently use OSS can be expected. It is more likely an organization will continue to use a software product once deployed and established than to abandon the product (Fig. 5).

The cities currently not using OSS provides a more interesting observation. Of the 259 IT managers and staff indicating their city is currently not using OSS, 82% indicated their city has no plans to use OSS in the future, 9% indicated their city did plan to use OSS in the future, and 9.7% (25) did not know.

The number of dedicated IT staff at the respondent cities may not be an influencing factor in decisions to use OSS in the future. While 74% of the cities not planning to use OSS in the future have IT staff numbering ten or less, 71% of cities currently using OSS also have ten or less IT staff.

The organizational support for using OSS appears to be a significant influencing factor for a city's future plans for OSS use. The survey design included questions regarding the respondent's perception of the Leadership, Management, and IT staff views of OSS. The subjects were asked if the city leadership, management, and IT staff support the use of OSS. For the cities not planning to use OSS in the future only 6% of the respondents indicated the city leadership supports the use of OSS, 8% of respondents indicated city management supports the use of OSS, and 33%

Fig. 4 Cities currently using
OSS

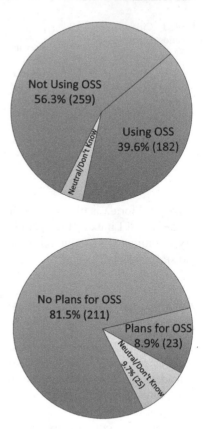

Fig. 5 OSS plans for cities
currently not using OSS

indicated city IT staff supports use of OSS. For cities currently using OSS the
responses were 22% leadership, 33% management, and 71% IT staff.

7.3   Current OSS Support by Leadership, Management,
and IT Staff

IT managers and staff report a significant difference in the perceived current support
of OSS and the support of OSS if using OSS would reduce IT operating costs. 11%
of IT managers and staff indicate they agree their city leadership currently supports
the use of OSS. The IT managers' and staff's perception of city management's
current support of OSS is similar, if somewhat higher, to their perception of city
leadership. Sixteen percent agree their city management currently supports OSS.
The IT managers' and staff's perception of city IT staff's current support of OSS,
that is their perception of themselves, was significantly higher than their perception
of city leadership and management support of OSS with 26% agreeing the city IT
staff supports the use of OSS (Fig. 6).

When asked about support of OSS if it meant saving money, 36% of the IT staff agrees their city leadership would support OSS if it would save money, a threefold increase. 41% agree their city management would support OSS if it would save money, a 150% increase. However, only 36% agreed the city IT staff would support OSS to save money, just a 50% increase. While these results indicate city leadership and management may be motivated to support OSS given the potential cost savings, IT staff may not share those same motivations (Fig. 7).

Of note is the drop in frequency (from ~70% to ~50%) of respondents indicating a neutral position or those who did not know. The possibility of reducing the costs of information technology is a significant influence on IT strategy and technology adoption.

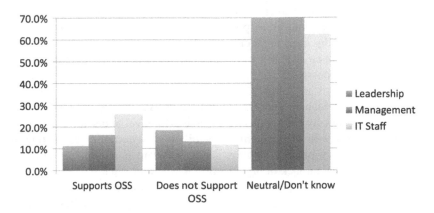

Fig. 6   Current support of OSS

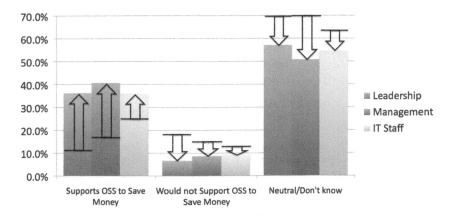

Fig. 7   Support if OSS would save money

7.4   Discrepancy of OSS Awareness: Self, Others

The survey data suggests a discrepancy between the subject's own awareness of OSS and their perception of city leadership, management, and IT staff's awareness of OSS. Within the sample set 69% of the respondents indicated they are aware of OSS. However, their responses regarding their city leadership, management, and IT staff's awareness of OSS show that most respondents perceive the leadership, management and IT staff as generally unaware of OSS. The high frequency of those individually aware of OSS could be attributed to the survey attracting individuals interested in OSS.

8   Conclusion

The results indicate cities in general do not have the necessary characteristics to successfully adopt OSS to deliver services and conduct city business on a comprehensive scale. The key indicators point to significant deficiencies in the three domains: capability, discipline, and cultural affinity.

While a majority of cities in the study show some characteristics that indicate the adoption of OSS is possible, and indeed on a trivial level (with a few notable exceptions) some cities are using OSS, still most cities lack key characteristics in the three domains to enable a successful comprehensive adoption of OSS.

The data suggest many cities may have an adequate level of discipline to support open source adoption with IT line items in the city budget and sufficient IT funding. However, a significant number of cities make software purchases on an ad hoc basis, indicating potential lack of organizational planning capability.

A city's Culture, with respect to IT decision making, appears to be a significant barrier to open source adoption. City leadership and management of cities that do not support the use of OSS are generally unaware of OSS as an alternative to commercial software. Cities currently using OSS are highly likely to continue to use OSS in the future while cities not presently using OSS have no future plans to use OSS.

Because the cities represented in this study in general do not exhibit the indicators in the three domains examined we conclude most cities do not have the capability, discipline, and cultural affinity to successfully adopt OSS on more than a trivia level.

References

1. Associated Press (14 Dec 2007) Dutch Government Ditches Microsoft, Moves to Open-Source Software. Foxnews.com. Retrieved 7 Feb 2008, from Fox News Web site: http://www.foxnews.com/story/0,2933,316841,00.html
2. R.J. Carey, *Department of the Navy Open Source Software Guidance (-)* Washington, DC, U.S. Navy. Retrieved 8 Feb 2008, from Open Source Software Institute Web site: http://oss-institute.org/Navy/DONCIO_OSS_User_Guidance.pdf
3. V. Henderson, Medium size cities. Reg. Sci. Urban Econ. **27**(6), 583–612 (1997)

4. D. Gichoya, Factors affecting the successful implementation of ICT projects in government. Electron. J. E Govern. **3**(4), 175–184 (2005)
5. J. Mtsweni, E. Biermann, An investigation into the implementation of open source software within the SA government: an emerging expansion model. In *Proceedings of the 2008 Annual Research Conference of the South African institute of Computer Scientists and information Technologists on IT Research in Developing Countries: Riding the Wave of Technology (Wilderness, South Africa, October 06–08, 2008)*. SAICSIT '08, vol. 338, ACM, New York, pp. 148–158
6. W. Castelnovo, M. Simonetta, The evaluation of e-government projects for small local government organisations. Electron. J. e-Government **5**(1), 21–28 (2007)
7. M. Välimäki, V. Oksanen, J. Laine, An empirical look at the problems of open source adoption in Finnish municipalities, in *Proceedings of the 7th International Conference on Electronic Commerce (Xi'an, China, August 15–17, 2005)*. ICEC '05, vol. 113, ACM, New York, pp. 514–520
8. H. Thorbergsso, T. Björgvinsson, Á. Valfells, Economic benefits of free and open source software in electronic governance, in *Proceedings of the 1st International Conference on Theory and Practice of Electronic Governance (Macao, China, December 10–13, 2007)*. ICEGOV '07, vol. 232, ACM, New York, pp. 183–186
9. S. Comino, F. Manenti, Free/open software: public policies in the software market (July 2004)
10. Z. Ebrahim, Z Irani, E-Government adoption: architecture and barriers. Business Process Manage. J. **11**(5), 580–611 (2005)
11. M. Murphy, Organisational change and firm performance. OECD Directorate for Science, Technology, and Industry Working Papers (2002), http:////www.oecd.org//sti
12. P. Alpar, Legal requirements and modeling or processes in government. Electron. J. E-Govern. **3**(3), 107–116 (2005)
13. D. Swedberg, Transformation by design: an innovative approach to implementation of e-government. Electron. J. E-Govern. **1**(1), 51–56 (2003)
14. J. Borras, International technical standards for e-government. Electron. J. E-Govern. **2**(2), 75–80 (2004)
15. D. Kaplowitz, T. Hadlock, R. Levine, A comparison of web and mail survey response rates. Public Opin. Quart. **68**(1), 94–101 (2004)
16. Apple Inc. (2008) Software License Agreement for MAC OS X, http://manuals.info.apple.com/en/Welcome_to_Leopard.pdf
17. L. Haber, ZDnet. (2002) City saves with Linux, thin clients, http://techupdate.zdnet.com/techupdate/stories/main/0,14179,2860180,00.html

Chapter 9
Pheromone-Balance Driven Ant Colony Optimization with Greedy Mechanism

Masaya Yoshikawa

Abstract Ant colony optimization (ACO), which has been based on the feeding behavior of ants, has a powerful solution searching ability. However, since processing must be repeated many times, the computation process also requires a very long time. In this chapter, we discuss a new ACO algorithm that incorporates adaptive greedy mechanism to shorten the processing time. The proposed algorithm switches two selection techniques adaptively according to generation. In addition, the new pheromone update rules are introduced in order to control the balance of the intensification and diversification. Experiments using benchmark data prove the validity of the proposed algorithm.

1 Introduction

Combinatorial optimization problems can be used for a variety of fields, and various solutions for these types of problems have been studied. Among these solutions, a heuristic solution, referred to as meta-heuristics, has attracted much attention in recent years. Meta-heuristics is a general term for an algorithm that is obtained by technological modeling of biological behaviors [1–10] and physical phenomena [11]. Using this approach, an excellent solution can be obtained within a relatively short period of time. One form of meta-heuristics is ant colony optimization (ACO), which has been based on the feeding behavior of ants. The fundamental elements of ACO are the secretion of pheromone by the ants and its evaporation.

M. Yoshikawa
Department of information engineering, Meijo university, 1-501 Shiogamaguchi, Tenpaku Nagoya 468-8502, Japan
e-mail: evolution_algorithm@yahoo.co.jp

S.-I. Ao et al. (eds.), *Machine Learning and Systems Engineering,*
Lecture Notes in Electrical Engineering 68,
DOI 10.1007/978-90-481-9419-3_9, © Springer Science+Business Media B.V. 2010

Using these two elements, ants can search out the shortest route from their nest to a feeding spot. Fig. 1 shows the actual method for searching the shortest route.

Figure 1a shows the case where two ants A and B have returned from a feeding spot to the nest via different routes. In this case, ants A and B have secreted pheromone on their routes from the nest to the feeding spot and from the feeding spot to the nest. At this point, a third ant C moves from the nest to the feeding spot, relying on the pheromone trail that remains on the route.

ACO is performed based on the following three preconditions: (1) the amount of pheromone secreted from all ants is the same, (2) the moving speed of all ants is the same, and (3) the secreted pheromone evaporates at the same rate. In the above case, since the moving distance from the nest to the feeding spot is longer for route A than for route B, a larger amount of pheromone will have evaporated along route A than along route B, as shown in Fig. 1b. Therefore, ant C will select route B because a larger amount of pheromone remains on this route. However, ant C does not go to the feeding spot by a route that is the same as route B; rather, ant C goes to the feeding spot by route C and then returns to the nest, as shown in Fig. 1c. Another ant D then goes to the feeding spot either by route B or by route C (on which a larger amount of pheromone remains). The fundamental optimization mechanism of ACO is to find a shorter route by repeating this process.

For this reason, ACO has a powerful solution searching ability. However, since processing must be repeated many times, the computation process also requires a very long time. In order to shorten the processing time, this study proposes a new ACO algorithm that incorporates adaptive greedy selection. The validity of the proposed algorithm is verified by performing evaluation experiments using benchmark data.

This chapter is organized as follows: Section 2 indicates the search mechanism of ACO, and describes related studies. Section 3 explains the proposed algorithm with modified pheromone update rules. Section 4 reports the results of computer simulations applied to the travelling salesman problem (TSP) benchmark data. We summarize and conclude this study in Section 5.

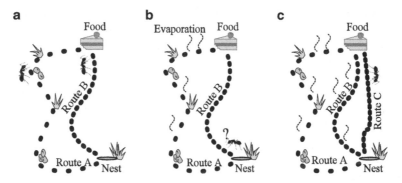

Fig. 1 Example of pheromone communication: (**a**) two routes on initial state, (**b**) positive feedback reinforcement using pheromone information, and (**c**) example of another route

2 Preliminaries

2.1 Ant Colony Optimization

ACO is a general term for the algorithm that can be obtained by technological modeling of the feeding behavior of ants. The basic model of ACO is the ant system (AS) [1] designed by M. Dorigo. The ant colony system (ACS) [2] is one of the modified algorithms of the AS. The ACS provided a better solution when the TSP was to be solved, compared to a genetic algorithm (GA) [9, 10] and a simulated annealing (SA) [11] method. Therefore, this study adopts ACS as its basic algorithm. Henceforth, ACO also denotes the ACS in this paper.

The processing procedure of ACO is explained using an example in which ACO is applied to the TSP. In ACO, each of several ants independently creates a travelling route (visiting every city just one time). At this time, each ant determines the next destination according to the probability p^k calculated from formula (1).

$$p^k(i,j) = \frac{[\tau(i,j)][\eta(i,j)]^\beta}{\sum\limits_{l \in n^k} [\tau(l,j)][\eta(l,j)]^\beta} \tag{1}$$

Where, the value $\eta\,(i,j)$ in formula (1) is called the static evaluation value. It is the reciprocal of the distance between city i and city j, and is a fixed value. On the other hand, $\tau\,(i,j)$ is referred to as the dynamic evaluation value. It expresses the amount of pheromone on the route between city i and city j, and this changes during the process of optimization. The term β represents a parameter and n^k represents a set of unvisited cities.

In ACO, the probability is high for selecting a route whose distance is short and whose pheromone amount is large. That is, selection using a roulette wheel is performed, as shown in Fig. 2.

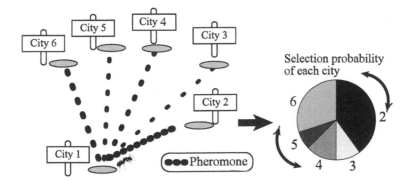

Fig. 2 Example of the selection probability

Here, the amount of pheromone on each partial route is determined by two types of rules: the local update rule and the global update rule. The local update rule is expressed by formula (2), and it is applied whenever each ant moves.

$$\tau(i,j) \leftarrow (1 - \psi)\tau(i,j) + \psi\tau_0 \tag{2}$$

In formula (2), ψ is a decay parameter in local update rule, τ_0 is the initial value of pheromone. The global update rule is expressed by formula (3), and it is applied after each ant completes a travelling route.

$$\tau(i,j) \leftarrow (1 - \rho)\tau(i,j) + \rho\Delta\tau(i,j) \tag{3}$$

$$\Delta\tau(i,j) = \begin{cases} 1/L^+ & \text{if} (i,j) \in T^+ \\ 0 & \text{otherwise} \end{cases} \tag{4}$$

In formula (3), ρ represents the evaporation rate of pheromone in the global update rule, T^+ represents the shortest travelling route among the completed travelling routes, and L^+ represents the length of the partial route that composes the shortest travelling route.

In ACO, the amount of pheromone is larger on a route that has been more often selected, and is smaller, due to evaporation, on a route that has been less often selected. In addition, a rule called pseudo-random-proportional rule is also introduced. For this rule, an ant is forced to move to a city, with the selection probability being the highest at a certain probability. In other words, a random number q is generated between 0 and 1. If q is smaller than threshold value q_t, an ant will move to a city whose value of numerator in formula (1) is the largest. If q is larger than the threshold value, the usual selection by formula (1) will be performed.

2.2 Related Studies

Many studies [1–8] have been reported for the use of ACO, such as those that have applied ACO to network routing problems [3] and to scheduling problems [4]. A number of studies [5, 6] have also been reported on pheromone, which is an important factor for controlling the intensification and diversification of a search. Hybridization with other algorithms, such as a GA [7] or SA [8] method, has also been studied. However, no study has yet been reported that incorporates adaptive greedy selection into an ACO, as is proposed in this study.

3   Hybrid ACO with Modified Pheromone Update Rules

In ACO, when x represents the number of cities, y represents the total number of ants, and z represents the number of processing repetitions, the number of calculations for formulae (1), (2), and (3) are expressed by $(x - 1) \times y \times z$, a $x \times y \times z$, and z, respectively. Since the number of processing repetitions in ACO is large, a processing time problem is inherent in ACO. In this study, incorporating a greedy selection mechanism into the ACO reduced the number of calculations in formulae (1) and (2). Consequently, the processing time was shortened.

A greedy selection approach only employs static evaluation values, with no use of dynamic evaluation values, when selecting the next destination city. That is, an ant moves from the present city to the nearest city. Since static evaluation values are constant in the optimization process, once the calculation is performed in the early stage, it is not necessary to re-calculate. Therefore, the processing time can be shortened by performing a greedy selection. However, since greedy selection favors the local optimal solution, the balance between formula (1) and the greedy selection becomes important.

In order to control this trade-off relationship, the greedy selection is adaptively used in this study. The adaptive greedy selection is explained using the TSP of six cities, as shown in Fig. 3.

In Fig. 3, when city A is set as the start city, the following three selections are performed using formula (1): (1) movement from city A to city B; (2) movement from city B to city C; and (3) movement from city C to city D. The greedy selection is also applied to the other movements; namely, from city D to city E and from city E to city F.

In contrast, when city D is set as the start city, the following three selections are performed using the greedy selection: (1) movement from city D to city E,

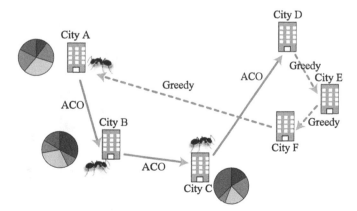

Fig. 3 Example of adaptive greedy selection

(2) movement from city E to city F, and (3) movement from city F to city A. Formula (1) is applied to the other movements; namely, from city A to city B and from city B to city C. In this study, the proportion of the greedy selection and the selection by formula (1) is changed according to generation, as shown in Fig. 4.

Figure 4 shows an example obtained when the total number of processing repetitions was set at 1,000, and the proportion of the greedy selection and the selection by formula (1) was changed every 200 repetitions. In this example, the selection by formula (1) was performed until the 200th repetition. From the 201st repetition to the 400th repetition, the selection by formula (1) was applied to 80% of cities, and the greedy selection was applied to 20% of cities. In other words, for the problem of 100 cities, the selection by formula (1) was performed on 80 cities and the greedy selection was performed on 20 cities. Similarly, from the 401st repetition to the 600th repetition, the selection by formula (1) was applied to 60% of cities, and the greedy selection was applied to 40% of cities. Thus, by changing the proportion of two selection methods according to generation, the trade-off relationship between the accuracy of the solution and the processing time can be controlled.

In optimization algorithms, controlling the trade-off relationship between the intensification and diversification of search is also important. In the adaptive greedy selection proposed in this study, the pressure of intensification is strong. Therefore, in order to strengthen the pressure of diversification, the local update rule and the global update rule are partially changed.

First, the local update rule is generally applied whenever an ant moves. However, it is not applied when the greedy selection is performed. Next, the global update rule is generally applied to the best travelling route; i.e., the shortest travelling route. It is also applied to the second shortest travelling route. By adding these two modifications, the proposed algorithm can realize a balance between the intensification and diversification of the search.

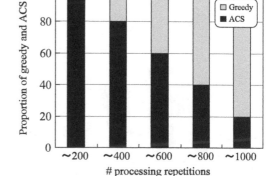

Fig. 4 Example of the proportion of the greedy selection and the selection by formula (1)

4 Experiments and Discussion

In order to verify the validity of the proposed algorithm, several comparison experiments were performed. Each experiment was performed 10 times. Table 1 shows the conditions used in the experiments. In the table, the designation ACS/ Greedy indicates that the city selection of the first half was performed using formula (1), and that of the second half was performed using the greedy selection. Similarly, the designation 80/20 indicates that 80% of the cities were selected using formula (1), and selection of 20% of the cities used the greedy selection. In each table, technique (a) represents the conventional ACO. Techniques (b) and (c) represent algorithms in which the proportion of the greedy selection and the selection by formula (1) was not changed according to generation. Techniques (d), (e), (f), and (g) represent algorithms in which the proportion was changed according to generation. These four algorithms are proposed in this study. Tables 2–4 show the experimental results.

As shown in Tables 2–4, the processing time was shortened when the greedy selection was used, regardless of its proportion. By changing the proportion of the greedy selection and the selection by formula (1) according to generation, the ability to search a solution also improved, when compared with the case where the proportion was fixed. In order to examine the optimal proportion, changes in solutions obtained using techniques (d), (e), (f), and (g) were plotted, as shown in Fig. 5. In this figure, the horizontal axis represents the number of processing

Table 1 Experimental conditions

Technique	Generation (# repetitions)				
	1–200	201–400	401–600	601–800	801–1000
(a) ACS			(N/A)		
(b) Greedy/ACS			50/50		
(c) ACS/Greedy			50/50		
(d) Greedy/ACS	0/100	20/80	40/60	60/40	80/20
(e) ACS/Greedy	100/0	80/20	60/40	40/60	20/80
(f) Greedy/ACS	80/20	60/40	40/60	20/80	0/100
(g) ACS/Greedy	20/80	40/60	60/40	80/20	100/0

Table 2 Results of $q_t = 0.25$

Technique	Best	Average	Worst	Time (s)
(a) ACS	21839	22049	22393	20.00
(b) Greedy/ACS	23196	23482	23690	14.73
(c) ACS/Greedy	22348	22599	22915	15.39
(d) Greedy/ACS	22103	22295	22619	15.64
(e) ACS/Greedy	21997	22264	22709	16.35
(f) Greedy/ACS	21967	22161	22535	15.93
(g) ACS/Greedy	21731	22109	22543	16.42

Table 3 Results of $q_t = 0.50$

Technique	Best	Average	Worst	Time (s)
(a) ACS	21700	21876	22056	18.90
(b) Greedy/ACS	23082	23490	23700	14.24
(c) ACS/Greedy	22348	22707	23122	14.85
(d) Greedy/ACS	21665	22114	22330	15.36
(e) ACS/Greedy	21866	22062	22386	15.72
(f) Greedy/ACS	21824	22026	22346	15.31
(g) ACS/Greedy	21817	22149	22721	15.81

Table 4 Results of $q_t = 0.75$

Technique	Best	Average	Worst	Time (s)
(a) ACS	21558	21768	22262	18.14
(b) Greedy/ACS	23218	23561	23773	13.80
(c) ACS/Greedy	22336	22978	23429	14.37
(d) Greedy/ACS	21787	22038	22503	14.78
(e) ACS/Greedy	21713	21888	22181	15.21
(f) Greedy/ACS	21826	21901	21977	14.73
(g) ACS/Greedy	21665	21975	22312	15.27

Fig. 5 Comparisons of selective techniques (d), (e), (f), and (g)

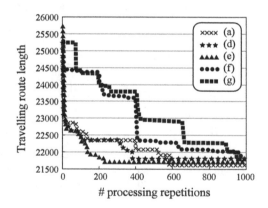

repetitions and the vertical axis represents the travelling route length of the optimal solution.

As shown in Fig. 5, when the number of processing repetitions was small as the termination condition, techniques (d) and (e) showed excellent performance. When the number of processing repetitions was sufficient, techniques (f) and (g) also exhibited excellent performance. Thus, the selection of technique according to the given experimental condition (termination condition) was clearly important.

5 Conclusion

In this chapter, we proposed a new technique that incorporates a greedy mechanism to shorten the processing time of an ACO. A hybrid technique that uses two types of selection techniques (i.e., the conventional selection technique and the greedy selection technique) is introduced to generate one solution. A technique was also developed for adaptively changing these two selection techniques according to generation. In order to control the balance of the intensification and diversification, the update rule of pheromone was improved in each phase of global update and local update. The validity of the proposed technique was verified by comparative experiments using benchmark data. As a future task, the proposed technique should be applied to practical problems other than the TSP.

References

1. M. Dorigo, V. Maniezzo, A. Colorni, Ant system: optimization by a colony of cooperating agents. IEEE Trans. Syst. Man Cybern. Part B. **26**(1), 29–41 (1996)
2. M. Dorigo, L.M. Gambardella, Ant colony system: a cooperative learning approach to the traveling salesman problem. IEEE Trans. Evol. Comput. **1**(1), 53–66 (1997)
3. T.H. Ahmed, Simulation of mobility and routing in ad hoc networks using ant colony algorithms. Proc. Int. Conf. Inf. Technol. Coding Comput. **2**, 698–703 (2005)
4. M. Yoshikawa, H. Terai, A Hybrid Ant Colony Optimization Technique for Job-Shop Scheduling Problems, Proc. of 4th IEEE/ACIS International Conference on Software Engineering Research, Management & Applications (SERA 2006), 95–100 (2006)
5. A. Hara, T. Ichimura, N. Fujita, T. Takahama, "Effective Diversification of Ant-Based Search Using Colony Fission and Extinction", Proc, of IEEE Congress on Evolutionary Computation (CEC 2006), 1028–1035 (2006)
6. Z. Ren, Z. Feng, Z. Zhang, GSP-ANT: an efficient ant colony optimization algorithm with multiple good solutions for pheromone update. Proc. IEEE Int. Conf. Intelligent Comput. Intelligent Syst. **1**, 589–592 (2009)
7. G. Shang, J. Xinzi, T. Kezong, "Hybrid Algorithm Combining Ant Colony Optimization Algorithm with Genetic Algorithm", Proc. of Chinese Control Conference (CCC 2007), 701–704 (2007)
8. Y.-H. Wang, P.-Z. Pan, A Novel Bi-Directional Convergence Ant Colony Optimization with SA for Job-Shop Scheduling, Proc. of International Conference on Computational Intelligence and Software Engineering (CiSE 2009), 1–4 (2009)
9. J. Holland, *Adaptation in Natural Artificial Systems*, 2nd edn. (The University of Michigan Press, MIT Press, 1992)
10. D.E. Goldberg, *Genetic Algorithms in Search Optimization, and Machine Learning* (Addison Wesley, Reading MA, 1989)
11. R.A. Rutenbar, Simulated annealing algorithms: an overview. IEEE Circuits Dev. Mag. **5**(1), 19–26 (1989)

Chapter 10
Study of Pitchfork Bifurcation in Discrete Hopfield Neural Network

R. Marichal, J. D. Piñeiro, E. González, and J. Torres

Abstract A simple two-neuron model of a discrete Hopfield neural network is considered. The local stability is analyzed with the associated characteristic model. In order to study the dynamic behavior, the Pitchfork bifurcation is examined. In the case of two neurons, one necessary condition for yielding the Pitchfork bifurcation is found. In addition, the stability and direction of the Pitchfork bifurcation are determined by applying the normal form theory and the center manifold theorem.

1 Introduction

The purpose of this paper is to present some results on the analysis of the dynamics of a discrete recurrent neural network. The particular network in which we are interested is the Hopfield network, also known as a Discrete Hopfield Neural Network in [1]. Its state evolution equation is

$$x_i(k+1) = \sum_{n=1}^{N} w_{in} f(x_i(k)) + \sum_{m=1}^{M} w'_{im} u_m(k) + w''_i \tag{1}$$

where
$x_i(k)$ is the ith neuron output
$u_m(k)$ is the mth input of the network
w_{in}, w'_{im} are the weight factors of the neuron outputs, network inputs and
w''_i is a bias weight
N is the neuron number

R. Marichal (✉)
System Engineering and Control and Computer Architecture Department, University of La Laguna, Avda. Francisco Sánchez S/N, Edf. Informatica, 38206 Tenerife, Canary Islands, Spain
e-mail: rlmarpla@ull.es

S.-I. Ao et al. (eds.), *Machine Learning and Systems Engineering*,
Lecture Notes in Electrical Engineering 68,
DOI 10.1007/978-90-481-9419-3_10, © Springer Science+Business Media B.V. 2010

M is the input number

$f(\cdot)$ is a continuous, bounded, monotonically increasing function, such as the hyperbolic tangent

This model has the same dynamic behavior as the Williams–Zipser neural network. The relationship between the Williams–Zipser states and Hopfield states is

$$X_h = W X_{W-Z}$$

where

X_h are Hopfield states

X_{W-Z} are the Williams-Zipser states

W is the weight matrix without the bias and input weight factor

We will consider the Williams−Zipser model in order to simplify the mathematical calculations.

The neural network presents different classes of equivalent dynamics. A system will be equivalent to another if its trajectories exhibit the same qualitative behavior. This is made mathematically precise in the definition of topological equivalence [2]. The simplest trajectories are those that are equilibrium or fixed points that do not change in time. Their character or stability is given by the local behavior of nearby trajectories. A fixed point can attract (sink), repel (source) or have directions of attraction and repulsion (saddle) of close trajectories [3]. Next in complexity are periodic trajectories, quasi-periodic trajectories or even chaotic sets, each with its own stability characterization. All of these features are similar in a class of topologically equivalent systems. When a system parameter is varied, the system can reach a critical point at which it is no longer equivalent. This is called a bifurcation, and the system will exhibit new behavior. The study of how these changes can be carried out will be another powerful tool in the analysis.

With respect to discrete recurrent neural networks as systems, several results on their dynamics are available in the literature. The most general result is derived using the Lyapunov stability theorem in [4], and establishes that for a symmetric weight matrix, there are only fixed points and period two limit cycles, such as stable equilibrium states. It also gives the conditions under which only fixed-point attractors exist. More recently Cao [5] proposed other, less restrictive and more complex, conditions. In [6], chaos is found even in a simple two-neuron network in a specific weight configuration by demonstrating its equivalence with a one-dimension chaotic system (the logistic map). In [7], the same author describes another interesting type of trajectory, the quasi-periodic orbits. These are closed orbits with irrational periods that appear in complex phenomena, such as frequency-locking and synchronization, which are typical of biological networks. In the same paper, conditions for the stability of these orbits are given. These can be simplified, as we shall show below.

Passeman [8] obtains some experimental results, such as the coexisting of the periodic, quasi-periodic and chaotic attractors. Additionally, [9] gives the position, number and stability types of fixed points of a two-neuron discrete recurrent network with nonzero weights.

There are some works that analyze the Hopfield continuous neural networks [10, 11], like [12–15]. These papers show the stability of Hopf-bifurcation with two delays.

Firstly, we analyze the number and stability-type characterization of the fixed points. We then continue with an analysis of the Pitchfork bifurcation. Finally, the simulations are shown and conclusions are given.

2   Determination of Fixed Points

For the sake of simplicity, we studied the two-neuron network. This allows for an easy visualization of the problem. In this model, we considered zero inputs so as to isolate the dynamics from the input action. Secondly, and without loss of generality with respect to dynamics, we used zero bias weights. The activation function is the hyperbolic tangent.

With these conditions, the network mapping function is

$$\begin{aligned}
x_1(k+1) &= \tanh(w_{11}x_1(k) + w_{12}x_2(k)) \\
x_2(k+1) &= \tanh(w_{21}x_1(k) + w_{22}x_2(k))
\end{aligned} \tag{2}$$

where $x(k)$ and $y(k)$ are the neural output of step k.

The fixed points are solutions of the following equations

$$\begin{aligned}
x_{1,p} &= \tanh(w_{11}x_{1,p} + w_{12}x_{2,p}) \\
x_{2,p} &= \tanh(w_{21}x_{1,p} + w_{22}x_{2,p})
\end{aligned} \tag{3}$$

The point (0, 0) is always a fixed point for every value of the weights. The number of fixed points is odd because for every fixed point $(x_{1,p}, x_{2,p})$, $(-x_{1,p}, -x_{2,p})$ is also a fixed point.

To graphically determine the configuration of fixed points, we redefine the above equations as

$$\begin{aligned}
x_{2,p} &= \frac{a\tanh(x_{1,p}) - w_{11}x_{1,p}}{w_{12}} = F(x_{1,p}, w_{11}, w_{12}) \\
x_{1,p} &= \frac{a\tanh(x_{2,p}) - w_{22}x_{2,p}}{w_{21}} = F(x_{2,p}, w_{22}, w_{21})
\end{aligned} \tag{4}$$

Depending on the diagonal weights, there are two qualitative behavior functions. We are going to determine the number of fixed points using the graphical representation of the above Eq. (4). First, we can show that the graph of the F function has a maximum and a minimum if $w_{ii} > 1$ or, if the opposite condition holds, is like the hyperbolic arctangent function (Fig. 1).

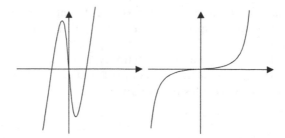

Fig. 1 The two possible behaviors of the F function. The left figure corresponds to the respective diagonal weight lower than unity. The right shows the opposite condition

The combination of these two possibilities with another condition on the ratio of slopes at the origin of the two curves (4) gives the number of fixed points. The latter condition can be expressed as

$$|W| = w_{11} + w_{22} - 1$$

where $|W|$ is the weight matrix determinant.

If $w_{11} > 1$, $w_{22} > 1$ and $|W| > w_{11} + w_{22} - 1$, then there can exist nine, seven or five fixed points. When this condition fails, there are three fixed points.

When a diagonal weight is less than one, there can be three or one fixed points.

3 Local Stability Analysis

In the process below, a two-neuron neural network is considered. It is usual for the activation function to be a sigmoid function or a tangent hyperbolic function.

Considering the fixed point Eq. (3), the elements of the Jacobian matrix at the fixed point (x, y) are

$$J = \begin{bmatrix} w_{11} f'(x_1) & w_{12} f'(x_1) \\ w_{21} f'(x_2) & w_{22} f'(x_2) \end{bmatrix}$$

The associated characteristic equation of the linearized system evaluated at the fixed point is

$$\lambda^2 - [w_{11} f'(x_1) + w_{22} f'(x_2)]\lambda + |W| f'(x_1) f'(x_2) = 0 \qquad (5)$$

where w_{11}, w_{22} and $|W|$ are the diagonal elements and the determinant of the matrix weight, respectively.

We can define new variables

$$\sigma_1 = \frac{w_{11} f'(x_1) + w_{22} f'(x_2)}{2}$$

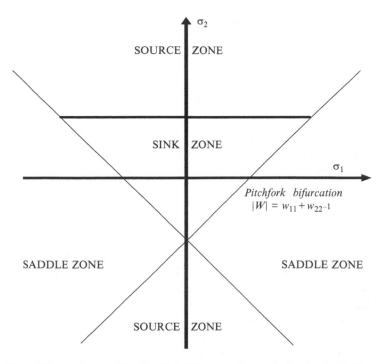

Fig. 2 The stability regions and the Pitchfork bifurcation line at the fixed point (0, 0)

$$\sigma_2 = |W| f'(x_1) f'(x_2)$$

The eigenvalues of the characteristic Eq. (5) are defined as

$$\lambda_\pm = \sigma_1 \pm \sqrt{\sigma_1^2 - \sigma_2}$$

The Pitchfork bifurcation appears when two complex conjugate eigenvalues reach the unit circle. It is easy to show that this limit condition is.

The boundaries between the regions shown in Fig. 2 are the bifurcations. At these limit zones the fixed point changes its character. The Pitchfork bifurcation is represented by the line $|W| = w_{11} + w_{22} - 1$ in Fig. 2.

4   Pitchfork Bifurcation Direction

In order to determine the direction and stability of the Pitchfork bifurcation, it is necessary to use the center manifold theory [2]. The center manifold theory demonstrates that the mapping behavior in the bifurcation is equivalent to the complex mapping below:

$$u(k+1) = u(k) + c(0)u(k)^3 + o(u(k)^4) \tag{6}$$

The parameter $c(0)$ is [2]

$$c(0) = \frac{1}{6}\langle p, C(q,q,q)\rangle \tag{7}$$

where C is the third derivative terms of the Taylor development, the notation $<.,.>$ represents the scalar product, and p, q are the eigenvector Jacobian matrix and its transpose, respectively. These vectors satisfy the normalization condition

$$\langle p,q\rangle = 1$$

The above coefficients are evaluated for the critical parameter of the system where the bifurcation takes place. The $c(0)$ sign determines the bifurcation direction. When $c(0)$ is negative, a stable fixed point becomes an unstable fixed point and two additional stable symmetrical fixed points appear. In the opposite case, $c(0)$ positive, an unstable fixed point becomes a stable fixed point and two additional unstable symmetrical fixed points appear.

In the neural network mapping, p and q are

$$q = \frac{d}{e+d}\left\{\frac{e}{w_{21}X_{2,0}}, -1\right\} \tag{8}$$

$$p = \left\{\frac{e}{w_{12}X_{1,0}}, -1\right\} \tag{9}$$

where

$$d = w_{11}X_{1,0} - 1$$

$$e = w_{22}X_{2,0} - 1$$

$$X_{1,0} = 1 - x_{1,0}^2$$

$$X_{2,0} = 1 - x_{2,0}^2$$

$x_{1,0}$ and $x_{2,0}$ are the fixed point coordinates where the bifurcation appears.

The Taylor development term is

$$C_i(q,q,q) = \sum_{j,k,l=1}^{2} \frac{\partial f_i}{\partial x_j \partial x_k \partial x_l} q_j q_k q_l$$

$$= f'''(0) \sum_{j,k,l=1}^{2} \delta_{ik}\delta_{il} w_{ij} w_{ik} w_{il} q_j q_k q_l$$

$$= f'''(0) \sum_{j=1}^{2} w_{ij}^3 q_j^2 q_j \tag{10}$$

where δ_{ij} is the Kronecker delta.

In order to determine the parameter $c(0)$, it is necessary to calculate the third derivate of the mapping (1) given by Eq. (10)

$$\frac{\partial f_i}{\partial x_j \partial x_k \partial x_l} = 2(1 - x_i^2)(3x_i^2 - 1)w_{ij}w_{ik}w_{il}$$

The Taylor development term is then

$$C_i(a,b,c) = 2 \sum_{j,k,l=1}^{2} (1 - x_i^2)(3x_i^2 - 1)w_{ij}w_{ik}w_{il}a_j b_k c_l.$$

Taking into account the previous equations and the q autovector Eq. (8)

$$\mathbf{C(q,q,q)} = 2 \begin{bmatrix} \frac{2X_{1,0}(3x_{1,0}^2-1)w_{12}^3}{d^3} \\ \frac{(1-3x_{2,0}^2)}{X_{2,0}^2} \end{bmatrix} \tag{11}$$

It can be shown in Eq. (3) that the zero is always a fixed point. Replacing the expressions for $C(q,q,q)$, q and p given by Eqs. (8) (9) and (11), respectively, and evaluating them at the zero fixed point, the previous $c(0)$ coefficient is

$$c(0) = \frac{1}{6}\langle p, C(q,q,q)\rangle = \frac{w_{12}^2(w_{22}-1)+(w_{11}-1)^3}{3(1-w_{11})^2(2-w_{11}-w_{22})}$$

The previous expression is not defined for the following cases

a. $w_{11} = 1$
b. $w_{11} + w_{22} = 2$.

In this paper we only consider condition (a) since condition (b) shows the presence of another bifurcation, known as Neimark-Sacker and which is analyzed in another paper [16].

Taking into account condition a) and the bifurcation parameter equation

$$|W| = w_{11} + w_{22} - 1$$

then

$$w_{12}w_{21} = 0$$

In this particular case, the eigenvalues match with the diagonal elements of the weight matrix

$$\lambda_1 = w_{11}$$
$$\lambda_2 = w_{22}$$

The new q and p eigenvectors are given by

$$q = \{1, 0\}$$
$$p = \left\{1, -\frac{w_{12}}{w_{22} - 1}\right\}$$

The $c(0)$ coefficient is

$$c(0) = \frac{1}{6}\langle p, C(q, q, q)\rangle = -\frac{1}{3}w_{11}^3 = -\frac{1}{3}.$$

Therefore, in this particular case, the coefficient of the normal form c (0) is negative, a stable fixed point becomes a saddle fixed point and two additional stable symmetrical fixed points appear.

5 Simulations

In order to examine the results obtained. The simulation shows the Pitchfork bifurcation (Fig. 3). The Pitchfork bifurcation is produced by the diagonal element weight matrix w_{11}. Figure 3a shows the dynamic configuration before the bifurcation is produced, with only one stable fixed point. Subsequently, when the bifurcation is produced (Fig. 3b), two additional stable fixed points appear and the zero fixed point changes its stability from stable to unstable (the normal form coefficient c is negative).

6 Conclusion

In this paper we considered the Hopfield discrete two-neuron network model. We discussed the number of fixed points and the type of stability. We showed the bifurcation Pitchfork direction and the dynamical behavior associated with the bifurcation.

The two-neuron networks discussed above are quite simple, but they are potentially useful since the complexity found in these simple cases might be carried

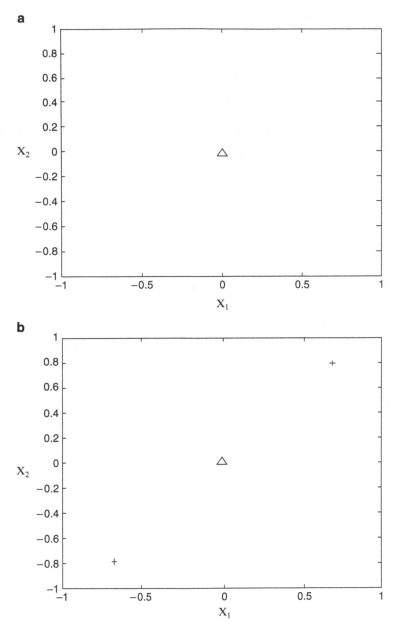

Fig. 3 The dynamic behavior when the Pitchfork bifurcation is produced. + and Δ are the saddle and source fixed points, respectively. (a) $w_{11} = 0.9$, $w_{12} = 0.1$, $w_{21} = 1$ and $w_{22} = 0.5$; (b) $w_{11} = 1.1$, $w_{12} = 0.1$, $w_{21} = 1$ and $w_{22} = 0.5$

over to larger Hopfield discrete neural networks. There exists the possibility of generalizing some of these results to higher dimensions and of using them to design training algorithms that avoid the problems associated with the learning process.

References

1. R. Hush, B.G. Horne, Progress in supervised neural network. IEEE Signal Process. Mag. 8–39 (1993)
2. Y.A. Kuznetsov, *Elements of Applied Bifurcation Theory*, 3rd edn. (Springer-Verlag, New York, 2004)
3. K.T. Alligood, T.D. Sauer, J.A. Yorke, *Chaos: An Introduction to Dynamical Systems* (Springer Verlag, New York, 1998)
4. C.M. Marcus, R.M. Westervelt, Dynamics of iterated-map neural networks. Phys. Rev. A **40**, 501–504 (1989)
5. Cao Jine, On stability of delayed cellular neural networks. Phys. Lett. A **261**(5–6), 303–308 (1999)
6. X. Wang, Period-doublings to chaos in a simple neural network: an analytical proof. Complex Syst. **5**, 425–441 (1991)
7. X. Wang, Discrete-time dynamics of coupled quasi-periodic and chaotic neural network oscillators. International Joint Conference on Neural Networks (1992)
8. F. Pasemann, Complex dynamics and the structure of small neural networks. Netw. Comput. Neural Syst. **13**(2), 195–216 (2002)
9. Tino et al., Attractive periodic sets in discrete-time recurrent networks (with emphasis on fixed-point stability and bifurcations in two-neuron networks). Neural Comput. **13**(6), 1379–1414 (2001)
10. J. Hopfield, Neurons with graded response have collective computational properties like those of two-state neurons. Procs. Nat. Acad. Sci. USA **81**, 3088–3092 (1984)
11. D.W. Tank, J.J. Hopfield, Neural computation by concentrating information in time, Proc. Nat. Acad. Sci. USA, **84**, 1896–1991 (1984)
12. X. Liao, K. Wong, Z. Wu, Bifurcation analysis on a two-neuron system with distributed delays. Physica D **149**, 123–141 (2001)
13. S. Guo, X. Tang, L. Huang, Hopf bifurcating periodic orbits in a ring of neurons with delays, Physica D, **183**, 19–44 (2003)
14. S. Guo, X. Tang, L. Huang, Stability and bifurcation in a discrete system of two neurons with delays. Nonlinear Anal. Real World. doi:10.1016/j.nonrwa.2007.03.002 (2007)
15. S. Guo, X. Tang, L. Huang, Bifurcation analysis in a discrete-time single-directional network with delays. Neurocomputing **71**, 1422–1435 (2008)
16. R. Marichal et al., Bifurcation analysis on hopfield discrete neural networks. *WSEAS* Trans. Syst. **5**, 119–124 (2006)

Chapter 11
Grammatical Evolution and STE Criterion

Statistical Properties of STE Objective Function

Radomil Matousek and Josef Bednar

Abstract Grammatical evolution (GE) is one of the newest among computational methods (Ryan et al. 1998; O'Neill and Ryan 2001). Basically, it is a tool used to automatically generate Backus-Naur-Form (BNF) computer programmes. The method's evolution mechanism may be based on a standard genetic algorithm (GA). GE is very often used to solve the problem of a symbolic regression, determining a module's own parameters (as it is also the case of other optimization problems) as well as the module structure itself. A Sum Square Error (SSE) method is usually used as the testing criterion. In this paper, however, we will present the original method, which uses a Sum epsilon Tube Error (STE) optimizing criterion. In addition, we will draw a possible parallel between the SSE and STE criteria describing the statistical properties of this new and promising minimizing method.

1 Introduction

The general problem of optimizing a nonlinear model or a specific problem of nonlinear regression in statistics may be seen as typical problems that can success-fully be approached using evolutional optimization techniques. A genetic algorithm method, for example, may be very efficient in determining the parameters of models representing a multimodal or non-differentiable behaviour of the goal function. On the other hand, there are numerous mathematical methods that may, under certain conditions, be successful in looking for a minimum or maximum. The classical optimization methods may differ by the type of the problem to be solved (linear, nonlinear, integer programming, etc.) or by the actual algorithm used. However, in all such cases, the model structure for which optimal parameters are

R. Matousek (✉)
Dept. of Mathematic and Applied Computer Science, Brno University of Technology, FME,
Technicka 2896/2, 61600 Brno, Czech Republic
e-mail: matousek@fme.vutbr.cz

S.-I. Ao et al. (eds.), *Machine Learning and Systems Engineering*,
Lecture Notes in Electrical Engineering 68,
DOI 10.1007/978-90-481-9419-3_11, © Springer Science+Business Media B.V. 2010

Table 1 Grammar G which was used with respect of the approximation tasks

Approximation task	Grammar and example of generating function
Trigonometric (data: ET20x50)	$G = \{+,-,*,/,\sin,\cos,\text{variables},\text{constants}\}$ notes: *unsigned integer constants* $\in [0,15]$, *real constants* $\in [0,1]$ $y = \sin(x) + \sin(2.5x)$
Polynomial (data: ET10x50)	$G = \{+,-,*,/,\text{variables},\text{constants}\}$ notes: *unsigned integer constants* $\in [0,15]$, *real constants* $\in [0,1]$ $y = 3x^4 - 3x + 1$

to be found is already known. But it is exactly the process of an adequate model structure design that may be of key importance in statistical regression and numerical approximation.

When a problem of regression is dealt with, it is usually split into two parts. First, a model is designed to fit the character of the empirically determined data as much as possible. Then its optimum parameters are computed for the whole to be statistically adequate.

In the event of a pure approximation problem, the situation is the same: a model is designed (that is, its structure such as a polynomial $P(x) = ax^2 + bx + c$) with its parameters being subsequently optimized (such as a, b, c) using a previously chosen minimization criterion, which is usually a Sum Square Error (SSE) one.

GE uses such data representation as to deal with the above problem in a comprehensive way, that is, combining the determination of a model structure and finding its optimum parameter in a single process. Such a process is usually called the problem of a symbolic regression. The GE optimisation algorithm usually uses a GA with the genotype-phenotype interpretation using a context-free Backus-Naur-Form (BNF) based grammar [1, 2]. The problem solved is then represented by a BNF structure, which a priori determines its scope – an automatic computer program generating tool. Further we will focus on issues of symbolic regression of trigonometric and polynomial data, that is, data that can be described using a grammar containing trigonometric functions and/or powers.

The advanced GE algorithm implemented was using a binary GA. The GA implemented the following operators: tournament selection, one- and two-point structural crossing-over, and structural mutations. Numerous modern methods of chromosome encryption and decryption were used [3, 4]. The grammar G was used with respect to the class to be tested (Table 1).

2 STE – Sum Epsilon Tube Error

To keep further description as simple as possible, consider a data approximation problem ignoring the influence of a random variable, that is, disregarding any errors of measurement. The least-square method is used to test the quality of a solution in an absolute majority of cases. Generally, the least-squares method is used to find a solution with the squared errors r summed over all control points being minimal. A control point will denote a point at which the approximation error is calculated using the data to be fitted, that is, the difference between the actual y value and the approximation $\hat{y}$, ($r=y-\hat{y}$).

$$SSE = \sum_i r_i^2 \tag{1}$$

However, in the event of a GE problem of symbolic regression, this criterion may not be sufficient. For this reason a new, Sum Epsilon-Tube Error (STE) evaluation criterion has been devised. For each value ε, this evaluation criterion may be calculated as follows:

$$STE_\varepsilon = \sum_i e_\varepsilon(r_i), \quad e_\varepsilon(r_i) = \begin{cases} 0 & r_i \notin [-\varepsilon, \varepsilon] \\ 1 & r_i \in [-\varepsilon, \varepsilon] \end{cases} \tag{2}$$

where ε is the radius of the epsilon tube, i is the index set of the control points, e is an objective function determining whether the approximated value lies within the given epsilon tube. The workings of this criterion may be seen as a tube of diameter epsilon that stretches along the entire approximation domain (Fig. 2). The axis of such a tube is determined by the points corresponding to the approximated values. Such points may be called control points.

The actual evaluating function then checks whether an approximation point lies within or out of the tube (2). The epsilon value changes dynamically during the optimization process being set to the highest value at the beginning and reduced when the adaptation condition is met.

The variable parameters of the STE method are listed in Table 2. At the beginning of the evolution process, the algorithm sets the diameter of the epsilon tube. If the condition (3), which indicates the minimum number of points that need to be contained in the epsilon cube, is met, the epsilon tube is adapted, that is, the

Table 2 Parameters of the STE minimization method

Parameter	Value	Description
steCP	Unsigned integer	The number of control points
steEpsStart	Positive real number	Epsilon initial value (ε-upper bound)
steEpsStop	Positive real number	Final epsilon value (ε-lower bound)
steEpsReduction	Unsigned integer	Percentage of value decrease on adapting the value
steAdaptation	Unsigned integer	Percentage indicating the success rate (control points are in the ε tube) at which epsilon is adapted
steModel	"Interest-Rate or Linear Model"	Model of the computation of a new epsilon value if adaptation condition is met
steIteration	Unsigned integer	This may replace the steEpsStop parameter, and then this parameter indicates the number of adaptations until the end of the computation

current epsilon value is reduced. This value can either be reduced using an interest-rate model with the current epsilon value being reduced by a given percentage of the previous epsilon value or using a linear model with the current epsilon value being reduced by a constant value.

The condition for the adaptation of the epsilon tube to be finished may either be chosen as a limit epsilon value (the steEpsStop parameter) or be given by the number of adaptive changes (the steIteration parameter).

$$(steCP/STE_\varepsilon) \cdot 100 \geq steAdaptation \tag{3}$$

3   STE – Empirical Properties

Practical implementations of the symbolic regression problem use STE as an evaluating criterion. The residua obtained by applying STE to the approximation problems implemented have been statistically analyzed. It has been found out that these residua are governed by the Laplace distribution (4) with $\mu = 0$ and $b = 1/$sqrt (2) in our particular case. The residua was standardised for requirement of our experiments (the data are standardized by subtracting the mean and dividing by the standard deviation).

$$f(x|\mu, b) = \frac{1}{2b} \exp\left(-\frac{|x - \mu|}{b}\right) = \frac{1}{2b} \begin{cases} \exp\left(-\dfrac{\mu - x}{b}\right) & \text{if } x < \mu \\ \exp\left(-\dfrac{x - \mu}{b}\right) & \text{if } x \geq \mu \end{cases} \tag{4}$$

A random variable has a Laplace (μ, b) distribution if its probability density function is (4). Here, μ is a location parameter and $b > 0$ is a scale parameter. If $\mu = 0$ and $b = 1$, the positive half-line is exactly an exponential distribution scaled by ½.

An intuitive approach to approximation shows better results for the STE method (Fig. 1) mainly because there are no big differences between the approximating functions (Figs. 3 and 4). The reason for this is obvious: very distant points influence the result of approximation a great deal for SSE while, for STE, the weight of such points are relatively insignificant.

Next, the main advantages and disadvantages are summarized of the minimization methods using the SSE and STE criteria.

3.1   SSE (Advantages, Disadvantages)

+ a classical and well-researched method both in statistical regression and numerical approximation.

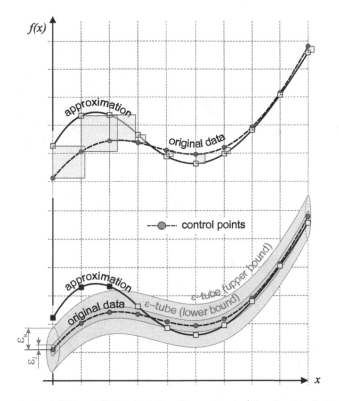

Fig. 1 Comparison of SSE and STE principles. Common principle of residual sum of squares (*above*). The principle of the Sum of epsilon-Tube Error (*below*)

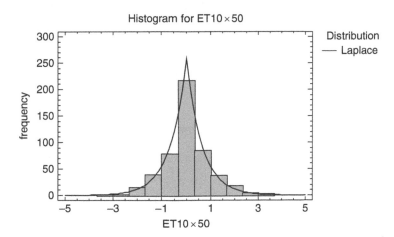

Fig. 2 Histogram of the standardized residua (from the experiment ET10x50) and the probability density function of the Laplace distribution

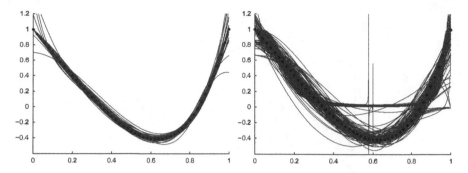

Fig. 3 The final approximations (symbolic regression) of polynomial data (see Table 1). STE method was used (*left*), respectively SEE method was used (*right*)

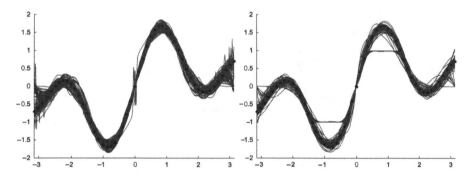

Fig. 4 The final approximations (symbolic regression) of trigonometric data (Table 1). STE method was used (*left*), respectively SEE method was used (*right*)

+ a metric is defined.
+ a $N(\mu = 0, \sigma^2)$ residua error distribution may be assumed with descriptive statistics for this distribution being available.
− more time-consuming.
− being used excessively, this method hardly provides an incentive for users to gain new insights [5].

3.2 STE (Advantages, Disadvantages)

+ less time consuming.
+ more intuitive and readable.
+ when using a GE-based symbolic regression, the Laplace distribution may be assumed (which is one of the original results of this paper).
+ better results are generated for the problem class.
− mathematical processing is worse with no metric defined.
− the method being new, it is not so well described.

4 Probabilistic Mapping of SSE to STE

This part shows how the results obtained by a SSE-based minimization method may be used to derive a result corresponding to the values of a STE-based minimization criterion. In other words, a procedure will be shown for using SSE to obtain a corresponding STE. The procedure is based on the following:

- It is assumed that, when applying the least-squares method, the residua are normally distributed with $N(\mu = 0, \sigma^2)$. Here, the standard deviation may be arbitrary, but a reduction to the standardized normal distribution $N(0, 1)$ is possible.
- In order to reach a sufficient statistical significance, the experiment simulated 10,000 instances of solution to an approximation problem (nRUN). This problem was discretized using 100 control points.
- Different sizes $eps = \{2, 1, 0.5, 0.25, 0.125\}$ of epsilon tubes were chosen to simulate possible adaptations of the epsilon tube by (3).
- The frequencies were calculated of the values lying within the given epsilon tubes.

This simulation and the subsequent statistical analysis were implemented using the Minitab software package. Table 3 provides partial information on the procedure used.

Where r is residua ($X \sim N(0,1)$), i is control point index ($i \in [1, 100]$ in our particular case), nRUN is given implementation (nRUN $\in [1, 10000]$ in our particular case).

Next, for each residuum value, the number of cases will be determined of these residua lying in the interval given by a particular epsilon tube. This evaluation criterion denoted by STE (Sum epsilon-Tube Error) is defined by (2) and, for the particular test values, corresponds to:

$$STE_\varepsilon = \sum_{i=1}^{100} e_\varepsilon(r_i), \ \varepsilon = \{2, 1, 0.5, 0.25, 0.125\} \tag{5}$$

By a statistical analysis, it has been determined that STE behaves as a random variable with a Binomial distribution

Table 3 A fragment of a table of the simulated residua values with standardized normal distribution and the corresponding $e_\varepsilon(r_i)$ values

r	i	nRUN	$\varepsilon = 2$	$\varepsilon = 1$	$\varepsilon = 0.5$	$\varepsilon = 0.25$	$\varepsilon = 0.125$
0.48593	1	1	1	1	1	0	0
−0.07727	2	1	1	1	1	1	1
1.84247	3	1	1	0	0	0	0
−1.29301	4	1	1	0	0	0	0
0.34326	5	1	1	1	1	0	0
...	...	...	...	...	...	...	...
0.02298	100	.10,000	1	1	1	1	1

Table 4 Table of probabilities p_s

ε	$P(N(0,1)< -\varepsilon)$	$P(N(0,1)<\varepsilon)$	p_ε
2	0.022750	0.977250	0.954500
1	0.158655	0.841345	0.682689
0.5	0.308538	0.691462	0.382925
0.25	0.401294	0.598706	0.197413
0.125	0.450262	0.549738	0.099476

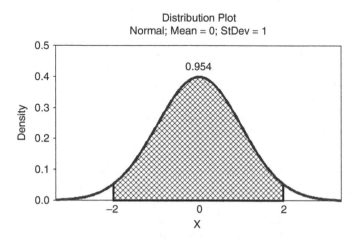

Fig. 5 Calculating probability p_ε with $\varepsilon = 2$

$$K \sim Bi(n, p_\varepsilon), \tag{6}$$

where

$$p_\varepsilon = P(X \sim N(0,1) \in [-\varepsilon, \varepsilon])$$

For the epsilons chosen, the STE characteristic has a binomial distribution given by (6) where the probabilities p_ε are calculated from Table 4. For the case of $\varepsilon = 2$, Fig. 5 shows the probability calculation.

The Fig. 6 shows the probability functions for epsilon tubes with value k indicating, with a given probability, the number of control points for which the $e(r)$ function takes on a value of one, that is, the number of control points contained in a given epsilon tube.

The correctness of the calculation was proved empirically. The STE values simulated (10,000 runs) were compared with distribution (6). The following figure (Fig. 7) displays an empirical probability density function obtained from Table 3 by (5) compared with the probability density function as calculated by (6) for $\varepsilon = 1$.

It can be seen in Fig. 7 that the empirical distribution is close the calculated one, which was verified using a goodness-of-fit test.

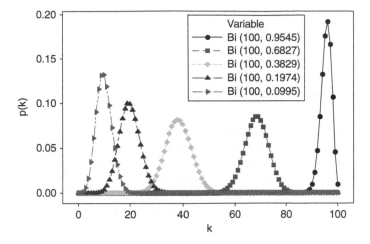

Fig. 6 STE probability density functions for ε-tubes with ε = {2, 1, 0.5, 0.25, 0.125}

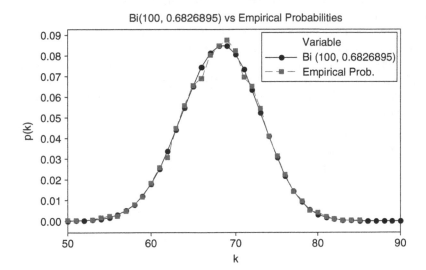

Fig. 7 Comparing empirical and calculate probability density functions of occurrences within the epsilon tube for ε = 1

5 Goodness-of-Fit Tests of Data Sets

Residua which we obtained from our symbolic regression experiments were analyzed for Laplace distribution coincidence. We have tested many residua data for this distribution. The examples of specific results which confirmed our hypothesis about Laplace distribution follows. There we denoted data sets ET10x50 (500 values ranging from −5.12124 to 3.27653) for polynomial problem and ET20x50

(1000 values ranging from -5.12124 to 5.11919) for goniometrical problem which is described by Table 1.

5.1   Uncensored Data – ET10x50

Fitted distribution is Laplace, where mean is -0.00198 and scale is 1.55075. This analysis shows the results of fitting a Laplace distribution to the data on ET10x50 (Tables 5 and 6). We can test whether the Laplace distribution fits the data adequately by selecting Goodness-of-Fit Tests from the list of Tabular Options. The visual fiiting of Laplace distribution is in the Fig. 2.

This pane shows the results of tests run to determine whether ET10x50 can be adequately modeled by a Laplace distribution. The chi-squared test divides the range of ET10x50 into nonoverlapping intervals and compares the number of observations in each class to the number expected based on the fitted distribution. The Kolmogorov-Smirnov test computes the maximum distance between the cumulative distribution of ET10x50 and the CDF of the fitted Laplace distribution. In this case, the maximum distance is 0.0540167. The other statistics compare the empirical distribution function to the fitted CDF in different ways.

Since the smallest P-value amongst the tests performed is greater than or equal to 0.05, we can not reject the idea that ET10x50 comes from a Laplace distribution with 95% confidence. Analogous procedure can be use in case of ET20x50 data sets and another.

Table 5 Table of goodness-of-fit tests for ET10x50: *chi-squared test

	Lower limit	Upper limit	Observed frequency	Expected frequency	Chi-squared
At or below		-3.0	2	2.39	0.06
	-3.0	-2.33333	1	4.33	2.57
	-2.33333	-1.66667	15	12.19	0.65
	-1.66667	-1.0	39	34.27	0.65
	-1.0	-0.333333	78	96.36	3.50
	-0.333333	0.333333	217	201.82	1.14
	0.333333	1.0	85	95.77	1.21
	1.0	1.66667	37	34.06	0.25
	1.66667	2.33333	18	12.11	2.86
	2.33333	3.0	5	4.31	0.11
above	3.0		3	2.38	0.16

*Chi-squared $= 13.1719$ with 8 d.f. P-Value $= 0.10607$

Table 6 Table of goodness-of-fit tests for ET10x50: *Kolmogorov-Smirnov test

	Laplace		Laplace
DPLUS	0.0397308	DN	0.0540167
DMINUS	0.0540167	*P-Value	0.108113

6 Probabilistic Relationship Between STE and SSE

Fundamental difference between STE and SSE optimal criteria is metric regarding Eqs. (1) and (2). The SSE contrary to the STE has define the Euclidean metric (in this case triangle inequality exist). Hence, exact relation form STE to SSE optimal criteria doesn't exist. But, if we try to use probabilistic relation, we can obtain relationship for transformation of Laplace distribution to Normal distribution. In other words, we can transform STE solution to SSE solution in probabilistic point of view.

We suppose (from caption IV and on the base chi-square goodness test) that the residua has Laplace distribution in case of STE criteria. As we can see in 0, Laplace distribution is a typical distribution for some numerical methods. Laplace distribution is also called double exponential distribution. It is the distribution of differences between two independent varieties with identical exponential distribution 0. The probability is given by (7)

$$P(x) = \frac{1}{2b} e^{-|x-\mu|/b} \tag{7}$$

and the moments about the mean μ_n are related to the moments about 0 by (8)

$$\mu_n = \sum_{j=0}^{n} \sum_{k=0}^{floor(j/2)} (-1)^{n-j} \binom{n}{j} \binom{j}{2k} b^{2k} \mu^{n-2k} \Gamma(2k+1) = \begin{cases} n! b^n & \text{for } n \text{ even} \\ 0 & \text{for } n \text{ odd} \end{cases} \tag{8}$$

Common uses moments are in the Table 7. Because the Mean Square Error (MSE) given by MSE = SSE/n is estimation of the variance σ^2 we can derive from Table 7 and given parameter b the Sum Square Error by Eq. (9).

$$SSE = \sigma^2 n = 2b^2 n \tag{9}$$

7 Conclusion

- A new STE evaluation criterion for GE was introduced using epsilon tubes. Based on the experiments conducted, this criterion seems to be more appropriate for the symbolic regression problem as compared with the SSE method.

Table 7 Common uses moments in normal and Laplacian probabilistic distributions

Relationship			Description
μ	=	μ	Mean
σ^2	=	$2b^2$	Variance
γ_1	=	0	Skewness
γ_2	=	3	Kurtosis

- Using statistical tools the method of transforming the results obtained by SSE minimisation into results obtained by STE has been shown.
- Using a particular class implemented using approximations obtained by an STE minimization, the empirically obtained residua were found to be Laplace distributed. It was confirmed on the base statistical approach of goodness-of-fit tests.
- The properties of such residua and transformation of the STE-based minimization criterion into a SSE-based was shown by means of probabilistic relationship.

Acknowledgement This work was supported by the Czech Ministry of Education in the frame of MSM 0021630529 and by the GACR No.: 102/091668.

References

1. M. O'Neill, C. Ryan, *Grammatical Evolution: Evolutionary Automatic Programming in an Arbitrary Language* (Kluwer, London, USA, 2003), ISBN 1-4020-7444-1
2. C. Ryan, M. O'Neill, J.J. Collins, Grammatical evolution: solving trigonometric identities, in *Proceedings of MENDEL '98* (Brno, Czech Republic, 1998), pp. 111–119, ISBN 80-214-1199-6
3. P. Osmera, R. Matousek, O. Popelka, T. Panacek, Parallel grammatical evolution for circuit optimization, in *Proceedings of MENDEL 2008* (Brno, Czech Republic, 2008), pp. 206–213, ISBN 978-80-214-3675-6
4. R. Matousek, GAHC: hybrid genetic algorithm, in *The Springer Book Series*, eds. by S.L. Ao, B. Rieger, S.S. Chen. *Lecture Notes in Electrical Engineering: Advances in Computational Algorithms and Data Analysis*, vol 14 (Springer, The Netherlands 2008), pp. 549–562, ISSN 1876-1100, ISBN 978-1-4020-8918-3
5. P. Popela, Numerical techniques and available software, in *Chapter 8 in Part II, Stochastic Modeling in Economics and Finance*, eds J. Dupacova, J. Hurt, J. St_ep_an (Kluwer Academic Publishers) pp.206–227
6. M. Abramowitz, I.A. Stegun (eds), *Handbook of Mathematical Functions with Formulas, Graphs, and Mathematical Tables, 9th printing* (Dover, New York, 1972)
7. E.W. Weisstein, Laplace Distribution. From *MathWorld* – A Wolfram. http://mathworld. wolfram.com/LaplaceDistribution.html
8. W.H. Pressm, S.A. Teukolsky, Numerical Recipes, *The Art of Scientific Computing*, 3rd edn. (Cambridge University Press, 2007) ISBN 978-0521880688
9. J. Roupec, GA-based parameters tuning in grammatical evolution, in *Proceedings of MENDEL 2007* (Prague, Czech Republic, 2007), pp. 66–71, ISBN 978-80-214-3473-8

Chapter 12
Data Quality in ANFIS Based Soft Sensors

S. Jassar, Z. Liao, and L. Zhao

Abstract Soft sensor are used to infer the critical process variables that are otherwise difficult, if not impossible, to measure in broad range of engineering fields. Adaptive Neuro-Fuzzy Inference System (ANFIS) has been employed to develop successful ANFIS based inferential model that represents the dynamics of the targeted system. In addition to the structure of the model, the quality of the training as well as of the testing data also plays a crucial role in determining the performance of the soft sensor. This paper investigates the impact of data quality on the performance of an ANFIS based inferential model that is designed to estimate the average air temperature in distributed heating systems. The results of the two experiments are reported. The results show that the performance of ANFIS based sensor models is sensitive to the quality of data. The paper also discusses how to reduce the sensitivity by an improved mathematical algorithm.

1 Introduction

A soft sensor computes the value of the process variables, which are difficult to directly measure using conventional sensors, based on the measurement of other relevant variables [1]. The performance of soft sensors is sensitive to the inferential model that represents the dynamics of the targeted system. Such inferential models can be based on one of the following modeling techniques:

- Physical model
- Neural network
- Fuzzy logic
- Adaptive Neuro-Fuzzy Inference System

S. Jassar (✉)
Ryerson University, 350 Victoria Street, Toronto M5B2K3, Canada
e-mail: sjassar@ee.ryerson.ca

S.-I. Ao et al. (eds.), *Machine Learning and Systems Engineering*,
Lecture Notes in Electrical Engineering 68,
DOI 10.1007/978-90-481-9419-3_12, © Springer Science+Business Media B.V. 2010

Recent research demonstrates the use of ANFIS in the development of an inferential model that estimated the average air temperature in a distributed heating system [2]. The estimated average air temperature allows for a closed-loop boiler control scheme, resulting in higher energy efficiency and improved comfort. In the current practice, due to the absence of economic and technically reliable method for measuring the overall comfort level in the buildings, the boilers are normally controlled to maintain the supply water temperature as a predefined level that normally does not reflect the heating demand of the buildings [3].

For ANFIS based inferential models, when estimation/prediction accuracy is concerned, it is assumed that both the data used to train the model and the testing data to make estimations are free of errors [4]. But rarely a dataset is clean before extraordinary effort having been made to clean the data. For this problem of average air temperature estimation, with the measurement errors in the input variables of the model, it is not unusual to have some uneven patterns in the dataset. The research presented in this chapter aims to analyze the impact of data quality of both training and testing datasets on the estimation accuracy of the developed model.

2 ANFIS Based Inferential Model

As an AI technique, "Soft Computing", integrates the powerful artificial intelligence methodologies such as neural networks and fuzzy inference systems. While fuzzy logic performs an inference mechanism under cognitive uncertainty, neural networks posses exciting capabilities such as learning, adaption, fault-tolerance, parallelism and generalization. Since Jang proposed ANFIS, its applications are numerous in various fields, including engineering, management, health, biology and even social sciences [5].

ANFIS is a multi-layer adaptive network-based fuzzy inference system. An ANFIS consists of a total of five layers to implement different node functions to learn and tune parameters in a fuzzy inference system (FIS) structure using a hybrid learning mode. In the forward pass of learning, with fixed premise parameters, the least squared error estimate approach is employed to update the consequent parameters and to pass the errors to the backward pass. In the backward pass of learning, the consequent parameters are fixed and the gradient descent method is applied to update the premise parameters. One complete cycle of forward and backward pass is called an epoch. Premise and consequent parameters will be identified for membership function (MF) and FIS by repeating the forward and backward passes. ANFIS has been widely used in prediction problems and other areas.

ANFIS based inferential model represents the dynamic relationship between the average air temperature, T_{avg}, and three easily measurable variables: external temperature, T_0, solar radiation, Q_{sol}, and energy consumed by the boilers, Q_{in} [6]. The FIS structure is generated by Grid partitioning method.

Grid partition divides the data space into rectangular sub-spaces using axis-paralleled partition based on pre-defined number of MFs and their types in each dimension. The wider application of grid partition in FIS generation is blocked by the curse of dimensions. The number of fuzzy rules increases exponentially when the number of input variables increases. For example, if there are m MFs for each input variable and a total of n input variables for the problem, the total number of fuzzy rules is m^n. It is obvious that the wide application of grid partition is threatened by the large number of rules. According to Jang, grid partition is only suitable for cases with small number of input variables (e.g. less than 6). In this research, the average air temperature estimation problem has three input variables. It is reasonable to apply the grid partition to generate FIS structure, ANFIS-GRID.

Gaussian type MF is used for characterizing the premise variables. Each input has 4 MFs, thus there are 64 rules. The developed structure is trained using hybrid learning algorithm [5]. The parameters associated with the MFs change through training process.

2.1 Training and Testing Data

Experimental data obtained from a laboratory heating system is used for training and testing of the developed model [7]. The laboratory heating system is located in Milan, Italy. The details of experimental data collection for the four variables, Q_{in}, Q_{sol}, T_0 and T_{avg}, are given by the authors [2, 6]. The dataset used for the training of ANFIS-GRID has 1800 input-output data pairs and is shown in Fig. 1.

The experimental data used for checking the performance of the developed model is shown in Fig. 2. The testing dataset has 7,132 data pairs, which is large enough as compared to training dataset used for the development of the model.

3 Impact of Data Quality

Data quality is generally recognized as a multidimensional concept [8]. While no single definition of data quality has been accepted by the researchers working in this area, there is agreement that data accuracy, currency, completeness, and consistency are important areas of concern [9]. This Chapter is primarily considering the data accuracy, defined as conformity between a recorded value and the actual data value.

Several studies have investigated the effect of data errors on the outputs of computer based models. Bansel et al. studied the effect of errors in test data on predictions made by neural network and linear regression models [10]. The training dataset applied in the research was free of errors. The research concluded that the error size had a statistically significant effect on predictive accuracy of both the linear regression and neural network models.

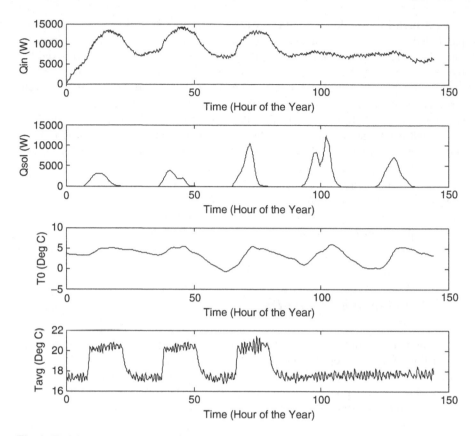

Fig. 1 Training dataset (February 2000: day 22 to day 27)

O'Leary investigated the effect of data errors in the context of a rule-based artificial intelligence system [11]. He presented a general methodology for analyzing the impact of data accuracy on the performance of an artificial intelligence system designed to generate rules from data stored in a database. The methodology can be applied to artificial intelligence systems that analyze the data and generate a set of rules of the form "if X then Y". It is often assumed that a subset of the generated rules is added to the system's rule base on the basis of the measure of the "goodness" of each rule. O'Leary showed that the data errors can affect the subset of rules that are added to the rule base and that inappropriate rules may be retained while useful rules are discarded if data accuracy is ignored.

Wei et al. analyzed the effect of data quality on the predictive accuracy of ANFIS model [12]. The ANFIS model is developed for predicting the injection profiles in the Daqing Oilfields, China. The research analyzed the data quality using TANE algorithm. They concluded that the cleaning of data has improved the accuracy of ANFIS model from 78% to 86.1%.

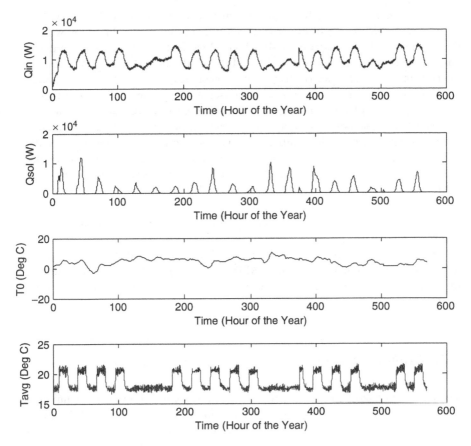

Fig. 2 Testing dataset (February 2000: day 1 to day 21)

In this chapter, the developed ANFIS-GRID based inferential model is trained and tested using the experimental data collected from a laboratory heating system [7]. The data collected has some uneven patterns. In this section we will discuss the experiments conducted to examine the impact of data quality on the predictive performance of the developed ANFIS-GRID model.

3.1 Experimental Methodology

Data errors may affect the accuracy of the ANFIS based models in two ways. First, the data used to build and train the model may contain errors. Second, even if training data are free of errors, once the developed model is used for estimation tasks a user may use input data containing errors to the model.

The research in this area has assumed that data used to train the models and data input to make estimation of the processes are free of errors. In this study we relax this assumption by asking two questions: (1) What is the effect of errors in the test data on the estimation accuracy of the ANFIS based models? (2) What is the effect of errors in the training data on the predictive accuracy of the ANFIS based models?

While many sources of error in a dataset are possible, we assume that the underlying cause of errors affect data items randomly rather than systematically. One source of inaccuracy that may affect a dataset in this way is the measurement errors caused by reading the equipment. This type of error may affect any data item in the dataset and may understate or overstate the actual data value. This study does not address the effect of systematic data errors on the estimations made by the ANFIS based models.

Two experiments are conducted to examine the research targets. Both the experiments used the same application (estimation of average air temperature) and the same dataset.

Experiment 1 examines the first question: What is the effect of errors in the test data on the estimation ability of the ANFIS based models? Experiment 2 examines the second question: How the errors of the training data affect the accuracy of the ANFIS based models?

3.2 Experimental Factors

There are two factors in each experiment: (1) fraction-error and (2) amount-error. Fraction-error is the percent of the data items in the appropriate part of the dataset (the test data in experiment 1 and the training data in experiment 2) that are perturbed. Amount-error is the percentage by which the data items identified in the fraction-error factor are perturbed.

3.2.1 Fraction-Error

Since fraction-error is defined as a percent of the data items in a dataset, the number of data items that are changed for a given level of fraction-error is determined by multiplying the fraction-error by the total number of data items in the dataset.

Experiment 1: The test data used in experiment 1, shown in Fig. 2, has 4 data items (1 value for each of the 4 input and output variables for 1 entry of the total 7,132 data pairs). This experiment examines all of the possible number of data items that could be perturbed. These four levels for fraction-error factor are: 25% (one data item perturbed), 50% (two data items perturbed), 75% (three data items perturbed), and 100% (four data items perturbed)

Experiment 2: The training data used in experiment 2 contains 1,800 data pairs (1 value for each of the 4 input and output variables for 1,800 entries). Four levels of the fraction-error factor are tested: 5% (90 data items are perturbed), 10% (180 data items are perturbed), 15% (270 data items are perturbed), and 20% (360 data items are perturbed).

3.2.2   Amount-Error

For both the experiments, the amount-error factor has two levels: (1) plus or minus 5% and (2) plus or minus 10%. The amount-error applied to the dataset can be represented by the following set of equations:

$$y' = y \pm 0.05 \times y \tag{1}$$

$$y' = y \pm 0.1 \times y \tag{2}$$

For Eqs. (1) and (2), y' is the value of the variable after adding or subtracting the noise error to the unmodified variable y.

3.3   Experimental Design

The experimental design is shown in Table 1. Both the experiments have four levels for the fraction-error factor and two levels for the amount-error. For each combination of fraction-error and amount-error, four runs with random combinations of the input and output variable are performed.

Although the levels of the fraction-error are different in the two experiments, the sampling procedure is the same. For each fraction-error level, the variables are randomly selected to be perturbed. This is repeated a total of four times per level. Table 2 shows the combinations of the variables for experiment 1.

Second, for each level of the amount-error factor, each variable is randomly assigned either a positive or negative sign to indicate the appropriate amount-error to be applied. Table 3 shows the randomly assigned amount-error levels in

Table 1 Experimental design

Experiment 1 (errors in the test data)	
Fraction-error levels (25%, 50%, 75%, and 100%)	4
Amount-error levels (5%, and 10%)	2
Number of random combinations of the variables considered within each fraction-error level	4
Total number of samples considered	7132
Experiment 2 (errors in the training data)	
Fraction-error levels (5%, 10%, 15%, and 20%)	4
Amount-error levels (5%, and 10%)	2
Number of random combinations of the variables considered within each fraction-error level	4
Total number of samples considered	1800

Table 2 Four combinations of the variables for each fraction-error level in Experiment 1

Fraction-error level (%)	Input and output variable combination			
	1	2	3	4
25	(Q_{in})	(Q_{sol})	(T_0)	(T_{avg})
50	(Q_{in}, T_0)	(Q_{in}, T_{avg})	(Q_{sol}, T_0)	(T_{avg}, T_0)
75	(Q_{in}, T_0, Q_{sol})	(Q_{in}, T_{avg}, T_0)	$(Q_{in}, T_{avg}, Q_{sol})$	(T_0, T_{avg}, Q_{sol})
100	$(Q_{in}, T_0, Q_{sol}, T_{avg})$	$(Q_{in}, T_0, Q_{sol}, T_{avg})$	$(Q_{in}, T_0, Q_{sol}, T_{avg})$	$(Q_{in}, T_0, Q_{sol}, T_{avg})$

Table 3 Randomly assigned percentage increase (+) or decrease (−) for a given amount-error level in Experiment 1

Fraction-error level (%)	Input and output variable combination			
	1	2	3	4
25	(Q_{in})	(Q_{sol})	(T_0)	(T_{avg})
	−	+	+	−
50	(Q_{in}, T_0)	(Q_{in}, T_{avg})	(Q_{sol}, T_0)	(T_{avg}, T_0)
	+,−	−,−	+,+	−,+
75	(Q_{in}, T_0, Q_{sol})	(Q_{in}, T_{avg}, T_0)	$(Q_{in}, T_{avg}, Q_{sol})$	(T_0, T_{avg}, Q_{sol})
	+,−,−	−,+,−	+,+,−	+,−,+
100	$(Q_{in}, T_0, Q_{sol}, T_{avg})$	$(Q_{in}, T_0, Q_{sol}, T_{avg})$	$(Q_{in}, T_0, Q_{sol}, T_{avg})$	$(Q_{in}, T_0, Q_{sol}, T_{avg})$
	−,+,−,+	+,+,+,+	−,−,+,+	−,−,−,+

experiment 1. The procedure for experiment 2 differs only in the number of variables that were randomly selected to be perturbed for the four levels of the fraction-error factor.

3.4 Experimental Result

For both the experiments, the measured average air temperature values and ANFIS-GRID estimated average air temperature values are compared using Root Mean Square Error (RMSE) as a measure of estimation accuracy.

3.4.1 Experiment 1 Results: Errors in the Test Data

Estimation accuracy results, using the simulated inaccuracies for amount-error and fraction-error for the average air temperature estimation are given in Fig. 3. It shows that as fraction-error increases from 25% to 100%, RMSE increases results in a decrease in predictive accuracy. As amount-error increases from 5% to 10%, RMSE increases also indicating a decrease in estimation accuracy. Both fraction-error and amount-error have an effect on predictive accuracy.

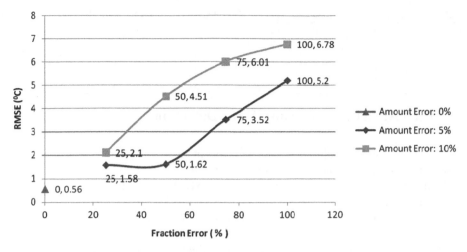

Fig. 3 RMSE (°C) values as error level in the test data varies

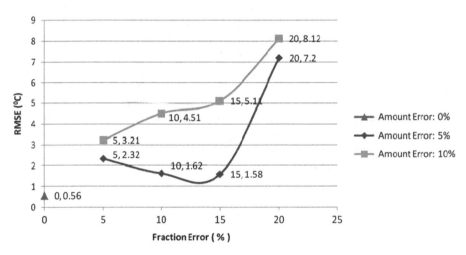

Fig. 4 RMSE (°C) values as error level in the training data varies

3.4.2   Experiment 1 Results: Errors in the Training Data

Predictive accuracy results, using the simulated inaccuracies for amount-error and fraction-error for the average air temperature estimation are given in Fig. 4. It shows that as fraction-error increases from 5% to 20%, RMSE increases indicating a decrease in predictive accuracy.

Table 4 Approximate functional dependencies detected using the TANE algorithm

Index	Approximate dependencies	Number of rows with conflicting tuples
1	$Q_{in}, Q_{sol}, T_0 \to T_{avg}$	42
2	$Q_{in}, T_0, T_{avg} \to Q_{sol}$	47
3	$Q_{in}, Q_{sol}, T_{avg} \to T_0$	43
4	$Q_{in}, T_0, T_{avg} \to Q_{sol}$	54

4 Tane Algorithm for Noisy Data Detection

Data quality analysis results show that the errors in the training data as well as in the testing data affect the predictive accuracy of the ANFIS based soft sensor models. This section discusses an efficient algorithm, TANE algorithm, to identify the noisy data pairs in the dataset.

The TANE algorithm, which deals with discovering functional and approximate dependencies in large data files, is an effective algorithm in practice [13]. The TANE algorithm partitions attributes into equivalence partitions of the set of tuples. By checking if the tuples that agree on the right-hand side agree on the left-hand side, one can determine whether a dependency holds or not. By analyzing the identified approximate dependencies, one can identify potential erroneous data in the relations.

In this research, relationship of the three input parameters (Q_{in}, Q_{sol}, and T_0) and the average air temperature (T_{avg}) is analyzed using TANE algorithm. For equivalence partition, all the four parameters are rounded off to zero decimal points.

After data pre-processing, four approximate dependencies are discovered, as shown in Table 4. Although all these dependencies reflect the relationships among the parameters, the first dependency is the most important one because it shows that the selected input parameters have consistent association relationship with the average air temperature except a few data pairs, which is a very important dependency for average air temperature estimation.

To identify exceptional tuples by analyzing the approximate dependencies, it is required to investigate the equivalence partitions of both left-hand and right-hand sides of an approximate dependency. It is non-trivial work that could lead to the discovery of problematic data. By analyzing the first dependency, conflicting tuples are identified. The total dataset has 7,132 data pairs. For the first approximate dependency from Table 4, 42 conflicting data pairs are present which needs were fixed using data interpolations for better performance of ANFIS-GRID model.

5 Results

The developed ANFIS-GRID model is validated using experimental results [7]. The model performance is measured using RMSE

$$RMSE = \sqrt{\frac{1}{N} \sum_{i=1}^{N} \left(T_{avg}(i) - \hat{T}_{avg}(i) \right)} \tag{3}$$

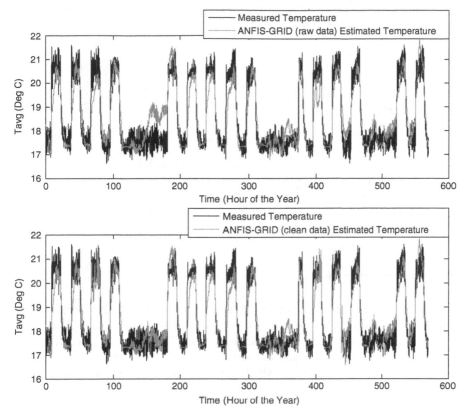

Fig. 5 Comparison of ANFIS-GRID estimated and the measured temperature values

For Eq. (3), N is the total number of data pairs, $\hat{T}_{avg}$ is the estimated and T_{avg} is the experimental value of average air temperature.

Initially, ANFIS-GRID model uses the raw data for both the training as well as the testing. Figure 5 compares ANFIS-GRID estimated average air temperature values with the experimental results. First plot in Fig. 5 shows that ANFIS-GRID estimated average air temperature values are in agreement with the experimental results, with RMSE 0.56°C. However there are some points at which estimation is not following the experimental results. For example, around 152–176 and 408–416 h of the year time, there is a significant difference between estimated and experimental results.

For checking the effect of data quality on ANFIS-GRID performance, the training and testing datasets are cleaned using TANE algorithm. The conflicting data pairs are replaced with the required data pairs. Then the cleaned dataset is applied for the training and the testing of ANFIS-GRID model. A comparison of the model output with clean data and the experimental results is shown in second plot of Fig. 5.

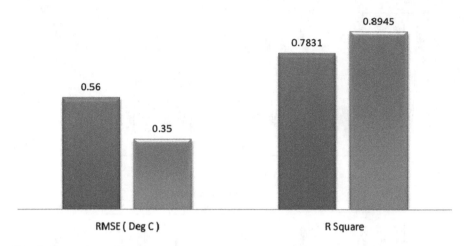

Fig. 6 Comparison of results

Figure 6 clearly shows the effect of data quality on predictive accuracy of ANFIS-GRID model. The RMSE is improved by 37.5–0.35°C. RMSE is considered as a measure of predictive accuracy. Less RMSE means less difference between the estimated values and the actual values. Finally, predictive accuracy is improved with decrease in RMSE. Figure 6 also shows the improvement in R^2 values.

6 Conclusion

ANFIS-GRID based inferential model has been developed to estimate the average air temperature in the distributed space heating systems. This model is simpler than the subtractive clustering based ANFIS model [2] and may be used as the air temperature estimator for the development of an inferential control scheme. Grid partition based FIS structure is used as there are only three input variables. The training dataset is also large enough as compared to the modifiable parameters of the ANFIS. As the experimental data is used for both the training as well as the testing of the developed model, it is expected that data can have some discrepancies. TANE algorithm is used to identify the approximate functional dependencies among the input and the output variables. The most important approximate dependency is analyzed to identify the data pairs with uneven patterns. The identified data pairs are fixed and again the developed model is trained and tested with the cleaned data. Figure 6 shows that the RMSE is improved by 37.5% and R^2 is improved by 12%. Therefore, it is highly recommended that the quality of datasets should be analyzed before they are applied in ANFIS based modelling.

References

1. M.T. Tham, G.A. Montague, A.J. Morris, P.A. Lant, Estimation and inferential control. J. Process Control **1**, 3–14 (1993)
2. S. Jassar, Z. Liao, L. Zhao, Adaptive neuro-fuzzy based inferential sensor model for estimating the average air temperature in space heating systems. Build. Environ. **44**, 1609–1616 (2009)
3. Z. Liao, A.L. Dexter, An experimental study on an inferential control scheme for optimising the control of boilers in multi-zone heating systems. Energy Build. **37**, 55–63 (2005)
4. B.D. Klein, D.F. Rossin, Data errors in neural network and linear regression models: an experimental comparison. Data Quality **5**, 33–43 (1999)
5. J.S.R. Jang, ANFIS: adaptive-network-based fuzzy inference system. IEEE Trans. Syst. Man Cybern. **2**, 665–685 (1993)
6. Z. Liao, A.L. Dexter, A simplified physical model for estimating the average air temperature in multi-zone heating systems. Build. Environ. **39**, 1013–1022 (2004)
7. BRE, ICITE, Controller efficiency improvement for commercial and industrial gas and oil fired boilers, *A craft project*, Contract JOE-CT98-7010, 1999–2001
8. J. Wang, B. Malakooti, A feed forward neural network for multiple criteria decision making. Comput. Oper. Res. **19**, 151–167 (1992)
9. Y. Huh, F. Keller, T. Redman, A. Watkins, Data quality. Inf. Softw. Technol. **32**, 559–565 (1990)
10. Bansal, R. Kauffman, R. Weitz, Comparing the modeling performance of regression and neural networks as data quality varies. J. Manage. Inf. Syst. **10**, 11–32 (1993)
11. D. O'Leary, The impact of data accuracy on system learning. J. Manage. Inf. Syst. **9**, 83–98 (1993)
12. M. Wei, et al., Predicting injection profiles using ANFIS. Inf. Sci. **177**, 4445–4461 (2007)
13. Y. Huhtala, J. Karkkainen, P. Porkka, H. Toivonen, TANE: an efficient algorithm for discovering functional and approximate dependencies. Comput. J. **42**, 100–111 (1999)

Chapter 13
The Meccano Method for Automatic Volume Parametrization of Solids

R. Montenegro, J.M. Cascón, J.M. Escobar, E. Rodríguez, and G. Montero

Abstract In this paper, we present significant advances of the novel meccano technique for simultaneously constructing adaptive tetrahedral meshes of 3-D complex solids and their volume parametrization. Specifically, we will consider a solid whose boundary is a surface of genus zero. In this particular case, the automatic procedure is defined by a surface triangulation of the solid, a simple meccano composed by one cube and a tolerance that fixes the desired approximation of the solid surface. The main idea is based on an automatic mapping from the cube faces to the solid surface, a 3-D local refinement algorithm and a simultaneous mesh untangling and smoothing procedure. Although the initial surface triangulation can be a poor quality mesh, the meccano technique constructs high quality surface and volume adaptive meshes. Several examples show the efficiency of the proposed technique. Future possibilities of the meccano method for meshing a complex solid, whose boundary is a surface of genus greater than zero, are commented.

1 Introduction

Many authors have devoted great effort to solving the automatic mesh generation problem in different ways [4, 15, 16, 28], but the 3-D problem is still open [1]. Along the past, the main objective has been to achieve high quality adaptive meshes of complex solids with minimal user intervention and low computational cost. At present, it is well known that most mesh generators are based on Delaunay triangulation and advancing front technique, but problems, related to mesh quality or mesh conformity with the solid boundary, can still appear for complex geometries. In addition, an appropriate definition of element sizes is demanded for

R. Montenegro (✉)
Institute for Intelligent Systems and Numerical Applications in Engineering (SIANI), University of Las Palmas de Gran Canaria, 35017 Las Palmas de Gran Canaria, Spain
e-mail: rmontenegro@siani.es

S.-I. Ao et al. (eds.), *Machine Learning and Systems Engineering*,
Lecture Notes in Electrical Engineering 68,
DOI 10.1007/978-90-481-9419-3_13, © Springer Science+Business Media B.V. 2010

obtaining good quality elements and mesh adaption. Particularly, local adaptive refinement strategies have been employed to mainly adapt the mesh to singularities of numerical solution. These adaptive methods usually involve remeshing or nested refinement.

We introduced the new meccano technique in [2, 3, 23, 24] for constructing adaptive tetrahedral meshes of solids. We have given this name to the method because the process starts with the construction of a coarse approximation of the solid, i.e. a meccano composed by connected polyhedral pieces. The method builds a 3-D triangulation of the solid as a deformation of an appropriate tetrahedral mesh of the meccano. A particular case is when meccano is composed by connected cubes, i.e. a polycube.

The new automatic mesh generation strategy uses no Delaunay triangulation, nor advancing front technique, and it simplifies the geometrical discretization problem for 3-D complex domains, whose surfaces can be mapped to the meccano faces. The main idea of the meccano method is to combine a local refinement/derefinement algorithm for 3-D nested triangulations [20], a parameterization of surface triangulations [8] and a simultaneous untangling and smoothing procedure [5]. At present, the meccano technique has been implemented by using the local refinement/derefinement of Kossaczky [20], but the idea could be implemented with other types of local refinement algorithms [17]. The resulting adaptive tetrahedral meshes with the meccano method have good quality for finite element applications.

Our approach is based on the combination of several former procedures (refinement, mapping, untangling and smoothing) which are not in themselves new, but the overall integration is an original contribution. Many authors have used them in different ways. Triangulations for convex domains can be constructed from a coarse mesh by using refinement/projection [25]. Adaptive nested meshes have been constructed with refinement and derefinement algorithms for evolution problems [7]. Mappings between physical and parametric spaces have been analyzed by several authors. Significant advances in surface parametrization have been done in [8, 10, 11, 22, 27, 29], but the volume parametrization is still open. Floater et al. [12] give a simple counterexample to show that convex combination mappings over tetrahedral meshes are not necessarily one-to-one. Large domain deformations can lead to severe mesh distortions, especially in 3-D. Mesh optimization is thus key for keeping mesh shape regularity and for avoiding a costly remeshing [18, 19]. In traditional mesh optimization, mesh moving is guided by the minimization of certain overall functions, but it is usually done in a local fashion. In general, this procedure involves two steps [13, 14]: the first is for mesh untangling and the second one for mesh smoothing. Each step leads to a different objective function. In this paper, we use the improvement proposed by [5, 6], where a simultaneous untangling and smoothing guided by the same objective function is introduced.

Some advantages of the meccano technique are that: surface triangulation is automatically constructed, the final 3-D triangulation is conforming with the object boundary, inner surfaces are automatically preserved (for example, interface between several materials), node distribution is adapted in accordance with the object geometry, and parallel computations can easily be developed for meshing the meccano pieces. However, our procedure demands an automatic construction of the meccano and an admissible mapping between the meccano boundary and the object surface must be defined.

In this paper, we consider a complex genus-zero solid, i.e. a solid whose boundary is a surface that is homeomorphic to the surface of a sphere, and we assume that the solid geometry is defined by a triangulation of its surface. In this case, it is sufficient to fix a meccano composed by a single cube and a tolerance that fixes the desired approximation of the solid surface. In order to define an admissible mapping between the cube faces and patches of the initial surface triangulation of the solid, we introduce a new automatic method to decompose the surface triangulation into six patches that preserves the same topological connections than the cube faces. Then, a discrete mapping from each surface patch to the corresponding cube face is constructed by using the parameterization of surface triangulations proposed by M. Floater in [8–11]. The shape-preserving parametrizations, which are planar triangulations on the cube faces, are the solutions of linear systems based on convex combinations.

In the near future, more effort should be made in developing an automatic construction of the meccano when the genus of the solid surface is greater than zero. Currently, several authors are working on this aspect in the context of polycube-maps, see for example [22, 27, 29]. They are analyzing how to construct a polycube for a generic solid and, simultaneously, how to define a conformal mapping between the polycube boundary and the solid surface. Although harmonic maps have been extensively studied in the literature of surface parameterization, only a few works are related to volume parametrization, for example a procedure is presented in see [21].

In the following Section we present a brief description of the main stages of the method for a generic meccano composed of polyhedral pieces. In Section 3 we introduce applications of the algorithm in the case that the meccano is formed by a simple cube. Finally, conclusions and future research are presented in Section 4.

2   The Meccano Method

The main steps of the general *meccano tetrahedral mesh generation algorithm* are summarized in this section. A detailed description of this technique can be analyzed in [2, 23, 24]. The input data are the definition of the solid boundary (for example by a given surface triangulation) and a given tolerance (corresponding to the solid surface approximation). The following algorithm describes the whole mesh generation approach.

Meccano tetrahedral mesh generation algorithm

1. Construct a meccano approximation of the 3-D solid formed by polyhedral pieces.
2. Define an admissible mapping between the meccano boundary faces and the solid boundary.
3. Build a coarse tetrahedral mesh of the meccano.

4. Generate a local refined tetrahedral mesh of the meccano, such that the mapping of the meccano boundary triangulation approximates the solid boundary for a given precision.
5. Move the boundary nodes of the meccano to the object surface with the mapping defined in 2.
6. Relocate the inner nodes of the meccano.
7. Optimize the tetrahedral mesh with the simultaneous untangling and smoothing procedure.

The first step of the procedure is to construct a meccano approximation by connecting different polyhedral pieces. Once the meccano approximation is fixed, we have to define an *admissible* one-to-one mapping between the boundary faces of the meccano and the boundary of the object. In step 3, the meccano is decomposed into a coarse and valid tetrahedral mesh by an appropriate subdivision of its initial polyhedral pieces. We continue with a local refinement strategy to obtain an adapted mesh which can approximate the boundaries of the domain within a given precision. Then, we construct a mesh of the solid by mapping the boundary nodes from the meccano faces to the true solid surface and by relocating the inner nodes at a reasonable position. After those two steps the resulting mesh is tangled, but it has an admissible topology. Finally, a simultaneous untangling and smoothing procedure is applied and a valid adaptive tetrahedral mesh of the object is obtained.

We note that the general idea of the meccano technique could be understood as the connection of different polyhedral pieces. So, the use of cuboid pieces, or a polycube meccano, are particular cases.

3   Application of the Meccano Method to Complex Genus-ZeroSolids

In this section, we present the application of the meccano algorithm in the case of the solid surface being genus-zero and the meccano being formed by a single cube. We assume as datum a triangulation of the solid surface.

We introduce an automatic parametrization between the surface triangulation of the solid and the cube boundary. To that end, we automatically divide the surface triangulation into six patches, with the same topological connection that cube faces, so that each patch is mapped to a cube face. These parametrizations have been done with GoTools core and parametrization modules from SINTEF ICT, available in the website http://www.sintef.no/math_software. This code implements Floater's parametrization in C++. Specifically, in the following application we have used the mean value method for the parametrization of the inner nodes of the patch triangulation, and the boundary nodes are fixed with chord length parametrization [8, 10].

We have implemented the meccano method by using the local refinement of ALBERTA. This code is an adaptive multilevel finite element toolbox [26] developed in C. This software can be used to solve several types of 1-D, 2-D or 3-D problems. ALBERTA uses the Kossaczky refinement algorithm [20] and requires an initial mesh topology [25]. The recursive refinement algorithm could not terminate for general meshes. The meccano technique constructs meshes that verify the imposed restrictions of ALBERTA in relation to topology and structure. The minimum quality of refined meshes is function of the initial mesh quality.

The performance of our novel tetrahedral mesh generator is shown in the following applications. The first corresponds to a Bust, the second to the Stanford Bunny and the third to a Bone. We have obtained a surface triangulation of these objects from internet.

3.1 Example 1: Bust

The original surface triangulation of the Bust has been obtained from the website *http://shapes.aimatshape.net*, i.e. AIM@SHAPE Shape Repository. It has 64,000 triangles and 32,002 nodes. The bounding box of the solid is defined by the points $(x, y, z)_{min} = (-120, -30.5, -44)$ and $(x, y, z)_{max} = (106, 50, 46)$.

We consider a cube, with an edge length equal to 20, as meccano. Its center is placed inside the solid at the point $(5, -3, 4)$. We obtain an initial subdivision of Bust surface in seven maximal connected subtriangulations by using the Voronoi diagram associated to the centers of the cube faces. In order to get a compatible decomposition of the surface triangulation, we apply an iterative procedure to reduce the current seven patches to six.

We map each surface patch Σ_S^i to the cube face Σ_C^i by using the Floater parametrization [8]. The definition of the one-to-one mapping between the cube and Bust boundaries is straightforward once the global parametrization of the Bust surface triangulation is built.

Fixing a tolerance $\varepsilon_2 = 0.1$, the meccano method generates a tetrahedral mesh of the cube with 147, 352 tetrahedra and 34, 524 nodes, see a cross section of the cube mesh in Fig. 1a. This mesh has 32, 254 triangles and 16, 129 nodes on its boundary and it has been reached after 42 Kossaczky refinements from the initial subdivision of the cube into six tetrahedra. The mapping of the cube external nodes to the Bust surface produces a 3-D tangled mesh with 8, 947 inverted elements, see Fig. 1b. The location of the cube is shown in this figure. The relocation of inner nodes by using volume parametrizations reduces the number of inverted tetrahedra to 285. We apply our mesh optimization procedure [5] and the mesh is untangled in two iterations. The mesh quality is improved to a minimum value of 0.07 and an average $\bar{q}_k = 0.73$ after 10 smoothing iterations.

We note that the meccano technique generates a high quality tetrahedra mesh (see Figs. 1c, d): only one tetrahedron has a quality lower than 0.1, 13 lower than 0.2 and 405 lower than 0.3.

The CPU time for constructing the final mesh of the Bust is 93.27 s on a Dell precision 690, 2 Dual Core Xeon processor and 8 Gb RAM memory. More precisely, the CPU time of each step of the meccano algorithm is: 1.83 s for the subdivision of the initial surface triangulation into six patches, 3.03 s for the Floater

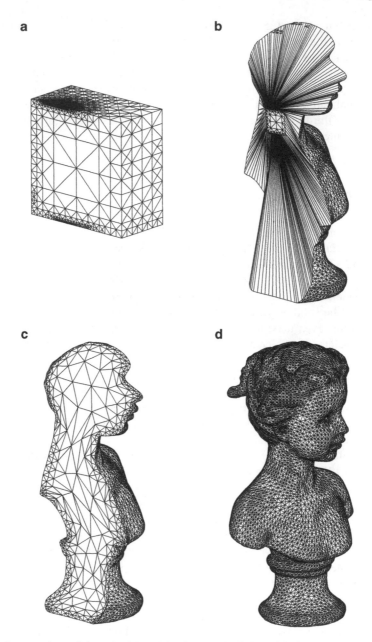

Fig. 1 Cross sections of the cube (**a**) and the Bust tetrahedral mesh before (**b**) and after (**c**) the application of the mesh optimization procedure. (**d**) Resulting tetrahedral mesh of the Bust obtained by the meccano method

parametrization, 44.50 s for the Kossaczky recursive bisections, 2.31 s for the external node mapping and inner node relocation, and 41.60 s for the mesh optimization.

3.2   Example 2: Bunny

The original surface triangulation of the Stanford Bunny has been obtained from the website *http://graphics.stanford.edu/data/3Dscanrep/*, i.e. the Stanford Computer Graphics Laboratory. It has 12, 654 triangles and 7, 502 nodes. The bounding box of the solid is defined by the points $(x, y, z)_{min} = (-10, 3.5, -6)$ and $(x, y, z)_{max} = (6, 2, 6)$.

We consider a unit cube as meccano. Its center is placed inside the solid at the point $(-4.5, 10.5, 0.5)$. We obtain an initial subdivision of the Bunny surface in eight maximal connected subtriangulations using Voronoi diagram. We reduce the surface partition to six patches and we construct the Floater parametrization from each surface patch Σ_S^i to the corresponding cube face Σ_C^i. Fixing a tolerance $\varepsilon_2 = 0.0005$, the meccano method generates a cube tetrahedral mesh with 54, 496 tetrahedra and 13, 015 nodes, see Fig. 2a. This mesh has 11, 530 triangles and 6, 329 nodes on its boundary and has been reached after 44 Kossaczky refinements from the initial subdivision of the cube into six tetrahedra.

The mapping of the cube external nodes to the Bunny surface produces a 3-D tangled mesh with 2, 384 inverted elements, see Fig. 2b. The relocation of inner nodes by using volume parametrizations reduces the number of inverted tetrahedra to 42. We apply eight iterations of the tetrahedral mesh optimization and only one inverted tetrahedron can not be untangled. To solve this problem, we allow the movement of the external nodes of this inverted tetrahedron and we apply eight new optimization iterations. The mesh is then untangled and, finally, we apply eight smoothing iterations fixing the boundary nodes. The resulting mesh quality is improved to a minimum value of 0.08 and an average $\bar{q}_k = 0.68$, see Figs. 2c, d. We note that the meccano technique generates a high quality tetrahedra mesh: only one tetrahedron has a quality below 0.1, 41 below 0.2 and 391 below 0.3.

The CPU time for constructing the final mesh of the Bunny is 40.28 s on a Dell precision 690, 2 Dual Core Xeon processor and 8 Gb RAM memory. More precisely, the CPU time of each step of the meccano algorithm is: 0.24 s for the subdivision of the initial surface triangulation into six patches, 0.37 s for the Floater parametrization, 8.62 s for the Kossaczky recursive bisections, 0.70 s for the external node mapping and inner node relocation, and 30.35 s for the mesh optimization.

3.3   Example 3: Bone

The original surface triangulation of the Bone has been obtained from *http://www-c. inria.fr/gamma/download/*..., and it can be found in the CYBERWARE Catalogue. This surface mesh contains 274, 120 triangles and 137, 062 nodes.

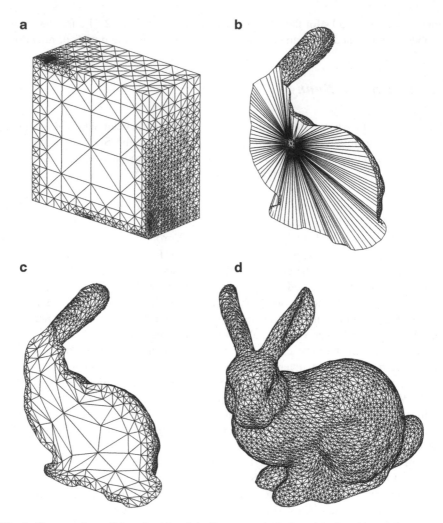

Fig. 2 Cross sections of the cube (**a**) and the Bunny tetrahedral mesh before (**b**) and after (**c**) the application of the mesh optimization procedure. (**d**) Resulting tetrahedral mesh of the Bunny obtained by the meccano method

Steps of the meccano technique are shown in Fig. 3. The resulting mesh has 47, 824 tetrahedra and 11, 525 nodes. This mesh has 11, 530 triangles and 5, 767 nodes on its boundary and it has been reached after 23 Kossaczky refinements from the initial subdivision of the cube into six tetrahedra. A tangled tetrahedra mesh with 1, 307 inverted elements appears after the mapping of the cube external nodes to the bone surface. The node relocation process reduces the number of inverted tetrahedra to 16. Finally, our mesh optimization algorithm produces a high quality tetrahedra mesh: the minimum mesh quality is 0.15 and the average quality is 0.64.

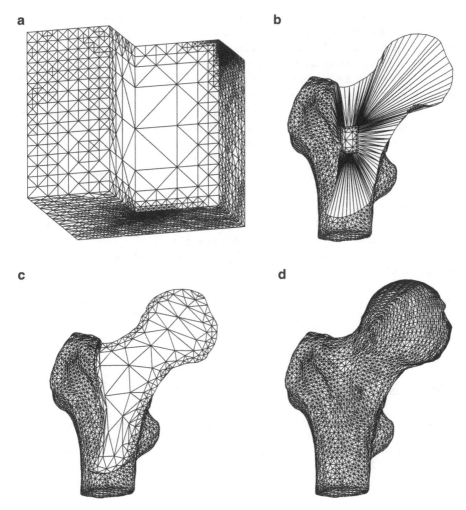

Fig. 3 Cross sections of the cube (**a**) and the Bone tetrahedral mesh before (**b**) and after (**c**) the application of the mesh optimization procedure. (**d**) Resulting tetrahedral mesh of the Bone obtained by the meccano method

4 Conclusions and Future Research

The meccano technique is a very efficient adaptive tetrahedral mesh generator for solids whose boundary is a surface of genus zero. We remark that the method requires minimum user intervention and has a low computational cost. The procedure is fully automatic and it is only defined by a surface triangulation of the solid, a cube and a tolerance that fixes the desired approximation of the solid surface. A crucial consequence of the new mesh generation technique is the resulting discrete parametrization of a complex volume (solid) to a simple cube (meccano).

We have introduced an automatic partition of the given solid surface triangulation for fixing an admissible mapping between the cube faces and the solid surface patches, such that each cube face is the parametric space of its corresponding patch.

The mesh generation technique is based on sub-processes (subdivision, mapping, optimization) which are not in themselves new, but the overall integration using a simple shape as starting point is an original contribution of the method and has some obvious performance advantages. Another interesting property of the new mesh generation strategy is that it automatically achieves a good mesh adaption to the geometrical characteristics of the domain. In addition, the quality of the resulting meshes is high.

The main ideas presented in this paper can be applied for constructing tetrahedral or hexahedral meshes of complex solids. In future works, the meccano technique can be extended for meshing a complex solid whose boundary is a surface of genus greater than zero. In this case, the meccano can be a polycube or constructed by polyhedral pieces with compatible connections. At present, the user has to define the meccano associated to the solid, but we are implementing a special CAD package for more general input solid.

Acknowledgements This work has been partially supported by the Spanish Government, "Secretaría de Estado de Universidades e Investigación", "Ministerio de Ciencia e Innovación", and FEDER, grant contract: CGL2008-06003-C03.

References

1. Y. Bazilevs, V.M. Calo, J.A. Cottrell, J. Evans, T.J.R. Hughes, S. Lipton, M.A. Scott, T.W. Sederberg, Isogeometric analysis: toward unification of computer aided design and finite element analysis. *Trends in Engineering Computational Technology* (Saxe-Coburg Publications, Stirling, 2008), pp. 1–16
2. J.M. Cascón, R. Montenegro, J.M. Escobar, E. Rodríguez, G. Montero, A new meccano technique for adaptive 3-D triangulations. *Proceedings of the 16th International Meshing Roundtable* (Springer, New York, 2007), pp. 103–120
3. J.M. Cascón, R. Montenegro, J.M. Escobar, E. Rodríguez, G. Montero, The Meccano method for automatic tetrahedral mesh generation of complex genus-zero solids. *Proceedings of the 18th International Meshing Roundtable* (Springer, New York, 2009), pp. 463–480
4. G.F. Carey, in *Computational Grids: Generation, Adaptation, and Solution Strategies* (Taylor & Francis, Washington, 1997)
5. J.M. Escobar, E. Rodríguez, R. Montenegro, G. Montero, J.M. González-Yuste, Simultaneous untangling and smoothing of tetrahedral meshes. Comput. Meth. Appl. Mech. Eng. **192**, 2775–2787 (2003)
6. J.M. Escobar, G. Montero, R. Montenegro, E. Rodríguez, An algebraic method for smoothing surface triangulations on a local parametric space. Int. J. Num. Meth. Eng. **66**, 740–760 (2006)
7. L. Ferragut, R. Montenegro, A. Plaza, Efficient refinement/derefinement algorithm of nested meshes to solve evolution problems. Comm. Num. Meth. Eng. **10**, 403–412 (1994)
8. M.S. Floater, Parametrization and smooth approximation of surface triangulations. Comput. Aid. Geom. Design **14**, 231–250 (1997)
9. M.S. Floater, One-to-one piece linear mappings over triangulations. Math. Comput. **72**, 85–696 (2002)

10. M.S. Floater, Mean value coordinates. Comput. Aid. Geom. Design **20**, 19–27 (2003)
11. M.S. Floater, K. Hormann, Surface parameterization: a tutorial and survey. *Advances in Multiresolution for Geometric Modelling, Mathematics and Visualization* (Springer, Berlin, 2005), pp. 157–186
12. M.S. Floater, V. Pham-Trong, Convex combination maps over triangulations, tilings, and tetrahedral meshes. Adv. Computat. Math. **25**, 347–356 (2006)
13. L.A. Freitag, P.M. Knupp, Tetrahedral mesh improvement via optimization of the element condition number. Int. J. Num. Meth. Eng. **53**, 1377–1391 (2002)
14. L.A. Freitag, P. Plassmann, Local optimization-based simplicial mesh untangling and improvement. Int. J. Num. Meth. Eng. **49**, 109–125 (2000)
15. P.J. Frey, P.L. George, in *Mesh Generation* (Hermes Sci. Publishing, Oxford, 2000)
16. P.L. George, H. Borouchaki, in *Delaunay Triangulation and Meshing: Application to Finite Elements* (Editions Hermes, Paris, 1998)
17. J.M. González-Yuste, R. Montenegro, J.M. Escobar, G. Montero, E. Rodríguez, Local refinement of 3-D triangulations using object-oriented methods. Adv. Eng. Soft. **35**, 693–702 (2004)
18. P.M. Knupp, Achieving finite element mesh quality via optimization of the Jacobian matrix norm and associated quantities. Part II-A frame work for volume mesh optimization and the condition number of the Jacobian matrix. Int. J. Num. Meth. Eng. **48**, 1165–1185 (2000)
19. P.M. Knupp, Algebraic mesh quality metrics. SIAM J. Sci. Comput. **23**, 193–218 (2001)
20. I. Kossaczky, A recursive approach to local mesh refinement in two and three dimensions. J. Comput. Appl. Math. **55**, 275–288 (1994)
21. X. Li, X. Guo, H. Wang, Y. He, X. Gu, H. Qin, Harmonic volumetric mapping for solid modeling applications. *Proceedings of the ACM Solid and Physical Modeling Symposium*, Association for Computing Machinery, Inc., 2007, pp. 109–120
22. J. Lin, X. Jin, Z. Fan, C.C.L. Wang, Automatic PolyCube-Maps. *Lecture Notes in Computer Science* **4975**, 3–16 (2008)
23. R. Montenegro, J.M. Cascón, J.M. Escobar, E. Rodríguez, G. Montero, Implementation in ALBERTA of an automatic tetrahedral mesh generator. *Proceedings of the 15th International Meshing Roundtable* (Springer, New York, 2006), pp. 325–338
24. R. Montenegro, J.M. Cascón, J.M. Escobar, E. Rodríguez, G. Montero, An automatic strategy for adaptive tetrahedral mesh generation. Appl. Num. Math. **59**, 2203–2217 (2009)
25. A. Schmidt, K.G. Siebert, in *Design of Adaptive Finite Element Software: The Finite Element Toolbox ALBERTA*. Lecture Notes in Computer Science and Engineering, vol. 42. (Springer, Berlin, 2005)
26. A. Schmidt, K.G. Siebert, ALBERTA – an adaptive hierarchical finite element toolbox. http://www.alberta-fem.de/
27. M. Tarini, K. Hormann, P. Cignoni, C. Montani, Polycube-Maps. ACM Trans. Graph. **23**, 853–860 (2004)
28. J.F. Thompson, B. Soni, N. Weatherill, in *Handbook of Grid Generation* (CRC Press, London, 1999)
29. H. Wang, Y. He, X. Li, X. Gu, H. Qin, Polycube splines. Comput. Aid. Geom. Design **40**, 721–733 (2008)

Chapter 14
A Buck Converter Model for Multi-Domain Simulations

Johannes V. Gragger, Anton Haumer, and Markus Einhorn

Abstract In this work a buck converter model for multi-domain simulations is proposed and compared with a state-of-the-art buck converter model. In the proposed model no switching events are calculated. By avoiding the computation of the switching events in power electronic models the processing time of multi-domain simulations can be decreased significantly. The proposed model calculates any operation point of the buck converter in continuous inductor current conduction mode (CICM) while considering the conduction losses and switching losses. It is possible to utilize the proposed modeling approach also for other dc-to-dc converter topologies. Laboratory test results for the validation of the proposed model are included.

1 Introduction

For the efficient utilization of multi-domain simulation software it is of high importance to have fast simulation models of power electronic components on hand. Especially in simulations of vast and complex electromechanical systems (e.g. power trains of hybrid electric vehicles [11] or drive systems in processing plants [7]) it is crucial to limit the processing effort to a minimum. Many times such electromechanical systems contain power electronic subsystems such as rectifiers, inverters, dc-to-dc converters, balancing systems (for energy sources), etc. When simulating these power electronic devices together with the other electrical and mechanical components of the application, computing the quantities of the power electronic models requires a large share of the available processing power if switching events are calculated in the power electronic models. Simulation models

J.V. Gragger (✉)
Electric Drive Technologies, Austrian Institute of Technology, Giefinggasse 2, 1210 Vienna, Austria
e-mail: johannes.gragger@ait.ac.at

S.-I. Ao et al. (eds.), *Machine Learning and Systems Engineering*,
Lecture Notes in Electrical Engineering 68,
DOI 10.1007/978-90-481-9419-3_14, © Springer Science+Business Media B.V. 2010

Fig. 1 Topology of a
conventional buck converter

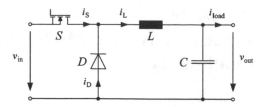

including power electronic devices with switching frequencies around 100 kHz
require at least four calculation points within simulation times of around 10 μs for
calculating the switching events. However, if the energy flow in an electromechanical system has to be investigated by simulation it is not necessary to calculate the
switching events in the power electronic model as long as the relevant losses are
considered.

In this work two different buck converter models are described. The first model,
model A, which is state-of-the-art describes the behavior of a conventional buck
converter, as shown in Fig. 1, including the calculation of switching events. This
means that in model A the switching of the semiconductors in the circuit is implemented with if-clauses. Therefore, model A directly calculates the ripple of the current
through the storage inductor, and the ripple of the voltage across the buffer capacitor.
Due to the if-clauses in model A the duration of the computing time is very high.

The second model in this work, indicated as model B, describes the behavior of
the buck converter without calculating the switching events with if-clauses. Only
the mean and RMS values of the voltages and currents are calculated. Therefore, the
computation times of model B are significantly shorter than the computation times
of model A.

In both models the conduction losses are considered by an ohmic resistance of
the storage inductor, the knee voltage and the on-resistance of the diode, and the on-resistance of the MOSFET. Linear temperature dependence is implemented for the
ohmic resistances of the storage inductor, the knee voltage and the on-resistance of
the diode and the on-resistance of the MOSFET in both buck converter models.

The switching losses are calculated assuming a linear dependency on the switching frequency, the blocking voltage and the commutating current between the
MOSFET and the diode. A controlled current source connected to the positive
and the negative pin of the supply side of the buck converter is used to model the
switching losses. This current source assures that the energy balance between the
supply side and the load side of the buck converter is guaranteed.

2 The Model for Calculating Switching Events

If the buck converter circuit in Fig. 1 is operated in continuous inductor current
conduction mode (CICM) the circuit can be in two different states. As long as
the MOSFET *S* is on and the diode *D* blocks the current, the buck converter is

in state 1. The corresponding equivalent circuit of the buck converter in state 1 is shown in Fig. 2.

v_{in} is the input voltage and v_{out} is the output voltage of the converter. R_S indicates the on-resistance of the MOSFET and i_S denotes the current through the MOSFET. R_L represents the ohmic contribution of the storage inductor L and i_L is the current through L. C stands for the buffer capacitor, i_{load} indicates the output current of the converter and i_D represents the current through the diode (in state 1, $i_D = 0$).

After S switched from on to off the diode begins to conduct. If S is off and D conducts, the circuit is in state 2. The corresponding equivalent circuit of the buck converter in state 2 is shown in Fig. 3 where R_D is the on-resistance and V_D represents the knee voltage of the diode.

Discontinuous inductor current conduction mode (DICM) could be considered in a third state where S and D are open at the same time. The buck converter is in DICM if S is open and the current passing through the diode becomes zero.

A buck converter model for calculating switching events can be implemented according to the pseudo code given in Alg. 1 where d stands for the duty cycle, f_s represents the switching frequency, and t indicates the time. $s_{control}$, the Boolean control signal of the MOSFET, is true during

$$t_{on} = dT_s \tag{1}$$

and false during

$$t_{off} = (1 - d)T_s \tag{2}$$

in a switching period $T_s = \frac{1}{f_s}$. In Alg. 1 only CICM is considered. However, it is easy to modify the model so that DICM can be simulated as well.

The basic principle of the modeling approach described in Alg. 1 is used in many state-of-the-art simulation tools. A disadvantage of such a model is the processing

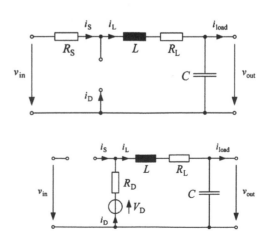

Fig. 2 Equivalent circuit of the buck converter in state 1. Switch S is on

Fig. 3 Equivalent circuit of the buck converter in state 2. Switch S is off

---

Algorithm 1 Pseudo code of a buck converter model for calculating switching events in CICM

Model:
```
BuckConverter
```
Parameter:
$L, C, R_S, R_L, R_D, V_D, f_s$
Real variables:
$v_{in}, v_{out}, i_S, i_L, i_D, i_{load}, t, d$
Boolean variables:
$S_{control}$
Equations:
if ($S_{control}$ = true),
```
consider equations corresponding to the equivalent circuit of state 1
   (Fig. 2)
```
else
```
consider equations corresponding to the equivalent circuit of state 2
   (Fig. 3)
```

---

effort that is caused by the if-clauses. Strictly speaking, the whole set of equations describing the circuit changes whenever the converter switches from state 1 to state 2 and vice versa. In such a model the relevant conduction losses are considered inherently. For the consideration of the switching losses a model expansion as described in Section 4 is necessary.

3 The Averaged Model

If the dynamic behavior of the buck converter is not of interest but the energy flow needs to be investigated it is possible to model the buck converter without calculating the switching events. Assuming the buck converter is in steady state the integral of the inductor voltage v_L over one switching period T_s equals zero [9]. Hence,

$$\int_0^{T_s} v_L dt = \int_0^{t_{on}} v_L dt + \int_{t_{on}}^{T_s} v_L dt = 0. \tag{3}$$

During the time t_{on} the equivalent circuit of state 1 describes the behavior of the buck converter. In the circuit in Fig. 2 the inductor voltage is given by

$$\bar{v}_{L,state\ 1} = v_{in} - \bar{v}_{out} - \bar{v}_{RL} - \bar{v}_{RS,state\ 1}, \tag{4}$$

where the voltage across R_L

$$\bar{v}_{RL} = \bar{i}_L R_L \tag{5}$$

and the voltage across R_S

$$\bar{v}_{RS,state\ 1} = \bar{i}_L R_S, \tag{6}$$

with the mean value of the inductor current

$$\bar{i}_{\mathrm{L}} = \bar{i}_{\mathrm{load}}. \tag{7}$$

The equivalent circuit of state 2 (shown in Fig. 3) describes the behavior of the buck converter during the time $t_{\mathrm{off}} = T_{\mathrm{s}} - t_{\mathrm{on}}$. In state 2 the inductor voltage

$$\bar{v}_{\mathrm{L,state\ 2}} = -\bar{v}_{\mathrm{out}} - \bar{v}_{\mathrm{RL}} - \bar{v}_{\mathrm{RD,state\ 2}} - V_{\mathrm{D}}, \tag{8}$$

where $\bar{v}_{\mathrm{RL}}$ is given by (5) and the voltage across R_{D}

$$\bar{v}_{\mathrm{RD,state\ 2}} = \bar{i}_{\mathrm{L}} R_{\mathrm{D}}. \tag{9}$$

Combining (3) with (4) and (8) one can derive

$$dT_{\mathrm{s}}\left[v_{\mathrm{in}} - \bar{v}_{\mathrm{RL}} - \bar{v}_{\mathrm{out}} - \bar{v}_{\mathrm{RS,state\ 1}}\right] + (1 - d)T_{\mathrm{s}}\left[-\bar{v}_{\mathrm{out}} - \bar{v}_{\mathrm{RL}} - \bar{v}_{\mathrm{RD,state\ 2}} - V_{\mathrm{D}}\right] = 0. \tag{10}$$

From (10) it is possible to find the mean value of the output voltage by

$$d(v_{\mathrm{in}} - \bar{v}_{\mathrm{RS,state\ 1}} + \bar{v}_{\mathrm{RD,state\ 2}} + V_{\mathrm{D}}) - (\bar{v}_{\mathrm{RL}} + \bar{v}_{\mathrm{RD,state\ 2}} + V_{\mathrm{D}}) = \bar{v}_{\mathrm{out}}. \tag{11}$$

$\bar{v}_{\mathrm{out}}$ is a function of the duty cycle d, the input voltage v_{in}, and the mean value of the load current $\bar{i}_{\mathrm{load}}$. Consequently, it is possible to calculate the average output voltage with considering the conduction losses if there are relations for d, v_{in}, and $\bar{i}_{\mathrm{load}}$ available in other models, which is usually the case. The result of (11) can be used as the input of a voltage source that is linked to the connectors of the load side of the buck converter model. Please note that with (11) only the influence of the conduction losses on the average output voltage is considered. In order to calculate the influence of the conduction losses on the supply current the conduction losses of the individual elements (MOSFET, diode, and inductor) need to be known. From the equivalent circuits in Figs. 2 and 3 it appears that by approximation (with the assumption that v_{out} only changes insignificantly in one switching period) in state 1 the inductor current rises with a time constant

$$\tau_{\mathrm{state\ 1}} = \frac{L}{R_{\mathrm{S}} + R_{\mathrm{L}}} \tag{12}$$

and in state 2 the inductor current decays exponentially with

$$\tau_{\mathrm{state\ 2}} = \frac{L}{R_{\mathrm{D}} + R_{\mathrm{L}}}. \tag{13}$$

Provided that the time constants $\tau_{\mathrm{state\ 1}}$ and $\tau_{\mathrm{state\ 2}}$ are much larger than the switching period T_{s} (which applies practically to all buck converters with proper design), the instantaneous current through the inductor can be assumed to have a triangular waveform such as

$$i_{\mathrm{L}} = \begin{cases} i_{\mathrm{L,state\ 1}} & \text{if } nT_{\mathrm{s}} < t \le (n+d)T_{\mathrm{s}} \\ i_{\mathrm{L,state\ 2}} & \text{if } (n+d)Ts < t \le (n+1)Ts \end{cases} \tag{14}$$

with $n = 0, 1, 2, 3, \ldots$ and

$$i_{\mathrm{L,state\ 1}} = \bar{i}_{\mathrm{load}} - \frac{\Delta I_{\mathrm{L}}}{2} + \frac{\Delta I_{\mathrm{L}}}{dT_{\mathrm{s}}} t \tag{15}$$

$$i_{\mathrm{L,state\ 2}} = \bar{i}_{\mathrm{load}} + \frac{\Delta I_{\mathrm{L}}}{2} - \frac{\Delta I_{\mathrm{L}}}{(1-d)T_{\mathrm{s}}} t, \tag{16}$$

where the current ripple

$$\Delta I_{\mathrm{L}} = \frac{\bar{v}_{\mathrm{out}} + V_{\mathrm{D}}}{L} (1-d)T_{\mathrm{s}}. \tag{17}$$

Considering the two states of the buck converter circuit and using (14)–(17) the waveform of the current through the MOSFET

$$i_{\mathrm{S}} = \begin{cases} i_{\mathrm{L,state\ 1}} & \text{if } nT_{\mathrm{s}} < t \le (n+d)T_{\mathrm{s}} \\ 0 & \text{if } (n+d)T_{\mathrm{s}} < t \le (n+1)T_{\mathrm{s}} \end{cases} \tag{18}$$

and the waveform of the current through the diode

$$i_{\mathrm{D}} = \begin{cases} 0 & \text{if } nT_{\mathrm{s}} < t \le (n+d)T_{\mathrm{s}} \\ i_{\mathrm{L,state\ 2}} & \text{if } (n+d)T_{\mathrm{s}} < t \le (n+1)T_{\mathrm{s}}. \end{cases} \tag{19}$$

For calculating the conduction losses of the individual elements in the converter, the RMS values of the current through the MOSFET $I_{\mathrm{S,rms}}$, the current through the diode $I_{\mathrm{D,rms}}$ and the inductor current $I_{\mathrm{L,rms}}$ have to be available. Applying the general relation

$$I_{\mathrm{rms}} = \sqrt{\frac{1}{T} \int_{t_0}^{t_0+T} i(t)^2 dt} \tag{20}$$

to (18) and (19) results in

$$I_{\mathrm{S,rms}} = \sqrt{d \left[I_{\mathrm{L,min}}^2 + I_{\mathrm{L,min}} \Delta I_{\mathrm{L}} + \frac{\Delta I_{\mathrm{L}}^2}{3} \right]}, \tag{21}$$

with

$$I_{L,min} = \bar{i}_{load} - \frac{\Delta I_L}{2} \tag{22}$$

and

$$I_{D,rms} = \sqrt{(1-d)\left[I_{L,max}^2 - I_{L,max}\Delta I_L + \frac{\Delta I_L^2}{3}\right]}, \tag{23}$$

with

$$I_{L,max} = \bar{i}_{load} + \frac{\Delta I_L}{2}. \tag{24}$$

Using (21) – (24) and considering

$$i_L = i_S + i_D \tag{25}$$

the RMS value of the inductor current can be written as

$$I_{L,rms} = \sqrt{I_{S,rms}^2 + I_{D,rms}^2}. \tag{26}$$

The conduction losses of the MOSFET $P_{S,con}$ and the storage inductor $P_{L,con}$ can be calculated by

$$P_{S,con} = R_S I_{S,rms}^2 \tag{27}$$

and

$$P_{L,con} = R_L I_{L,rms}^2. \tag{28}$$

When calculating the conduction losses of the diode also the portion of the power emission contributed by the knee voltage has to be taken into account. Since the knee voltage is modeled as a constant voltage source the mean value of the current through the diode

$$\bar{i}_D = (1-d)\bar{i}_{load} \tag{29}$$

has to be used to calculate the respective contribution to the conduction losses. The total conduction losses in the diode can be written as

$$P_{D,con} = R_D I_{D,rms}^2 + V_D \bar{i}_D. \tag{30}$$

Using (27), (28), and (30) the total amount of conduction losses can be calculated by

$$P_{\text{tot,con}} = P_{\text{S,con}} + P_{\text{D,con}} + P_{\text{L,con}}. \tag{31}$$

4 Consideration of Switching Losses

Models on different levels of detail for switching loss calculation have been published. In many MOSFET models the parasitic capacitances are considered and in some also the parasitic inductances at the drain and at the source of the MOSFET are taken into account. In [1] a model considering the voltage dependence of the parasitic capacitances is proposed. A model in which constant parasitic capacities as well as parasitic inductances are considered is suggested in [2] and in [4] voltage dependent parasitic capacities together with the parasitic inductances are used for the calculation.

In data sheets such as [6] an equation combining two terms is used. In the first term constant slopes of the drain current and the drain source voltage are assumed and in the second the influence of the output capacitance is taken into account. Also for this approach the parasitic capacities as well as the switching times (or at least the gate switch charge and the gate current) have to be known. In [10] is stated that the approach in [6] leads to an overestimation of the switching losses in the MOSFET.

A general approach for switching loss calculation in power semiconductors using measurement results with linearization and polynomial fitting is presented in [3]. In [12] the switching losses are considered to be linear dependent on the blocking voltage, the current through the switch, and the switching frequency. This approach was initially developed for modeling switching losses in IGBTs but it can also be applied to the calculation of MOSFET switching losses. In [8] a modified version of the model proposed in [12] is presented. The difference is that in [8] the switching losses are dependent on the blocking voltage, the current through the switch, and the switching frequency with higher order.

In the presented work the approach described in [12] is used to model the switching losses. The two buck converter models in Section 2 and 3 can be expanded with switching loss models using

$$P_{\text{switch}} = P_{\text{ref,switch}} \frac{f_{\text{s}}}{f_{\text{ref,s}}} \frac{i_{\text{load}}}{i_{\text{ref,load}}} \frac{v_{\text{in}}}{v_{\text{ref,in}}}, \tag{32}$$

where P_{switch} represents the sum of the actual switching losses in the MOSFET and the diode of the buck converter, f_{s} denotes the actual switching frequency, i_{load} is the actual commutating current between the diode and the MOSFET, and v_{in} is the actual blocking voltage of the diode and the MOSFET. $P_{\text{ref,switch}}$ represents a measured value of the switching losses at a reference operation point defined by $f_{ref,s}$, $i_{ref,load}$, and $v_{ref,in}$.

With this approach no knowledge of the parasitic capacitances and inductances is needed. Neither the switching energy nor the switching times need to be known. For the accuracy of the model given in (32) the precision of the measurement results at the reference operation point is very important.

5   Implementation of the Simulation Models

The buck converter models described in Sections 2 and 3 got implemented with *Modelica* modeling language [5] using the Dymola programming and simulation environment. *Modelica* is an open and object oriented modeling language that allows the user to create models of any kind of physical object or device which can be described by algebraic equations and ordinary differential equations.

In both models the conduction losses with linear temperature dependence and the switching losses (calculated according to Section 4) are considered.

In Fig. 4 the scheme of model A, the model calculating the switching events (as explained in Section 2) is shown.

The conduction losses are inherently considered in Alg. 1 and the switching losses are considered by means of a controlled current source with

$$i^*_{\text{model A}} = \frac{P_{\text{switch}}}{v_{\text{in}}} \tag{33}$$

whereas P_{switch} is calculated by (32).

Figure 5 illustrates the scheme of the averaged buck converter model (as explained in Section 3) with consideration of the switching and conduction losses. The basic components of model B are the current source controlled with $i^*_{\text{model B}}$, the voltage source controlled with $v^* = \bar{v}_{\text{out}}$, and the power meter measuring the averaged output power $P_{\text{out}} = \bar{v}_{\text{out}} \bar{i}_{\text{load}}$.

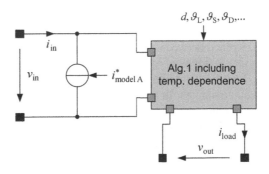

Fig. 4 Model A (buck converter model)

Fig. 5 Model B (buck
converter model)

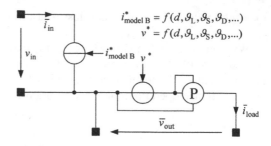

In model B the control signal of the voltage source v^* is computed according to
(11) and the control signal of the current source $i^*_{\text{model B}}$ is calculated by

$$i^*_{\text{model B}} = \frac{P_{\text{switch}} + P_{\text{tot,con}} + P_{\text{out}}}{v_{\text{in}}}. \tag{34}$$

In (34) P_{switch} is given by (32), $P_{\text{tot,con}}$ is calculated from (31), and P_{out} is the
output signal of the power meter in Fig. 5.

6   Simulation and Laboratory Test Results

The approach applied in model A is well established. Therefore the results of model
A are used as a reference for the verification of model B. For the comparison of the
two models a buck converter ($f_s = 100$ kHz) supplied with a constant voltage of
30 V and loaded with a constant load current of 40 A was simulated using model A
and model B. In the two simulations the duty cycle was decreased step by step with
$\Delta d = 0.1$ every 0.03 s starting from $d = 0.8$ to $d = 0.2$.

The purpose of model B is to calculate the efficiency and the electric quantities
in steady state. The supply current signals and the load voltage signals in Fig. 6
show that after the transients decay both models reach the same operation point.
Please note that in Fig. 6 the instantaneous supply current signal computed with
model A is averaged over a switching period.

Both simulations were computed on a state-of-the-art PC with 3 GHz dual core
and 3 GB RAM. It took only 2.8 s to process the results of the simulation with
model B whereas the CPU time for the simulation with model A was 36 s. The large
difference between the CPU times indicates that it is much more efficient to use
model B if the energy flow through a converter is the focus of the simulation.

For the validation of the two simulation models several laboratory tests have
been conducted. In order to avoid core losses an air-cored coil was implemented as
the storage inductor. As the passive and the active switch two IRFPS3810 power
MOSFETs were chosen whereas the body diode of one of the MOSFETs was used
as the freewheeling diode. The temperatures of the two MOSFETs and the air-cored
coil were measured with type-K thermocouples.

Fig. 6 Supply current (*top*) and load voltage (*bottom*) simulated with model A and B

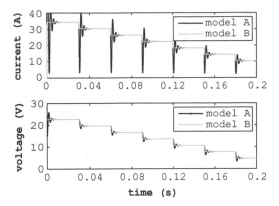

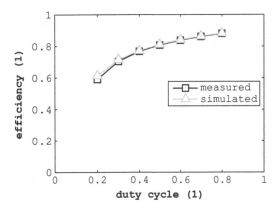

Fig. 7 Measured and simulated efficiency of the buck converter with $v_{in} = 30$ V and $i_{load} = 40$ A

The buck converter used in the laboratory tests was operated with a similar duty cycle reference signal as used for the results in Fig. 6. However, the step time of the duty cycle signal in the laboratory test was significantly longer compared to the signal used for the results in Fig. 6. Because of this, the temperatures of the semiconductors increased significantly. Figure 7 shows the measured efficiency of the circuit under test and the respective (steady state) simulation results of model A and B. The measured and simulated results show satisfactory coherence.

In Fig. 8 the measured losses of the buck converter operated with $d = 0.2$, and $d = 0.8$ during warm-up tests are compared with the results of a simulation carried out with model B. In the top diagram it can be seen that the conduction losses decrease with increasing time and temperature. This is because the knee voltage of the freewheeling diode has a negative temperature coefficient and at $d = 0.2$ the freewheeling diode conducts 80 % of the time in a switching period. In the bottom diagram of Fig. 8 the conduction losses raise with increasing time and temperature. The reason for this is the positive linear temperature coefficient of the on-resistance of the MOSFET and the longer duration in which the MOSFET conducts during a

Fig. 8 Warm-up test results at $d = 0.2$ (*top*) and $d = 0.8$ (*bottom*). Measured and simulated losses of the buck converter with $v_{in} = 30$ V and $i_{load} = 40$ A

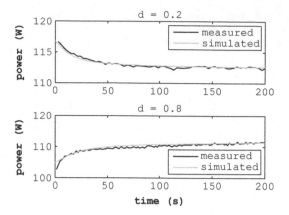

switching period. Please note that the MOSFET dissipates more energy and reaches higher temperatures if the buck converter is operated with high duty cycles.

7 Conclusion

An analytical approach to calculate the steady state behavior of a buck converter including the consideration of conduction losses is described. The presented model B is generated from the derived equations and expanded so that switching losses and temperature dependence of the conduction losses are considered. For the verification of the described modeling approach two simulation models (models A and B) are programmed with *Modelica* language. In steady state model A, the model calculating switching events, matches the behavior of model B, the model based on the approach of system averaging. When comparing the CPU times of models A and B it appears that model B can be computed more than 10 times faster than model A. Consequently, it is recommended to preferably use model B in simulations whenever only the steady state values of the electrical quantities in the buck converter are of interest. This is for instance the case in energy flow analyzes and in simulations for core component dimensioning of electromechanical systems. The simulation results of model A and B show satisfying conformity with the laboratory test results.

References

1. L. Aubard, G. Verneau, J.C. Crebier, C. Schaeffer, Y. Avenas, Power MOSFET switching waveforms: an empirical model based on a physical analysis of charge locations. *The 33rd Annual Power Electronics Specialists Conference, IEEE PESC*, **3**, 1305–1310 (2002)

2. Y. Bai, Y. Meng, A.Q. Huang, F.C. Lee, A novel model for MOSFET switching loss calculation. *The 4th International Power Electronics and Motion Control Conference, IPEMC*, **3**, 1669–1672 (2004)
3. U. Drofenik, J.W. Kolar, A general scheme for calculating switching- and conduction-losses of power semiconductors in numerical circuit simulations of power electronic systems. *Proceedings of the International Power Electronics Conference, IPEC*, 2005
4. W. Eberle, Z. Zhang, Y.-F. Liu, P. Sen, A simple switching loss model for buck voltage regulators with current source drive. *The 39th Annual Power Electronics Specialists Conference, IEEE PESC*, 2008, pp. 3780–3786
5. P. Fritzson, *Principles of Object-Oriented Modeling and Simulation with Modelica 2.1*. (IEEE Press, Piscataway, NJ, 2004)
6. Datasheet of the IRF6603, N-Channel HEXFET Power MOSFET. *International Rectifier*, 2005
7. H. Kapeller, A. Haumer, C. Kral, G. Pascoli, F. Pirker, Modeling and simulation of a large chipper drive. *Proceedings of the 6th International Modelica Conference*, 2008, pp. 361–367
8. K. Mainka, J. Aurich, M. Hornkamp, Fast and reliable average IGBT simulation model with heat transfer emphasis. *Proceedings of the International Conference for Power Conversion, Intelligent Motion and Power Quality, PCIM*, 2006
9. N. Mohan, T.M. Undeland, W.P. Robbins, *Power Electronics – Converters, Applications, and Design*, 2nd edn. (Wiley, New York, 1989)
10. Z.J. Shen, Y. Xiong, X. Cheng, Y. Fu, P. Kumar, Power MOSFET switching loss analysis: A new insight. *Conference Record of the IEEE Industry Applications Conference, 41st IAS Annual Meeting*, **3**, 1438–1442 (2006)
11. D. Simic, T. Bäuml, F. Pirker, Modeling and simulation of different hybrid electric vehicles in modelica using Dymola. *Proceedings of the International Conference on Advances in Hybrid Powertrains, IFP*, 2008
12. D. Srajber, W. Lukasch, The calculation of the power dissipation of the IGBT and the inverse diode in circuits with the sinusoidal output voltage. *Conference Proceedings of Electronica*, 1992, pp. 51–58

Chapter 15
The Computer Simulation of Shaping in Rotating Electrical Discharge Machining

Jerzy Kozak and Zbigniew Gulbinowicz

Abstract The effect of the tool electrode wear on the accuracy is very important problem in the Rotating Electrical Discharge Machining (REDM). Two mathematical models of REDM are presented: the first one considers machining with the face of the end tool electrode and the second one considers EDM with the lateral side of the electrode. The software for computer simulation of EDM machining with the side and face of the electrodes has been developed. This simulation model for NC contouring EDM using rotating electrode may also be applied for tool electrode path optimization. The experimental results confirm the validity of the proposed mathematical models and the simulation software.

1 Introduction

Today's manufacturing industries are facing challenges from advanced difficult-to-machine materials (tough super alloys, ceramics, and composites), stringent design requirements (high precision, complex shapes, and high surface quality), and machining costs. The greatly-improved thermal, chemical, and mechanical properties of the material (such as improved strength, heat resistance, wear resistance, and corrosion resistance) are making ordinary machining processes unable to machine them economically. The technological improvement of manufacturing attributes can be achieved by high efficiency Rotating Electrical Discharge Machining (REDM: Electrical Discharge Grinding-EDG or Electrical Discharge Milling), Abrasive Electrodischarge Grinding (AEDG), Rotary Electrochemical Arc/Discharge Machining – RECAM/RCDM or grinding using metallic bond diamond wheels. These machining processes use a rotating tool.

J. Kozak (✉)
Warsaw University of Technology, ul. Narbutta 85, 02-524 Warsaw, Poland
e-mail: jkozak64@wp.pl

S.-I. Ao et al. (eds.), *Machine Learning and Systems Engineering*,
Lecture Notes in Electrical Engineering 68,
DOI 10.1007/978-90-481-9419-3_15, © Springer Science+Business Media B.V. 2010

The Electrical Discharge Machining process is widely used to machine complicated shapes with high accuracy on hard and advanced materials including hardened steels, super-alloys, tungsten carbides, electrically conductive engineering ceramics, polycrystalline diamonds etc. The material is removed by the thermal energy generated by a series of electrical discharges between the tool electrode and workpiece immersed in the dielectric. There is no direct contact between the tool and workpiece during the machining process. The workpiece can be formed, either by replication of a shaped tool electrode or by 3D movement of a simple electrode like in milling.

The fabrication of complex shaped multiple electrodes used in the die-sinking micro and macro Electrical Discharge Machining (EDM) is expensive and time consuming [1]. Moving a simple shaped electrode along designed tool paths has been proposed as a solution to some of these problems [2–5]. A numerical control system enables the workpiece or tool electrode to move a previously programmed path. In this way, very complex 2 or 3 dimensional shape according to the design requirement can be shaped.

However, the tool wear during machining adversely affects the accuracy of the machined components. The problem of tool wear in 3D micro and macro EDM using simple shaped electrodes has been addressed by applying the Uniform Wear Method to maintain the electrode shape unchanged and compensate for the longitudinal tool wear [5–8].

Tool wear in EDM is characterized by the relative tool-electrode wear (RTW) which is generally defined as ratio tool wear rate (*TWR*, which is the volume of tool material removed per unit time) to material removal rate (*MRR*, which is the volume of workpiece material removed per unit time):

$$v = \frac{TWR}{MRR} \tag{1}$$

Depending upon the operating parameters of REDM, the relative wear may be 0.01–2. Changes in dimensions of tool due to wear during machining is expected to reflect in the actual depth of cut and finally in the profile and dimensional accuracy of machined parts.

Results of investigation of EDM with rotating electrode reported in studies [9, 10] show a slope curvilinear profile of bottom surface of machined groove due to the wear of disk electrode. Controlling the path of electrode can reduce shape error. In the paper [3] preliminary analysis of shape error indicates that one of the main factors leading to shape errors is wheel wear. More extended study of this problem based on mathematical modeling and experiments is reported in articles [11, 12], where the general differential equation described relationship between tool wear, initial and final shape of machined surface has been derived. Effect of wheel wear on dimensional accuracy of grinding is known from theory of tolerances. During machining of parts loaded together in a pocket, and in one pass of the rotary tool, height of parts achieved is random variable in the machined set. Based on the

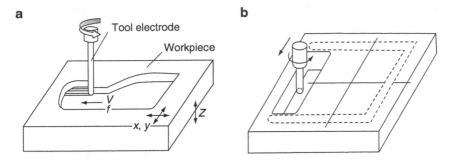

Fig. 1 Example of EDM using rotating tool electrode: (**a**) machining with the face of end electrode, (**b**) machining by the side of rotating electrode

Fig. 2 Principal scheme for mathematical modeling of REDM by end tool electrode

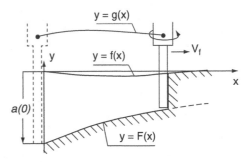

assumption of constant wear rate, uniform probability density function (PDF) has been obtained for height of machined parts [11].

Two cases of machining operations are taken for mathematical modeling: the first one considers machining with the face of the tool electrode and the second one considers EDM with the side of the electrode (Fig. 1).

In first case, the technique of integrating Uniform Wear Method with CAD/CAM software has been successful in generating very complex 3D cavities, it involves a time consuming, empirical approach for selecting tool paths and machining parameters. Therefore, it is necessary to develop a theoretical model which accounts for the effect of tool wear on the surface profile generation during each pass. As the number of passes can be very large, a corresponding computer simulation software simulation also needs to be developed.

2   Mathematical Modelling of Redm Shaping by End Tool Electrode

The principal scheme of shaping using the end tool electrode process is presented in Fig. 2. The initial profile of workpiece is given by function $y = f(x)$. The electrode is controlled to move along the tool head path $y = g(x)$. However, the longitudinal tool wear results in the profile of machined surface $y = F(x)$, which is different from $g(x)$.

The purpose of this mathematical modeling and computer simulation is to determine surface profile $y = F(x)$, taking into account the change in tool length which occurs due to the wear of electrode. The final profile depends on input parameters, such as depth of cut a_0, initial profile $y = f(x)$, diameter of tool d_0 and tool head path $y = g(x)$.

Let us consider the case of machining presented in Fig. 2, when the initial surface is $y = f(x)$ and $g(x) = $ constant.

In determining the profile of machined surface, the following assumptions are made:

– Changes in tool shape are neglected because the Uniform Wear Method is applying.
– The gap between tool electrode and workpiece is neglected.
– Material removal rate MRR is equal:

$$MRR = (f(x) - F(x)) \cdot d_0 \cdot V_f \tag{2}$$

– Tool wear rate TWR, defined as the volume of tool material removed per unit time, for $g(x) = const.$ is:

$$TWR = \frac{\pi \cdot d_0^2}{4} \frac{dF}{dt} \tag{3}$$

or

$$TWR = \frac{\pi \cdot d_0^2}{4} \frac{dF}{dx} \frac{dx}{dt} = \frac{\pi \cdot d_0^2}{4} \frac{dF}{dx} V_f \tag{4}$$

where V_f is feed rate.

After substituting (2) and (4) in (1) for v, the equation describing the profile of machined surface takes the following form:

$$\frac{dF}{dx} + mF = mf(x) \tag{5}$$

with initial condition $F(0) = -a(0)$ (Fig. 2), where $m = 4v/\pi \cdot d_0$ is the wear factor.

In many cases it is possible to assume, that relative wear during machining is constant ($m = const.$), i.e. it is not dependent on the actual depth of cut. For this condition and for $f(x) = 0$, the solution of (5) becomes:

$$F(x) = -a_0 \exp(-mx) \tag{6}$$

While considering accuracy, the profile is conveniently described in coordinates relative to required allowance, $-a_0$ i.e.:

$$h(x) = F - (-a_0) = a_0[1 - \exp(-mx)] \tag{7}$$

For more universal application of (6) a non-dimensional form of notations is used. The non-dimensional variables are defined as $\bar{h} = h/a_0$ and $\bar{x} = x/L$, where L is the length of workpiece. In non-dimensional form, (7) can be written as:

$$\bar{h} = 1 - \exp(-A \cdot \bar{x}) \tag{8}$$

where $A = m \cdot L = 4v\,L/(\pi \cdot d_0)$.

Profiles of machined surface obtained for different values of A are presented in Fig. 3. As result of wear, the depth of cut is decreasing and the tool wear rate (*TWR*) is changing during machining. For comparison in Fig. 3 the dashed line (for $A = 1$) represents the profile neglecting changes in *TWR* during machining.

In general case when $g = g(x)$, mathematical model of REDM process is described as follows:

$$\frac{dF}{dx} + mF = mf(x) + \frac{dg}{dx} \tag{9}$$

with initial condition $F(0) = -a_0$.

The machining accuracy may be improved by control motion of tool electrode. For example, the tool wear can be compensated by moving tool or workpiece along y-axis, to obtain flat surface. In the case when initial surface is flat $f(x) = 0$ and the depth of cut is equal a_0, the required flat shape is $F(x) = -a_0$. Therefore, based on solution (9) the linear path of tool is needed, which can be described as follows:

$$y = g(x) = H_0 - a_0 \cdot m \cdot x \tag{10}$$

where H_0 is initial position of tool head.

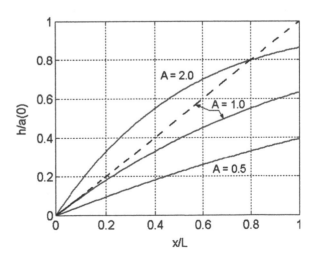

Fig. 3 Profile of machined surfaces for different values of A

For machining with constant feed rate V_f along x-axis, the compensation of the tool wear can be obtained by adding the relative motion of the tool/workpiece with constant feed rate V_y along y-axis equal:

$$V_y = -a_0 \cdot m \cdot V_f = -\frac{4v \cdot a_0}{\pi \cdot d_0} V_f \qquad (11)$$

This theoretical conclusion about linear path of tool electrode for compensation of wear has been confirmed by the experiments [5].

3 Mathematical Modelling of REDM Shaping by Lateral Surface of Tool Electrode

The principal scheme of shaping process using the lateral side of tool electrode is presented in Fig. 4. The purpose of this mathematical modeling and computer simulation is to determine the profile of generated surface $y = F(x)$, taking into account the change in tool diameter which occurs due to the tool wear of rotary tool during machining (Fig. 4). The final surface profile depends on the input data, such as depth of cut a_0, initial surface profile $y = f(x)$, diameter of the tool d_0 and the curvilinear path of the center of tool $y = g(x)$.

For the modeling purpose the following assumptions were made:

- The inter electrode gap s is constant and included to the effective radius of tool i.e. $R = R(electrode) + s$.
- Actual depth of cut, a, is determined by the position of points A and B, which are point tangency of the tool electrode to the generated profile and intersection point of the tool and the initial profile $y = f(x)$, respectively (Fig. 4).
- Feed rate along axis y is significant lower in compare to the feed rate V_f i.e. the value of $dg/dx \ll 1$.
- Changes in the tool shape along the axis of rotation are neglected.

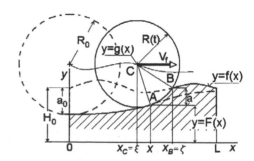

Fig. 4 Scheme of machining with curvilinear path of rotating tool

After machining for certain time t, the center of the tool has reached the coordinate $x_C = \xi$, the feed rate is $V(t) = V_f$ and the effective tool radius due to the wear is $R(t)$.

The material removal rate MRR at $dg/dx \ll 1$ is:

$$MRR = a \cdot b \cdot V_f = [(f(\zeta) - F(x)] \cdot b \cdot V_f \tag{12}$$

where b is width of workpiece and ξ is function of x, $\xi = \xi(x)$.

The coordinates of intersection point $B(\xi, f(\xi))$ can be calculated from the relation $AC = BC$ (Fig. 4):

$$(x - \xi)^2 + [F(x) - g(\xi)]^2 = (\varsigma - \xi)^2 + [f(\varsigma) - g(\xi)]^2 \tag{13}$$

The tool wear rate TWR, is given by:

$$TWR = -2\pi \cdot b \cdot R(t) \cdot \frac{dR}{dt} \tag{14}$$

Since

$$\frac{dR}{dt} = \frac{dR}{d\xi} \frac{d\xi}{dt} = \frac{dR}{d\xi} V_f \tag{15}$$

and according to the relation from Fig. 4,

$$R^2(t) = [g(\xi) - F(x)]^2 + (x - \xi)^2 \tag{16}$$

the rate of change of tool radius can be expressed as following:

$$\frac{dR}{dt} = \frac{1}{R} [g(\xi) - F(x)] \left(\frac{dg}{d\xi} - \frac{dF}{dx} \frac{dx}{d\xi} \right) + (x - \xi) \left(\frac{dx}{d\xi} - 1 \right) \tag{17}$$

Substituting (12) and (17) into (1) and performing transformation, the profile of machined surface can be described by (18):

$$\frac{dF}{dx} + \frac{v}{2\pi} \frac{F}{[g(\xi) - F]} \frac{d\xi}{dx} =$$
$$= \frac{v}{2\pi} \frac{f(\varsigma)}{[g(\xi) - F]} \frac{d\xi}{dx} + \frac{dg}{dx} - \frac{(x - \xi)}{[g(\xi) - F]} \left(1 - \frac{d\xi}{dx} \right) \tag{18}$$

with initial condition $F = F(0) = g(0) - R(0)$.

Generally, the value of relative tool wear is depending on the depth of cut $a = f$ (ξ)-$F(x)$ i.e. $v = v(a)$, and this function can be only determined experimentally.

An additional equation to (18) and (13) can be derived from the condition for tangency of the tool to the profile machined surface at the point A as follows:

$$[g(\xi) - F(x)]\frac{dF}{dx} = x - \xi \tag{19}$$

The described above mathematical model can be used in following processes: REDM/EDG (side EDM milling), AREDM, RECAM/RCDM and grinding (in this case from definition of the G-ratio, $v = 1/G$).

Based on the presented mathematical model, computer simulation of evolution of workpiece profile can be carried out by using in-house developed software and two main tasks can be formulated:

1. The tool path, $g(x)$, and initial shape of surface, $f(x)$, are known but the resulting shape of the machined surface, $F(x)$, and needs to be predicted.
2. For a required shape of the machined surface, $F(x)$, the tool path, $g(x)$, needs to be determined in order to compensate for the tool wear.

These tasks have been solved numerically using the Finite Difference Method and iterative procedure.

In many cases the changes in the radius and the differences between coordinates x and ξ (for points A and B, respectively) are usually small when compared to the initial radius. Therefore $d\xi/dx \cong 1$ and (18) will evolve in:

$$\frac{dF}{dx} + \frac{v}{2\pi}\frac{F}{[g(x) - F]} = \frac{v}{2\pi}\frac{f(x)}{[g(x) - F]} + \frac{dg}{dx} \tag{20}$$

This simplified mathematical model can be successfully used to plan REDM operations and in solving the mentioned above two main tasks.

4 Software for Computer Simulation

The developed software supports process design for the REDM. The software is easy to use, menu driven application allowing evaluating the effect of various parameters on process performance.

Programming of kinematics of the tool electrode can be achieved by supplying program with tool path in x direction described. Geometry of the surface at the beginning of machining can be defined in two ways:

– By supplying software with the surface equation, $y = f(x)$
– By supplying software with coordinates of the surface (for example, a surface just obtained a machining process cycle can be used as a starting surface for the next machining operation)

The important feature of the software is the capability to define relative tool wear as function of depth of cut.

Input date is inserted in the overlap Data. The windows for input data are shown in Fig. 5.

Results of simulation can be viewed in two ways: tables of coordinates of points of the workpiece surface and 2D graphs.

Fig. 6 shows an example of a graph of REDM machined surface defined by function $y = a/(1 + b\sin(2x/))$. The tool path was given by $g(x) = H_0 - a_0 + R_0 = $ constant (in Fig. 7 is shown shifted tool path to axis x).

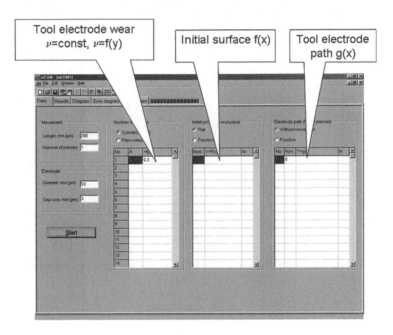

Fig. 5 Overlap data

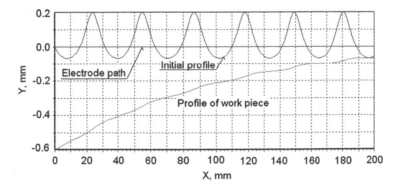

Fig. 6 The profiles from software screen ($R_0 = 10$ mm, $a_0 = 0.6$, $v = 0.3$)

In Fig. 7 is shown simulation of REDM process with compensation of tool wear. An effect of periodical changes in the depth on final shape is appeared on machined surface.

5 Experimental Verification

The theoretical model and simulation results were verified in Rotary Electrical Discharge Machining (REDM) on M35K Mitsubishi, EDM – NC machine tool using electrode with diameter of 8.2 and 18 mm. The range for workpiece movement was 0–160 mm for x-axis. Experiments were performed on tool steel material P233330 (HRC 56) with copper electrode using the following process parameters: Pulse voltage of U: 40 V, Pulse current I: 55 and 120 A, Pulse on time t_p = pulse off

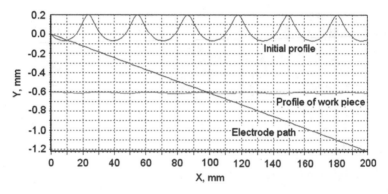

Fig. 7 The surface profile and tool path at EDM with compensation of tool wear ($R_0 = 10$ mm, $a_0 = 0.6$, $v = 0.3$)

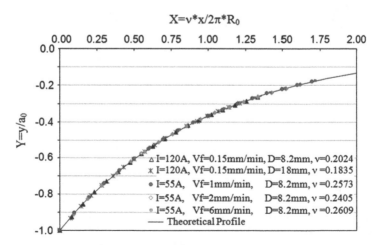

Fig. 8 Comparison of theoretical and experimental results in non-dimensional system coordinates. Setting parameters: $t_p = t_0 = 120$ µs, $n = 200$ rpm

Table 1 The machining conditions for experiments carried out on EDIOS-16

Setting parameters							
Test	t (min)	a_0 (mm)	I (A)	U (V)	n (rev/min)	TWR (mm^3/min)	MRR (mm^3/min)
a	25	1	6	90	600	0,033	9,077
b	19	1,5	10	50	2,700	0,254	2780
c	18	0,5	5	100	6,000	1,954	3,262

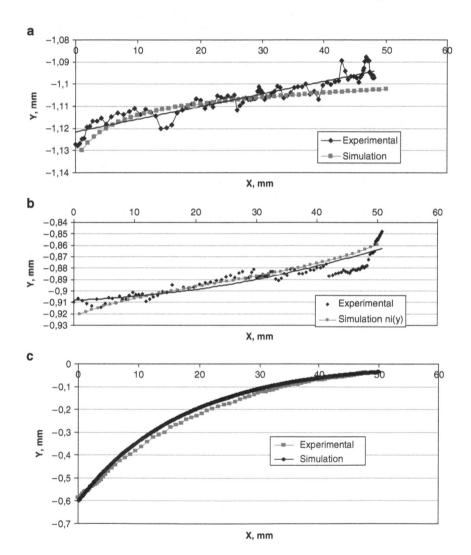

Fig. 9 Comparison of theoretical and experimental results

time t_0: 60 and 120 μs, Depth of cut a_0: 1.2 and 1.5 mm, Feed rate V_f: 0.15, 1.0, 2.0, and 6.0 mm/min, Rotation speed n: 200 rpm. For different combination of setting parameters for verification used non-dimensional system coordinates $X = v \cdot x/(2\pi \cdot R_0)$ and $\bar{Y} = y/a_0$. The results are presented in Fig. 8.

Further experiments were carried out on EDIOS-16 machine tool using copper electrode with diameter of 4 mm, pulse on time 160 μs pulse off time 10 μs and the maximal travel distances in x axis were 0–50 mm.

The machining conditions are shown in the Table 1.

Experimental results are compared with simulation prediction in Fig. 9.

Experimental verifications show high accuracy of developed mathematical model and computer simulation. An overall average of 6% (of the initial depth of cut) deviation was found between simulation and experimental results.

6 Conclusion

The study showed a good agreement of theoretical and experimental results of modeling of REDM process. The developed software can be useful for analysis of the REDM process, parameter optimization and surface prediction. Computer simulation of REDM has a significant potential to be used in industry.

Acknowledgements This work was supported in part by the European Commission project "Micro-Technologies for Re-Launching European Machine Manufacturing SMEs (LAUNCH-MICRO)".

References

1. M. Kunieda, B. Lauwers, K.P. Rajurkar, B.M. Schumacher, Advancing EDM through fundamental insight into the process. Ann. CIRP **54**(2), 599–622
2. P. Bleys, J.P. Kruth, B. Lauwers, Milling EDM of 3D shapes with tubular electrodes, *Proceedings of ISEM XIII*, 2001, pp. 555–567
3. P. Bleys, J. P. Kruth, Machining complex shapes by numerically controlled EDM. Int. J. Electric. Mach. **6**, 61–69 (2001)
4. J.P. Kruth, B. Lauwers, W. Clappaert, A study of EDM pocketing, *Proceedings of ISEM X*, 1992, pp. 121–135
5. Z.Y. Yu, T. Masuzawa, M. Fujino, Micro-EDM for three-dimensional cavities, development of uniform wear method. Ann. CIRP **47**(1), 169–172 (1998)
6. K.P. Rajurkar, Z.Y. Yu, 3D micro EDM using CAD/CAM., Ann. CIRP **49**(1), 127–130 (2000)
7. Z.Y. Yu, T. Masuzawa, M. Fujino, A basic study on 3D Micro-EDM. Die Mould Technol. **11** (8), 122–123 (1996) (in Japanese)
8. Z.Y. Yu, J. Kozak, K.P. Rajurkar, Modeling and simulation of micro EDM process. Ann. CIRP **52**(1), 143–146 (2003)
9. Y. Uno, A. Okada, M. Itoh, T. Yamaguchi, EDM of groove with rotating disk electrode, Int. J. Electr.Mach. **1**, 13–20 (1996)

10. J. Quian, H. Ohmori, T. Kato, I. Marinescu, Fabrication of micro shapes of advanced materials by ELID- Grinding. Trans. NAMRI/SME **27**, 269–278 (2000)
11. J. Kozak, Effect of the wear of rotating tool on the accuracy of machined parts, in *Proceedings of the 2nd International Conference on Advances in Production Engineering APE-2*, vol. I. Warsaw, 2001, pp. 253–262
12. J. Kozak, Z. Gulbinowicz, D. Gulbinowicz, Computer simulation of rotating electrical machining (REDM). Arch. Mech. Eng. **XL**(1), 111–125 (2004)

Chapter 16
Parameter Identification of a Nonlinear Two Mass System Using Prior Knowledge

C. Endisch, M. Brache, and R. Kennel

Abstract This article presents a new method for system identification based on dynamic neural networks using prior knowledge. A discrete chart is derived from a given signal flow chart. This discrete chart is implemented in a dynamic neural network model. The weights of the model correspond to physical parameters of the real system. Nonlinear parts of the signal flow chart are represented by nonlinear subparts of the neural network. An optimization algorithm trains the weights of the dynamic neural network model. The proposed identification approach is tested with a nonlinear two mass system.

1 Introduction

If the system is unknown and no prior knowledge is available the identification process can start with an oversized neural network. To find a model of the system as simple as possible a pruning algorithm deletes unnecessary parts of the network. This approach reduces the complexity of the network. As shown in [2, 3] this identification approach is able to remove weights from an oversized general dynamic neural networks (short GDNN, in GDNNs the layers have feedback connections with time delays, see Fig. 1).

The approach presented in this article is completely different. The assumption is that prior knowledge is available in form of a continuous signal flow chart. The model is derived in two steps. First the structure is altered in such a way, that all backward paths contain integrator blocks. In the second step the integrator blocks are discretized by Implicit and Explicit Euler approximation. In this article the

C. Endisch (✉)
Institute for Electrical Drive Systems, Technical University of Munich, Arcisstraße 21, 80333 München, Germany
e-mail: christian.endisch@tum.de

S.-I. Ao et al. (eds.), *Machine Learning and Systems Engineering*,
Lecture Notes in Electrical Engineering 68,
DOI 10.1007/978-90-481-9419-3_16, © Springer Science+Business Media B.V. 2010

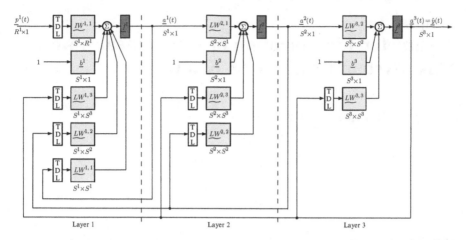

Fig. 1 Example of a three-layer GDNN with feedback connections in all layers – the output of a tapped delay line (TDL) is a vector containing delayed values of the TDL input. Below the matrix-boxes and below the arrows the dimensions are shown. R^m and S^m respectively indicate the dimension of the input and the number of neurons in layer m. $\underline{\hat{y}}$ is the output of the GDNN

parameters are trained with the powerful Levenberg–Marquardt (LM) algorithm [5]. Real time recurrent learning (RTRL) is used to calculate the necessary Jacobian matrix.

In [6, 7] a so called structured recurrent neural network is used for identifying dynamic systems. On a first glance the structured recurrent network seems to be similar to the approach in this article. However, the approach presented in this article does neither depend on a state observer nor on system-dependent derivative calculations. No matter what model is used for the identification process, the derivative calculations are conducted for the GDNN-model in general [9–11].

The second section presents the general dynamic network (GDNN). For implementing a special model structure administration matrices are introduced. Section 3 explains the optimization method used to train the network parameters throughout this article. Section 4 describes the mechatronic system to be identified. The structured GDNN is constructed in Section 5. In Section 6 the identification approach is tested with a nonlinear dynamic system. Finally, Section 7 summarizes the results.

2   General Dynamic Neural Network

Figure 1 shows an example of a three-layer GDNN with feedback connections. De Jesus described the GDNN-model in his doctoral thesis [10]. The sophisticated formulations and notations of the GDNN-model allow an efficient computation of the Jacobian matrix using real time recurrent learning (RTRL) [4, 9, 11, 16]. In this article we follow these conventions suggested by De Jesus. The simulation equation for layer m is calculated by

$$\underline{n}^m(t) = \sum_{l \in L_m^f} \sum_{d \in DL^{m,l}} LW^{m,l}(d) \cdot \underline{a}^l(t-d) + \sum_{l \in I_m} \sum_{d \in DI^{m,l}} LW^{m,l}(d) \cdot \underline{p}^l(t-d) + \underline{b}^m,$$

(1)

where $\underline{n}^m(t)$ is the summation output of layer m, $\underline{p}^l(t)$ is the l-th input to the network, $IW^{m,l}$ is the input weight matrix between input l and layer m, $LW^{m,l}$ is the layer weight matrix between layer l and layer m, $\underline{b}^m$ is the bias vector of layer m, $DL^{m,l}$ is the set of all delays in the tapped delay line between layer l and layer m, $DI^{m,l}$ is the set of all input delays in the tapped delay line between input l and layer m, I_m is the set of indices of input vectors that connect to layer m and L_m^f is the set of indices of layers that directly connect forward to layer m. The output of layer m is

$$\underline{a}^m(t) = \underline{f}^m(\underline{n}^m(t)),$$

(2)

where $\underline{f}^m(\cdot)$ are either nonlinear *tanh*- or linear activation functions. At each point in time the Eqs. 1 and 2 are iterated forward through the layers. Time is incremented from $t = 1$ to $t = Q$. (See [10] for a full description of the notation used here.) In order to construct a flexible model-structure, it is necessary that only particular weights in the weight matrices do exist. This is realized by the introduction of administration matrices.

2.1 Administration Matrices

For each weight matrix there exists one weight administration matrix to mark which weights are used in the GDNN-model. The layer weight administration matrices $\underline{AL}^{m,l}(d)$ have the same dimensions as the layer weight matrices $LW^{m,l}(d)$, the input weight administration matrices $\underline{AI}^{m,l}(d)$ have the same dimensions as the input weight matrices $IW^{m,l}(d)$ and the bias weight administration vectors $\underline{Ab}^m$ have the same dimensions as the bias weight vectors $\underline{b}^m$. The elements of the administration matrices can have the boolean values 0 or 1, indicating if a weight is valid or not. If e.g. the layer weight $lw^{m,l}(d) = [LW^{m,l}(d)]_{k,i}$ from neuron i of layer l to neuron k of layer m with $d^{k,i}$th-order time-delay is valid, then $\left[\underline{AL}^{m,l}(d)\right]_{k,i} = al_{k,i}^{m,l}(d) = 1$. If the element in the administration matrice equals to zero, the corresponding weight has no influence on the GDNN. With these definitions the kth output of layer m can be computed by

$$n_k^m(t) = \sum_{l \in L_m^f} \sum_{d \in DL^{m,l}} \left(\sum_{i=1}^{S^l} lw_{k,i}^{m,l}(d) \cdot \alpha l_{k,i}^{m,l}(d) \cdot a_i^l(t-d) \right)$$

$$+ \sum_{l \in I_m} \sum_{d \in DI^{m,l}} \left(\sum_{i=1}^{R^l} iw_{k,i}^{m,l}(d) \cdot \alpha i_{k,i}^{m,l}(d) \cdot p_i^l(t-d) \right) + b_k^m \cdot \alpha b_k^m,$$

$$a_k^m(t) = f_k^m(n_k^m(t)), \tag{3}$$

where S^l is the number of neurons in layer l and R^l is the dimension of the lth input. By setting certain entries of the administration matrices to one a certain GDNN-structure is generated. As this model uses structural knowledge from the system, it is called Structured Dynamic Neural Network (SDNN).

2.2 Implementation

For the simulations throughout this paper the graphical programming language *Simulink (Matlab)* is used. SDNN and the optimization algorithm are implemented as S-function in C++.

3 Parameter Optimization

First of all a quantitative measure of the network performance has to be defined. In the following we use the squared error

$$E(\underline{w}_k) = \frac{1}{2} \cdot \sum_{q=1}^{Q} (\underline{y}_q - \hat{\underline{y}}_q(\underline{w}_k))^T \cdot (\underline{y}_q - \hat{\underline{y}}_q(\underline{w}_k))$$

$$= \frac{1}{2} \cdot \sum_{q=1}^{Q} \underline{e}_q^T(\underline{w}_k) \cdot \underline{e}_q(\underline{w}_k), \tag{4}$$

where q denotes one pattern in the training set, $\underline{y}_q$ and $\hat{y}_q(\underline{w}_k)$ are the desired target and the actual model output of the q-th pattern respectively. The vector $\underline{w}_k$ is composed of all weights in the SDNN. The cost function $E(\underline{w}_k)$ is small if the training process performs well and large if it performs poorly. The cost function forms an error surface in a $(N+1)$-dimensional space, where N is equal to the number of weights in the SDNN. In the next step this space has to be searched in order to reduce the cost function.

3.1   Levenberg–Marquardt Algorithm

All Newton methods are based on the second-order Taylor series expansion about the old weight vector $\underline{w}_k$:

$$E(\underline{w}_{k+1}) = E(\underline{w}_k + \Delta\underline{w}_k)$$
$$= E(\underline{w}_k) + \underline{g}_k^T \cdot \Delta\underline{w}_k + \frac{1}{2} \cdot \Delta\underline{w}_k^T \cdot \underset{\sim}{H}_k \cdot \Delta\underline{w}_k. \tag{5}$$

If a minimum on the error surface is found, the gradient of the expansion Eq. 2 with respect to $\Delta\underline{w}_k$ is zero:

$$\nabla E(\underline{w}_{k+1}) = \underline{g}_k + \underset{\sim}{H}_k \cdot \Delta\underline{w}_k = 0. \tag{6}$$

Solving Eq. 6 for $\Delta\underline{w}_k$ results in the Newton method

$$\Delta\underline{w}_k = -\underset{\sim}{H}_k^{-1} \cdot \underline{g}_k^T,$$
$$\underline{w}_{k+1} = \underline{w}_k - \underset{\sim}{H}_k^{-1} \cdot \underline{g}_k. \tag{7}$$

The vector $-\underset{\sim}{H}_k^{-1} \cdot \underline{g}_k^T$ is known as the Newton direction, which is a descent direction, if the Hessian matrix $\underset{\sim}{H}_k$ is positive definite. The LM approach approximates the Hessian matrix by [5]

$$\underset{\sim}{H}_k \approx \underset{\sim}{J}^T(\underline{w}_k) \cdot \underset{\sim}{J}(\underline{w}_k) \tag{8}$$

and it can be shown that

$$\underline{g}_k = \underset{\sim}{J}^T(\underline{w}_k) \cdot \underline{e}(\underline{w}_k), \tag{9}$$

where $\underset{\sim}{J}(\underline{w}_k)$ is the Jacobian matrix:

$$\underset{\sim}{J}(\underline{w}_k) = \begin{bmatrix} \dfrac{\partial e_1(\underline{w}_k)}{\partial w_1} & \dfrac{\partial e_1(\underline{w}_k)}{\partial w_2} & \cdots & \dfrac{\partial e_1(\underline{w}_k)}{\partial w_N} \\ \dfrac{\partial e_2(\underline{w}_k)}{\partial w_1} & \dfrac{\partial e_2(\underline{w}_k)}{\partial w_2} & \cdots & \dfrac{\partial e_2(\underline{w}_k)}{\partial w_N} \\ \vdots & \vdots & \ddots & \vdots \\ \dfrac{\partial e_Q(\underline{w}_k)}{\partial w_1} & \dfrac{\partial e_Q(\underline{w}_k)}{\partial w_2} & \cdots & \dfrac{\partial e_Q(\underline{w}_k)}{\partial w_N} \end{bmatrix} \tag{10}$$

The Jacobian matrix includes first derivatives only. N is the number of all weights in the neural network and Q is the number of evaluated time steps. With Eqs. 4, 5 and 6 the LM method can be expressed with the scaling factor μ_k

$$\underline{w}_{k+1} = \underline{w}_k - \left[\underset{\sim}{J}^T(\underline{w}_k) \cdot \underset{\sim}{J}(\underline{w}_k) + \mu_k \cdot \underset{\sim}{I}\right]^{-1} \cdot \underset{\sim}{J}^T(\underline{w}_k) \cdot \underline{e}(\underline{w}_k), \qquad (11)$$

where $\underset{\sim}{I}$ is the identity matrix. As the LM algorithm is the best optimization method for small and moderate networks (up to a few hundred weights), this algorithm is used for all simulations in this paper.

LM optimization is usually carried out offline. In this paper we use a sliding time window that includes the information of the last Q time steps. With the last Q errors the Jacobian matrix $\underset{\sim}{J}(\underline{w}_k)$ from Eq. 10 is calculated quasi-online. In every time step the oldest training pattern drops out of the time window and a new one (from the current time step) is added – just like a first in first out (FIFO) buffer. If the time window is large enough, it can be assumed that the information content of the training data is constant. With this simple method we are able to implement the LM algorithm quasi-online. For the simulations in this paper the window size is set to $Q = 25000$ using a sampling time of 1ms.

3.2 Jacobian Calculations

To create the Jacobian matrix, the derivatives of the errors have to be computed, see Eq. 10. The GDNN has feedback elements and internal delays, so that the Jacobian cannot be calculated by the standard backpropagation algorithm. There are two general approaches to calculate the Jacobian matrix for dynamic systems: By backpropagation through time (BPTT) [15] or by real time recurrent learning (RTRL) [16]. For Jacobian calculations the RTRL algorithm is more efficient than the BPTT algorithm [11]. According to this the RTRL algorithm is used in this paper. The interested reader is referred to [2, 8, 11] for further details.

4 Two-Mass-System

The considered plant (shown in Fig. 2) is a nonlinear two-mass flexible servo system (TMS), which is a common example for an electrical drive connected to a work machine via flexible shaft. Figure 3 displays the signal flow chart of the TMS,

Fig. 2 Laboratory setup of the TMS

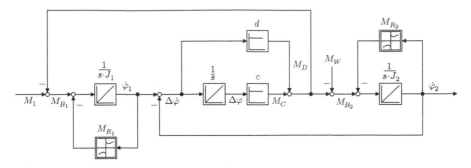

Fig. 3 Signal flow chart of the TMS with friction

where the spring constant c and the damping d model the shaft between the two machines [14]. $\dot{\varphi}_1$ and $\dot{\varphi}_2$ denote the rotation speed of the main engine and the work machine respectively. The torque of inertia of the machines are depicted by J_1 and J_2. The motor torque is M_1 and the torque at the work machine is M_2. M_{B1} and M_{B2} are the acceleration torques of the main engine and the work machine respectively. The difference of the rotation speeds is denoted by $\Delta\dot{\varphi}$ and $\Delta\varphi$ is the difference of the angles. The torque of the spring and the damping torque are depicted by M_C and M_D respectively. The friction torques of the engine and the working machine are M_{R1} and M_{R2} respectively. The objective in this paper is to identify the linear TMS-parameters and the characteristics of the two friction torques.

5   Structured Dynamic Neural Networks

To construct a structured network with the help of GDNNs it is necessary to redraw the signal flow chart from Fig. 3 because the implementation of algebraic loops is not feasible [1]. All feedback connections must contain at least one time delay, otherwise the signals cannot be propagated through the network correctly. This goal is accomplished by inserting integrator blocks in the feedback loops. Figure 4 displays the redrawn version of Fig. 3. By using the Euler approximation it is possible to discretize the redrawn signal flow chart. The Implicit Euler approximation $y(t) = y(t - 1) + x(t) \cdot T$ replaces all integrator blocks in the forward paths and the Explicit Euler approximation $y(t) = y(t - 1) + x(t - 1) \cdot T$ replaces all integrator blocks in the feedback paths. This approach ensures that all feedback connections contain the necessary time delays. The resulting discrete signal flow chart, which can be implemented as a SDNN, is displayed in Fig. 5. z^{-1} denotes a first order time delay and T is the sampling time. All other denotations are the same as in Fig. 3. In total the SDNN consists of 16 layers. The summing junctions depict the neurons of the network. Every summing junction is marked with a number, which denotes the layer of the neuron and its position within the layer.

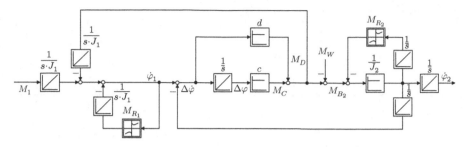

Fig. 4 Redrawn signal flow chart of the TMS from Fig. 3

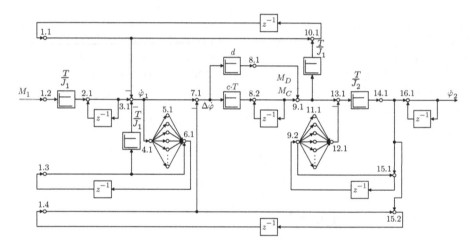

Fig. 5 Structured recurrent network of a nonlinear TMS

For instance, 15. 1 marks the first neuron of the 15th layer. The friction of the engine and the work machine can be modeled by an optional number of neurons in the 5th layer and in the 11th layer respectively. These are the only neurons with tanh-transfer functions. All other neurons have linear transfer functions. The connections in Fig. 5 which do neither belong to a linear parameter (depicted as box) nor to a friction-subpart are initialized with 1 or −1. The optimization algorithm is able to tune the parameters corresponding to the spring constant c, the damping d and the torque of inertia J_2 and the friction weights of the work machine. As it is not possible to identify the two torques of inertia as well as the two characteristic curves of the TMS simultaneously, the engine parameters are determined in a first no-load-identification which is conducted in idle running, see Section 6.

6   Identification

6.1   Excitation Signal

The system is excited by an APRBS-signal (Amplitude Modulated Pseudo Random Binary Sequence [13]) combined with a bias produced by a relay. The APRBS-signal has an amplitude range between –7 and 7 Nm and an amplitude interval between 10 and 250 ms. The relay output switches to –4 Nm if the rotation speed of the TMS is greater than $10\frac{rad}{s}$ and it switches to 4 Nm if the rotation speed is smaller than $-10\frac{rad}{s}$. The suggested excitation signal ensures that the TMS, which is globally integrating, remains in a well defined range for which the SDNN is able to learn the friction. Moreover, the output of the relay is multiplied by 0.2 for a rotation speed in the range of -2 to $2\frac{rad}{s}$. Thus, the SDNN receives more information about the friction in the region of the very important zero crossing. The resulting output of the TMS can be regarded in upper panel of Fig. 8. This excitation signal is used in all the following identification processes.

6.2   Engine Parameters

For a successful TMS-identification it is necessary to identify the engine parameters in idle mode first. The obtained values are used as fix parameters in the identification of the whole TMS in Chapter 6. The upper left drawing of Fig. 6 displays the signal flow chart of the nonlinear engine, where M_1 is the torque, $\dot{\varphi}_1$ is the rotation speed, and J_1 denotes the torque of inertia. In order to be able to discretize the signal flow chart, we insert the integrator block in the backward path. The resulting signal flow chart is shown in the upper right drawing of Fig. 6.

As explained above, for discretizing the signal flow chart the Implicit Euler approximation has to replace the integrator in the forward path, whereas the Explicit Euler approximation replaces the integrator in the backward path. The sampling time T is incorporated in the gain corresponding to the torque of inertia J_1. The lower drawing of Fig. 6 displays the resulting discrete signal flow chart, which can be implemented as a SDNN in which all the summing junctions are regarded as neurons. The neurons in layer 5 model the friction of the engine.

For the identification process the engine is excited with the signal explained in Section 6. The quasi-online calculated cost function $E(\underline{w}_k)$ Eq. 4 with $Q = 25000$ is minimized by the LM optimization algorithm Eq. 11. The sampling time is set to $T = 1$ms and the identification process starts after 5 s, seen in the upper panel of Fig. 7. The SDNN-model of Fig. 6 has three neurons in the fifth layer with six weights to model the friction. These weights are initialized randomly between –0.5 and 0.5. Table 1 shows two different initial values for the torque of inertia J_1 and the corresponding results. The calculated results are mean values between the last 25, 000 optimization steps. Figure 7 displays the characteristic curve of the friction at

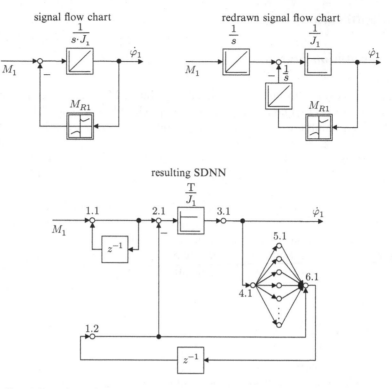

Fig. 6 Signal flow chart of the engine (*upper left side*), redrawn signal flow chart of the engine (*upper right side*) and resulting SDNN (*lower drawing*)

$t = 100$ s identified by the SDNN. We observe that the network is able to model the jump due to the static friction with just three neurons. The following identification of the whole TMS uses this friction curve from Fig. 7 and the result $J_1 = 0.1912$ from Table 1 for the torque of inertia.

6.3 TMS Parameters

To identify the parameters of the whole TMS we excite the plant with the torque signal described in Section 6 and use the SDNN-model constructed in Chapter 5. The torque of inertia of the engine J_1 and its friction are initialized according to the results of Section 6. These weights remain unchanged during the whole identification process, whereas the weights corresponding to the torque of inertia of the work machine J_2, the spring constant c, the damping d and the work machine friction are trained. The work machine friction is modeled by three neurons with tanh functions in the 11th layer, see Fig. 5. The six weights of this nonlinear subpart are initialized

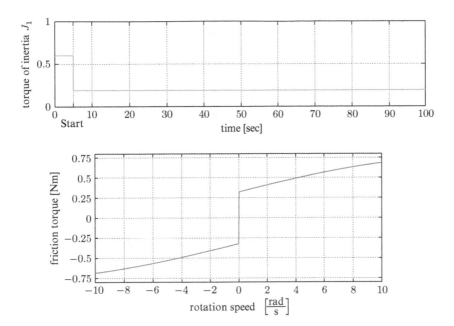

Fig. 7 Identification of the engine parameters – Torque of inertia and friction curve identified by the nonlinear subpart in the 5th layer

Table 1 Initial values and final values of the torque of inertia of the engine J_1

	J_1 [kg m^2]	E (Mean value)
Initial value	0. 6	
Result (mean value)	0.1912	$5.629 \cdot 10^{-2}$
Initial value	0.01	
Result (mean value)	0.1912	$4.387 \cdot 10^{-2}$

randomly between –0.5 and 0.5. The upper panel of Fig. 8 displays the outputs of the TMS and the SDNN-model during the identification process for the first set of initial values of Table 2. The lower panel of this figure shows only 5 s for a detailed view. The identification process starts after 5 s. The sampling time is set to $T = 1$ ms. The quasi-online calculated cost function $E(\underline{w}_k)$ Eq. 4 with $Q = 25000$ is minimized by the LM optimization algorithm Eq. 11 and is depicted in the middle panel of Fig. 8. Due to the quasi-online approach the cost function value increases until $t = 25$ s, until the training data window is completely filled up. The results in Table 2 are mean values of the last 25000 optimization steps.

Figure 9 displays the developing of the damping, the spring constant and the torque of inertia during the identification process for the first initialization of Table 2. Figure 10 shows the characteristic curve of the work machine friction

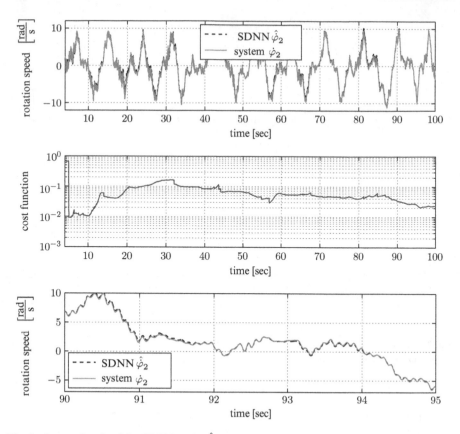

Fig. 8 Output signals of the SDNN model $\hat{\dot{\varphi}}_2$ and the real TMS $\dot{\varphi}_2$ with resulting cost function

Table 2 Initial values and final values for the identification of the TMS

	J_2 [kgm^2]	$d\left[\dfrac{\text{Nms}}{\text{rad}}\right]$	$c\left[\dfrac{\text{Nm}}{\text{rad}}\right]$	Error (mean value)
Initial value	0.7	0.4	100	
Result (mean value)	0.3838	0.2406	477.2	$3.792 \cdot 10^{-2}$
Initial value	0.1	0.7	800	
Result (mean value)	0.3881	0.0919	477.6	$6.400 \cdot 10^{-2}$

identified by the SDNN after 100 s. In addition to that Table 2 shows the results of a second identification run with another initialization. The resulting torque of inertia and spring constant are almost equal. Only the damping shows different final values. The higher mean error (compared to the first identification) implies that the second optimization process ended in a local minimum.

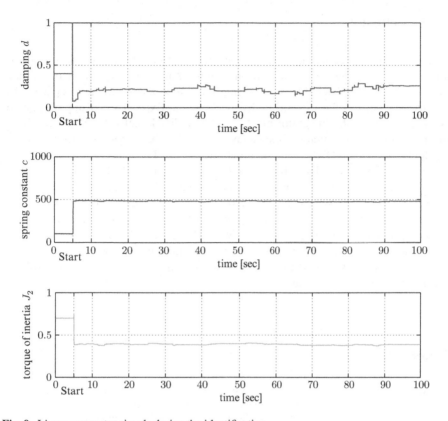

Fig. 9 Linear parameter signals during the identification

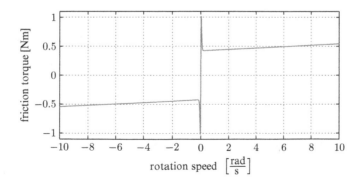

Fig. 10 Friction curve identified by the nonlinear subpart in the 11th layer

7   Conclusion

The system identification approach in this article is based on a given continuous signal flow chart. Before discretizing the signal flow chart, integrator blocks are inserted in all feedback loops. The Implicit Euler approximation replaces all integrator blocks in the forward paths, whereas the Explicit Euler approximation is used to replace the integrator blocks in the feedback paths. Since the resulting discrete flow chart contains time delays in all feedback connections it can be implemented as a SDNN without algebraic loops. Physical parameters of the system are represented by certain weights of the SDNN. Nonlinear subparts of the network model nonlinearities of the system. The suggested identification approach is tested with a nonlinear TMS. The results verify that the suggested approach enables us to identify the torques of inertia, the spring constant, the damping and the two friction curves of the TMS. Problems may occur if the optimization algorithm gets stuck in a local minimum. In this case the identification process has to be restarted with a different set of initial values. With the help of the mean error the success of the identification process can be validated.

References

1. M.J. Brache, Identification of dynamic systems using previous knowledge, Diploma Theses, Lehrstuhl für Elektrische Antriebssysteme, Technische Universität München, 2008
2. C. Endisch, C. Hackl, D. Schröder, Optimal brain surgeon for general dynamic neural networks, in *Lecture Notes in Artificial Intelligence (LNAI) 4874* (Springer, Berlin, 2007), pp. 15–28
3. C. Endisch, C. Hackl, D. Schröder, System identification with general dynamic neural networks and network pruning. Int. J. Computat. Intel. **4**(3), 187–195 (2008)
4. C. Endisch, P. Stolze, C. Hackl, D. Schröder, Comments on backpropagation algorithms for a broad class of dynamic networks. IEEE Trans. Neural Network **20**(3), 540–541 (2009)
5. M. Hagan, B.M. Mohammed, Training feedforward networks with the marquardt algorithm. IEEE Trans. Neural Networks **5**(6), 989–993 (1994)
6. C. Hintz, B. Angerer, D. Schrder, Online identification of mechatronic system with structured recurrent neural networks. *Proceedings of the IEEE-ISIE 2002*, L'Aquila, Italy, pp. 288–293
7. C. Hintz, Identifikation nichtlinearer mechatronischer Systeme mit strukturierten rekurrenten Netzen. Dissertation, Lehrstuhl fr Elektrische Antriebssysteme, Technische Universität München, 2003
8. O. De Jesus, M. Hagan. Backpropagation Algorithms Through time for a general class of recurrent network. IEEE International Joint Conference Neural Network, Washington, 2001, pp. 2638–2643
9. O. De Jesus, M. Hagan, Forward perturbation algorithm for a general class of recurrent network. IEEE International Joint Conference Neural Network, Washington, 2001, pp. 2626–2631
10. O. De Jesus, Training general dynamic neural networks. Ph.D. dissertation, Oklahoma State University, Stillwater, OK, 2002
11. O. De Jesus, M. Hagan backpropagation algorithms for a broad class of dynamic networks. IEEE Trans Neural Networks **18**(1), 14–27 (2007)

13. O. Nelles, in *Nonlinear System Identification* (Springer, Berlin, 2001).
14. D. Schröder, in *Elektrische Antriebe – Regelung von Antriebssystemen*, 2nd edn. (Springer, Berlin, 2001)
15. P.J. Werbos, Backpropagation through time: what it is and how to do it. Proc. IEEE **78**(10), 1550–1560 (1990)
16. R.J. Williams, D. Zipser, A learning algorithm for continually running fully recurrent neural networks. Neural Comput. **1**, 270–280 (1989)

Chapter 17
Adaptive and Neural Learning for Biped Robot Actuator Control

Pavan K. Vempaty, Ka C. Cheok, Robert N.K. Loh,
and Micho Radovnikovich

Abstract Many robotics problems do not take the dynamics of the actuators into account in the formulation of the control solutions. The fallacy is in assuming that forces/torques can be instantaneously and accurately generated. In practice, actuator dynamics may be unknown. This paper presents a Model Reference Adaptive Controller (MRAC) for the actuators of a biped robot that mimics a human walking motion. The MRAC self-adjusts so that the actuators produce the desired torques. Lyapunov stability criterion and a rate of convergence analysis is provided. The control scheme for the biped robot is simulated on a sagittal plane to verify the MRAC scheme for the actuators. Next, the paper shows how a neural network (NN) can learn to generate its own walking gaits using successful runs from the adaptive control scheme. In this case, the NN learns to estimate and anticipate the reference commands for the gaits.

1 Introduction

Biped walking dynamics is highly non-linear, has many degrees of freedom and requires developing complicated model to describe its walking behavior. Many novel approaches have emerged in the field of biped walking to address this complicated control mechanism. Existing biped walking methods [3, 6, 7, 10] give precise stability control for walking bipeds. However these methods require highly precise biped walking dynamics. In recent years, biped walking through imitation has been a promising approach, since it avoids developing complex kinematics of the human walking trajectory and gives the biped a human like

P.K. Vempaty (✉)
Department of Electrical and Computer Engineering, Oakland University, Rochester, Michigan 48309, USA
e-mail: pvempaty@oakland.edu

S.-I. Ao et al. (eds.), *Machine Learning and Systems Engineering*,
Lecture Notes in Electrical Engineering 68,
DOI 10.1007/978-90-481-9419-3_17, © Springer Science+Business Media B.V. 2010

walking behavior. These methods combine the conventional control schemes to develop the walking mechanism for the bipeds. Examples include, imitation based on intelligent control methods like genetic algorithm [3, 14], fuzzy logic [15], neural network approach [8], and other methods such as adaptation of biped locomotion [10], learning to walk through imitation [8] and reinforcement learning [9, 12, 15]. But these methods cannot adapt their behavior to the changes in the dynamics of the process and the character of the disturbances [7]. Therefore, adaptive control approaches [1, 2, 5, 11] are useful.

To address the problem of poorly known actuator dynamic characteristics and unpredictable variations of a biped system, we propose a Lyapunov based model reference adaptive control system (MRAC) method for the biped walking control. The controlled plant (biped actuator) adapts itself to the reference model (desired dynamics for the actuators). Lyapunov's stability criterion and convergence analysis are shown for parameter tuning. Through this scheme, a robot can learn its behavior through its reference models. The paper further describes a NN scheme for estimating and anticipating reference commands needed for walking gaits. Data from successful adaptive control biped runs were used to train the NN.

2　Problem Description

2.1　Objective

Consider the objective of controlling a biped robot so that it imitates the movement of a person. Figure 1 shows the basic idea where the human movement is represented by $\mathbf{y}_d$ and the biped movement by $\mathbf{y}$. The biped motion is determined by the actuators which are controlled by the inputs $\mathbf{u}_a$. The overall objective is to find the adaptive $\mathbf{u}_a$ such that $\mathbf{y} \rightarrow \mathbf{y}_d$.

The actuator dynamics have uncertainties including non-linearities, unknown parameter values and delays, which have not been widely addressed. Figure 2 shows the adaptive actuator objective where the actuator output moment $\mathbf{M}$ is made to follow a required $\mathbf{M}_d$, which will be computed from the desired requirement that $\mathbf{y}$ tracks $\mathbf{y}_d$. Figure 2 also shows a NN scheme for generating an estimate $\hat{\mathbf{y}}_d$ of

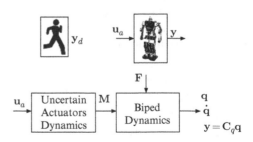

Fig. 1 Human movements, biped robot, its actuators

Fig. 2 MRAC scheme
for the biped walker

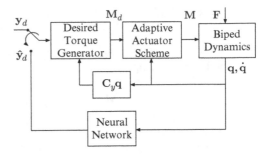

the reference command signal. This paper deals with the formulation and simulation aspects of the MRAC actuator and neural network learning schemes.

2.2 Biped Dynamics

Equations for describing the dynamics of a biped robot were introduced in [4], and can be summarized as follows.

$$\mathbf{A}_q(\mathbf{q})\ddot{\mathbf{q}} = B(\mathbf{q}, \dot{\mathbf{q}}, \mathbf{M}, \mathbf{F}) \tag{1}$$

where $\mathbf{q}$ is the generalized coordinates of the robot, $\dot{\mathbf{q}}$ and $\ddot{\mathbf{q}}$ are the first and second derivatives, $\mathbf{M}$ the moments/torques applied to the joints in the robot and $\mathbf{F}$ the reaction forces at the contact of the robot's feet and ground surface.

2.3 Uncertain Actuator Dynamics

The literature often assumes that $\mathbf{M}$ can be readily generated without considering the dynamics of the actuators. For example, if a set of desired torques are calculated as $\mathbf{M}_d$, then it would assume that $\mathbf{M} = \mathbf{M}_d$ and applied directly as inputs to the robot. However, this is not a valid assumption since in practice the moments $\mathbf{M}$ will be generated by actuators which normally have unknown parameters, time delays and non-linearities. The moments $\mathbf{M}$ can be modelled as the states of

$$\begin{aligned}
\dot{\mathbf{x}}_a(t) &= \mathbf{A}_a\mathbf{x}_a(t) + \mathbf{B}_a\mathbf{u}_a(t) + \mathbf{d}_a(\mathbf{q}(t), \dot{\mathbf{q}}(t), \tau_{ext}(t)) \\
\mathbf{M} &= \mathbf{C}_a\mathbf{x}_a
\end{aligned} \tag{2}$$

where $\mathbf{u}_a$ are inputs of the actuators, and $\mathbf{d}_a(\mathbf{q}, \dot{\mathbf{q}}, \tau_{ext})$ represents disturbance torques to the actuators due to robot movements. τ_{ext} is an external disturbance torque. We assume that the moments/torques $\mathbf{M}$ can be measured; for example, by measuring the currents in motors or pressure in hydraulics. In pre-tuned actuators, we can assume that $\mathbf{M} = \mathbf{x}_a$, i.e., $\mathbf{C}_a = I$.

2.4   Desired Moments $\mathbf{M}_d$

The desired moments $\mathbf{M}_d$ can be derived as the output of a controller that operates on $\mathbf{y}_d$ and $\mathbf{y}$. For example,

$$\mathbf{M}_d(s) = \mathbf{G}_c(s)\left[\mathbf{y}_d(s) - \mathbf{y}(s)\right] \tag{3}$$

where s is the Laplace variable, $\mathbf{G}_c(s)$ is the controller transfer function. The controller $\mathbf{G}_c$ is designed to generate the desired moments $\mathbf{M}_d$, required for the adaptive actuator scheme by using the information from $\mathbf{y}$ and $\mathbf{y}_d$.

2.5   Adaptive Control Approach

In the presence of actuator unknown parameters and uncertainties in (2). We would like the actuator output $\mathbf{M}$ to follow $\mathbf{M}_d$. Figure 2 shows the adaptation scheme.

3   Solution

3.1   Reference Model for Actuator

To apply the MRAC approach to the actuator (2), a reference model for the actuator is needed as follows. The computed desired moments $\mathbf{M}_d$ (3) will be represented as the states of the reference model

$$
\begin{aligned}
\dot{\mathbf{x}}_m(t) &= \mathbf{A}_m \mathbf{x}_m(t) + \mathbf{B}_m \mathbf{u}_m(t) \\
\mathbf{M}_d(t) &= \mathbf{x}_m(t)
\end{aligned}
\tag{4}
$$

where $\mathbf{A}_m$ and $\mathbf{B}_m$ represent the desired dynamics for the actuator to follow. $\mathbf{u}_m(t)$ represents the command input to the reference model of the actuator and is required for the MRAC.

3.2   Inverse Reference Model

However, we do not know the input $\mathbf{u}_m$. So the approach here is to estimate the unknown $\mathbf{u}_m$ knowing $\mathbf{x}_m$. The unknown $\mathbf{u}_m$ can be represented as the output of an waveform shaping model, i.e.,

$$
\begin{aligned}
\dot{\mathbf{x}}_u &= \mathbf{A}_u \mathbf{x}_u + \mathbf{w}_u \\
\mathbf{u}_m &= \mathbf{C}_u \mathbf{x}_u
\end{aligned}
\tag{5a}
$$

where $\mathbf{A}_u$ and $\mathbf{C}_u$ represent approximate waveform characteristics and $\mathbf{w}_u$ is a sparse and small shaping input [13].

An estimate of the of $\mathbf{u}_m$ can be found using an observer of the form

$$\dot{\hat{\mathbf{x}}}_{mu} = \mathbf{A}_{mu}\hat{\mathbf{x}}_{mu} + \mathbf{K}_{mu}\left[\mathbf{x}_m - \mathbf{C}_{mu}\hat{\mathbf{x}}_{mu}\right]$$
$$\hat{\mathbf{u}}_m = \begin{bmatrix} \mathbf{0} & \mathbf{C}_u \end{bmatrix}\hat{\mathbf{x}}_{mu}$$

(5b)

where, $\mathbf{x}_{mu} = \begin{bmatrix} \mathbf{x}_m \\ \mathbf{x}_u \end{bmatrix}$, $\mathbf{A}_{mu} = \begin{bmatrix} \mathbf{A}_m & \mathbf{B}_m\mathbf{C}_u \\ 0 & \mathbf{A}_u \end{bmatrix}$, $\mathbf{B}_{mu} = \begin{bmatrix} 0 \\ \mathbf{I} \end{bmatrix}$ and $C_{mu} = \begin{bmatrix} I & 0 \end{bmatrix}$; 0 and I are null and identity matrix. $\mathbf{K}_{mu}$ is chosen such that $\mathbf{A}_{mu} - \mathbf{K}_{mu}\,C_{mu}$ has exponentially stable eigenvalues. We refer (5b) as the inverse reference model, $\mathbf{x}_m$ is given and input $\mathbf{u}_m$ is estimated by the inverse model as the output $\hat{\mathbf{u}}_m$.

3.3 MRAC Scheme

3.3.1 Configuration of MRAC Actuator

The adaptive actuator scheme is shown in Fig. 3, where the reference and control models are specified by

$$\dot{\mathbf{x}}_m = \mathbf{A}_m\mathbf{x}_m + \mathbf{B}_m\hat{\mathbf{u}}_m$$
$$\mathbf{u}_a = \mathbf{L}\mathbf{x}_a + \mathbf{N}\hat{\mathbf{u}}_m$$

(6)

The adaptation algorithm in the MRAC will adjust the gains $\mathbf{L}$ and $\mathbf{N}$ based on Lyapunov stability criteria as illustrated in Fig. 3.

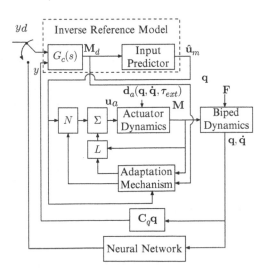

Fig. 3 The MRAC with the input predictor

3.3.2   Error Dynamics

The error $\mathbf{e} = \mathbf{x}_m - \mathbf{x}_a$ between the actuator torque $\mathbf{x}_a = \mathbf{M}$ and the desired torque $\mathbf{x}_m = \mathbf{M}_d$ behave according to

$$\dot{\mathbf{e}} = \mathbf{A}_m\mathbf{e} + [\mathbf{A}_m - \mathbf{A}_a - \mathbf{B}_a\mathbf{L}]\mathbf{x}_a \\ + [\mathbf{B}_m - \mathbf{B}_a\mathbf{N}]\hat{\mathbf{u}}_m - \mathbf{d}_a(\mathbf{q}, \dot{\mathbf{q}}, \boldsymbol{\tau}_{ext}) \tag{7}$$

3.3.3   Lyapunov Stability Analysis

Define a candidate for a Lyapunov function as

$$v = \mathbf{e}^T\mathbf{P}\mathbf{e} \\ + trace\left[(\mathbf{A}_m - \mathbf{A}_a - \mathbf{B}_a\mathbf{L})^T\mathbf{Q}(\mathbf{A}_m - \mathbf{A}_a - \mathbf{B}_a\mathbf{L})\right] \\ + trace\left[(\mathbf{B}_m - \mathbf{B}_a\mathbf{N})^T\mathbf{R}(\mathbf{B}_m - \mathbf{B}_a\mathbf{N})\right] \tag{8}$$

where $\mathbf{P} = \mathbf{P}^T > 0, \mathbf{Q} = \mathbf{Q}^T > 0$ and $\mathbf{R} = \mathbf{R}^T > 0$ are positive definite matrices. Then

$$\dot{v} = \dot{\mathbf{e}}^T\mathbf{P}\mathbf{e} + \mathbf{e}^T\mathbf{P}\dot{\mathbf{e}} \tag{9}$$

Based on the analysis of $\dot{v}$, we choose

$$\mathbf{B}_a\dot{\mathbf{L}} = \mathbf{Q}^{-1}\mathbf{P}\mathbf{e}\mathbf{x}_a^T \\ \mathbf{B}_a\dot{\mathbf{N}} = -\mathbf{R}^{-1}\mathbf{P}\mathbf{e}\hat{\mathbf{u}}_m^T \tag{10}$$

so that $\dot{v} = \mathbf{e}^T[\mathbf{P}\mathbf{A}_m + \mathbf{A}_m{}^T\mathbf{P}]\mathbf{e} - 2\mathbf{e}^T\mathbf{A}_m{}^T\mathbf{d}_a(\mathbf{q}, \dot{\mathbf{q}}, \boldsymbol{\tau}_{ext})$. We next choose an $\mathbf{S} = \mathbf{S}^T > 0$ and solve $\mathbf{P}$ from

$$\mathbf{P}\mathbf{A}_m + \mathbf{A}_m^T\mathbf{P} = -\mathbf{S} \tag{11}$$

We arrive at $\dot{v} = -\mathbf{e}^T\mathbf{S}\mathbf{e} + 2\mathbf{e}^T\mathbf{A}_m{}^T\mathbf{d}_a(\mathbf{q}, \dot{\mathbf{q}}, \boldsymbol{\tau}_{ext})$, where $\dot{v}$ is negative under the assumption $\mathbf{e}^T\mathbf{S}\mathbf{e} > 2\mathbf{e}^T\mathbf{A}_m{}^T\mathbf{d}_a(\mathbf{q}, \dot{\mathbf{q}}, \boldsymbol{\tau}_{ext})$, which implies that the magnitude of error should be larger than the disturbance. In practice this means that the system should be persistently excited. Hence we conclude that the overall dynamic system comprising of (4) and (7) has a candidate function that satisfies the Lyapunov stability criterion.

4   MRAC for Walking Biped Actuators

4.1   Dynamics of Walking Biped

The bipedal model has five links with four pin joints as shown in Fig. 4. One link represents the upper body and two links are for each lower limb. The biped has two hip joints, two knee joints and two ankles at the tips of the lower limbs. There is an actuator located at each joint and all of the joints are considered rotating only in the sagittal plane. The system contains five links and seven degrees of freedom, selected according to Fig. 4a and given by

$$\mathbf{q} = [x_0, y_0, \alpha, \beta_L, \beta_R, \gamma_L, \gamma_R]^T \tag{12}$$

The coordinates (x_0, y_0) fix the position of the center of mass of the torso, and the rest of the coordinates describe the joint angles. The link lengths are denoted by (h_0, h_1, h_2) and masses by (m_0, m_1, m_2). The centers of mass of the links are located at the distances (r_0, r_1, r_2) from the corresponding joints.

The model is actuated with four moments; two of them acting between the torso and both thighs, and two at the knee joints, as illustrated in Fig. 4b.

$$\mathbf{M} = [M_{L1}, \; M_{R1}, \; M_{L2}, \; M_{R2}]^T \tag{13}$$

The walking surface is modelled using external forces that affect the leg tips, as shown in Fig. 4b.

$$\mathbf{F} = [F_{Lx}, \; F_{Ly}, \; F_{Rx}, \; F_{Ry}]^T \tag{14}$$

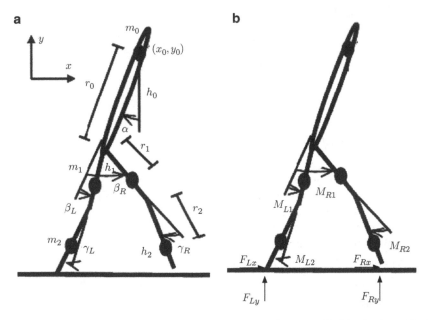

Fig. 4 (a) Biped robot with the corresponding seven coordinates **q**. (b) Biped robot with the moments **M** and reaction forces **F**

When the leg should touch the ground, the corresponding forces are switched on to support the leg. As the leg rises, the forces are zeroed.

Using Lagrangian mechanics, the dynamic equations for the biped system can be derived as shown in (1) where $\mathbf{A(q)} \in \mathfrak{R}^{7\times7}$ is the inertia matrix and $\mathbf{bq}, \dot{\mathbf{q}}, \mathbf{M}, \mathbf{F} \in \mathfrak{R}^{7\times1}$ is a vector containing the right hand sides of the seven partial differential equations. The closed form formulas for both $\mathbf{A}$ and $\mathbf{b}$ are presented in [4].

4.2   Computation of Desired Moments

The desired human movement angles will be measured and represented by

$$\mathbf{y}_d = [\alpha_m \quad (\beta_L - \beta_R) \quad \gamma_{Lm} \quad \gamma_{Rm}]^T \tag{15}$$

An output $\mathbf{y}$ is constructed from the biped feedback information $\mathbf{q}$ as

$$\mathbf{y} = [\alpha \quad (\beta_L - \beta_R) \quad \gamma_L \quad \gamma_R]^T = C_y \mathbf{q} \tag{16}$$

The desired torque is derived as the output of a controller that operates on the error between $\mathbf{y}_d$ and $\mathbf{y}$. That is

$$\mathbf{M}_d(s) = \mathbf{G}_C(s)(\mathbf{y}_d(s) - \mathbf{y}(s))$$
$$\mathbf{M}_d = [M_{d1} \quad M_{d2} \quad M_{d3} \quad M_{d4}]^T \tag{17}$$

4.3   Dynamics of Actuators

We note that the actuator states $\mathbf{x}_a = \mathbf{M}$ are the torques that will drive the biped robot. The goal is to find $\mathbf{u}_a$ such that $\mathbf{M} \to \mathbf{M}_d$. We assume that DC motors are used as actuators. It follows that (2) can be decoupled into individual motors representing the first-order stator-rotor dynamics (a_{ai}, b_{ai}) that generates output torque (x_{ai}) while subject to disturbance (d_{ai}), that is

$$\mathbf{x}_a = [x_{a1} \quad x_{a2} \quad x_{a3} \quad x_{a4}]^T, \quad \mathbf{u}_a = [u_{a1} \quad u_{a2} \quad u_{a3} \quad u_{a4}]^T$$
$$\mathbf{A}_a = diag\{a_{a1} \quad a_{a2} \quad a_{a3} \quad a_{a4}\}, \quad \mathbf{B}_a = diag\{b_{a1} \quad b_{a2} \quad b_{a3} \quad b_{a4}\}$$
$$\mathbf{d}_a(\mathbf{q}, \dot{\mathbf{q}}, \tau_{ext}) = [d_{a1} \quad d_{a2} \quad d_{a3} \quad d_{a4}]^T$$

where $\mathbf{A}_a$, $\mathbf{B}_a$ and $\mathbf{d}_a$ are the uncertain parameter values.

4.4 Configuration of MRAC Actuator

The control specified by

$$\mathbf{u}_a = \mathbf{L}\mathbf{x}_a + \mathbf{N}\hat{\mathbf{u}}_m$$
$$\mathbf{L} = diag\{l_1,\ l_2,\ l_3,\ l_4\}, \tag{18}$$
$$\mathbf{N} = diag\{n_1,\ n_2,\ n_3,\ n_4\}$$

It follows from the Lyapunov design that the gains are adjusted according to

$$l_i = \frac{1}{b_{ai}q_i}p_i(M_{di} - x_{ai})x_{ai}, \quad n_i = -\frac{1}{b_{ai}r_i}p_i(M_{di} - x_{ai})\hat{u}_{mi} \tag{19}$$

4.5 Convergence Analysis of MRAC

The convergence of the MRAC depends on the following dynamics

$$\dot{e}_i = a_{mi}e_i + b_{ai}x_{ai}\tilde{l}_i - b_{ai}\hat{u}_{mi}\tilde{n}_i - d_{ai}(\mathbf{q}, \dot{\mathbf{q}}, \tau_{ext})$$
$$\dot{\tilde{l}}_i = -\frac{p_i x_{ai}}{b_{ai}q_i}e_i, \qquad \tilde{n}_i = \frac{p_i \hat{u}_{mi}}{b_{ai}r_i}e_i \tag{20}$$

where $e_i = x_m - x_{ai}$, $\tilde{l}_i = l_i^* - \hat{l}_i$, $\tilde{n}_i = n_i^* - \hat{n}_i$, and l_i^* and n_i^* are the true values of parameters l_i and n_i, whose dynamics are characterized by a third order polynomial with coefficients given by $-a_{mi}, \frac{p_i x_{ai}}{q_i} + \frac{p_i \hat{u}_{mi}}{r_i}$ and 0. We would assign the first two coefficients as the convergence characteristics. The first coefficient in turn defines a_{mi}. We next choose a large value for s_i to ensure that the Lyapunov rate is satisfied,

$$\dot{v}_i = -s_i e_i^2 - 2a_{mi}e_i d_{ai}(\mathbf{q}, \dot{\mathbf{q}}, \tau_{ext}) < 0 \tag{21}$$

We can now compute p_i, q_i, and r_i for the algebraic Lyapunov function as

$$p_i = -\frac{s_i}{2a_{mi}}, \quad q_i = \frac{2p_i x_a^2}{c_2}, \quad r_i = \frac{2p_i u_m^2}{c_2}$$

The analysis here was used to assist tuning of the parameters.

4.6 Neural Network Learning

The purpose of the NN is to eventually replace the necessity of human input ($\mathbf{y}_d$). After the MRAC converges, the NN is trained to generate an estimate $\hat{\mathbf{y}}_d$ of $\mathbf{y}_d$ using y and $\dot{\mathbf{y}}$ as inputs. The basis for the mapping comes from the closed-loop system formed by (1), (3) and The closed-loop characteristic can be expressed in a generic form by $f(\mathbf{q}, \dot{\mathbf{q}}, \ddot{\mathbf{q}}, \mathbf{x}_a, \mathbf{y}_d, \mathbf{F}) = 0$. The NN is an approximate mapping of $f_1(\mathbf{y}, \dot{\mathbf{y}}, \mathbf{y}_d)$ $= 0$, where f_1 is a reduced version of f.

5   Simulation Results

The biped movements from (16) are measured and used to generate the desired torques from (17), (18) and (19) were implemented as the MRAC scheme. Three sets of simulation runs are shown below. The first simulation does not include the disturbance torque $\mathbf{d}_a(\mathbf{q}, \dot{\mathbf{q}}, \boldsymbol{\tau}_{ext})$. We introduced the disturbance in the second simulation [13]. The third simulation shows the estimation performance of the NN in predicting the reference command.

5.1   First Simulation (Without Disturbance)

5.1.1   Results of the Biped Moments M and Desired Biped Moments $\mathbf{M}_d$

Figure 5 shows $\mathbf{M}_{R1}$ converging to $\mathbf{M}_{d2}$ for one step cycle. Similarly, it can be noted that the rest of the biped moments converge to the reference model's behavior. Therefore, this gives the biped walker a human-like gait as described by $\mathbf{M}_d$.

5.1.2   Results of the Biped Walking Simulation

Figure 6 plots the stable torso height y_0 (see Fig. 4) of the biped walking as the result of the adapted torques $\mathbf{M}$.

5.2   Second Simulation (Disturbance)

5.2.1   Results of the Biped Walking with Torque Disturbance

Equation (2) includes disturbance torques $\mathbf{d}_a(\mathbf{q}, \dot{\mathbf{q}}, \boldsymbol{\tau}_{ext})$ which represents torque feedback and other external moments. This causes the dynamics of the actuator to vary. Figure 7 shows the biped torso height y_0, recovering from the impact due to

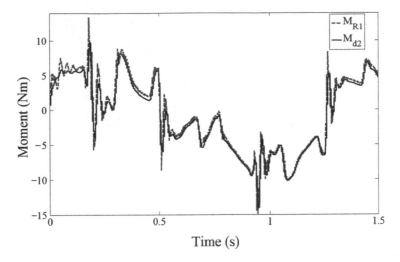

Fig. 5 M_{R1} tracking M_{d2}

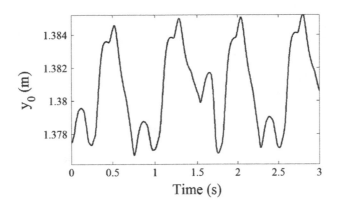

Fig. 6 Height of the biped torso y_0

the external disturbance to the biped walker, introduced by four impulses at 0.05, 0.3, 1.34, and 2.2 s with a magnitude of 5 and 6 Nm.

5.3   Third Simulation (Neural Network Estimation)

Figure 8 shows an 8-15-1 neural network model with **y** & **ẏ** as the inputs and estimate $\hat{y}_{d3}$ of the reference command signal $\mathbf{y}_{d3}$ as the output.

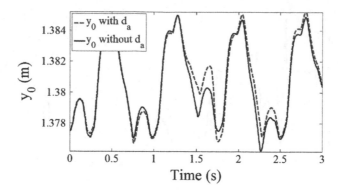

Fig. 7 Height of the biped torso without disturbance and with disturbance torques $d_a(\mathbf{q}, \dot{\mathbf{q}}, \tau_{ext})$, applied at $t = 0.05$, $t = 0.3$, $t = 1.34$, and $t = 2.2$s

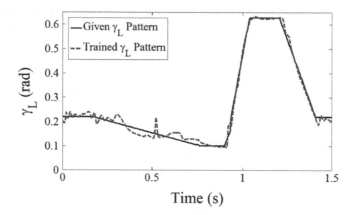

Fig. 8 Tracking performance of trained γ_L pattern to the reference pattern

6 Conclusions

In this paper, we presented an MRAC technique to ensure that the actuators reliably produce desired torques necessary for a walking robot. An observer was used to predict an anticipated state of the desired torque, thus causing the adaptive actuators to anticipate motion torques. We provided a proof to show that the MRAC scheme results in a stable system in the sense of Lyapunov when errors between the desired and actuator torques are significant. Also, the convergence analysis for tuning p, q, and r is provided. Simulation results verify that the system is robust when tested with various parameters and unknown coefficients. This paper also investigates and presents a neural network estimation of the human gait reference inputs. In the future, we plan to implement the MRAC and neural network schemes to control a real-time biped walker with transport delay.

References

1. E. Bobasu, D. Popescu, Adaptive nonlinear control algorithms of robotic manipulators. *Proceedings of the 7th WSEAS International Conference on Automation & Information*, 2006, pp. 83–88
2. R. Chalodnan, D.B. Grimes, R.P.N. Rao, Learning to walk through imitation. *Proceedings of the 20th International Joint Conference on Artificial Intelligence*, 2007, pp. 2084–2090
3. S. Ha, Y. Han, H. Hahn, Adaptive gait pattern generation of biped robot based on humans gait pattern analysis. Proc. Int. J. Mech. Sys. Sci. Engr. 1(2), 80–85 (2008)
4. O. Haavisto, H. Hyotyniemi, Simulation tool of biped walking robot model. Master's thesis, Helsinki University of Technology, Espoo, Finland, 2004
5. J. Hu, J. Pratt, C.-M. Chew, H. Herr, G. Pratt, Adaptive virtual model control of a bipedal walking robot. Int. J. Artif. Intel. Tools 8(3), 337–348(1999)
6. S. Kajita, F. Kanehiro, K. Kaneko, K. Fijiwara, K. Yokoi, H. Hirukawa, A realtime pattern generator for biped walking. *Proceedings of the IEEE International Conference on Robotics & Automation*, vol. 1, 2002, pp. 31–37
7. S. Kajita, F. Kanehiro, K. Kaneko, K. Fujiwara, K. Harada, K. Yokoi, H. Hirukawa, Biped walking pattern generation by using preview control of zero-moment point. *Proceedings of the IEEE International Conference on Robotics & Automation*, vol. 2, 2003, pp. 1620–1626
8. P. Manoonpong, T. Geng, T. Kulvicius, B. Porr, F. Worgotter, Adaptive, fast walking in a biped robot under neuronal control and learning. PLos Computat. Biol. 3(7), 1305–1320(2007)
9. J. Morimoto, G. Cheng, C.-G. Atkeson,G. Zeglen, A simple reinforcement learning algorithm for biped walking. *Proceedings of the IEEE Conference on Robotics and Automation*, vol. 3, 2004, pp. 3030–3035
10. J. Nakanishi, J. Morimoto, G. Endo, G. Cheng, S. Schaal, M. Kawato, Learning from demonstration and adaptation of biped locomotion with dynamical movement primitives. *Proceedings of the 4th IEEE/RAS International Conference on Humanoid Robotics*, vol. 2, 2004, pp. 925–940
11. J. Pratt, C.-M. Chew, A. Torres, P. Dilworth, G. Pratt, Virtual model control: an intuitive approach for bipedal locomotion. Int. J. Robot. Res. 20(2), 129–143 (2001)
12. A. Takanobu, S. Jun, K. Shigenobu, Reinforcement learning for biped walking using human demonstration towards the human like walking of biped robot. Chino Shisutemu Shinpojiumu Shiryo, 32, 393–398 (2005)
13. P.K. Vempaty, K.C. Cheok, R.N.K. Loh, Model reference adaptive control for actuators of a biped robot locomotion. *Proceedings of the World Congress on Engineering and Computer Science*, San Francisco, USA, vol. 2, 2009, pp. 983–988
14. K. Wolff, P. Nordin, Evolutionary learning from first principles of biped walking on a simulated humanoid robot. *Proceedings of the Business and Industry Symposium of the Simulation Technologies Conference*, 2003, pp. 31–36
15. C. Zhou, Q. Meng, Dynamic balance of a biped robot using fuzzy reinforcement learning agents. Fuzzy Sets Syst. Arch. 134, 169–187 (2003)

Chapter 18
Modeling, Simulation, and Analysis
for Battery Electric Vehicles

Wei Zhan, Make McDermott, Behbood Zoghi, and Muhammad Hasan

Abstract Steady state and dynamic vehicle models are derived for analyses of requirements on motors and batteries for battery electric vehicles. Vehicle level performance requirements such as driving range, maximum cruise speed, maximum gradeability, and maximum acceleration are used to analyze the requirements on motors and batteries including motor power and torque, battery weight and specific energy. MATLAB simulation tools are developed to allow validation of these vehicle level performance requirements for a given set of motor/battery.

1 Introduction

Even though the internal combustion engine (ICE) is currently still the dominant power source for automobiles, the cost of fuel and more stringent government regulations on greenhouse gas emissions have led to more active interest in hybrid and electric vehicles. Hybrid vehicles have better gas mileage than ICE-powered vehicles. But they still have greenhouse gas emissions and the dual power sources make them more complex and expensive. Battery electric vehicles (BEV) have zero emission with a single power source that makes their design, control, and maintenance relatively simple compared to hybrid vehicles. In addition, the wide use of BEVs will reduce dependence on foreign oil, lower the cost per mile of driving, and can potentially reduce the cost of electricity by using the vehicle-to-grid power capability.

W. Zhan (✉)
Department of Engineering Technology and Industrial Distribution, Texas A&M University, College Station, TX 77843-3367, USA
e-mail: zhan@entc.tamu.edu

S.-I. Ao et al. (eds.), *Machine Learning and Systems Engineering*, Lecture Notes in Electrical Engineering 68, DOI 10.1007/978-90-481-9419-3_18, © Springer Science+Business Media B.V. 2010

The main limitations on BEVs lie in the battery technology. The low energy and power densities of batteries compared to hydrocarbon fuels significantly reduce the driving range and other vehicle level performances. Initial cost is another factor that slows the commercialization of electric vehicles. However, the latest developments in battery technologies are making BEVs more and more attractive. From lead-acid batteries to nickel metal-hydride (NiMH) batteries, lithium-ion cell technology [1], and the latest nano-technology based batteries [2], the energy and power densities have improved drastically.

The U.S. government is investing heavily to support battery technologies and infrastructure for electric vehicles. It has set a target of one million electric vehicles on U.S. roads by 2012. Tax credits up to $7,500 for buyers of plug-in electric vehicles are offered by the U.S. government. The private sector is also investing billions of dollars in electric vehicles. GM and Ford are both planning to roll out plug-in electric vehicles in 2010. All these developments point to a trend toward electric vehicles in the auto industry.

Not only are an increasing number of new BEVs being manufactured, there is also significant interest in converting existing ICE-powered vehicles to electric power. A Google search for "conversion to EV" results in millions of websites and books, many of them providing Do it Yourself kits with focus on removal and addition of components [3]. There is increasing interest in academia in the development of BEVs; for example, an undergraduate senior design project developed an electric vehicle conversion [4]. However, many of these efforts lack consideration of detailed system design requirements. Most of the components for conversions are selected to provide similar output to that of the ICE or by using the vehicle weight to determine the energy and power requirements. The conversion to electric propulsion is a complex process and requires analysis that can be very different from that of an ICE-powered vehicle [5]. Many performance objectives impose conflicting demands. If not designed carefully, the resulting electric vehicle can have many problems such as driving range shorter than expected, battery and motor lacking enough power for desired acceleration, and safety-related design problems.

In this paper, first principle models are derived for electric vehicles and used to establish quantitative design requirements. Software tools are developed in MATLAB [6, 7] to allow users to quickly determine expected vehicle level performances for a given set of motors and batteries. They can also be used to conduct trade-off studies for many design parameters.

2 Steady State Analysis

One of the system level requirements for BEV is the driving range at constant speed. This requirement can be used to derive motor power and battery power and energy requirements. Since the vehicle is assumed to be moving at a constant speed, steady state analysis can be use to study this problem.

Fig. 1 Road load force
components

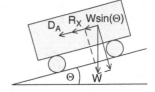

The energy required to move a vehicle is determined by the distance it travels
and the force it has to overcome. The road load force the vehicle must overcome to
move the given distance has three components [8, 9], as illustrated in Fig. 1:

- The component of the gravity force in the direction of travel, if it is an inclined
  path (Wsin(θ))
- The aerodynamic drag (D_A)
- The rolling resistance (R_x)

2.1   Projected Gravity Force

The gravity force is decomposed into two components, one in the direction of travel
and the other in the direction perpendicular to the surface. In order to move the
vehicle up the inclined surface, the vehicle must overcome the gravity force
component in the direction of travel. This is given by

$$W_x = W sin(\theta) \tag{1}$$

where W is the gravity force, θ is the angle of the inclined surface, and W_x is the
component of the gravity force in the direction of travel.

2.2   Aerodynamic Drag

The drag is a function of speed for any given vehicle. At low speed the drag force is
negligible. At high speed, the drag becomes a significant factor. For simplicity, a
semi-empirical model is used here [9]

$$D_A = \frac{1}{2}\rho V^2 C_D A \tag{2}$$

where V is the vehicle speed (ft/s), A is the frontal area of the vehicle (ft^2), C_D is the
aerodynamic drag coefficient, D_A is the aerodynamic drag (lb), ρ is the air density

(lb-s^2/ft^4). A nominal value of $\rho = 0.00236$ (lb-s^2/ft^4) and an estimated value of 0.45 for C_D are used in this paper.

2.3 The Rolling Resistance

Rolling resistance of the tires is a major vehicle resistance force. It is the dominant motion resistance force at low speed (<50 mph). The rolling resistance can be modeled as the *load* on the tires multiplied by the coefficient of rolling resistance f_r:

$$R_x = f_r W \tag{3}$$

The coefficient of rolling resistance is affected by tire construction, tire temperature, vehicle speed, road surface, and tire pressure. For instance, the rolling resistance coefficient changes as the temperature changes. To simplify our analysis, the coefficient of rolling resistance is assumed constant. We use 0.015 as the nominal value.

2.4 Power Required

Based on the above analysis, the power required to drive the vehicle at a given speed V (mph) is given by the total road load forces multiplied by the vehicle speed, i.e.,

$$HP = 0.00267(D_A + R_x + W_x)V \tag{4}$$

where W_x can be calculated using (1), D_A is given by (2), R_x can be calculated using (3) with $f_r = 0.015$, and 0.00267 is the conversion factor to horsepower, HP. To calculate these quantities, we need the following inputs:

- Vehicle speed (mph)
- Vehicle weight (including trailer if there is one) (lb)
- Frontal area of the vehicle, (including trailer if there is one) (ft^2)
- Aerodynamic drag coefficient (including trailer if there is one)
- Coefficient of rolling resistance
- Surface incline angle (degree)

2.5 Energy Required

Energy is power integrated over time. If the total distance traveled is long enough, the initial acceleration and final deceleration have negligible effect on the total energy calculation. Also, since this is a steady state analysis the aerodynamic drag is constant. Noting that $W_x = W\sin(\theta)$ and $V = dx/dt$, it follows that

$$\int W_x \frac{dx}{dt} dt = \int W\sin(\theta)dx = \int W\sin(\theta)dx = W\Delta h$$

where Δh is the change in elevation between the starting and ending points. Thus, the energy required to move a vehicle for a distance of d (miles) at a speed V (mph) with a change in elevation of Δh (miles) is given by

$$\begin{aligned} E(\text{kWh}) &= 0.00267[(D_A + R_x)d + W\Delta h] \times 0.746(\text{kW})/HP \\ &= 0.002[(D_A + R_x)d + W\Delta h]. \end{aligned} \tag{5}$$

Define $\theta*$ as the average slope; i.e.,

$$\sin\theta^* = \Delta h/d$$

and

$$W_x{}^* = W\sin\theta^*$$

Then the trip energy becomes

$$E(\text{kWh}) = 0.002(D_A + R_x + W_x{}^*)d. \tag{6}$$

If speed is not constant, Eqs. (5) and (6) do not apply and the power consumed to overcome drag must be evaluated as an integral.

The energy calculations in (6) can be converted to MJ (1 kWh = 3.6 MJ):

$$E(\text{MJ}) = 0.0072(D_A + R_x + W_x{}^*)d. \tag{7}$$

2.6   Battery Specific Energy

The total energy required for driving a vehicle at constant speed over a given range can be used to derive the requirement on battery specific energy D_{se} (MJ/kg).

Let the battery weight be W_b (lb) and the battery/motor delivery efficiency be η. The total available energy E_T (MJ) from the battery/electric motor is determined by

$$E_T(\text{MJ}) = 0.455(\text{kg/lb}) \times D_{se}(\text{MJ/Kg}) \times Wb(\text{lb}) \times \eta. \tag{8}$$

This amount must be greater than or equal to the total energy required as given in (7), i.e.,

$$0.0072(D_A + Rx + Wx^*)d \le 0.455\,\eta D_{se}W_b. \tag{9}$$

From this, one can solve for D_{se} required to travel a given distance

$$D_{se} \geq 0.0158 \frac{D_A + R_x + W_x^*}{\eta W_b} d. \tag{10}$$

Alternatively, we can calculate the maximum distance d_{max} the vehicle can travel when the specific energy is given

$$d_{max} = 63.29 \frac{\eta W_b}{D_A + R_x + W_x^*} D_{se}. \tag{11}$$

Note that the battery weight W_b is part of the vehicle weight W, which is used in the calculation of R_x and W_x. Denoting the vehicle weight without the battery by W_0 (lb), we have

$$W = W_0 + W_b \tag{12}$$

Combining (9) and (12), we get

$$W_b[63.29\eta D_{se} - d(\sin \theta^* + f_r)] \geq d[D_A + (\sin \theta^* + f_r)W_0]. \tag{13}$$

Since the right hand side is positive, and the battery weight must be positive, we must have

$$d < \frac{63.29\eta D_{se}}{\sin \theta^* + f_r} \tag{14}$$

The right hand side of (14) provides a theoretical upper bound for the distance a vehicle can travel with infinite battery weight (energy) regardless of the speed, vehicle weight, and aerodynamic drag.

Under the assumption that (14) holds, one can determine the weight of a battery in order to travel a distance d

$$W_b \geq d \frac{D_A + (\sin\theta + f_r)W_0}{63.29\eta D_{se} - d(\sin\theta + f_r)} \tag{15}$$

One can conclude that, as long as (14) holds, a sufficiently heavy battery would always enable the vehicle to travel a given distance d. In practice, there are other constraints such as the volumetric limitation for the battery and vehicle load capacity.

Using (12) in (10) yields

$$D_{se} \geq 0.0158 \left[\frac{D_A}{W_b} + (f_r + \sin(\theta^*)) \left(1 + \frac{W_0}{W_b} \right) \right] \frac{d}{\eta} \tag{16}$$

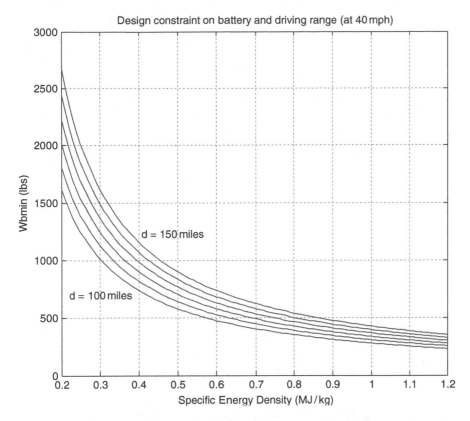

Fig. 2 Design constraint among specific energy, driving range, and battery weight for a given speed

The relationship between minimum battery weight and specific energy is plotted in Fig. 2 for different values of driving range. The efficiency for battery and motor is assumed to be 75% [10] and the speed is 40 mph. For a given driving range, the battery weight and specific energy must be chosen so that the point is above the curve corresponding to the driving range.

Similarly, one can plot the design constraints for any fixed driving range and various speed.

2.7 Maximum Cruise Speed

The maximum cruise speed that the vehicle is required to maintain imposes requirements on the power delivered by the motor and batteries. The example below is for a cruise speed of 80 mph.

Using the following vehicle parameters in the steady state model: vehicle speed = 80 mph; vehicle weight (including motors, but without engine, transmission, battery) = 4,000 lb; frontal area of the vehicle = 34 ft^2; air temperature = 59°F; atmospheric

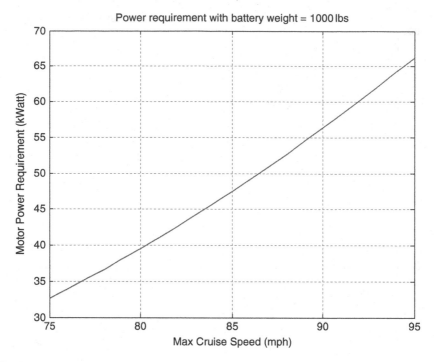

Fig. 3 Motor power as a function of max vehicle speed

pressure = 29.92 in-Hg; aerodynamic drag coefficient = 0.45; coefficient of rolling resistance = 0.015; efficiency = 75%; surface incline angle = 0°; battery weight = 1,000 lbs, one can calculate the requirement on motor power for a maximum cruise speed of 80 mph to be greater than 40 kW. The power requirement on the motor is plotted as a function of maximum cruise speed in Fig. 3. If two motors are used, then each motor should have rated power greater than 20 kW. The corresponding power requirement on the batteries can be derived by dividing the motor power by the efficiency. With a maximum cruise speed of 80 mph, the battery power can be calculated as greater than 53 kW. This can be translated to the current/voltage requirements on the batteries. For example, if the battery voltage output is 300 V, then the current must be greater than 53,000/300 = 177 A.

Figure 3 shows that the power is very sensitive to the maximum cruise speed. One can also find the impact of the battery weight on the power requirement based on maximum cruise speed. Simulation result shows that a 200% increase of battery weight from 500 to 1,500 lbs only results in about 5% change in the battery power requirement. A 53 kW total motor power can cover all realistic battery weights. A similar conclusion holds for battery power required vs. battery weight: the impact of the battery weight on the power required based on maximum speed is insignificant. It can be calculated that a 71 kW battery power is sufficient for any realistic battery weight for an 80 mph cruise speed.

The maximum cruise speed scenario is used to determine the rated power for batteries and motors. In other words, the battery and motor are required to provide the power over a long period of time.

3   Dynamic Analysis

When the vehicle is driven over a short period of time or the time spent in accelerating/decelerating is a significant portion of the total time, steady state analysis is not adequate. Instead, dynamic analysis is needed. The dynamic model developed in this section will be used to derive requirements on motor and battery outputs.

The forces acting on the vehicle are illustrated in Fig. 4, where

- W is the gravity force.
- R_{xf} and R_{xr} are front and rear rolling resistant forces and $R_{xf} + R_{xr} = R_x$.
- W_f and W_r are front and rear normal forces.
- D_A is the aerodynamic drag.
- L_A is the aerodynamic lift.
- F_x is the tractive force (rear wheel drive is assumed).

Newton's Second Law is applied in the direction of the vehicle movement and the direction perpendicular to the road surface.

$$F_x - W \sin \theta - D_A - \frac{W}{g} a - R_x = 0 \tag{17}$$

$$W_f + W_r + L_A - W \cos \theta = 0 \tag{18}$$

where the aerodynamic lift force is given by

$$L_A = \frac{1}{2} \rho V^2 C_L A \tag{19}$$

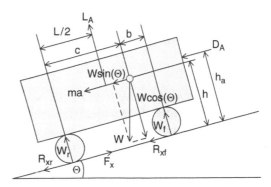

Fig. 4   Vehicle dynamics model

The typical range [6] of values for the aerodynamic lift coefficient is $C_L = 0.0-0.5$. The lift force is applied at the center of the wheel base.

A moment equation about the contact point at the front wheels can also be written.

$$W_r(b + c) - D_A h_a - (Wa/g + W \sin \theta)h + L(L_A/2) - W \cos(\theta)b = 0 \qquad (20)$$

There are two scenarios for the traction force F_x:

- Limited by the motor output (power limited)
- Limited by the road surface friction coefficient (traction limited)

3.1 Power Limited

During acceleration, the motor, transmission, and wheel dynamics have significant impact on the vehicle acceleration.

Newton's Second Law can be applied to the motor to get

$$T_m - T_t = I_m \alpha_m \qquad (21)$$

where T_m is the total motor torque (Nm) (if two motors are used, the torque from each motor needs to be multiplied by 2), T_t is the torque input to the drive train (transmission and differential) (Nm), I_m is the motor total rotational inertia (kg m^2), and α_m is the motor angular acceleration (rad/s^2).

The torque delivered at the output of the drive train is amplified by the gear ratio of the transmission times the gear ratio of the differential, but is decreased by the torque required to accelerate the gears and shafts. If the drive train inertia is characterized by its value on the input side, we have the following equation:

$$T_t - T_w/G = I_t \alpha_m \qquad (22)$$

where G is the gear ratio of the combined transmission and differential, i.e., the ratio between the angular velocities of the motor shaft and the driven wheels, T_w is the torque at the output of the drive train, and I_t is the rotational inertia for the drive train.

Applying Newton's Second Law to the driven wheels, one gets

$$T_w - F_x r = I_w \alpha_w \qquad (23)$$

where F_x is the total tractive force for the two driven wheels, r is the tire radius, I_w is the rotational inertia of the wheels (I_w includes the rotational inertia of everything down stream of the drive train), and is the wheel acceleration (rad/s^2) of the wheels. By the definition of gear ratio, we have

$$G \alpha_w = \alpha_m \qquad (24)$$

Equations (21)–(24) could be combined to solve for the tractive force available at the ground. Recognizing that for power limited operation the vehicle acceleration, a, is the wheel rotational acceleration, α_w, times the tire radius, yields:

$$F_x = \frac{G}{r}T_m - \frac{(I_m + I_t)G^2 + I_w}{r^2}a \tag{25}$$

where r is the radius of the tire.

The effect of mechanical losses can be approximated by including an efficiency factor, η_t, in the first term in the right hand side of (26), as a result we have

$$F_x = \frac{G}{r}\eta_t T_m - \frac{(I_m + I_t)G^2 + I_w}{r^2}a \tag{26}$$

In order to see more clearly the effect of the rotational inertia on the vehicle acceleration, (17) and (26) are combined to get the following

$$(M + M_r)a = \frac{G}{r}\eta_t T_m - W\sin\theta - D_A - R_x \tag{27}$$

where $M = W/g$ and M_r is the equivalent mass of the rotating components, given by

$$M_r = \frac{(I_m + I_t)G^2 + I_w}{r^2} \tag{28}$$

Clearly, the effect of the rotating components on the vehicle acceleration is equivalent to increasing the vehicle mass by the amount in (28).

3.2 Traction Limited

In this case, the tractive force calculated by (26) exceeds the maximum tractive force the road surface and the tire can generate, which is determined by

$$F_x = \mu W_r \tag{29}$$

where μ is the surface friction coefficient. For truck tires, on dry asphalt, μ is approximately 1.

From (17), (20), and (29), we can solve for the maximum tractive force F_{xmax}

$$F_{xmax} = \frac{\mu}{L - \mu h}\left[D_A h_a - (D_A + R_x)h - \frac{L}{2}L_A + Wb\cos\theta\right] \tag{30}$$

In the model, these two cases can be combined by calculating the minimum of the two forces in (26) and (30).

3.3   0–60 mph

The requirement on 0–60 mph time determines the maximum outputs from the
batteries and the motors. A 10 s 0–60 mph time is used as a vehicle acceleration
requirement. During the 10 s while the vehicle accelerates to 60 mph from a
standing start, the maximum power generated by the motor and battery can be
significantly higher than the rated values.

Figure 5 shows the characteristic of a typical AC induction motor. This can be
used together with the vehicle dynamics model to derive the maximum power
requirements on the motor and battery. The combined model can be used
to determine the 0–60 mph time. Based on the simulation result, one can find if a
specific motor/battery combination meets the requirement of 0–60 mph time.
In addition to the parameters used in the derivation of power requirements for
maximum speed, the following parameters are used:

Gear ratio = 6; Motor inertia = 0.36 kg m^2 = 3.18 lb in s^2; Rotational inertia of
drive train = 0.045 kg m^2 = 0.40 lb in s^2; Rotational inertia of each driven wheel =
1.69 kg m^2 = 14.92 lb in s^2; Drive train efficiency = 96%; Wheel base = 126 in;
CG height = 3.3 ft; Height of aerodynamic drag = 3.5 ft; Radius of loaded wheel =
13.82 in; Weight percentage on front wheels = 55%; Motor torque and motor
efficiency curves given in Fig. 5; battery power limit = 140 kW; surface mu = 0.95;

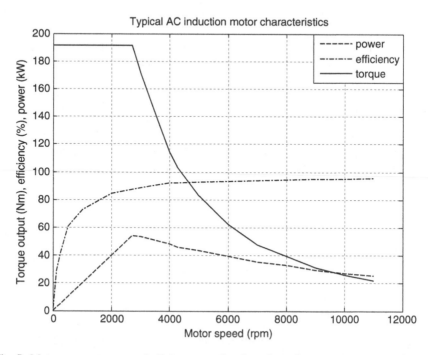

Fig. 5 Motor power, torque, and efficiency as a function of speed

where the gear ratio is the ratio between the angular velocities of the motor and the driven wheels.

The simulation result in Fig. 6 shows that the vehicle can accelerate from 0 to 60 mph in 9.5 s.

By varying the battery weight and repeating the simulation, the 0–60 mph time as a function of the battery weight is plotted in Fig. 7. It can be seen that the battery weight has a significant impact on the 0–60 mph time.

3.4 Maximum Gradeability

Maximum gradeability is defined as the largest surface incline that a vehicle can overcome. It is an important vehicle level performance requirement. Lower level requirements on motor torque and battery power can be developed based on the maximum gradeability requirement.

In the analysis of maximum gradeability, the following conditions hold:

$$a \geq 0; D_A = 0; L_a = 0$$

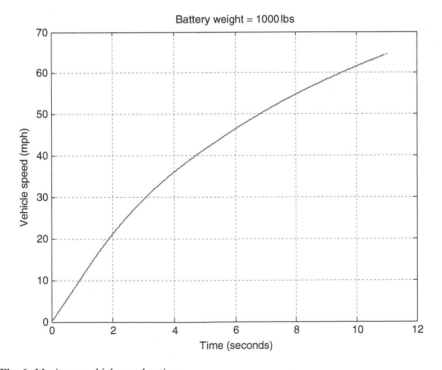

Fig. 6 Maximum vehicle acceleration

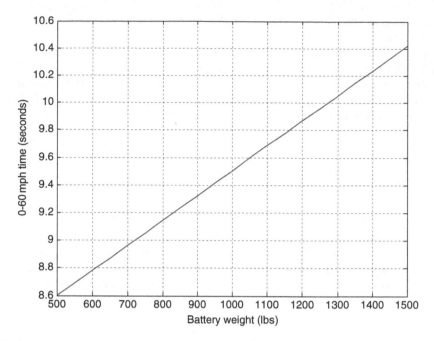

Fig. 7 Impact of battery weight on maximum acceleration

There are two cases one must consider: traction limited and power limited. When the vehicle is traction limited, from (17), (20), and (29) we have

$$F_x = \frac{\mu}{L} W(h \sin\theta + b \cos\theta) \geq W \sin\theta + R_x \tag{31}$$

From the above inequality, we can solve for the maximum surface incline when it is traction limited

$$\theta \leq \tan^{-1}\left(\frac{\mu b - L f_r}{L - \mu h}\right) \tag{32}$$

When the vehicle is power limited, from (17) and (26), we have

$$\frac{G}{r} T_m \geq W \sin\theta + R_x \tag{33}$$

Note that $\eta_t = 1$ is assumed for maximum gradability analysis. Combining (32) and (33), one can plot the maximum grade as a function of motor torque.

4 Conclusion

Several design requirements for electric vehicles are discussed in this paper. System level design requirements are used to derive requirements on motor power and torque and battery power, weight and specific energy based on simulation tools developed using first principle models. A design constraint involving battery weight, specific energy, and driving range is derived. The maximum cruise speed requirement results in requirements on the rated motor torque, power, and battery power. For a given motor and battery, the maximum acceleration can be simulated to see if the 0–60 mph time requirement is met. The maximum gradeability requirement can be used to derive requirements on maximum motor torque/battery current. A software package EVSim [11] has been developed based on the analysis in this paper. The simulation tool allows further investigation of design trade-offs among different parameters. Future research includes the optimization of vehicle performance and cost using the simulation tools developed in this paper, sensitivity studies for the design parameters in the simulation model, and incorporating motor dynamics [12] into our model.

References

1. G.A. Nazri, G. Pistoia, *Lithium Batteries: Science and Technology* (Kluwer/Plenum, Boston, 2004)
2. C.K. Chan, H. Peng, G. Liu, K. McIlwrath, X.F. Zhang, R.A. Huggins, Y. Cui, High-performance lithium battery anodes using silicon nanowires. Nat. Nanotechnol. **V**(3) 31–35 (2008)
3. T. Lucas, F. Riess, *How to Convert to an Electric Car* (Crown Publishers, New York, 1980)
4. J. Dave, J. Dong, Conversion of an existing car to a rechargeable electric vehicle, *ASEE Annual Conference*, 2009
5. I. Husain, *Electric and Hybrid Vehicles* (CRC Press, Boca Raton, FL, 2003)
6. R. Pratap, *Getting Started with MATLAB 7: A Quick Introduction for Scientists and Engineers* (Oxford University Press, New York, 2006)
7. The MathWorks, Inc. *MATLAB®7 Getting Started Guide* (2008)
8. M. Ehsani, Y. Gao, S.E. Gay, A. Emadi, *Modern Electric, Hybrid Electric, and Fuel Cell Vehicles* (CRC Press LLC, London, 2005)
9. T.D. Gillespie, *Fundamentals of Vehicle Dynamics* (Society of Automotive Engineers, Inc., Warrendale, PA, 1992)
10. Formula Hybrid Rules, SAE International, August 26, 2008. Available at http://www.formula-hybrid.org/pdf/Formula-Hybrid-2009-Rules.pdf. Last accessed on 29 Dec 2009
11. W. Zhan, M. McDermott, M. Hasan, *EVSim Software Design* (Aug 2009)
12. W. Gao, E. Solodovnik, R. Dougal, Symbolically-aided model development for an induction machine in virtual test bed. IEEE Trans. Energy Conver. **19**(1) 125–135 (March 2004)

Chapter 19
Modeling Confined Jets with Particles and Swril*

Osama A. Marzouk, and E. David Huckaby

Abstract We present mathematical modeling and numerical simulations, using the finite volume method, of a coaxial particle-laden airflow entering an expansion in a vertical pipe. An Eulerian approach is used for the gas (air) phase, which is modeled by the unsteady Favre-averaged Navier–Stokes equations. A Lagrangian model is used for the dispersed (particles) phase. The results of the simulations using three implementations of the $k-\varepsilon$ turbulence model (standard, renormalization group – RNG, and realizable) are compared with measured axial profiles of the mean gas-phase velocities. The standard model achieved the best overall performance. The realizable model was unable to satisfactorily predict the radial velocity; it is also the most computationally-expensive model. The simulations using the RNG model predicted extra recirculation zones.

1 Overview

The use of computational fluid dynamics (CFD) to accurately model energy production systems is a challenging task [1]. Of current interest, due to ever-increasing energy demands, are coal-based energy systems such as pulverized coal (PC) boilers and gasifiers with an emphasis on systems which provide for carbon capture and storage (e.g. PC-oxyfuel). Turbulence and particle sub-models

*The support of this work by U.S. DOE Existing Plants Emissions and Capture program is gratefully acknowledged. The first author was supported in part by an appointment to the U.S. Department of Energy (DOE) Postgraduate Research Program at the National Energy Technology Laboratory administered by the Oak Ridge Institute for Science and Education.

O.A. Marzouk (✉)
United States Department of Energy, National Energy Technology Laboratory, 3610 Collins Ferry Road, Morgantown, WV 26507-0880, USA
e-mail: omarzouk@vt.edu

S.-I. Ao et al. (eds.), *Machine Learning and Systems Engineering,*
Lecture Notes in Electrical Engineering 68,
DOI 10.1007/978-90-481-9419-3_19, © Springer Science+Business Media B.V. 2010

are one of many sub-models which are required to calculate the behavior of these gas–solid flow systems.

The particle-laden swirling flow experiment of Sommerfeld and Qiu [2] was selected as a test-case to assess the performance of three implementations of the $k-\varepsilon$ turbulence model for gas–solid flows. Previous numerical investigations of this experiment include, Euler–Lagrange (EL)/Reynolds-averaged Navier–Stokes (RANS-steady) [3], EL/large eddy simulations (LES) [4], Euler–Euler (EE)/RANS-unsteady [5], and EE/LES [6]. The extensive experimental measurements make this experiment a good test-case for gas–solid CFD.

A schematic of the experiment is shown in Fig. 1. The coaxial flow consists of a primary central jet, laden with particles at a loading of 0.034 kg-particles/kg-air and an annular secondary jet with a swirl number of 0.47 based on the inlet velocity. Coaxial combustors have a similar configuration to this system. Generating a swirling flow is an approach used to stabilize combustion and maintain a steady flame [7]. Swirl entrains and recirculates a portion of the hot combustion products. It also enhances the mixing of air and fuel. The inlet swirl number was calculated as the ratio between the axial flux of angular momentum to the axial flux of linear momentum

$$S = \frac{2\int_0^{R_{sec}} \rho\, U_\theta\, U_x\, r^2\, dr}{D_{cyl} \int_0^{R_{sec}} \rho\, U_x^2\, r\, dr} \tag{1}$$

where U_x and U_θ are the axial and tangential (swirl) velocities, $R_{sec} = 32$mm is the outer radius of the swirling secondary jet, and $D_{cyl} = 197$mm is the inner diameter

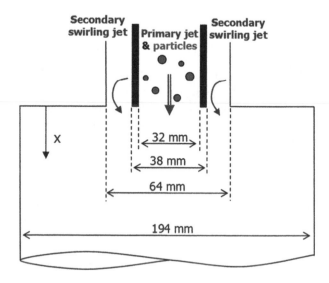

Fig.1 Illustration of the coaxial flow (*with swirl and particles*)

of the pipe into which the jets enter. The inlet Reynolds number (Re) is approximately 52,400 based on the outer diameter of the secondary jet, thus

$$Re = \frac{\rho \, \hat{U}_x \, (2R_{sec})}{\mu} \tag{2}$$

where ρ is the density, μ is the dynamic viscosity, and $\hat{U}_x$ is an equivalent axial velocity that accounts for the total volume flow rate from both primary and secondary jets. The particles were small spherical glass beads, with a density of 2,500kg/m^3. The beads injected according to a log-normal distribution with a mean number diameter of 45 μm.

2   Gas Phase and Turbulence Models

The continuity and momentum equations for the resolved gas-phase fields are expressed and solved in Cartesian coordinates as

$$\frac{\partial \rho}{\partial t} + \frac{\partial (\rho \, U_j)}{\partial x_j} = 0 \tag{3}$$

$$\frac{\partial (\rho \, U_i)}{\partial t} + \frac{\partial (\rho \, U_i U_j)}{\partial x_j} = -\frac{\partial p}{\partial x_i} + \frac{\partial (\sigma_{ij} + \tau_{ij})}{\partial x_j} + \rho \, g_i + S_p \tag{4}$$

where U_i is the velocity vector, p is the pressure, σ_{ij} and τ_{ij} are the viscous and Reynolds (or turbulent) stress tensors, g_i is gravitational vector (we only have $g_1 = 9.81$m/s^2), and S_p is a source term accounting for the momentum from the dispersed phase. As for Newtonian fluids, σ_{ij} is calculated as

$$\sigma_{ij} = 2\mu \, S_{ij}^{dev}$$

where S_{ij}^{dev} is the deviatoric (traceless) part of the strain-rate tensor S_{ij}

$$S_{ij} = \frac{1}{2}\left(\frac{\partial U_i}{\partial x_j} + \frac{\partial U_j}{\partial x_i}\right)$$

$$S_{ij}^{dev} = \frac{1}{2}\left(\frac{\partial U_i}{\partial x_j} + \frac{\partial U_j}{\partial x_i} - \frac{2}{3}\delta_{ij}\frac{\partial U_k}{\partial x_k}\right)$$

The tensor τ_{ij} is not resolved directly. Instead, its effects are approximated using the gradient transport hypothesis

$$\tau_{ij} = 2\mu_t \left(S_{ij}^{dev}\right) - \frac{2}{3}\rho \, k \, \delta_{ij} \tag{5}$$

where μ_t is the turbulent (or eddy) viscosity. The spherical tensor on the right-hand side is not considered here [3, 8]. Different eddy-viscosity turbulence models propose different strategies to calculate μ_t. In the case of k–ε models, μ_t is calculated as

$$\mu_t = C_\mu \, \rho \frac{k^2}{\epsilon} \tag{6}$$

where k is the turbulent kinetic energy per unit mass and ε is its dissipation rate. They are calculated by solving two coupled transport equations. The forms of these equations vary depending on the model implementation. We consider here three implementations, which are described in the following subsections.

2.1 Standard k–ε Model

The standard k–ε model refers to the Jones–Launder form [9], without wall damping functions, and with the empirical constants given by Launder and Sharma [10]. The k and ε equations are

$$\frac{\partial(\rho k)}{\partial t} + \frac{\partial(\rho U_j k)}{\partial x_j} = \frac{\partial}{\partial x_j}\left[\left(\mu + \frac{\mu_t}{\sigma_k}\right)\frac{\partial k}{\partial x_j}\right] + P - \rho \epsilon \tag{7}$$

$$\frac{\partial(\rho \epsilon)}{\partial t} + \frac{\partial(\rho U_j \epsilon)}{\partial x_j} = \frac{\partial}{\partial x_j}\left[\left(\mu + \frac{\mu_t}{\sigma_\epsilon}\right)\frac{\partial \epsilon}{\partial x_j}\right] + \frac{\epsilon}{k}(C_{\epsilon 1}G - C_{\epsilon 2}\,\rho \epsilon)$$
$$- \left(\frac{2}{3}C_{\epsilon 1} + C_{\epsilon 3}\right)\rho \epsilon \frac{\partial U_k}{\partial x_k} \tag{8}$$

where P is the production rate of kinetic energy (per unit volume) due to the gradients in the resolved velocity field

$$P = \tau_{ij}\frac{\partial U_i}{\partial x_j}$$

which is evaluated as

$$P = G - \frac{2}{3}\rho k \frac{\partial U_k}{\partial x_k}$$

with

$$G = 2\mu_t S_{ij}^{dev}\frac{\partial U_i}{\partial x_j} = 2\mu_t\left(S_{ij}S_{ij} - \frac{1}{3}\left[\frac{\partial U_k}{\partial x_k}\right]^2\right)$$

The addition of $C_{\varepsilon 3}$ in the last term on the right-hand side of (19.8) is not included in the standard model. It was proposed [8, 11] for compressible turbulence. However, we will refer to this implementation as the standard model. The model constants are

$$C_\mu = 0.09, \; \sigma_k = 1.0, \; \sigma_\epsilon = 1.3, \; C_{\epsilon 1} = 1.44, \; C_{\epsilon 2} = 1.92, \; C_{\epsilon 3} = -0.33$$

2.2   Renormalization Group (RNG) k−ε Model

The RNG model was developed [12, 13] using techniques from the renormalization group theory with scale expansions for the Reynolds stress. The k and ε equations have the same form in (7) and (8), but the constants have different values. In addition, the constant $C_{\varepsilon 1}$ is replaced by $C^*_{\varepsilon 1}$, which is no longer a constant, but is determined from an auxiliary function as

$$C^*_{\epsilon 1} = C_{\epsilon 1} - \frac{\eta(1 - \eta/\eta_0)}{1 + \beta\,\eta^3}$$

where

$$\eta = \frac{k}{\epsilon}\sqrt{2\,S_{ij}\,S_{ij}}$$

is the expansion parameter (ratio of the turbulent time scale to the mean-strain time scale). The model constants are

$$C_\mu = 0.0845, \; \sigma_k = 0.7194, \; \sigma_\epsilon = 0.7194, \; C_{\epsilon 1} = 1.42$$
$$C_{\epsilon 2} = 1.68, \quad C_{\epsilon 3} = -0.33, \quad \eta_0 = 4.38, \qquad \beta = 0.012$$

2.3   Realizable k−ε Model

The realizable $k-\varepsilon$ model was formulated [14] such that the calculated normal (diagonal) Reynolds stresses are positive definite and shear (off-diagonal) Reynolds stresses satisfy the Schwarz inequality. Similar to the RNG model, the form of the k equation is the same as the one in (7). In addition to altering the model constants, the two main modifications lie in replacing the constant C_μ used in calculating the eddy viscosity in (6) by a function, and in changing the right-hand side (the production and destruction terms) of the ε equation. The last term in (8) is dropped. With this, the ε equation becomes

$$\frac{\partial(\rho\,\epsilon)}{\partial t}+\frac{\partial(\rho\,U_j\,\epsilon)}{\partial x_j}=\frac{\partial}{\partial x_j}\left[\left(\mu+\frac{\mu_t}{\sigma_\epsilon}\right)\frac{\partial\epsilon}{\partial x_j}\right]+C_1\,\rho\,\epsilon S-C_{\epsilon 2}\,\rho\,\frac{\epsilon^2}{k+\sqrt{(\mu/\rho)\epsilon}}\quad(9)$$

where

$$C_1=max\left(0.43,\frac{\eta}{\eta+5}\right);\quad C_\mu=\frac{1}{A_0+A_S(U^*k/\epsilon)}$$

$$U^*=\sqrt{S_{ij}S_{ij}+\Omega_{ij}\Omega_{ij}};\quad \Omega_{ij}=\frac{1}{2}\left(\frac{\partial U_i}{\partial x_j}-\frac{\partial U_j}{\partial x_i}\right)$$

$$A_S=\sqrt{6}\cos(\phi);\quad \phi=\frac{1}{3}\arccos\left(\sqrt{6}\,W\right)$$

$$W=\min\left[\max\left(2\sqrt{2}\,\frac{S_{ij}S_{jk}S_{ik}}{S^3},-\frac{1}{\sqrt{6}}\right),\frac{1}{\sqrt{6}}\right]$$

and Ω_{ij} is rate-of-rotation tensor. The model constants are

$$\sigma_k=1.0,\ \sigma_\epsilon=1.2,\ C_{\epsilon 2}=1.9,\ A_0=4.0$$

3 Dispersed Phase

The Lagrangian equations of motion of a particle (in vector form) are

$$\frac{d\mathbf{x}}{dt}=\mathbf{u}\qquad(10a)$$

$$m\frac{d\mathbf{u}}{dt}=\mathbf{f}\qquad(10b)$$

where m is the mass of the particle, $\mathbf{u}$ is the particle velocity, and $\mathbf{f}$ is the force acting on the particle. In this study; the drag, gravity, and buoyancy are considered, thus the force $\mathbf{f}$ has the following form [15]:

$$\mathbf{f}=-\frac{\pi d^2}{8}\rho\,C_D|\mathbf{u}-\mathbf{U}^*|(\mathbf{u}-\mathbf{U}^*)+m\mathbf{g}-\rho\forall\mathbf{g}\qquad(11)$$

where d is the particle diameter, C_D is the drag coefficient (which is a function of the particle Reynolds number, Re_d, as will be described later), $\forall$ is the particle volume and $\mathbf{U}^*$ is the instantaneous fluid velocity

$$\mathbf{U}^* \equiv \mathbf{U} + \mathbf{U}' \tag{12}$$

The vector $\mathbf{U}$ is the resolved velocity of the fluid (interpolated at the particle location) which is calculated after solving the governing equations of the flow, coupled with the turbulence model. The fluctuating velocity, $\mathbf{U}'$, is estimated using the discrete random walk algorithm [16, 17]. In this algorithm, uncorrelated eddies are generated randomly, but the particle trajectory is deterministic within an eddy. The fluctuating velocity affects the particle over an interaction time, $T_{interac}$, which is the shortest of the eddy life time (Lagrangian integral time scale of turbulence), T_{eddy}, and the residence or transit time, T_{resid}. The latter is the time needed by the particle to traverse the eddy. These characteristic times are calculated as

$$T_{eddy} = \frac{k}{\epsilon} \tag{13a}$$

$$T_{resid} = C_{resid} \frac{k^{3/2}}{\epsilon |\mathbf{u} - \mathbf{U} - \mathbf{U}'|} \tag{13b}$$

$$T_{interac} = min\left(T_{eddy}, T_{resid}\right) \tag{13c}$$

In (19.13b), $\mathbf{U}'$ is lagged from the previous time step, and $C_{resid} = 0.09^{3/4} = 0.16432$. The turbulence information, thus the characteristic times in (13), are updated every time step to account for the fact that the turbulence encountered by a particle in its trajectory is not homogeneous.

The drag coefficient for the spherical particles is determined from the following two-region formula [11], which is very similar to the Schiller–Naumann[18] expression for $Re_d \leq 1,000$, and uses a constant Newton drag coefficient for $Re_d > 1,000$

$$C_D = \begin{cases} \dfrac{24}{Re_d}\left(1 + \dfrac{1}{6} Re_d^{2/3}\right), & Re_d \leq 1000 \\ 0.424, & Re_d > 1000 \end{cases} \tag{14}$$

where the particle Reynolds number is defined as

$$Re_d = \frac{\rho |\mathbf{u} - \mathbf{U}^*| d}{\mu} \tag{15}$$

Combining (19.10a) and (19.11), and using $m = \rho_p \pi d^3/6$ (ρ_p is the particle density), the particle equations of motion become

$$\frac{d\mathbf{x}}{dt} = \mathbf{u} \tag{16a}$$

$$\frac{d\mathbf{u}}{dt} = -\frac{\mathbf{u} - \mathbf{U}^*}{\tau} + \left(1 - \frac{\rho}{\rho_p}\right)\mathbf{g}$$

where

$$\tau = \frac{4}{3}\frac{\rho_p\, d}{\rho\, C_D|\mathbf{u} - \mathbf{U}^*|} = \frac{\rho_p\, d^2}{18\,\mu}\frac{24}{Re_d\, C_D}$$

is the momentum relaxation time of the particle.

The particle position is tracked using the algorithm described by Macpherson et al. [19]. The algorithm consists of a series of substeps in which the particle position, $\mathbf{x}$, is tracked within a cell using a forward (explicit) Euler scheme followed by integrating the particle momentum equation using a backward (implicit) Euler scheme to update the particle velocity, $\mathbf{u}$. When calculating the resultant force due to the particle on the fluid phase, the algorithm takes into account the particle residence time in each cell. Interaction with the wall is represented through elastic frictionless collisions.

4   Simulation Settings

The problem is treated as axisymmetric (although the results are mirrored in some figures for better visualization). The domain starts at the expansion location with $x = 0$ in Fig. 1 and extends to $x = 1.0$m. The domain is a 3D wedge (full-opening angle $5°$), with a Frontal Area of 1.0m, 0.097m, and 240 and 182 mesh points in the axial and radial directions, respectively. The mesh is nonuniform both axially and radially, with finer resolution near walls and between the two jets. The mesh has 40,080 cells. The inlet condition for the velocity in the primary (inner) jet is specified in terms of the mass flow rate (9.9g/s). For the secondary (outer) jet, the inlet velocity is specified using the experimental velocity profile. A zero-gradient condition is applied to the pressure at the inlet. The specific turbulent kinetic energy, k, is set to 0.211m^2/s^2 and 0.567m^2/s^2 in the primary and secondary jets, respectively; and the dissipation rate, ε, is set to 0.796m^2/s^3 and 3.51m^2/s^3 in the primary and secondary jets, respectively. The inlet k was estimated assuming 3% turbulence intensity (the experimental value was not specified, but 3% is a reasonable medium-turbulence level [20]) and the inlet ε was then estimated from [21]

$$\varepsilon = C_\mu^{3/4}\, k^{1.5}/l \tag{17}$$

where the standard value 0.09 is used for C_μ, and l is the turbulence length scale, which is approximated as $\approx 10\%$ of the pipe diameter ($l = 0.02$m). At the outlet, zero-gradient conditions are applied for all variables except the pressure, where a constant value of 10^5 N/m^2 is imposed. At the walls, the wall-function treatment is used for the turbulence, and a zero-gradient condition is used for the pressure.

The PISO (pressure implicit splitting of operators) scheme is used to solve the governing flow equations. A variable time step is adjusted dynamically to limit the maximum convective CFL to 0.3. The backward Euler scheme is used for the time integration of the flow equations. Upwind differencing is used for the convective terms. Linear (second-order central difference) interpolation is used to find the mass fluxes at the face centers of the cells from the cell-center values, and is also used for the diffusion terms.

The particle mass flow rate is 0.34 g/s, which corresponds to 0.00472 g/s for our case of 5° wedge. The particles are injected at a velocity of 12.5m/s, which is the nominal axial inlet velocity in the primary jet.

We used version 1.5 of the finite volume open source code OpenFOAM [22, 23] to perform all the simulations presented here. Similar particles are grouped into a parcel (or a particle cloud), where they have a common velocity, to reduce the amount of computation for the dispersed phase. The number of particles per parcel is not uniform.

5 Results

The simulated flow time is 0.6 s for the results presented in this chapter. We have found that this time interval is sufficient for all particles to traverse the domain and for the gas phase to achieve stationary conditions. The last 0.1 s is used to obtain the mean gas-phase velocities.

Figure 2 shows three snapshots of the parcels after 0.05, 0.1, and 0.15 s using the standard k–ε model (the diameters of the particles associated with a parcel are evenly scaled by a factor of 100). This figure illustrates that the model captures well the expected dynamics of the dispersed phase. The larger particles (with larger inertia) maintain their axial motion and penetrate the central recirculation bubble. They are not affected strongly by the swirl and radial velocity of the gas phase. Smaller particles are entrained due to smaller relaxation times, and directed to the walls.

The mean axial, radial, and tangential velocities of the gas phase are shown in Fig. 3. The negative mean axial velocity along the centerline and the walls identify the regions of recirculation. The strong variations in all velocities are confined to a distance of 150 mm after the inlet. The axial and tangential velocities exhibit initial decay, whereas the radial velocity increases at the upstream boundary of the central recirculation bubble.

A comparison between the mean streamlines obtained with the three turbulence models is given in Fig. 4. Besides the central bubble, there are secondary recirculation zones (due to the sudden expansion) at the top of the pipe. The standard model gives the shortest recirculation bubble, with best agreement with the experimental results. The realizable model gives a longer bubble, but with a qualitatively similar structure. The RNG model resolves, in addition to the central

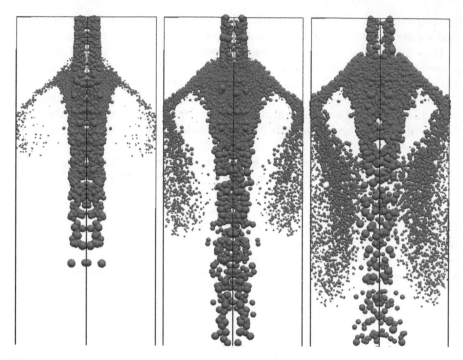

Fig. 2 Parcel motion with the standard $k-\varepsilon$ model at $t = 0.05$s (*left*), $t = 0.10$s (*middle*), and $t = 0.15$s (*right*)

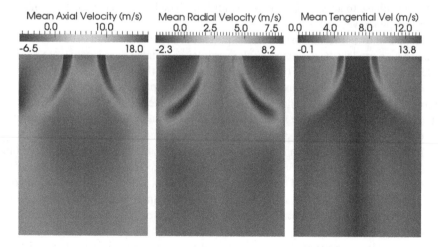

Fig. 3 Mean gas-phase velocities near the inlet with the standard $k-\varepsilon$ model

and two secondary recirculation zones, two noticeable tertiary recirculation zones at the beginning (top) of the central bubble. This feature was not reported in the experimental results.

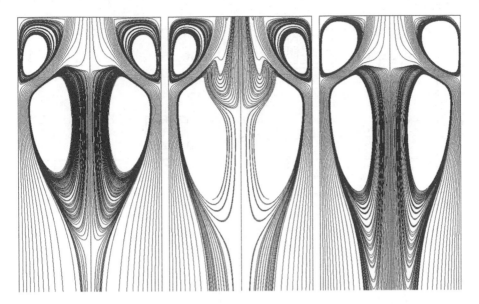

Fig. 4 Comparison of the mean streamlines with 3 k–ε models: standard (*left*), RNG (*middle*), and realizable (*right*)

The mean gas-phase velocity fields are sampled at different axial stations. We compare the mean velocities obtained using the three turbulence models with the measured values at two stations in Fig. 5, located at $x = 3$ and 112 mm. The first station is located close to the inlet, thus upstream of the central bubble, whereas the second one spans the bubble. At $x = 3$ mm, all models predict axial and tangential velocities that are in agreement with the measured ones. The radial velocity using the realizable model shows considerable disparity, with excessively-negative (inward) velocity in the region away from the jets, followed by an outward flow near the wall, which is opposite to the inward flow observed experimentally. The standard model has a slightly better agreement with the measurements than the RNG model at this axial location. At $x = 112$ mm, the RNG predictions deviate considerably from the measurements. This is a direct consequence of the tertiary recirculation shown in Fig. 4. As in the earlier station, the standard and realizable models provide similar results except for the radial velocity, with the realizable model failing to capture the measured peak at $r \approx 80$ mm. The standard model provides better prediction of the axial velocity in the vicinity of the wall than the other two models.

On a computing machine with two processors: quad core Intel Xeon L5335 2.00 GHz, a simulation interval of 0.5 s required 17.13 h CPU time for the standard model, 19.44 h for the RNG model, and 24.81 h for the realizable model. The standard model has the lowest computational demand due to its relative simplicity. The realizable model is the most computationally-expensive implementation, with CPU time equal to 145% and 128% of the CPU times in the case of the standard and RNG models, respectively.

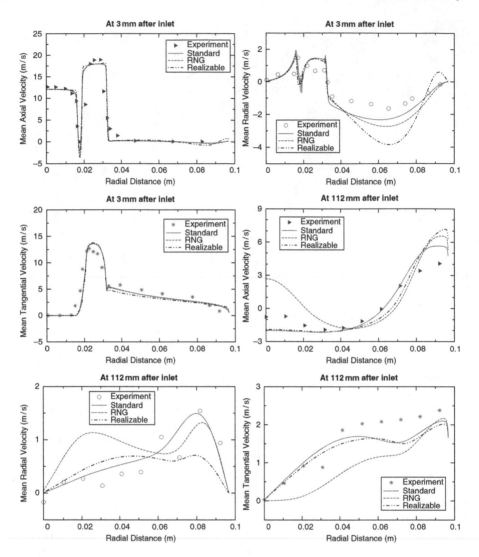

Fig. 5 Comparison of the mean gas-phase velocities at $x = 3$ and 112 mm

6 Conclusions

We simulated a coaxial airflow with a particle-laden primary jet and a swirling annular jet, entering a sudden expansion. Three implementations of the $k-\varepsilon$ model: standard, renormalization group (RNG), and realizable were applied. The standard model had the best overall performance based on the mean gas-phase velocities. The RNG model showed considerable deviations from the measurements in some regions. The main drawback of the realizable model was its erroneous prediction of the radial velocity. The main differences in the predicted velocity profiles were related to the different flow structures and mean streamlines. We should point out that our finding that the more complex models did not outperform the simpler one should not be considered a universal statement. In fact, more work is needed to investigate the reasons of this unexpected outcome. This includes examining the effect of $C_{\varepsilon3}$, and the effect of dropping the spherical component in the modeled Reynolds stresses (which causes violation of one of the realizability constraints).

References

1. T.F. Wall, Combustion processes for carbon capture. P. Combust. Inst. **V31**(N1), pp. 31–47 (2007)
2. M. Sommerfeld, H.-H. Qiu, Detailed measurements in a swirling particulate two-phase flow by a phase-Doppler Anemometer. Int. J. Heat Fluid Fl. **V12**(N1), pp. 20–28 (1991)
3. M. Sommerfeld, A. Ando, D. Wennerberg, Swirling, particle-laden flows through a pipe expansion. J. Fluids Eng. **V114**(N4), pp. 648–656 (1992)
4. S.V. Apte, K. Mahesh, P. Moin, J.C. Oefelein, Large-eddy simulation of swirling particle-laden flows in a coaxial-jet combustor. Int. J. Multiphase Flow **V29**, pp. 1311–1331 (2003)
5. Y. Yu, L.X. Zhou, C.G. Zheng, Z.H. Liu, Simulation of swirling gas-particle flows using different time scales for the closure of two-phase velocity correlation in the second-order moment two-phase turbulence model. J. Fluids Eng. **V125**(N2), pp. 247–250 (2003)
6. M. Boileau, S. Pascaud, E. Riber, B. Cuenot, L.Y.M. Gicquel, T.J. Poinsot, M. Cazalens, Investigation of two-fluid methods for large eddy simulation of spray combustion in gas turbines. Flow Turbul. Combust. **V80**(N3), pp. 291–321 (2008)
7. G. Monnot, in *Principles of Turbulent Fired Heat*, Editions Technip, 1985
8. F.P. Kärrholm, Numerical modelling of diesel spray injection, turbulence interaction and combustion. Ph.D. Dissertation, Chalmers University of Technology, Göteborg, Sweden, 2008
9. W.P. Jones, B.E. Launder, The prediction of laminarization with a two-equation model. Int. J. Heat Mass Tran. **V15**(N2), pp. 301–314 (1972)
10. B.E. Launder, B.I. Sharma, Application of the energy-dissipation model of turbulence to the calculation of flow near a spinning disk. *Lett Heat Mass Tran.*, **V1**(N2), pp. 131–138 (1974)
11. P.A.N. Nordin, Complex chemistry modeling of diesel spray combustion. Ph.D. Dissertation, Chalmers University of Technology, Göteborg, Sweden, 2001
12. V. Yakhot, S.A. Orszag, Renormalization group analysis of turbulence: 1. Basic theory. J. Sci. Comput. **V1**(N1), pp. 3–51 (1986)
13. V. Yakhot, S.A. Orszag, S. Thangam, T.B. Gatski, C.G. Speziale, Development of turbulence models for shear flows by a double expansion technique. Phys. Fluids A. **V4**(N7), pp. 1510–1520 (1992)
14. T.-H. Shih, W.W. Liou, A. Shabbir, J. Zhu, A new $k-\varepsilon$ eddy-viscosity model for high reynolds number turbulent flows – model development and validation. Comput. Fluids **V24**(N3), pp. 227–238 (1995)
15. S. Elghobashi, On predicting particle-laden turbulent flows. Appl. Sci. Res. **V52**(N4), pp. 309–329 (1994)
16. D. Milojević, Lagrangian stochastic-deterministic (LSD) predictions of particle dispersion in turbulence. Part Part Syst. Charact., **V7**(N1-4), pp. 181–190 (1990)
17. D. Lakehal, On the modelling of multiphase turbulent flows for environmental and hydrodynamic applications. Int. J. Multiphase Flow **V28**(N5), pp. 823–863 (2002)
18. L. Schiller, A. Naumann, Über die grundlegenden Berechungen bei der Schwerkraftaufbereitung. Zeit Ver Deut Ing **V77**, pp. 318–320 (1933)

19. G.B. Macpherson, N. Nordin, H.G. Weller, Particle tracking in unstructured, arbitrary poly-
    hedral meshes for use in CFD and molecular dynamics. Commun. Numer. Meth. Eng. **V25**
    (N3), pp. 263–273 (2009)
20. C. Sicot, P. Devinanta, S. Loyera, J. Hureaua, Rotational and turbulence effects on a wind
    turbine blade. Investigation of the stall mechanisms. J. Wind Eng. Ind. Aerod. **V96**(N8-9), pp.
    1320–1331 (2008)
21. J.H. Ferziger, M. Perić, in *Computational Methods for Fluid Dynamics*, 3rd edn. (Springer,
    2002)
22. OpenFOAM: the open source CFD toolbox. Available from:http://www.openfoam.org
23. H.G. Weller, G. Tabor, H. Jasak, C. Fureby, A tensorial approach to computational continuum
    mechanics using object-oriented techniques. Comput. Phys. **V12**(N6), pp. 620–631 (1998)

Chapter 20
Robust Tracking and Control of MIMO Processes with Input Saturation and Unknown Disturbance

Ajiboye Saheeb Osunleke and Mingcong Deng

Abstract In this chapter, the design of robust stabilization and output tracking performance of multi-input multi-output processes with input saturations and unknown disturbance are considered. The proposed control technique is the robust anti-windup generalized predictive control (RAGPC) scheme for multivariable processes. The proposed control scheme embodies both the optimal attributes of generalized predictive control and the robust performance feature of operator-based theoretic approach. As a result, a strongly robust stable feedback control system with disturbance rejection feature and good tracking performance is achieved.

1 Introduction

The control of multivariable systems is paramount in nearly all industrial and allied processes. This is so because of the need to optimize outputs at the expense of limited available materials. In designing MIMO control systems, systems and control engineers have to deal with some challenges which impose limitations on the performance of MIMO systems.

In recent past, several techniques for control of MIMO systems have been proposed and good reviews are available [1–3]. Among these techniques, the predictive control approach is distinctive in its application to control of MIMO processes which exhibit some of these performance limited characteristics such as input saturation and disturbance. Predictive control strategy is now widely regarded as one of the standard control paradigms for industrial processes and has continued to be a central

A.S. Osunleke (✉)
Division of Industrial Innovation Sciences, Graduate School of Natural Science and Technology, Okayama University, 3-1-1 Tsushima-Naka, 3-chome, Okayama 700-8530, Japan
e-mail: osunleke.boye@suri.sys.okayama-u.ac.jp

S.-I. Ao et al. (eds.), *Machine Learning and Systems Engineering*,
Lecture Notes in Electrical Engineering 68,
DOI 10.1007/978-90-481-9419-3_20, © Springer Science+Business Media B.V. 2010

research focus in recent times [1, 4, 5. In practice, control systems have to deal with disturbances of all kinds, such as stepwise load disturbances, high frequency sensor or thermal noise, input saturations etc. These undesirable effects are systematically dealt with using predictive control.

The new proposed control strategy under the generalized predictive control scheme is known as multivariable robust anti-windup generalized predictive control (MRAGPC). It is the multivariable extension of the robust anti-windup generalized predictive control (RAGPC) scheme for SISO systems in earlier work [8]. Like RAGPC scheme, MRAGPC also shares the optimal and robust performance attributes characteristic of a good control strategy. This work proposes a design procedure for Multivariable systems having both input saturation and disturbance. The proposed MRAGPC design procedure can be implemented in four different modes depending on the nature of problem at hand and the design control objective. The design application entails firstly, the design of compensator to compensate the effect of minimal phase and strictly properness as these are characteristics of MIMO industrial processes. Secondly is to construct stable coprime factors for the compensated system. Subsequently, anti-windup controllers are designed to achieve a stable feedback closed loop control system. This procedure is completed by the design of a tracking operator using operator-based theoretic approach. In order to estimate the unknown disturbance in the system, unknown disturbance estimation mechanism for MIMO system is also proposed. Based on the outlined procedure, the robust performance of the closed-loop system in the presence of input saturation and disturbances is ensured.

In this proposed design, the coupling effect of interactions between components of the MIMO systems is curtailed by the appropriate choice of MRAGPC design parameters. The improved performance of the proposed scheme is confirmed by simulation of the model of a two-input two-output MIMO system.

2   MRAGPC Design Scheme

In order to control MIMO processes with all the design limitations, MRAGPC scheme is implemented in three different modes depending on the type and nature of problems at hand.

Mode 1 – known as Anti-windup mode, is implementable for the case of input saturation but with no disturbances load on the input to the system.

Mode 2 – otherwise called compensation mode, is implementable for the case of input saturation, non-minimal phase and disturbance load in the input to the system.

Mode 3 – referred to as the fault detection mode, for system with both input saturation and unknown disturbance.

In this paper, all these modes of MRAGPC application are implemented.

2.1   MRAGPC Problem Formulation

The proposed linear multivariable system model is given by Deng et al. [4].

$$\mathbf{A}[p]\underline{Y}(p) = \mathbf{B}[p]\underline{U}(p) + \mathbf{C}[p]\underline{V}(p) \tag{1}$$

where $\underline{Y}(p)$, $\underline{U}(p)$, $\underline{V}(p)$ are $m \times 1$ output, $q \times 1$ input and $p \times 1$ disturbance vectors respectively. $\boldsymbol{B}[p]$ is a $m \times q$ polynomial matrix while $\mathbf{A}[p]$ and $\mathbf{C}[p]$ are $m \times m$ diagonal polynomial matrices. As before, the polynomial matrices $\mathbf{A}[p]$ and $\boldsymbol{B}[p]$, $\mathbf{A}[p]$ and $\mathbf{C}[p]$ are coprime. The elements of $\mathbf{C}[p]$ are with a degree of one less or equal to that of the corresponding elements of $\mathbf{A}[p]$ and are chosen by the designer. The model in (1) corresponds to a Left Matrix Fraction Description (LMFD) described as

$$\mathbf{H}_0(p) = \mathbf{A}^{-1}[p]\mathbf{B}[p] \tag{2}$$

Equation (2) can be equivalently expressed as

$$\mathbf{H}_0(p) = \mathbf{D}_0^{-1}(p)\mathbf{N}_0(p) \tag{3}$$

where $\mathbf{P}_0(p)$ is the nominal multivariable plant, $\mathbf{D}_0(p)$ and $\mathbf{N}_0(p)$ are coprime factors of $\mathbf{P}_0(p)$ and are defined as

$$\mathbf{N}_0(p) = \tilde{\mathbf{D}}^{-1}[p]\tilde{\mathbf{B}}[p], \quad \mathbf{D}_0(p) = \tilde{\mathbf{D}}^{-1}[p]\tilde{\mathbf{A}}[p] \tag{4}$$

where $\tilde{\mathbf{D}}(p)$ is a Hurwitz polynomial matrix with degree equal to that of $\tilde{\mathbf{A}}(p)$. The polynomial matrix $\tilde{\mathbf{B}}(p)$ is the same as $\mathbf{B}(p)$ for non-strictly proper and minimal system. The control input $\mathbf{u}_j(t)$ is subject to the following constraint (Fig. 1)

$$u_{\min,j} \leq u_j(t) \leq u_{\max,j} \quad (j = 1, 2, \ldots, m) \tag{5}$$

The constraint (7) is equivalently expressed as

$$\mathbf{u}(t) = \sigma(\mathbf{u}_1) = \begin{pmatrix} \sigma(\bar{u}_1(t)) \\ \vdots \\ \sigma(\bar{u}_m(t)) \end{pmatrix}, \tag{6}$$

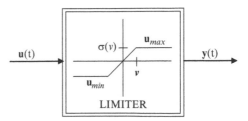

Fig. 1 Input saturation model

where $\mathbf{u}_1$ is the ideal controller output vector. The objective is to design stable controllers using coprime factorization and Youla-Kucera parameterization [5] for the above process.

2.2 Controllers Parameterization

The proposed design scheme for the MIMO systems follows four steps as highlighted below.

2.2.1 Design of Stable Feedback MCGPC Controllers V (p) and U (p) with Input Constraint Part (Fig. 2)

By adapting the MCGPC design method to the anti-windup design approach, the proposed controllers are given by Youla-Kucera parameterization as follows.

$$\mathbf{U}(p) = \mathbf{X}(p) + \mathbf{Q}(p)\mathbf{D}(p) \tag{7}$$

$$\mathbf{V}(p) = \mathbf{Y}(p) - \mathbf{Q}(p)\mathbf{N}(p) \tag{8}$$

where $Q(p) \in RH_\infty$ is a design parameter matrix for ensuring a strongly stable feedback controllers in the above and is given by

$$\mathbf{Q}(p) = \mathbf{U}_d^{-1}(p)\mathbf{U}_n(p) \tag{9}$$

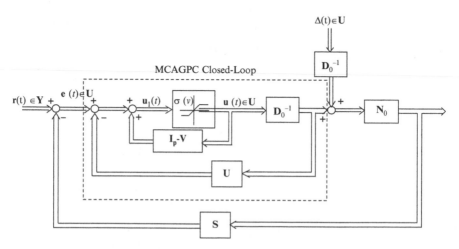

Fig. 2 Proposed MIMO control structure

$N(p)$, $D(p)$ are the coprime factor representation of $\mathbf{D}_0^{-1}(p)$, $X(p)$, $Y(p) \in RH_\infty$ are operators chosen to satisfy the Bezout Identity

$$\mathbf{X}(p)\mathbf{N}(p) + \mathbf{Y}(p)\mathbf{D}(p) = \mathbf{I}_m \tag{10}$$

$$\hat{\mathbf{N}}(p)\hat{\mathbf{X}}(p) + \hat{\mathbf{D}}(p)\hat{\mathbf{Y}}(p) = \mathbf{I}_m \tag{11}$$

$$\hat{\mathbf{X}}(p),\ \hat{\mathbf{Y}}(p),\ \mathbf{X}(p),\ \mathbf{Y}(p)\ \in \mathbf{RH}_\infty \tag{12}$$

where the coprime factorization $N(p)$ and $D(p)$, $\hat{N}(p)$ and $\hat{D}(p)$ of the CAGPC Plant can be chosen as follows.

$$\mathbf{H}_0(p) = \mathbf{D}_0^{-1}(p) = \tilde{\mathbf{D}}[p]\mathbf{A}^{-1}[p] = \mathbf{A}^{-1}[p]\tilde{\mathbf{D}}[p] \tag{13}$$

From (13), defining the stable closed-loop characteristic polynomial matrices of the process without input constraint as $\mathbf{T}_c[p]$ and $\hat{\mathbf{T}}_c[p]$, then we choose $\mathbf{N}(p)$ and $\mathbf{D}(p)$, $\hat{\mathbf{N}}(p)$ and $\hat{\mathbf{D}}(p)$ as

$$\mathbf{N}(p) = \hat{\tilde{\mathbf{D}}}[p]\hat{\mathbf{T}}_c^{-1}[p]\hat{\tilde{\mathbf{D}}}[p],\ \hat{\mathbf{N}}(p) = \mathbf{T}_c^{-1}[p]\tilde{\mathbf{D}}[p] \tag{14}$$

$$\mathbf{D}(p) = \hat{\mathbf{A}}[p]\hat{\mathbf{T}}_c^{-1}[p]\hat{\tilde{\mathbf{D}}}[p],\ \mathbf{D}(p) = \mathbf{T}_c^{-1}[p]\mathbf{A}[p] \tag{15}$$

$$\mathbf{X}(p) = \mathbf{K}_e\mathbf{R}_e + \mathbf{K}_e\mathbf{C}_e(p)\mathbf{F}(p) \tag{16}$$

$$\mathbf{Y}(p) = \mathbf{I}_m + \mathbf{K}_e\mathbf{C}_e(p)\mathbf{G}(p) \tag{17}$$

$\mathbf{T}_c[p]$ is the closed-loop characteristic polynomial matrix for the control system in Fig. 2 and is given by

$$\mathbf{T}_c[p] = \mathbf{A}[p] + \mathbf{K}_e\mathbf{L}(p) + \mathbf{K}_e\mathbf{R}_e\tilde{\mathbf{D}}[p] \tag{18}$$

where K_e, C_e, R_e, $G[p]$ and $F[p]$ are given in [4].

2.2.2   Robust Stability of $\mathbf{D}_0^{-1}(p)$ with Input Constraint

The predictive anti-windup part is quasi-linear with respect to the effect of input constraint. The closed loop transfer function of this part of the control structure is always stable and invertible. Therefore, it can be seen that the plant retains a robust *rcf* and it is stable. Then, we can design a Bezout operator $S(p)$ to satisfy the following Bezout Identity [7]

$$\mathbf{R}(p)\mathbf{D_0}(p) + \mathbf{S}(p)\mathbf{N_0}(p) = \mathbf{I_m} \tag{19}$$

2.2.3   Tracking Operator Design

Here a tracking operator $\mathbf{M}(p)$ for reference signal $\mathbf{r(t)}$ such that plant output $\mathbf{y}(t)$ tracks to the setpoint signal $\mathbf{r}(t)$ as depicted in Fig. 3. Based on the proposed lemma and proof in [7], operator $\mathbf{M}(p)$ is designed to satisfy

$$\mathbf{N_0}(p)(\mathbf{M}(p)\mathbf{r}(p) + \mathbf{R}(p)\Delta(p)) = \mathbf{I_m}\mathbf{r}(p) \tag{20}$$

2.2.4   Decoupling of the Control System

In the proposed control scheme, the possible interactions between different control loops are identified as input-input, disturbance-input and input-output. In order to minimize the effect of these interactions, the following design steps are proposed.

- Choosing larger value for control orders and/or smaller value for the maximum prediction horizons and/or using reference models for the output in the design step 1.
- Design a decoupling matrix $\mathbf{G_D}(p)$ such that loop interactions are decoupled. The elements of $\mathbf{G_D}(p)$ can be selected in a number of ways to achieve optimally decoupled control system. This is further investigated in the future works.

3   Additional Design Schemes for MRAGPC

In this section, we present two other design features to enhance the performance of the proposed method for MIMO processes similarly to that obtained for SISO [8] namely, the robust parallel compensator and unknown fault detection schemes.

3.1   Robust Parallel Compensator (RPC) Scheme for MIMO Processes

In an extension to the design of robust parallel compensator designed for SISO system [9], we present in this section, an equivalent form of this compensator for

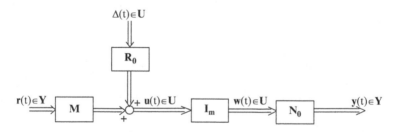

Fig. 3   Robust tracking structure for MIMO system

MIMO processes. Here, a RPC, $\mathbf{F}(p)$ (see Fig. 4) is designed for all classes of system which exhibit non-minimal phase and non-strictly properness. The condition for this design to satisfy static output feedback is the same as in SISO [10] that is:

- The process must be inversely stable.
- Its relative degree must be zero i.e. non-strictly properness.
- The leading coefficient must be positive.

For the process represented by (2), we propose the following GRPC design scheme.

$$\mathbf{F}(p) = \mathbf{F_d^{-1}}[p]\mathbf{F_n}[p]\mathbf{P_p^{\delta}} \tag{21}$$

where
$F_n[p]$ and $F_d[p]$ are $(\gamma - i)$th and $(\gamma - i + 1)$th order monic stable polynomial matrices respectively and are as given as

$$F_n[p] = \beta_{\gamma-k,i,j}p^{\gamma-k} + \cdots + \beta_{0,i,j}, \quad i = 1, 2, \ldots, m; \quad j = 1, 2, \ldots, q \tag{22}$$

$$F_d[p] = \alpha_{\gamma-k+1,i,j}p^{\gamma-k+1} + \cdots + \alpha_{0,i,j}, \quad i = 1, 2, \ldots, m; \quad j = 1, 2, \ldots, q \tag{23}$$

and P_p^{δ} is a diagonal matrix $m \times q$ of element p^{δ} where δ is the degree of polynomial in p.

Let $\tilde{\mathbf{H}}(p)$ be the augmented system matrix defined as

$$\tilde{\mathbf{H}}(p) = \mathbf{H}(p) + \mathbf{F}(p) \tag{24}$$

Equivalently,

$$\tilde{\mathbf{H}}(p) = \tilde{\mathbf{A}}^{-1}[p]\tilde{\mathbf{B}}[p] \tag{25}$$

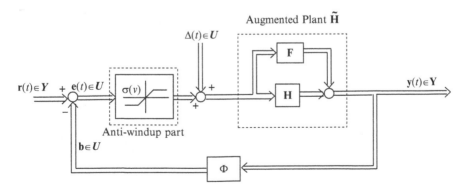

Fig. 4 Robust parallel compensator design for MIMO system

In summary, the coefficients $\beta_{k,i,j} > 0$, $\alpha_{k,i,j} > 0$ in (25) above are determined such that

1. $\tilde{\mathbf{B}}[p]$, $\tilde{\mathbf{A}}[p] \in$ RH∞ for sufficiently large $\beta_{0,i,j}$.
2. The augmented plant $\tilde{\mathbf{H}}(p)$ has the same steady state gain as the nominal plant $\mathbf{H}(p)$.
3. There exists positive constants $\alpha_{k,i,j}$ such that the augmented system $\tilde{\mathbf{H}}(p)$ in (25) will satisfy the sufficient condition of SOF for any small positive integer value of δ and $\alpha_{0,i,j} > 0$.

For the RAGPC design scheme, it is just sufficient to choose $\delta = 1$ and γ as minimum as possible while the unknown coefficients are designed satisfying the above stated conditions using Kharitonov's theorem [11]. Without loss of generality, it is obvious from (25) that $\tilde{\mathbf{H}}(p)$ have a relative degree equal zero and a positive leading coefficient thereby satisfying all the conditions of $\tilde{\mathbf{H}}(p)$ being SOF stabilized and almost strictly positive real (ASPR).

3.2 Unknown Disturbance Estimation Scheme for MIMO Processes

As noted in earlier work [2], the unknown disturbance of the system is estimated as fault signals acting on the process input. If stable low-order operators S_0 and R_0 can be designed such that the pair $[S_0, R_0] \approx [S, R]$ (see Fig. 5). Then the following equations enable us to detect the fault signal [2]

$$\left.\begin{array}{l} \mathbf{u_0} = \mathbf{R_0}\mathbf{u_d}(t) + \mathbf{S_0}(\mathbf{y_a})(t) \\ \mathbf{y_d} = \mathbf{D_0}(\mathbf{u_0})(t) \end{array}\right\} \qquad (26)$$

Then as shown in Fig. 3, the fault signal is estimated as

$$\tilde{\Delta} = abs(\mathbf{R}^{-1}(\mathbf{u_0}(t) - \mathbf{r_m}(t))) \qquad (27)$$

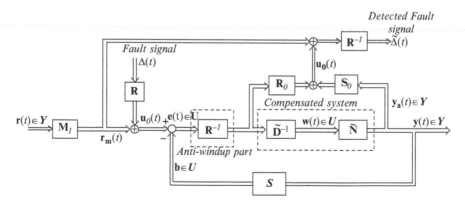

Fig. 5 Unknown disturbance estimation structure for MIMO system

4   Simulation Examples

The Mode 1 and Mode 2 implementation of MRAGPC are considered in this section. The process models used are the same. The objective is to control this MIMO system with and without disturbance load into the system.

4.1   Control of MIMO System Without Disturbance

The parameters of the 2-input 2-output system are given in Table 1 below.

4.2   Control of MIMO System with Disturbance

The process model (4.1) is considered here with disturbance load in the input. The process and control parameters are as shown in Table 2.

Table 1 Simulation parameters for MIMO process without disturbance

System parameters:		
$A_{11} = p^2 + p + 0.1$	$A_{22} = p^2 + 0.5p + 0.06$	$B_{11} = -0.2p + 10$
$B_{12} = 0.5p + 2$	$B_{21} = -0.1p + 1.2$	$B_{22} = -0.25p + 6$
$C_{11} = C_{22} = p + 0.8$	$m = q = 2$	
Controller parameters:		
$N_{y1} = N_{y2} = 10$	$N_{u1} = N_{u2} = 3$	$r_{m1} = r_{m2} = 0.5$
$U_{n11} = 0.3$	$U_{n22} = 0.2$	$u_{1,1} = u_{1,2} = 0$
$u_{2,1} = u_{2,2} = 0.1$	$r_1 = 10, r_2 = 2$	$\lambda_1 = \lambda_2 = 0.1$
$U_d = I_q$	$T_{1,1} = T_{1,2} = 0$	$T_{2,1} = T_{2,2} = 6$

Table 2 Simulation parameters for MIMO process with disturbance

System parameters:			
$A_{11} = p^2 + p + 0.1$	$A_{22} = p^2 + 0.5p + 0.06$	$B_{11} = -0.2p + 10$	
$B_{12} = 0.5\,p + 2$	$B_{21} = -0.1\,p + 1.2$	$B_{22} = -0.25\,p + 6$	
$C_{11} = C_{22} = p + 0.8$	$m = q = 2$	$\Delta_1 = \Delta_2 = 20$	
		$\sin(1.8t)$	
Compensator parameters:			
$\delta = 1, \gamma = 1$	$\beta_{0,11} = 1.21$	$\beta_{0,12} = 0.64$	$\beta_{0,21} = 0.64 \quad \beta_{0,22} = 0.06$
$\alpha_{0,11} = 1.476$	$\alpha_{0,12} = 4.51$	$\alpha_{0,21} = 0.88$	$\alpha_{0,22} = 0.004$
$\alpha_{1,11} = 3.105$	$\alpha_{1,12} = 1.25$	$\alpha_{1,21} = 1.25$	$\alpha_{1,22} = 0.112$
Controller parameters:			
$N_{y1} = N_{y2} = 12$	$N_{u1} = N_{u2} = 3$	$r_{m1} = r_{m2} = 1$	$U_{n11} = U_{n22}$
			$= 0.45$
$u_{1,1} = u_{1,2} = 0$	$u_{2,1} = u_{2,2} = 0.1$	$r_1 = 10, r_2 = 2$	$\lambda_1 = \lambda_2 = 0.1$
$U_d = I_q$	$T_{1,1} = T_{1,2} = 0$	$T_{2,1} = T_{2,2} = 6$	

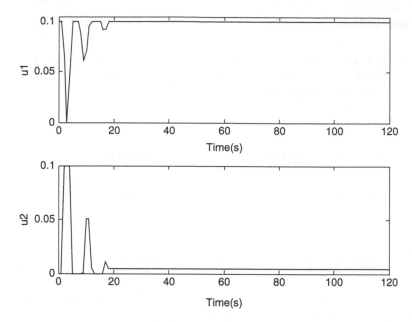

Fig. 6 Controlled inputs step responses for MIMO system without disturbance load

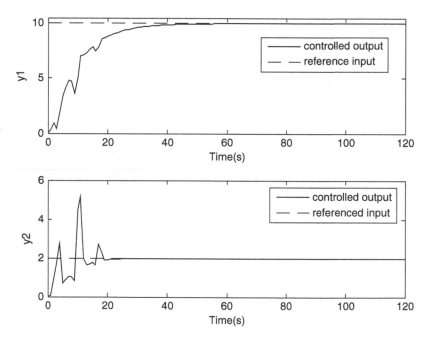

Fig. 7 Step responses of controlled outputs to step references of +10 and +2 respectively for MIMO process without disturbance load

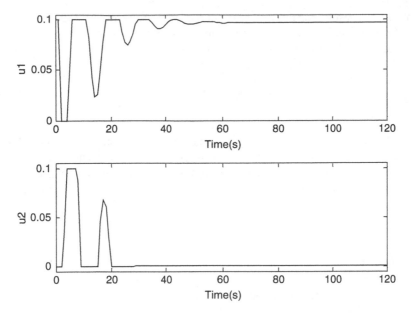

Fig. 8 Controlled inputs step responses for MIMO system with disturbance load

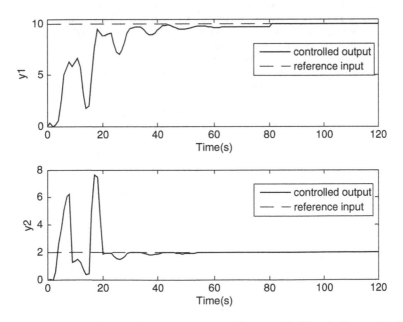

Fig. 9 Step responses of controlled outputs to step references of +10 and +2 respectively for MIMO process with disturbance load

Simulation results for the two cases considered are shown in Figs. 6 – 9. The control inputs saturate at 0.1 and the controlled outputs tracks to the given step reference inputs +10 and +2 for the two outputs respectively.

5 Conclusion

It is evident that the present control performance is better than the earlier work [12] and that the coupling effects have been considerably checked by chosen appropriate design parameters. Also, the designing of unknown disturbance estimation scheme confers another important feature on MRAGPC in serving as a fault detection scheme for MIMO processes.

References

1. H. Demircioglu, P.J. Gawthrop, Multivariable continuous-time generalized predictive control (MCGPC) Automatica **28**(4), 697–713 (1992)
2. M. Deng, A. Inoue, K. Edahiro, Fault detection in a thermal process system with input constraints using a robust right coprime factorization approach. Proc. IMechE Part I Syst. Control Eng. **221**(part I),819–831 (2007a)
3. A. Nassirhand, S.A. Hosseini, Tracking and decoupling of multivariable non-linear systems. Int. J. Model. Identification Contr. **6**(4): 341–348 (2009)
4. M. Deng, A. Inoue, N. Ishibashi, A. Yanou, Application of an anti-windup multivariable continuous-time generalized predictive control to a temperature control of an aluminum plate. Int. J. Model. Identification Contr. **2**(2), 130–1733 (2007b)
5. K. Zhou, J.C. Doyle, K. Glover, *Robust and Optimal Control* (Prentice-Hall, London, 1996)
6. A.S. Osunleke, M. Deng, A. Inoue, A robust anti-windup generalized predictive controller design for systems with input constraints and disturbances. Int. J. Innovative Comput. Inf. Contr. **5**(10B), 3517–3526 (2009a)
7. M. Deng, A. Inoue, K. Ishikawa, Operator-based nonlinear feedback control design using robust right coprime factorization. IEEE Trans. Automatic Contr. **51**(4), 645–648 (2006)
8. A.S. Osunleke, M. Deng, A. Yanou, A design procedure for control of strictly proper non-minimum phase processes with input constraints and disturbance. Intl. J. Model. Identification Contr. (2010) (in press)
9. A.S. Osunleke, M. Deng, A. Inoue, A. Yanou, Adaptive robust anti-windup generalized predictive control (RAGPC) of non-minimum phase systems with input constraint and disturbance, in *Proceedings of IEEE AFRICON 2009* (Nairobi, Kenya, 2009c)
10. M. Deng, Z. Iwai, I. Mizumoto, Robust parallel compensator design for output feed-back stabilization of plants with structured uncertainties. Syst. Contr. Lett. **36**, 193–198 (1999)
11. V.L. Kharitonov, Asymptotic stability of an equilibrium position of a family of systems of differential equations. Differentsialnye uravneniya **14**, 2086–2088 (1978)
12. A.S. Osunleke, M. Deng, A. Yanou, MRAGPC control of MIMO processes with input constraints and disturbance, Lecture notes in engineering and computer science, *World Congress on Engineering and Computer Science*, 20–22 Oct 2009b, pp. 872–877

Chapter 21
Analysis of Priority Rule-Based Scheduling in Dual-Resource-Constrained Shop-Floor Scenarios

Bernd Scholz-Reiter, Jens Heger, and Torsten Hildebrandt

Abstract A lot of research on scheduling manufacturing systems with priority rules has been done. Most studies, however, concentrate on simplified scenarios considering only one type of resource, usually machines. In this study priority rules are applied to a more realistic scenario, in which machines and operators are dual-constrained and have a re-entrant process flow. Interdependencies of priority rules are analyzed by long-term simulation. Strength and weaknesses of various priority rule combinations are determined at different utilization levels. Further insights are gained by additionally solving static instances optimally by using a mixed integer linear program (MILP) of the production system and comparing the results with those of the priority rules.

1 Introduction

Shop-scheduling has attracted researchers for many decades and is still of big interest, because of its practical relevance and its np-complete character. Multi- or dual-resource problems are significantly less analyzed, despite being more realistic. Scheduling with priority rules is quite appealing for various reasons (see Section 2.2) and often used, especially when considering more realistic and complex manufacturing systems.

This work analyses the quality of priority rules in the dual-resource constrained case. Different combinations of rules are tested in order to analyze their interdependencies. The analysis combines simulation with the optimal solution of static instances. To evaluate the results of the simulation, a mixed integer linear program (MILP) has been developed to calculate optimal solutions. The paper is organized

B. Scholz-Reiter (✉)
Bremer Institut für Produktion und Logistik (BIBA) at the University of Bremen, Hochschulring 20, D-28359 Bremen, Germany
e-mail: bsr@biba.uni-bremen.de

S.-I. Ao et al. (eds.), *Machine Learning and Systems Engineering*,
Lecture Notes in Electrical Engineering 68,
DOI 10.1007/978-90-481-9419-3_21, © Springer Science+Business Media B.V. 2010

as follows: Section 2 gives a short literature review about shop scheduling, priority rules and dual constrained scheduling. A problem description follows in Section 3. Section 4 analyses small instances in comparison with optimal solutions and Section 5 describes a long term simulation study. The paper ends with a conclusion and description of future research.

2 Literature Review

In this chapter a literature review is given considering the job shop problem, priority rules and dual-constrained scheduling.

2.1 Shop Scheduling

Haupt [1] gives a definition of the scheduling or sequencing problem as "the determination of the order in which a set of jobs (tasks) ($i|i = 1,\ldots, n$) is to be processed through a set of machines (processors, work stations) ($k|k = 1,\ldots, m$)." Usual abstractions of real-world scheduling problems are the job-shop and flow-shop problem. With both problem types the number of operations and their sequence are known in advance and fixed. In a Section scenario each job can have its own route whereas in the classical flow-shop all jobs share a common routing. Due to their high complexity optimal solutions can only be calculated for small problems. New, flexible approaches and different solutions need to be applied. Decentralized planning and scheduling as well as smart autonomous items are, in our opinion, a very promising approach to this (e.g. [2]). Decentralized decisions can be based on local decision rules. This is where priority rules come into play.

2.2 Priority Rules

Priority rules are applied to assign a job to a resource. This is done each time the resource gets idle and there are jobs waiting or a new job arrives. The priority rule assigns a priority to each job. This priority can be based on attributes of the job, the resource or the system. The job with the highest priority is chosen to be processed next.

Priority-scheduling rules have been developed and analyzed for many years [1, 3]. They are widely used in industry, especially in complex manufacturing systems, e.g. semiconductor manufacturing. Their popularity is derived from the fact that they perform reasonably well in a wide range of environments, are relatively easy to understand and need minimal computational time.

Depending on the manufacturing system and the various objectives (e.g. flow time, tardiness, etc.) no single rule has been found, which outperforms all others [4]. As a result, there are approaches to find new priority rules according to the system's current state (e.g. [3, 5]).

2.3 *Multi/dual-Resource Constrained Scheduling*

Most research has been done on the machine-only constrained problem, where machines are the only limiting resource. Nevertheless, closer to the 'real' world are multi- or dual-constrained (DRC) problems, where more than one resource restricts the output of the system and impacts shop performance. Gargeya and Deane [6] defined the multiple resource constrained job shop as "a job shop in which two or more resources are constraining output. The resources may include machines, labor and auxiliary resources. Dual-constrained job shops are constrained by two resources (e.g. machine and laborers). Dual-constrained job shops are thus a specific type of multiple resource constrained job-shop."

To solve the multi-resource constrained problem different approaches were proposed. Mati and Xie [7] developed a greedy heuristic. This heuristic is guided by a genetic algorithm in order to identify effective job sequences. Dauzère-Pérès et al. [8, 9] developed a disjunctive graph representation of the multi-resource problem and proposed a connected neighborhood structure, which can be used to apply a local search algorithm such as taboo search. Patel et al. [10] proposed a genetic algorithm for dual resource constrained manufacturing and compared different priority rules against different performance measures. In the study of Chen et al. [11], an integer optimization formulation with a separable structure is developed where both machines and operators are modeled as resources with finite capacities. By relaxing resource capacity constraints and portions of precedence constraints, the problem is decomposed into smaller sub-problems that are effectively solved by using a dynamic programming procedure.

3 Problem Description

ElMaraghy et al. [12] defined the machine/worker/job scheduling problem as: "Given process plans for each part, a shop capacity constrained by machines and workers, where the number of workers is less than the number of machines in the system, and workers are capable of operating more than one machine, the objective is to find a feasible schedule for a set of job orders such that a given performance criteria is optimized." An interesting object of investigation in DRC is the interaction effect of the resource constraints and how they impact the performance measure like the mean flow time.

The experimental design used in this study corresponds to the MiniFab scenario [13, 14], which is shown schematically in Fig. 1. For practical reasons the model has been simplified, yet the main properties remain.

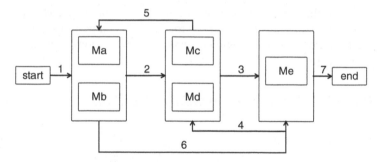

Fig. 1 Process flow of the 5 Machine 6 step production model

Table 1 Processing and (un)loading times used

	Machine	Loading	Processing	Unloading
Step1	Ma/Mb	15	55	25
Step2	Mc/Md	20	25	15
Step3	Me	15	45	15
Step4	Ma/Mb	20	35	25
Step5	Mc/Md	15	65	25
Step6	Me	10	10	10

The MiniFab scenario is a simplification of a semiconductor manufacturing system, but still contains the characteristics of such a system. It consists of five machines with 6-step re-entrant processes. Operators are used to load and unload machines. Machines Ma and Mb as well as Mc and Md are parallel identical machines sharing a common buffer. To be able to get optimal solutions, we do not consider sequence-dependent setup times, parallel batching and machine break downs.

Operations at the machines are divided into three parts: loading, processing and unloading. Processing starts directly after the machines are loaded. Unloading begins after processing, but if there is no operator available for unloading, the machine stays idle. In Table 1 processing, loading and unloading times are listed.

The processing and (un)loading times as well as the machine and sequence settings are fixed. The machines select a job out of the waiting queue and call an operator to load them (point in time a in Fig. 2). If no operator is available the machines wait until eventually they are loaded (b-c) and proceed with processing (c-d).

If an operator is available, unloading is performed immediately (e-f), otherwise there is an additional waiting period (d-e). The operators are not needed during processing and can work on other machines during that time period. We vary the number of operators from 0 to 3 to create scenarios constraining the available capacity in different ways. The four resulting scenarios are:

- NO_OPERATOR: operators are not considered at all in this case, operations require machines only (five machines, zero operators).
- MACHINE_CONSTRAINED: three operators are available, making machines the more critical resource. (five machines, three operators).

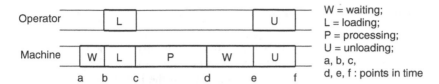

Fig. 2 Gantt chart illustration of machine-operator assignment

- OPERATOR_CONSTRAINED: there is just one operator; therefore operators are much more critical than machines. In fact in this scenario the operator becomes the capacity-constraining factor, reducing the maximum long-term system capacity to 6.857 instead of 13.714 orders per day (five machines, one operator).
- DUAL_CONSTRAINED: there are two operators in the model, resulting in machines being roughly as critical as operators. (five machines, two operators).

We consider FIFO (First In buffer First Out), FSFO (First in System first Out), Rnd (Random) and SPT (Shortest Processing Time first) as sequencing rules for the machines. Besides FIFO, FSFO, Rnd and SPT we consider two additional decision rules for operators: MQL (longest Machine Queue Length first) and SSPT (Shortest Step Processing Time first). With MQL, priority is given to machines with a long queue of waiting jobs. SSPT is similar to SPT, but instead of considering the processing time of the complete operation, only the length of the required load or unload processing steps are used. Furthermore, we vary the tie-breaker rule which is used as a secondary criterion if priorities of the primary rule are equal. Using FIFO, FSFO and Rnd no equal priorities can occur, so there is no need for a tie-breaker rule in these cases.

To decide which job to process next, we use a two-step procedure, first using the machine rule to select a job out of the waiting jobs in front of a machine. This sends a processing request to the operator pool. The operator rule is then used to decide between requests of different machines. If no operator is available immediately, the machine is kept in a waiting state.

In this study we only consider the common performance measure mean flow time. Our experiments are divided into two parts. Firstly we have experiments with only a few jobs, which enable us to compare optimal schedules with the schedules the priority rules provide (Section 4). Secondly we perform a long term simulation study, with a time span of ten years to analyze the priority rules in detail (Section 5).

4 Experiments with Static Instances

To determine not only the differences between priority rules and their combinations, it is interesting to analyze their performance in comparison to optimal solutions and evaluate their performance in general.

4.1 Experimental Design

We simulated instances from the same scenario with only 2–50 jobs in the system. All jobs were released at the same time and known in advance. This simplification makes it possible to calculate optimal solutions for the instances with few jobs in the system. For larger instances feasible schedules (solutions of the MILP) could be found, but they were not proven optimal. Due to the complexity of the model, there still was a gap between the found solution (upper bound) and the theoretically possible solution (lower bound).

The optimum results are calculated with CPLEX 11 solving a MILP-formulation, which is an extension of the advanced job-shop formulation used by Pan and Chen [15]. Operators and re-entrant processes were added to their flexible job-shop MILP, so that the MiniFab model as described above can be solved and optimal schedules are calculated. The notation used in the model is as follows:

F	Mean flow time	Z	Binary variables
m	Number of machines	$r_{i,j,g}$	1 if operation $O_{i,j}$ requires machine group g, 0 otherwise
n	Number of jobs	S_i	1 if $O_{i,j}$ is processed on machine k, 0 otherwise, binary
J_i	Job number i	s_i	Starting time of $O_{i,j}$
M_k	Machine number k	F_g	Group g of parallel machines
N	Number of operations	$QQ, QQ2$	Variables for optim. expressions
a	Operator action: 1 = load., 2 = unload.	$OS_{i,j,k,a}$	Starting time of operator h doing a with operation $O_{i,j}$
H	Number of operators	$OO_{i,j,k,a}$	1 if operator h does a with $O_{i,j}$
$O_{i,j}$	operation j of J_i	$bigM$	A very large positive number

$$Minimize\, \overline{F} \tag{1}$$

$$(OS_{i,j,h,2} + ut_j) - bigM(1 - oo_{i,j,h,2}) \le s_{i,j+1}, \tag{2a}$$
$$i = 1, 2...n; j = 1, 2...N - 1; h = 1..H$$

$$s_{i,j} + lt_{i,j} + p_{i,j} \le os_{i,j,h,2} + bigM(1 - oo_{i,j,h,2}), \tag{2b}$$
$$i = 1, 2...n; j = 1, 2...N; h = 1..H$$

$$bigM(3 - v_{i,j,k} - v_{ii,jj,k} - oo_{ii,jj,h,2} + bigM(1 - Z_{i,j,ii,jj}) \tag{3}$$
$$+ s_{ii,jj} - os_{i,j,h,2} - ut_{i,j} = Q_{i,j,ii,jj,k,h},$$
$$1 \le i \le ii \le n$$

$$Q_{i,j,ii,jj,k,h} \le bigM(7 - 2v_{i,j,k} - 2v_{ii,jj,k} - 2oo_{i,j,h,2}) + s_{ii,jj} \tag{4}$$
$$- os_{i,j,h,2} + ut_{i,j} + s_{i,j} - os_{ii,jj,h,2} - ut_{ii,jj}$$
$$1 \le i \le ii \le n, j, jj = 1, 2...N; h = 1..H, k = 1..m$$

$$\sum_{k \in F_g} v_{i,j,k} = r_{i,j,g} \tag{5}$$
$$i = 1, 2..n; j = 1, 2...N$$

$$os_{i,j,h,1} \ge s_{i,j} - (1 - oo_{i,j,h,1})bigM, \tag{6}$$
$$i = 1, 2..n; j = 1, 2...N; h = 1..H$$

$$os_{i,j,h,1} \le s_{i,j} + (1 - oo_{i,j,h,1})bigM, \tag{7}$$
$$i = 1, 2..n; j = 1, 2...N; h = 1..H$$

$$\sum_{h=1}^{H} (oo_{i,j,h,a}) = 1, \tag{8}$$
$$i = 1, 2..n; j = 1, 2...N; h = 1..H; a = 1..2$$

$$\tag{9}$$

(continued)

$$os_{i,j,h,a} \leq oo_{i,j,h,a}bigM,$$
$$i = 1,2..n; j = 1,2...N; h = 1..H; a = 1..2$$

$$os_{i,j,h,1} + lt_{i,j} - (3 - Y_{i,j,ii,jj,h,1}$$
$$-oo_{ii,jj,h,1} - oo_{i,j,h,1})bigM - os_{ii,jj,h,1} = QQ_{i,j,ii,jj,h} \qquad (10)$$
$$1 \leq i \leq ii \leq n; j, jj = 1,2...N; h = 1..H;$$

$$QQ_{i,j,ii,jj,h} \leq bigM(5 - 2oo_{ii,jj,h,1} - 2oo_{i,j,h,1}) - lt_{ii,jj} - lt_{i,j} \qquad (11)$$
$$1 \leq i \leq ii \leq n; j, j = 1,2...N; h = 1..H;$$

$$oo_{i,j,h,1} + lt_{i,j} - (3 - Y_{i,j,ii,jj,h,2} \qquad (12)$$
$$-oo_{ii,jj,h,2} - oo_{i,j,h,1})bigM - os_{ii,jj,h,2} = QQ2_{i,j,ii,jj,h}$$
$$1 \leq i \leq ii \leq n; j, jj = 1,2...N; h = 1..H;$$

$$QQ2_{i,j,ii,jj,h} \leq bigM(5 - 2oo_{ii,jj,h,2} - 2oo_{i,j,h,1}) - lt_{i,j} - ut_{i,j} \qquad (13)$$
$$1 \leq i \leq ii \leq n; j, jj = 1,2...N; h = 1..H;$$

$$\overline{F} = \sum_{i=1}^{n} (os_{i,N,h,2} + ut_{i,N} - bigM(1 - oo_{i,N,h,2}))/n \qquad (14)$$
$$h = 1..H;$$

[mean flow time in minutes]		number of jobs										
		2	3	4	5	6	7	8	9	10	20	50
NO_OPERATOR	optimum	483	520	568	622	678	733	785	870	918		
	gap (lower/upper bound)	0%	0%	0%	0%	0%	8%	0%	27%	37%		
	best rule	483	520	568	628	684	739	799	861	908	1424	2989
	worst rule	483	520	568	637	716	794	881	956	1040	1940	4616
	span best-worst	0%	0%	0%	1%	5%	7%	10%	11%	15%	36%	54%
	FSFO SSPT[FSFO]	483	520	568	628	684	739	799	861	908	1424	2989
MACH_CONSTRAINED	optimum	483	520	568	622							
	gap (lower/upper bound)	0%	0%	0%	0%							
	best rule	483	520	568	628	684	739	809	869	919	1437	3039
	worst rule	483	520	568	637	716	804	905	998	1095	2085	5050
	span best-worst	0%	0%	0%	1%	5%	9%	12%	15%	19%	45%	66%
	FSFO SSPT[FSFO]	483	520	568	628	684	739	809	869	930	1437	3042
DUAL_CONSTRAINED	optimum	483	524	569	640	746	838	930				
	gap (lower/upper bound)	0%	1%	0%	3%	9%	19%	16%				
	best rule	483	525	589	655	708	781	854	912	967	1527	3173
	worst rule	483	530	620	724	823	916	999	1135	1229	2325	5500
	span best-worst	0%	1%	5%	11%	16%	17%	17%	24%	27%	52%	73%
	FSFO SSPT[FSFO]	483	528	598	656	708	789	875	924	979	1547	3173
OP_CONSTRAINED	optimum	483	520	568	929	1088	1272					
	gap (lower/upper bound)	0%	0%	0%	33%	38%	47%					
	best rule	535	643	756	917	1013	1118	1225	1351	1456	2524	5695
	worst rule	548	685	842	1041	1243	1423	1616	1807	2022	4024	10010
	span best-worst	2%	6%	11%	13%	23%	27%	32%	34%	39%	59%	76%
	FSFO SSPT[FSFO]	540	643	803	918	1013	1118	1229	1351	1486	2541	5695

Fig. 3 Results of static scenarios in minutes

In our study, we were able to calculate some optimal solutions, which in most cases were only slightly better than the best rule combination. In Fig. 3 we list the solver results with the remaining gap and the performance of the best and worst rule combination taken from any of the 72 combinations (see Section 3). For a detailed analysis of the static runs, we chose the FSFO ShortestOpStepLength [FSFO] rule, which seems to be the best performing one. In Fig. 3 the corresponding graphs are plotted. The FSFO ShortestOpStepLength [FSFO] rule is indicated by the black line. The grey area defines the corridor between best and worst rule solutions. Black dots correspond with the solver results, gaps between found solutions and lower bounds are printed as vertical lines.

4.2 Analyses of Static Instances

About four or five jobs need to be in the system and first differences between good and bad rule combinations can be found. An example is the DUAL_ CONSTRAINED case with five jobs in the system: The rule combination FSFO (machine) ShortestOpStepLength [FSFO] (operator) has a mean flow time of 656 min and the combination FIFO (machine) FSFO (operator) has 724 min. This is already a difference of more than 10%. The difference to the solver solution is only around 2%.

The results of the scenario with NO_OPERATORS and the MACHINE_RES-TRICTED scenario are very similar as Figs. 3 and 4 show. When there are more

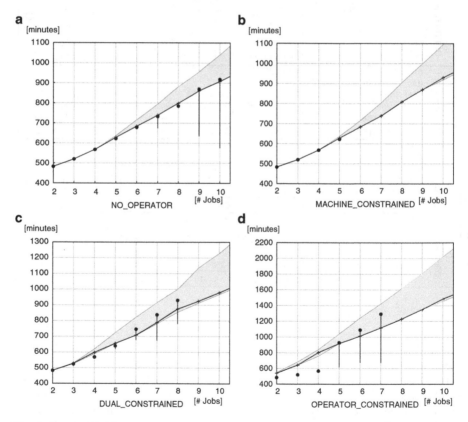

Fig. 4 Results for all static cases. Black line: FSFO ShortestOpStepLength [FSFO] rule, grey area: best/worst rule combination result, black points: Solver results with gap (line)

jobs in the system the performance differences between the rule combinations arise. In the DUAL_CONSTRAINED case, the performance differences between rule combinations are much higher and arise in smaller scenarios. This result was expected, because in the dual constrained case the interdependencies between the machine priority rule and the operator priority rule are the strongest and amplify each other.

In the OPERATOR_CONSTRAINED case the shop is restricted the most, which leads very quickly to high mean flow times. The differences between the rule combinations are also very high in this case. The solution the solver provided for the cases with three and four jobs indicate that either the rule performance or the order of selection (machines first, operators second) do not perform very well in this scenario. It seems more likely that the order of selection is responsible for this effect. The operator is clearly the bottleneck and the scheduling should be arranged in a way that utilizes him as best as possible, which is clearly not the case.

In instances with more jobs we were not able to proof the same effect, because the solver provided no optimal solutions even after days of calculations. The solutions found were comparable to those of the used rules.

5   Long-Term Simulation

To validate the results from our static analyses, an extensive simulation study was performed with a time horizon of 10 years.

5.1   Long-Term Simulation

To assess the performance of the various rule combinations in the long term, we simulate the system under three load conditions for each of the four scenarios: 70%, 80% and 90% bottleneck utilization. Given the scenario and load level, we determine the appropriate arrival rate based on a static analysis of system capacity. Inter-arrival times follow an exponential distribution. We simulate a time span of 10 years, ignoring data from the first year. Altogether we determine system performance for 864 different parameter settings (4 scenarios, 3 load levels, 72 rule combinations). For the different parameter settings we use common random numbers as a variance reduction technique [16]. The results presented are the averages of mean flow times achieved in 20 independent replications.

5.2   Analysis of Long-Term Simulations

Figure 5 shows graphically the scenario, load level setting and the best rule result achieved in this case. The four lines correspond to our scenarios:

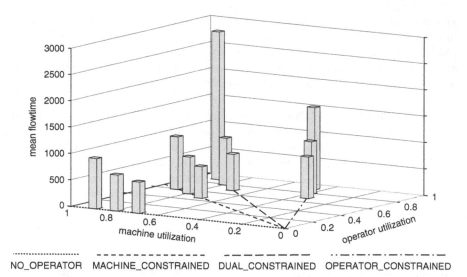

Fig. 5 Best rule performance for the four scenarios and three load level settings

- NO_OPERATOR (dotted line)
- MACHINE_CONSTRAINED (dashed line with short dashes)
- DUAL_CONSTRAINED (dashed line with long dashes)
- OPERATOR_CONSTRAINED (dot-and-dashed line)

Selecting one of the scenarios fixes the achievable combination of machine and operator utilization along the respective line. To give an example: if MACHINE_-CONSTRAINED is chosen and a load level of 90% bottleneck utilization (in this case: machine utilization), then this results in an operator utilization of 60%. The grey bar at this point shows the best flow time achieved, i.e. out of the 72 rule combinations investigated.

The best results can obviously be achieved if no operator constraints are present at all. This corresponds to the dotted line (NO_OPERATOR) in our simulation; we also have the lowest flow times in this scenario, compared to the other scenarios with the same load level. The DUAL_CONSTRAINED case (long dashes) on the other hand is the most difficult, operator and machines are equally critical.

Figure 6 lists our simulation results in more detail. Mean flow time as measured in our simulation experiment is expressed there as a flow factor, i.e. flow time divided by total processing time. The best and worst rule performances for a given scenario and load level are highlighted. To summarize our results, FSFO is the best machine rule irrespectively of the scenario and load level. Results concerning the best operator rule are a bit more complex: the MACHINE_CONSTRAINED and OPERATOR_CONSTRAINED scenarios where the operator rule SSPT[FSFO], i.e. Shortest Step Processing Time first with FSFO is used as a tie breaker, the best results are gained if used together with FSFO as a machine rule.

machine rule [tie breaker]	operator rule [tie breaker]	NO_OPERATOR (0 Operator) utilisation of bottleneck			MACH_CONSTRAINED (1 Operator) utilisation of bottleneck			OP_CONSTRAINED (2 Operator) utilisation of bottleneck			DUAL_CONSTRAINED (3 Operator) utilisation of bottleneck			Mean (flow factor)
		70%	80%	90%	70%	80%	90%	70%	80%	90%	70%	80%	90%	
FIFO	FIFO	1,42	1,74	2,71	1,49	1,94	3,71	2,26	3,40	8,26	2,22	6,28	so	3,22
FIFO	Rnd	1,42	1,74	2,71	1,50	1,95	3,81	2,29	3,48	8,93	2,27	6,90	so	3,36
FIFO	SSPT[FSFO]	1,42	1,74	2,71	1,46	1,83	3,08	1,88	2,50	4,44	1,76	2,70	so	2,32
FSFO	SSPT[FIFO]	1,32	1,52	2,13	1,34	1,56	2,25	1,83	2,31	3,68	1,54	2,00	so	1,95
FSFO	SSPT[FSFO]	1,32	1,52	2,13	1,34	1,56	2,25	1,76	2,21	3,51	1,58	2,27	so	1,95
FSFO	SPT[FIFO]	1,32	1,52	2,13	1,35	1,59	2,41	1,92	2,68	7,26	1,56	2,61	so	2,39
Rnd	Rnd	1,42	1,73	2,69	1,49	1,94	3,87	2,27	3,45	8,91	2,29	7,99	so	3,46
Rnd	SSPT[FSFO]	1,42	1,73	2,69	1,45	1,81	3,05	1,87	2,49	4,44	1,77	2,79	so	2,32
SPT[FIFO]	MQL[FSFO]	1,37	1,65	2,44	1,40	1,70	2,59	2,02	2,77	5,23	1,73	2,53	13,14	2,31
SPT[FIFO]	SPT[FSFO]	1,37	1,65	2,44	1,41	1,73	2,69	2,40	3,94	12,06	2,15	5,23	so	3,37
SPT[FSFO]	MQL[FSFO]	1,37	1,65	2,44	1,40	1,70	2,59	1,90	2,63	5,10	1,73	2,55	13,20	2,28
SPT[FSFO]	SPT[FSFO]	1,37	1,65	2,44	1,41	1,73	2,69	2,40	3,94	12,06	2,15	5,23	so	3,37
SPT[Rnd]	MQL[FSFO]	1,37	1,65	2,44	1,40	1,70	2,59	1,90	2,64	5,11	1,73	2,54	13,71	2,28
SPT[Rnd]	SPT[FIFO]	1,37	1,65	2,44	1,41	1,72	2,69	2,35	3,77	10,48	2,12	5,00	so	3,18
flowfactor best rule		1,32	1,52	2,13	1,34	1,56	2,25	1,76	2,21	3,51	1,54	2,00	6,56	1,95
flowfactor worst rule		1,42	1,74	2,71	1,50	1,95	3,87	2,40	3,94	12,06	2,29	7,99	14,65	3,46
flowtime span ((worst-best)/best)		8,2%	14,7%	27,3%	12,2%	24,8%	72,5%	36,0%	78,0%	243,0%	48,6%	299,6%	123,5%	77,4%

Fig. 6 Results for selected (worst/best) rule combinations investigated. Except for the last row (span) all values are flow factors, i.e. mean flow time divided by the processing time (445 min). The last column contains a mean performance averaged over all scenario/load level combinations (but excluding DUAL_CONSTRAINED/90%, see text); so (system overflow) more incoming than outgoing jobs

The FIFO rule performs quite badly in our experiments. In most cases it does not yield better results than random selection, i.e. choosing an arbitrary job to process next.

The DUAL_CONSTRAINED scenario SSPT[FIFO] is the best, i.e. SSPT with tie breaker FIFO, for the 70% and 80% load levels. This changes if load is further increased to 90%. Most rule combinations are not able to handle this high load for both machines and operators - the system is running full, meaning more jobs are entering the system than leaving it. Cases are marked with "so" in the table. Only combinations with MQL as an operator rule are able to handle this high load. The best results in this case are produced by the combination of the FSFO machine rule and MQL[FIFO] as an operator rule. The MQL - rule prefers machines with long queues, which means that operators go to highly loaded machines first. Although this rule combination gives by far the best results for this case, the increase in average flow time is very high.

Comparing the results for MACHINE_CONSTRAINED and NO_OPERATOR, the increase in mean flow time is only small, especially for lower load levels. If only flow time estimates are of interest, it seems viable to simplify the model by ignoring the operator requirements.

The experiments with small instances show that in the OPERATOR_CONSTRAINED cases, the optimal solutions are much better than the results produced by our heuristics. In the long term simulation, the best rule results are higher for scenarios with moderate load levels of 70% and 80% compared to the DUAL_CONSTRAINED. Only the 90% load level results are smaller than its DUAL_CONSTRAINED equivalent, but still about 56% higher than in the MACHINE_CONSTRAINED scenario. This seems to be an effect of the heuristic procedure used (machines choose next job first, then operators choose between the machines that are ready to process). In future research more simulation runs with a different heuristic setup will be performed to analyze this effect in more detail.

In summary, all four scenarios show that the performance of priority rules differs tremendously in some cases. The interdependencies of the rules, especially in the DUAL_CONSTRAINED scenarios, lead to high performance differences. Other system factors, e.g. the utilization rate, also affect the results.

6 Conclusion and Further Research

The paper analyses a priority-rule based approach to shop-floor scheduling considering both machines and operators. Our dual-constrained scenario was derived from semiconductor manufacturing. To gain insights into priority rule performance we use two approaches: simulation of small problem instances and its comparison with optimal solutions and simulation of long term system behavior.

Especially in shop environments, where more than one resource is constraining the system's performance, proper selection and combination of priority rules is crucial. The long term simulation has also shown that in high utilization scenarios, many rule combinations are not able to handle the workload at all.

Comparison of known optimal solutions and best rule results indicates only a small difference between the optimum and the best rule result. The differences between well performing rule combinations and unsuitable combinations is much higher than the differences to optimal solutions. They show that well adjusted priority rules provide good solutions.

It is interesting to see, that the rule combinations' performance depends on the utilization level of the system. This brings up the idea to develop a control system, which selects or develops priority rules automatically, considering system parameters and adopting to changing shop floor conditions.

Acknowledgements This research is funded by the German Research Foundation (Deutsche Forschungsgemeinschaft DFG), under grant SCHO 540/17-1.

References

1. R. Haupt, A survey of priority rule-based scheduling. OR Spektrum, **11**(1), 3–16 (1989)
2. B. Scholz-Reiter, M. Görges, T. Philipp, Autonomously controlled production systems-influence of autonomous control level on logistic performance. CIRP Ann. Manuf. Technol. **58**(1), 395–398 (2009)
3. C.D. Geiger, R. Uzsoy, Learning effective dispatching rules for batch processor scheduling. Int. Journal Prod. Res. **46**(6), 1431–1454 (2008)
4. C. Rajendran, O. Holthaus, A comparative study of dispatching rules in dynamic flowshops and jobshops. Eur. J. Oper. Res. **116**(1), 156–170 (1999)
5. K.-C. Jeong, Y.-D. Kim, A real-time scheduling mechanism for a exible manufacturing system: using simulation and dispatching rules. Int. J. Prod. Res. **36**(9), 2609–2626 (1998)
6. V.B. Gargeya, R.H. Deane, Scheduling research in multiple resource constrained job shops: a review and critique. Int. J. Prod. Res. **34**, 2077–2097 (1996)

7. Y. Mati, X. Xie, A genetic-search-guided greedy algorithm for multi-resource shop scheduling with resource flexibility, IIE Trans. **40**(12), 1228–1240 (2007)
8. S. Dauzère-Pérès, W. Roux, J.B. Lasserre, Multi-resource shop scheduling with resource flexibility. Eur. J. Oper. Res. **107**(2), 289–305 (1998)
9. S. Dauzère-Pérès, C. Pavageau, Extensions of an integrated approach for multi-resource shop scheduling, IEEE Transactions on Systems, Man. and Cybernetics, Part C: Applications and Reviews (2003)
10. V. Patel, H.A. ElMaraghy, I. Ben-Abdallah, Scheduling in dual-resources constrained manufacturing systems using genetic algorithms, 7th IEEE International Conference on Emerging Technologies and Factory Automation. Proceedings. ETFA 99 (1999)
11. D. Chen, P.B.c Luh, Optimization-based manufacturing scheduling with multiple resources, setup requirements, and transfer lots. IIE Trans. **35**, 973–985 (2003)
12. H. ElMaraghy, V. Patel, I.B. Abdallah, Scheduling of manufacturing systems under dual-resource constraints using genetic algorithms. J. Manuf. Syst. **19**(3), 186–201 (2000)
13. G. Feigin, J. Fowler, R. Leachman Masm test data sets (2009). http://www.eas.asu.edu/masmlab. Last accessed on 28 May 2009
14. M.K. El Adl, A.A. Rodriguez, K.S. Tsakalis Hierarchical modeling and control of re-entrant semiconductor manufacturing facilities, Proceedings of the 35th Conference on Decision and Control Kobe, Japan (1996)
15. J.C.-H. Pan, J.-S. Chen, Mixed binary integer programming formulations for the reentrant job shop scheduling problem. Comput. Oper. Res. **32**(5), 1197–1212 (2005)
16. A.M. Law, *Simulation Modeling and Analysis*, 4th edn. (McGraw-Hill, Boston, MA, 2007)

Chapter 22
A Hybrid Framework for Servo-Actuated Systems Fault Diagnosis

Seraphin C. Abou, Manali Kulkarni, and Marian Stachowicz

Abstract A hybrid fault diagnosis method is proposed in this paper which is based on analytical and fuzzy logic theory. Analytical redundancy is employed by using statistical analysis. Fuzzy logic is then used to maximize the signal- to-threshold ratio of the residual and to detect different faults. The method was successfully demonstrated experimentally on hydraulic actuated system test rig. Real data and simulation results have shown that the sensitivity of the residual to the faults is maximized, while that to the unknown input is minimized. The decision of whether 'a fault has occurred or not?' is upgraded to 'what is the severity of that fault?' at the output. Simulation results show that fuzzy logic is more sensitive and informative regarding the fault condition, and less sensitive to uncertainties and disturbances.

1 Introduction

Hydraulic systems are very commonly used in industry. Like any other system these systems too are prone to different types of faults. Proportional valves are much less expensive in hydraulic control applications; they are more suitable for industrial environments. Since proportional valves do not contain sensitive, precision components, they offer various advantages over servo valves because they are less prone to malfunction due to fluid contamination. However, these advantages are offset by their nonlinear response characteristics. Since proportional valves have less precise manufacturing tolerances, they suffer from performance degradation. Larger tolerances on spool geometry result in response nonlinearities, especially in the vicinity of neutral spool position. Proportional valves lack the smooth flow properties of "critical center" valves, a condition closely approximated by servo

S.C. Abou (✉)
Mechanical and Industrial Engineering Department, University of Minnesota Duluth, 1305 Ordean Ct, Duluth, MN 55812, USA
e-mail: sabou@d.umn.edu

S.-I. Ao et al. (eds.), *Machine Learning and Systems Engineering*,
Lecture Notes in Electrical Engineering 68,
DOI 10.1007/978-90-481-9419-3_22, © Springer Science+Business Media B.V. 2010

284   S.C. Abou et al.

valves at the expense of high machining cost. As a result, small changes in spool
geometry (in terms of lapping) may have large effects on the hydraulic system
dynamics [1]. Especially, a closed-center spool (overlapped) of proportional valve,
which usually provides the motion of the actuator in a proportional hydraulic
system, may result in the steady state error because of its dead-zones characteristics
in flow gain [1]. Figure 1 illustrates the characteristics of proportional valve.
Continuous online monitoring of fault in hydraulic system becomes increasingly
important day-by-day.

The characteristics of the proportional valve with dead-zones, $g(u)$ is described
as follow (Fig. 1):

$$g(u) = \begin{cases} a(u-b) & \text{if } u \geq b \\ 0 & \text{if } -b \leq u \leq b \\ a(u+b) & \text{if } u < -b \end{cases} \tag{1}$$

where b, $a \geq 0$; a represents the slope of the response outside the dead-zone, while
the width of the dead-zone equals 2b.

Remarkable efforts have been devoted to develop controllers. However, PID
controllers are not robust to the parameter variation to the plants being controlled.
Moreover, it takes time for the automatically self tuned PID controllers to adapt
themselves up to their final stable state (steady state). The fault detection problem
can be solved using different approaches like Wald's Sequential Test, as in [1]
which is a conventional approach or using innovative approaches like genetic
algorithms as in [2], neural networks as in [3, 4], fuzzy logic as in [5] etc. each
having its own advantages and disadvantages.

Human experts play central roles in troubleshooting or fault analysis. In power
systems, it is required to diagnose equipment malfunctions as well as disturbances.
The information available to perform equipment malfunction diagnosis is most of
the time incomplete. In addition, the conditions that induce faults may change with
time. Subjective conjectures based on experience are necessary. Accordingly, the

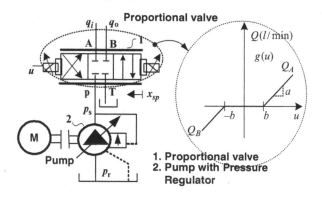

Fig. 1 Characteristics of 4/3 closed-center proportional valve

expert systems approach has proved to be useful. As stated previously, fuzzy theory can lend itself for representation of knowledge and the building of an expert system. In this paper we used fuzzy logic to detect the severity of fault at the output. The concept of fuzzy logic was first introduced by Professor Lofti Zadeh [6], who represented the vagueness of human concepts in terms of linguistic variables. After the introduction of fuzzy sets, their applications to solve real world problems were concentrated [7, 8]. Reference [9] concentrates on robust fault detection on an aircraft flight control system. A model based fault diagnosis method for an industrial robot is proposed in [10]. Residuals are calculated by the observer using a dynamic robot model and later evaluated using fuzzy logic. In this paper we demonstrate a similar *model based* approach for evaluating of severity of fault in the hydraulic actuator using fixed threshold approach [11]. The *objective knowledge* on the system is represented by mathematical modeling (calculating the residuals using nonlinear observer) [1], while the *subjective knowledge* is represented using fuzzy logic (fuzzy rules and membership functions).

2   System Under Consideration

The schematic of the hydraulic system under consideration, the mathematical model and the design of nonlinear observer can be found in [1]. General nonlinear dynamical systems can be described as follows:

$$\begin{cases} f(x, y, u, t) = A_0 x + Bh(x) + \eta(x) \\ y = C^T x \end{cases} \tag{2}$$

where $h(x) = l(x) + \varphi(x)u + \beta(t - T)\phi(x)$; φ and l are unknown smooth functions. $x = (x_1, x_2, x_3, ..., x_n)$ is the state vector, $x \in R^n$, u and $y \in R$; $z = \hat{y}$ is the observer output vector; $\eta(x)$ represents the uncertainty in the system dynamics that may include parameter perturbations, external disturbances, noise, etc. All long this study, we consider abrupt fault at time T. As a result, $\beta(t - T)$ is the time profile of failures as shown in Fig. 2.

$\phi(x)$ is another function which represents a failure in the system. Mathematical description of the nonlinear observer as follows:

$$\begin{cases} \dot{\hat{x}} = A\hat{x} + B\hat{h}(x) + K(y - C\hat{x}) \\ z = \hat{y} = C^T \hat{x} \end{cases} \tag{3}$$

where $\hat{h}(x|\theta_f) = \hat{l}(x|\theta_f) + \hat{\varphi}(x|\theta_f)u + \phi(x|\theta_f)$; K is the observer gain matrix selected as follows $A = A_0 - KC^T$, which is strictly a Hurwitz matrix.

In the discrete time system, consider from (2) and (3) $e(k) = y(k) - z(k)$. The actual state of the system $y(k)$ is known through the sensors. The residual $e(k)$ is calculated as follows:

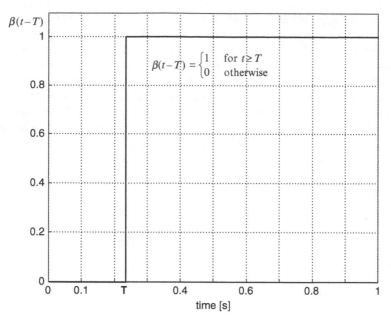

Fig. 2 Function $\beta(t - T)$ in (2)

$$e(k) = M_p y(k) - M_p z(k) \tag{4}$$

M_p is an identity matrix of size $m \times n$, $M_p = I_{4 \times 1}$

It is perceived that the performance of the actuated system is selected based on four parameters having a range of value from zero (0) to one (1). The elements of the state vector $z \approx [v \; P_i \; P_o \; x_{sp}]^T$ are: velocity $\dot{x} = v$, input pressure P_i, output pressure P_o and x_{sp} spool displacement. The residual of these four can be measured. In this paper we have concentrated on the *velocity residual* and the identity matrix $M_p = [1 \; 0 \; 0 \; 0]$.

Theoretically, these residuals should be zero under no fault condition. However, in practical context, due to noise, inexact mathematical modeling and system nonlinearity, this residual is never zero even under no fault condition. Reference [1] uses a conventional method called *Wald's Sequential Test* to detect fault. In this method, the cumulative residual error is calculated over a period of time and fault is detected using the fixed threshold concept. This conventional method has some disadvantages. A value just below the threshold is not considered as a fault while some value just above the threshold will be considered as a fault. This can also lead to missing alarms and false triggers. This information could be potentially misguiding to the operators working on the hydraulic system. This is the drawback of binary logic.

The conventional method is rigid and does not consider a smooth transition between the faulty and the no fault condition. The probability assignment procedure is heuristic and depends on the number of Zeros/Ones in the failure signature. This does not give any information about the fault in between the thresholds. In order to take care of this condition we try to replace this binary logic by multi-valued one using fuzzy logic. Evaluating these residuals using fuzzy logic replaces the yes/no decision of fault by the severity of fault at the output.

3   Role of Fuzzy Logic

From the point of view of human-machine cooperation, it is desirable that faults classification process would be interpretable by humans in such a way that experts could be able to evaluate easily the classifier solution. Another interest of an intelligent interface lies in the implementation of such a system in a control room. Operators have to be informed very quickly if a fault is occurring. They have to understand what exactly the process situation is, in order to make the right counteraction if possible or to stop the system if necessary. For instance, as shown in Fig. 3, the fuzzification vector can be assigned to an index of a color map, representing a color code, by defuzzification. Figure 3 depicts the overall architecture of the hydraulic system fault detection where u(t) is the control input.

The mapping of the inputs to the outputs for the fuzzy system is in part characterized by a set of *condition → action* rules (if-Then) form:

$$\textit{If} \text{ premise } \textit{then} \text{ consequent} \tag{5}$$

The inputs of the fuzzy system are associated with the *premise*, and the outputs are associated with the *consequent*. As already seen, the difference between the expected state z(k) and the actual state of the system y(k) gives the *residual* e(k).

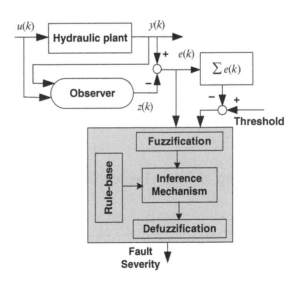

Fig. 3 The structure of Fuzzy fault diagnosis

The value of residual is added over a period of time which gives the cumulative residual $\sum e(k)$. This value is subtracted from the predicted threshold and is called *cumulative residual difference*.

The lower the value of this cumulative residual difference, higher is the fault severity, indicating that the cumulative residual is approaching the threshold and vice versa. The threshold is determined through observations. It will vary depending upon the fault tolerance of the application in which the hydraulic system is used. Even if there is no fault, the modeling errors or noise drive several residuals beyond their threshold. This is usually indicated by all suspect residuals being weak. The *residual* is bounded between the upper and the lower threshold. As soon as it approaches these thresholds, the fault severity increases. Thus, the *residual* and the *cumulative residual difference* are given as two inputs to the fuzzy logic controller. Based on these two inputs, the controller decides the fault severity at the output. One of the equations of fuzzy equality can be written as:

$$
S_{A,B} = \frac{\sum\limits_{i=1}^{n} \min\{\mu_A(i), \mu_B(i)\}}{\max\left\{\sum\limits_{i=1}^{n} \mu_A(i), \sum\limits_{i=1}^{n} \mu_B(i)\right\}}
\tag{6}
$$

where n is the dimension number of the discrete space, μ_A is the membership function of fuzzy set A, μ_B is the membership function of fuzzy set B.

Suppose that a fuzzy rule set to be detected F represents the current working class of the actuated hydraulic system, and the other fuzzy reference rule set F_i stands for one working class of the system. Since both the fuzzy reference rule sets and fuzzy rule set to be detected have the same hypersphere subspaces, the Eq. (6) can be used for their contrasts. As a result, in this study, the approximate measurement of the fuzzy reference rule set F_i can be expressed by:

$$
s_i(k) = \frac{|e_i(k)|}{e_i}
\tag{7}
$$

where $e_i(k)$ *is* the ith residual; $s_i(k)$ is the residual-to-threshold ratio.

Obviously $s_i(k)$ is greater than or equal to 1 if the test is fired on the residual and $s_i(k)$ is less than 1 if it did not.

4   Design of Fuzzy Logic Controller

On one hand, note that the fuzzy reference rule sets almost cover all the faults; while on the other, the fuzzy rule set to be detected may bring forward the undefined symptoms which cannot be distinguished from fuzzy reference rule sets, as shown in Fig. 3. *How can we solve this problem?*

To solve this problem, we include in the evaluation procedure an additional credit degree of unknown classes which is expressed as follows:

$$\varepsilon_{ei} = 1 - \max\left(S_{A,B}\right)_{i=1}^{n}$$

$$\text{Obviously, } 0 \leq S_{A,B} \leq 1 \tag{8}$$

4.1 Inputs

Figure 4 illustrates the actual and the estimate velocities. The difference is due to the error introduced in the actual system by adding random noise to the *velocity* during simulation.

The plot of residual, cumulative residual, cumulative residual difference along with the thresholds can be seen in Figs. 5 and 6. As seen earlier, the residual and the cumulative residual difference are the two inputs to the fuzzy logic controller. Fault isolation thresholds are very important parameters; their values are also decided by the statistical analysis of the fault credit degrees [13]. As for unknown fault type, consulting fault isolation thresholds are selected upon our knowledge on the system. The detection results of the normal data of the space propulsion

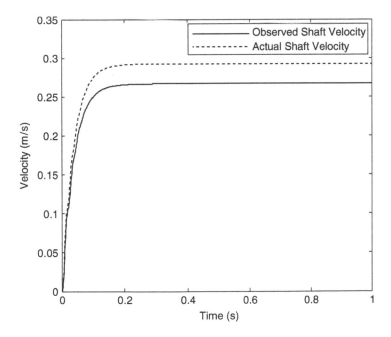

Fig. 4 Graph showing Actual velocity and observed velocity vs time

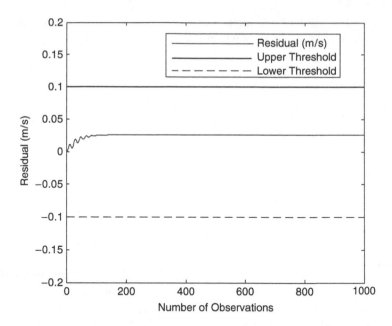

Fig. 5 Graph showing 'Residual' along with the upper and lower thresholds vs 'number of observations'

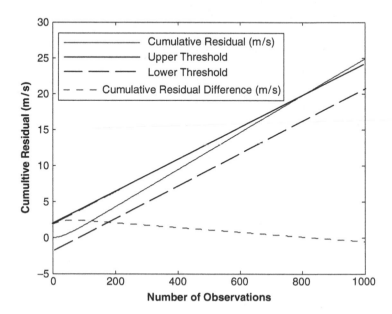

Fig. 6 Cumulative residual and the cumulative residual difference along with the upper and lower thresholds vs the number of observations

system are shown in Fig. 5. Because normal credit degree does not exceed the threshold, the detection results are that no fault exists, and working conditions are normal.

4.2   Membership Functions

The first input which is residual is divided into seven membership functions namely, Big Negative (BN), Negative (N), Small Negative (SN), Zero (Z), Small Positive (SP), Positive (P) and Big Positive (BP) shown below Fig. 7. Similarly, we developed five membership functions for the second input which is cumulative residual difference. They are Large Negative (LNeg), Medium Negative (MNeg), Small Negative (SNeg), Zero (Zero) and Positive (POS) as seen in the following Fig. 8. As already seen there are four parameters which can be used to calculate the residuals. Among them the *velocity* is the most concerned parameter in this case of study.

Hence, the *velocity residual* is selected to determine the fault severity at the output. The membership functions for the output i.e. fault severity are F0, F1, F2, F3, F4, F5 and F6 where F0 represents the lowest fault severity and F6 represents the highest fault severity Fig. 9.

The shapes of the membership functions which are triangular and trapezoidal were selected based on the simple guidelines suggested in [12]. This can be seen in Fig. 9.

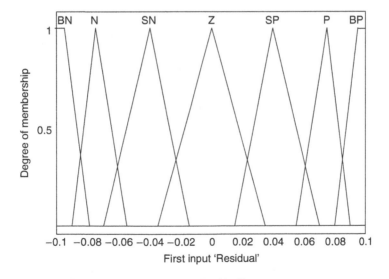

Fig. 7 Membership functions for the first input 'Residual'

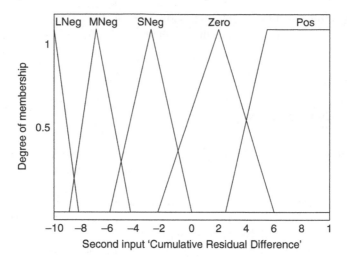

Fig. 8 Membership functions for the second input 'Cumulative Residual Difference'

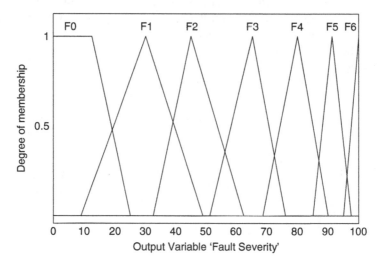

Fig. 9 Membership functions for the output 'Fault Severity'

4.3   Rule-Based Inference

Inference rules were developed which relate the two inputs to the output. They are summarized in the Table 1. As seen from the table, there are in all 35 rules. For example, if the residual is Big Positive (BP) and the cumulative residual difference is Large Negative (LNeg) then the output fault severity is the highest (F6). Similarly, if the residual is Zero (Z) and the cumulative residual difference is Positive (Pos) then the output fault severity is the lowest (F0).

Table 1 Rule based inference

		BN	NEG	SN	Z	SP	POS	BP
					Residual			
Cumulative residual difference	POS	F3	F2	F1	F0	F1	F2	F3
	Zero	F4	F3	F2	F1	F2	F3	F4
	SNeg	F5	F4	F3	F2	F3	F4	F5
	MNeg	F6	F5	F4	F3	F4	F5	F6
	Lneg	F6	F6	F5	F4	F5	F6	F6

4.4 Defuzzification

After converting the crisp information into fuzzy the last step is to reverse that. Converting the fuzzy information to crisp is known as defuzzification. The center of area/centroid method was used to defuzzify these sets which can be represented mathematically as follows:

$$Defuzzified\ value = \frac{\Sigma f_i \cdot \mu(f_i)}{\Sigma \mu(f_i)} \tag{9}$$

Where f_i is the fault severity at the output and $\mu(f_i)$ is the output membership function.

4.5 Rule Viewer

The rules can also be seen from the rule viewer using the fuzzy logic toolbox in MATLAB software. When the residual is 0.01, it is far away from both the upper and lower thresholds (almost at the center) and hence, has lower fault severity. Also, the cumulative residual difference is nine which means the difference between the actual value of cumulative residual and threshold is high i.e. cumulative residual is far away from the threshold. Hence, the fault severity should be low. A combination of these values of residual and cumulative residual gives fault severity percentage of 9.96% which is low. Similarly, when the residual is 0.089 it indicates that it is very close to the threshold. A cumulative residual difference of -9 indicates that the threshold has been already crossed by the cumulative residual (hence it is negative). Both of these conditions lead to a very high fault severity of 98.4%. This can be seen with the help of the rule viewer facility in the fuzzy logic toolbox. These examples are shown in Figs. 10 and 11 respectively with the help of rule viewer.

5 Simulation

This simulation was carried out in MATLAB SIMULINK using fuzzy logic controller from the fuzzy logic toolbox as shown in Figs. 10 and 11. The upper subsystem represents the actual system (actual state of the hydraulic system) and

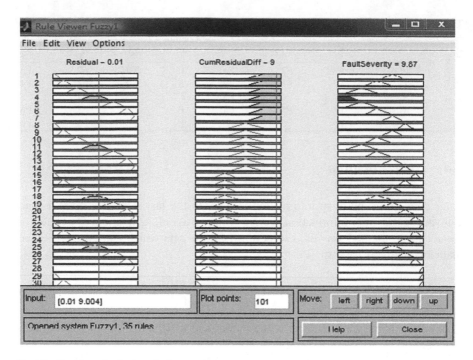

Fig. 10 Test Results for low fault severity

the lower subsystem is the nonlinear observer (which predicts the state of the system). The SIMULINK diagram is the implementation of the block diagram shown in Fig. 1. The simulation is carried out for a unit step input. Fault is introduced in the actual system by adding noise to the *velocity* in the actual system and different fault severities are tested at the output.

6 Conclusion

With the looming reliability challenges in future new technologies, the ability to provide practicing engineers online information about the systems health is vital for decision-making. The approach proposed in this study showed that fuzzy logic, when used in combination with analytical methods like non linear observer, can enhance the output. Simulation results clearly demonstrated what we all know: whatever fault type the plant generates, its symptoms always depart from the characteristics of the fuzzy reference rule set standing for the normal working condition. Moreover while fuzzy rule sets are set up, the fuzzy reference rule set generated is used for representing the normal working condition, which is supposed to be the first fuzzy reference rule set. Thus, with the credit degree representing the normal working condition, we judged whether the plant working condition is

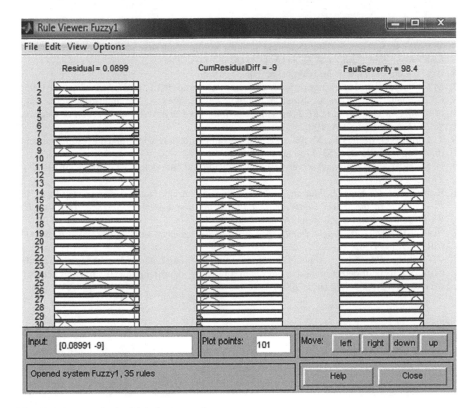

Fig. 11 Test results for high fault severity

normal, further obtain the fault degree. This study helped to assure that the plant fault existed and to report the system conditions. Future work will be developed to identify the fault type and predict the equipment remaining life.

Acknowledgement The present work has been performed in the scope of the activities of Grant In Aid project. The financial support of the University of Minnesota, Grant In Aid FY 2008 is gratefully acknowledged.

References

1. H. Khan, S.C. Abou, N. Sepehri, Nonlinear observer-based fault detection technique for electro hydraulic servo positioning systems. Mechatronics **15**(9), 1037–1059 (Nov 2005)
2. D. Xuewu, L. Guangyuan, L. Zhengji, Discrete-time robust fault detection observer design: a genetic algorithm approach, in *IEEE Proceedings of the 7th World Congress on Intelligent Control and Automation*, Chongqing, June 2008
3. G. Qian-jin Guo, Y. Hai-bin Yu, X. Ai-dong, Modified Morlet wavelet neural networks for fault detection. Int. Conf. Control Automat. (ICCA 2005) **2**, 1209–1214 (June 2005)

4. Q. Sheng, X.Z. Gao, Z. Xianyi, State-of-the-art in soft computing-based-motor fault diagnosis. IEEE **2**, 1381–1386 (June 2003)
5. V. Pedro, R. Jover, A. Antero, Detection of stator winding fault in induction motor using fuzzy logic. Appl. Soft Comput. **8**(2), 1112–1120 (Mar 2007)
6. L.A. Zadeh, Fuzzy sets, Inf. Control **8**, 338–35, 1965
7. M.M. Jerry, Fuzzy logic systems for engineering: a tutorial. IEEE **83**(3), 345–377 (Mar 1995)
8. C. Stephen, J. C. John, C. Sitter. E. Fooks, Perspectives on the industrial application of intelligent control, in *IEEE Proceedings of the 34th Conference on Decision & Control*, vol. 1. New Orleans, LA, 1995
9. T. Curry, E.G. Collins Jr., M. Selekwa, Robust fault detection using robust l_1 estimation and fuzzy logic. IEEE, **2**, 1753–1758 (June 2001)
10. H. Shneider, P.M. Frank, Observer-based supervision and fault detection in robots. IEEE **4**(3), 274–282 (May 1996)
11. K. Manali, S.C. Abou, S. Marian, Fault detection in hydraulic system using fuzzy logic, In *Proceedings of the World Congress on Engineering and Computer Science 2009 Vol. II; WCECS 2009*, San Francisco, USA, 20–22 Oct 2009
12. Y. John, L. Reza, *Fuzzy Logic: Intelligence, Control and Information*. (Prentice Hall, Englewood Cliffs, NJ, 1999), pp. 22–27
13. C.T. Shing, P.L. Chee, Application of an adaptive neural network with symbolic rule extraction to fault detection and diagnosis in a power generation plant. IEEE Trans. Energy Convers. **19**(2), 369–377 (June 2004)

Chapter 23
Multigrid Finite Volume Method for FGF-2 Transport and Binding

Wensheng Shen, Kimberly Forsten-Williams, Michael Fannon, Changjiang Zhang, and Jun Zhang

Abstract A multigrid finite volume method has been developed to accelerate the solution process for simulating complex biological transport phenomena, involving convection, diffusion, and reaction. The method has been applied to a computational model which includes media flow in a cell-lined cylindrical vessel and fibroblast growth factor-2 (FGF-2) within the fluid capable of binding to receptors and heparan sulfate proteoglycans (HSPGs) on the cell surface. The differential equations were discretized by the finite volume method and solved with the multigrid V-cycle algorithm. Our work indicates that the multigrid finite volume method may allow users to investigate complex biological systems with less CPU time.

1 Introduction

Basic fibroblast growth factor (FGF-2) is the prototypical member of the family of fibroblast growth factors [14]. It has been demonstrated to be important in normal physiology and pathologies in cancer development process [4]. In addition to its target cell surface tyrosine kinase receptor, FGF-2 also binds with high affinity to heparan sulfate proteoglycan (HSPG), which consists of a protein core with O-linked carbohydrate chains that are polymers of repeating disaccharides [6]. HSPGs are present on almost all cell surfaces and generally in much higher numbers than FGF surface receptors. FGF-2 is present in circulation and its presence in elevated levels in blood is used in clinical settings as criteria for treatment strategies such as interferon alpha therapy for infantile hemangioma [5]. Although FGF-2 binding interactions have been the subject of numerous studies in static systems, far less is known about their behavior in flow.

W. Shen (✉)
SUNY College at Brockport, Brockport, NY 14420, USA
e-mail: wshen@brockport.edu

S.-I. Ao et al. (eds.), *Machine Learning and Systems Engineering*,
Lecture Notes in Electrical Engineering 68,
DOI 10.1007/978-90-481-9419-3_23, © Springer Science+Business Media B.V. 2010

The interaction between biochemical reaction and mass transfer is of particular interest for the real time analysis of biomolecules in microflow systems. Glaser [12] proposed a two-dimensional (2D) computer model, which was greatly simplified by assuming that the flow is parallel and diffusion is perpendicular to the surface, to study antigen-antibody binding and mass transport in a microflow chamber. Myszka et al. [15] used a computer model similar to that of [12], to simulate the binding between soluble ligands and surface-bound receptors in a diagnostic device, BIACORE, for both transport-influenced and transport-negligible binding kinetics. A 2D convection and diffusion equation was used by Chang and Hammer [2] to analytically investigate the effect of relative motion between two surfaces on the bindings of surface-tethered reactants. We take a different approach by solving the 2D convection diffusion equation directly.

2 Methods

2.1 Mathematical Model

A coupled nonlinear convection-diffusion-reaction model for simulating growth factor binding under flow conditions has recently been developed by [10, 18] and is used here. This model can be applied to predict the time-dependent distribution of FGF-2 in a capillary with complicated binding kinetics on the tube surface.

For simplicity, we consider a basic model, where FGF-2 is the only ligand in the solution, and cells which line the capillary express both FGF-2 receptors (FGFRs) and HSPGs. As illustrated in Fig. 1, FGF-2 binds to FGFR and HSPG to form complexes of FGF-2-FGFR and FGF-2-HSPG. The resulting complexes may continue binding to produce either their dimers or FGF-2-FGFR-HSPG triads. The FGF-2-FGFR-HSPG triads may further bind to each other to generate FGF-2-FGFR-HSPG dimers. Consequently, the concentration of FGF-2 in the solution is affected due to the molecular binding process. The 2D time-dependent equations for

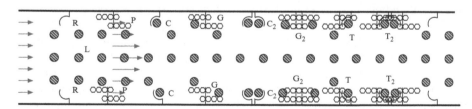

Fig. 1 Sketch of growth factor binding to receptors and HSPGs and the formation of various compounds on the surface of a capillary. The symbols in the sketch are as follows: L=FGF-2, R=FGFR, P=HSPG, C=FGF-2-FGFR complex, G=FGF-2-HSPG complex, C_2=FGF-2-FGFR dimer, G_2=FGF-2-HSPG dimer, T=FGF-2-FGFR-HSPG complex, and T_2=FGF-2-FGFR-HSPG dimer. The arrows represent velocity vectors, which are uniform in the entrance and evolve to parabolic later on

fluid mass and momentum transfer in conservative form for incompressible flow can be written as:

$$\frac{\partial \rho}{\partial t} + \frac{1}{r}\frac{\partial(r\rho v_r)}{\partial r} + \frac{\partial(\rho v_z)}{\partial z} = 0, \tag{1}$$

$$\frac{\partial(\rho v_r)}{\partial t} + \frac{1}{r}\frac{\partial(r\rho v_r v_r)}{\partial r} - \frac{1}{r}\frac{\partial}{\partial r}\left(r\mu\frac{\partial v_r}{\partial r}\right) + \frac{\partial(\rho v_r v_z)}{\partial z} - \frac{\partial}{\partial z}\left(\mu\frac{\partial v_r}{\partial z}\right) = -\frac{\partial p}{\partial r} - \mu\frac{v_r}{r^2}, \tag{2}$$

$$\frac{\partial(\rho v_z)}{\partial t} + \frac{1}{r}\frac{\partial(r\rho v_r v_z)}{\partial r} - \frac{1}{r}\frac{\partial}{\partial r}\left(r\mu\frac{\partial v_z}{\partial r}\right) + \frac{\partial(\rho v_z v_z)}{\partial z} - \frac{\partial}{\partial z}\left(\mu\frac{\partial v_z}{\partial z}\right) = -\frac{\partial p}{\partial z}, \tag{3}$$

where ρ is the density, μ the dynamic viscosity, p the dynamic pressure, and v_r and v_z are velocity components in the radial and axial directions, respectively. In the above equation set, the independent variables are time t, radial coordinate r, and axial coordinate z.

The mass of each species must be conserved. If binding or reactions occur within the fluid, the coupling of mass transport and chemical kinetics in a circular pipe can be described by the following equations:

$$\frac{\partial(\rho\phi_i)}{\partial t} + \frac{1}{r}\frac{\partial(\rho r u\phi_i)}{\partial r} + \frac{\partial(\rho v\phi_i)}{\partial x} = \frac{1}{r}\frac{\partial}{\partial r}\left(K_r r\frac{\partial(\rho\phi_i)}{\partial r}\right)$$
$$+ \frac{\partial}{\partial x}\left(K_x\frac{\partial(\rho\phi_i)}{\partial x}\right) + F_i(\phi_1\cdots\phi_n), \quad 1 \le i \le n, \tag{4}$$

where ϕ_i is the concentration of species i, u and v are the radial and longitudinal components of velocity, K_r and K_z the molecular diffusion coefficients, and F_i the rate of change due to kinetic transformations for each species i. The basic model however has only FGF-2 within the fluid (ϕ_i is simply FGF-2) and, thus, reactions (i.e., binding and dissociation) occur only on the capillary surface. That is to say that F_i is valid merely on the tube surface. The reactants and products involved in the chemical kinetics include FGF-2, FGFR, HSPG, FGF-FGFR complex and its dimer, FGF-HSPG complex and its dimer, FGF-HSPG-FGFR complex and its dimer, with a total of nine species ($n = 9$) [10].

2.2 Collocated Finite Volume Discretization

A cell-centered finite volume approach is applied to discretize the partial differential equations. Eqs. 1–4 can be re-organized in the form of convection diffusion equations and written as

$$\frac{\partial}{\partial t}(\rho\phi) + \frac{1}{y}\frac{\partial}{\partial y}\left(y\rho v\phi - y\Gamma\frac{\partial\phi}{\partial y}\right) + \frac{\partial}{\partial x}\left(\rho u\phi - \Gamma\frac{\partial\phi}{\partial x}\right) = S, \tag{5}$$

where x and y are axial and radial coordinates, u and v are the velocity components in the axial and radial directions, respectively. It is worth noticing that the mass conservation equation is a special case of Eq. 5 in which ϕ, Γ, and S are taken as $\phi = 1$, $\Gamma = 0$, and $S = 0$.

To achieve higher order temporal accuracy, we use a quadratic backward approximation for the time derivative term. Such arrangement gives us second order temporal accuracy. Integrating Eq. 5, the corresponding finite volume equations can be derived,

$$\frac{3(\rho\phi)_P^{n+1} - 4(\rho\phi)_P^n + (\rho\phi)_P^{n-1}}{\Delta t} + (J_e - J_w) + (J_n - J_s) = S_C + S_P\phi_P + S_{sym},$$

(6)

where $J_{e, w, n, s}$ is the convection-diffusion flux at each of the four interfaces of the control volume P, with $J_{e,w} = F_{e,w} - D_{e,w}$, $J_{n,s} = F_{n,s} - D_{n,s}$, S_C and S_P are the results of source term linearization, and S_{sym} is the contribution from axisymmetric coordinates, with $S_{sym} = -\Gamma\frac{v}{y^2}$ for the momentum equation of v. The approximation of convective flux is critical for an accurate solution. We use a so-called deferred correction technique [8]. This technique calculate higher-order flux explicitly using values from the previous iteration. Using deferred correction, which is a combination of the first order upwind differencing and the second order central differencing, the convective flux is written as a mixture of upwind and central differences, $F_{e,w} = F_{e,w}^u + \lambda(F_{e,w}^c - F_{e,w}^u)^o$, where $F_e^u = \max((\rho u)_e\Delta r_j, 0.)\phi_P + \min((\rho u)_e\Delta r_j, 0.)\phi_E$, $F_w^u = \max((\rho u)_w\Delta r_j, 0.)\phi_W + \min((\rho u)_w\Delta r_j, 0.)\phi_P$, $F_e^c = (\rho u)_e\Delta r_j(1 - \alpha_e)\phi_P + (\rho u)_e\Delta r_j\alpha_e\phi_E$, $F_w^c = (\rho u)_w\Delta r_j(1 - \alpha_w)\phi_W + (\rho u)_w\Delta r_j\alpha_w\phi_P$, λ is a parameter set as $\lambda = 0 \sim 1$, and the superscript $(^o)$ indicates taking the value from the previous iteration, which will be taken to the right hand side and treated as a part of the source term. The same can be applied to convective flux in radial direction. The interpolation factors are defined as $\alpha_e = \frac{x_e - x_P}{x_E - x_P}$, $\alpha_w = \frac{x_P - x_w}{x_P - x_W}$, $\alpha_n = \frac{r_n - r_P}{r_N - r_P}$, and $\alpha_s = \frac{r_P - r_s}{r_P - r_S}$. The diffusion fluxes are $D_e = \frac{K_x y_j(\phi_E - \phi_P)}{x_E - x_P}$, $D_w = \frac{K_x y_j(\phi_P - \phi_W)}{x_P - x_W}$, $D_n = \frac{K_r x_i(\phi_N - \phi_P)}{r_N - r_P}$, and $D_s = \frac{K_r x_i(\phi_P - \phi_S)}{r_P - r_S}$. The notations of spatial discretization in Eq. 6 is illustrated in Fig. 2, where the uppercase letters indicate the center of the control volumes, and the lowercase letters indicate the interfaces between neighboring control volumes.

Substituting numerical fluxes into Eq. 6 and collecting terms, a set of algebraic equations are obtained of the following form

$$A_S\phi_S + A_W\phi_W + A_P\phi_P + A_E\phi_E + A_N\phi_N = b.$$

(7)

The coefficients of Eq. 7 consist of a pentadiagonal matrix, which is solved using an efficient linear system solver that will be discussed later. The coefficients and the right-hand side in Eq. 7 are given by

$$A_S = -\max((\rho v)_s\Delta x_i, 0.) - \frac{K_r\Delta x_i r_s}{r_P - r_S},$$

(8a)

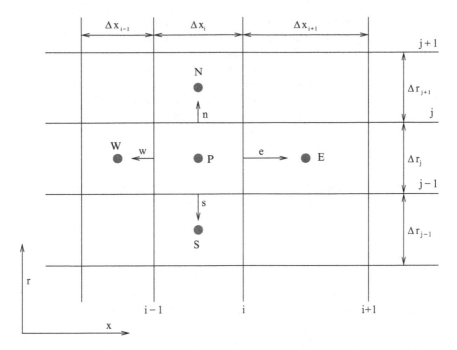

Fig. 2 Finite volume notation of control volumes in axisymmetrical coordinates

$$A_W = -\max((\rho u)_w \Delta r_j, 0.) - \frac{K_x \Delta r_j r_P}{x_P - x_W}, \tag{8b}$$

$$A_N = \min((\rho v)_n \Delta x_i, 0.) - \frac{K_r \Delta x_i r_n}{r_N - r_P}, \tag{8c}$$

$$A_E = \min((\rho u)_e \Delta r_j, 0.) - \frac{K_x \Delta r_j r_P}{x_E - x_P}, \tag{8d}$$

$$A_P = \frac{3\rho r_P \Delta x_i \Delta r_j}{2\Delta t} - (A_W + A_S + A_E + A_N), \tag{8e}$$

$$b = (S_C + S_P) r_P \Delta x_i \Delta r_j + \left(\frac{4.0\rho \phi_P^n}{2.0\Delta t} - \frac{\rho \phi_P^{n-1}}{2.0\Delta t}\right) r_P \Delta x_i \Delta r_j$$
$$- \lambda(F_e^c - F_e^u - F_w^c + F_w^u + F_n^c - F_n^u - F_s^c + F_s^u). \tag{8f}$$

2.3 Multigrid Methods

With the finite volume method on a structured grid, the multigrid version in 2D can be constructed such that each coarse grid control volume is composed of four control volumes of the next finer grid [8]. To do this, a grid generator that is able to generate multiple grids is used. The grid generator takes input of the number of grid levels, the number of lines along each boundary, the coordinates of starting and ending points for each line, the line type, etc., for the coarsest grid. The data of the rest levels of grids are computed automatically by subdividing each control volume in finer ones and saved as binary files to be loaded by the flow solver.

The algebraic Eq. 7 can be written more abstractly as

$$A\phi = b, \tag{9}$$

where A is a square matrix, ϕ the unknown vector, and b the source vector. After the kth outer iterations on a grid with spacing h, the intermediate solution satisfies the following equation,

$$A_h^k \phi_h^k - b^k = r_h^k, \tag{10}$$

where r_h^k is the residual vector after the kth iteration. Once the approximate solution at the kth iteration is obtained, we restrict the approximate solution as well as the residual to the the next coarse grid by the restriction operators I_h^{2h} and $\hat{I}_h^{2h}$. An approximate solution to the coarse grid problem can be found by solving the following system of equations,

$$A_{2h}\phi_{2h}^1 = A_{2h}(I_h^{2h}) - \hat{I}_h^{2h}r_h^k. \tag{11}$$

After the solution on the coarse grid is obtained, the correction $\Delta\phi = \phi_{2h}^1 - \phi_{2h}^0$ is transfered to the fine grid by interpolation (prolongation), where $\phi_{2h}^0 = I_h^{2h}\phi_h^k$. The difference of I_h^{2h} and $\hat{I}_h^{2h}$ is as follows, I_h^{2h} takes the mean value of states in a set of cells, but $\hat{I}_h^{2h}$ performs a summation of residuals over a set of cells. The value of ϕ_h^k is updated by

$$\phi_h^{k+1} = \phi_h^k + I_{2h}^h\Delta\phi, \tag{12}$$

where I_{2h}^h is a prolongation operator. This procedure is repeated until the approximate solution on the finest grid converges using the multigrid V-cycle algorithm [16]. The relaxation technique used in the multigrid method is Stone's strong implicit procedure (SIP), which is a modification from the standard ILU decomposition [19]. This paper has adapted the multigrid method for incompressible Navier–Stokes equations provided by [16], and extended it to include the mass transport. Since the concentration of ligand is very small, in the order of 10^{-11} to 10^{-10} M, we may assume that the momentum and mass transfer equations are independent to each other. Consequently the momentum transfer equation can be solved first to obtain the velocity distribution, which is then put into the mass transfer equation.

3   Results

The flow model in the current paper is the well-known incompressible Navier–Stokes equation, which has been widely used to solve low speed fluid mechanics problems. A convection-diffusion equation is applied to address the mass transport of growth factor in solution. A similar mass transport model has been used by [12] to describe the biological interaction of antigen and antibody, and by [15] to investigate the interactions of a variety of biomolecules in BIACORE, where the flow cell was of rectangular geometry. However, application of the 2D Navier–Stokes equation to the study of growth factor binding to cell surface receptors and HSPGs lining a cylindrical capillary has not, to the best of our knowledge, been reported. The numerical procedure and simulation results for a single grid is presented in a previous paper [18], while the main purpose of the current paper is to demonstrate a multigrid solution to the convection-diffusion-reaction model of mass transport and chemical kinetics of protein ligands.

The dimensions of the capillary under consideration were length $L = 0. 1$ m and radius $R = 0. 00035$ m, corresponding to that of a Cellmax Cartridge System bioreactor (FiberCell Systems, Inc., Frederick, MD, USA). The ratio of length over diameter is 286. To simulate the capillary flow, four types of boundary conditions are used in the numerical simulation: inlet boundary to the left-hand side, outlet boundary to the right-hand side, symmetry boundary at the bottom, and impermeable wall boundary at the top. For inlet boundary, all quantities are prescribed, and the incoming convective flux is calculated as well. For outlet boundary, zero gradient along the grid line is applied. A three-level multigrid method is considered, as shown in Fig. 3, and the number of control volumes are 40×5, 80×10, and 160×20, respectively. Each time the grid is refined, one control volume is divided into four smaller control volumes. The ligand interactions within the capillary were modeled as a problem of mass transport with a reactive boundary condition. For the mass transport equation, the boundary conditions are: given concentration on the west, zero flux on the east, symmetrical condition on the south, and reaction boundary on the north. An existing model of biochemical reactions on the capillary surface from [10] is used in our simulations (illustrated in Fig. 1). The reaction rates for all species are represented by a set of ordinary differential equations (ODEs), as shown in Table 1, and the parameters are primarily determined by experiments [10]. The system of ordinary differential equations are solved by a variable-coefficient ODE solver VODE for stiff and nonstiff systems of initial value problems [1]. The code is developed to solve the time-dependent Navier–Stokes and convection-diffusion-reaction equations with a particular application in growth factor transport and binding. The computation is performed on a Sun-Blade-100 machine with a single 500 MHz SPARC processor and 2 GB memory. A typical numerical solution is plotted in Fig. 4, where the concentration distribution of fibroblast growth factor inside the capillary is shown at an instant time of $t = 10$ min. The solution in Fig. 4 corresponds to the finest grid arrangement in the multigrid system. Note that the ligand concentration in the figure is non-

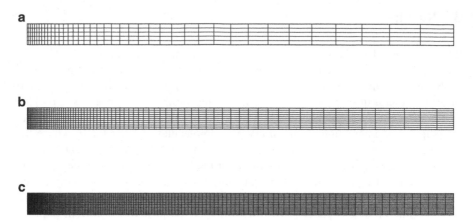

Fig. 3 Computational grids for capillary flow and protein transport: (**a**) first level grid, (**b**) second level grid, (**c**) third level grid

dimensionalized with respect to the inlet concentration on the west boundary. To demonstrate the impact of surface binding on ligand transport, two numerical experiments have been conducted, one without surface binding, Fig. 4(a), and the other with surface binding to receptors and heparan sulfate proteoglycans, Fig. 4(b). As expected, without surface binding, displayed in Fig. 4(a), the ligand moves with the flow by convection and diffusion, and its concentration has a uniform distribution along the radial direction in a portion of the capillary from $x = 0$ to roughly $x = 0.05$ m. One may further predict that a uniform concentration distribution in the whole capillary will be obtained after $t = 20$ min. It clearly shows in Fig. 4(b) that the concentration of ligand in the capillary is spatially reduced by a great margin down along the capillary as well as in the radial direction due to biochemical reactions on the surface. In Fig. 4, the front of ligand transport is dissipative, which may be explained by the relative importance between convection and diffusion. The Peclet number is defined as the ratio of convective transport to diffusive transport, $Pe = uL/D$, where u is the velocity, L is the characteristic length, and D is the diffusion coefficient. The characteristic length can have a significant impact on the value of Pe. Under current simulation, the velocity is $u = 0.0867$ mm/s, the diffusion coefficient is $D = 9.2 \times 10^{-7}$ cm^2/s. Using the capillary radius as the characteristic length, the resulting Peclet number is $Pe = 330$. This value indicates a convective dominated process however the value is somewhat intermediate indicating diffusion does make a not insignificant contribution to the overall process.

Figure 5(a) displays the convergence history for capillary flow in a single grid with the finest grid spacing. In the current calculation, the energy equation is not considered. The results clearly show that the residuals are reduced effectively in the first 200 outer iterations, and after that, they cannot be further eliminated no matter how many more outer iterations are performed. The results also indicate that the pressure equation has a relatively large residual (Fig. 5(a)).

Table 1 The rate of concentration change due to protein interactions (the initial concentration of surface variables are $R_0 = 1.6 \times 10^4$ receptors/cell and $P_0 = 3.36 \times 10^5$ sites/cell)

Reaction rate	Parameters
$\dfrac{d[R]}{dt} = -k_f^R[L][R] + k_r^R[C] + k_r^T[T]$ $- k_c[R][G] - k_{int}[R] + V_R$	$k_f^R = 2.5 \times 10^8 \,\mathrm{M}^{-1}\,\mathrm{min}^{-1}, k_r^R = 0.048 \,\mathrm{min}^{-1}$ $k_r^T = 0.001 \,\mathrm{min}^{-1}, k_c = 0.001 \,\mathrm{min}^{-1}(\#/\mathrm{cell})^{-1}$ $k_{int} = 0.005 \,\mathrm{min}^{-1}, V_R = 80 \,\mathrm{sites}\,\mathrm{min}^{-1}$
$\dfrac{d[P]}{dt} = -k_f^P[L][P] + k_r^P[G] + k_r^T[T]$ $- k_c[C][P] - k_{int}[P] + V_P$	$k_f^P = 0.9 \times 10^8 \,\mathrm{M}^{-1}\,\mathrm{min}^{-1}, k_r^P = 0.068 \,\mathrm{min}^{-1}$ $k_r^T = 0.001 \,\mathrm{min}^{-1}, k_c = 0.001 \,\mathrm{min}^{-1}(\#/\mathrm{cell})^{-1}$ $k_{int} = 0.005 \,\mathrm{min}^{-1}, V_P = 1,680 \,\mathrm{sites}\,\mathrm{min}^{-1}$
$V\dfrac{d[L]}{dt} = -k_f^R[L][R] + k_r^R[C] + k_r^T[T]$ $- k_f^P[L][P] - k_r^R[G]$	$k_f^R = 2.5 \times 10^8 \,\mathrm{M}^{-1}\,\mathrm{min}^{-1}$ $k_f^P = 0.9 \times 10^8 \,\mathrm{M}^{-1}\,\mathrm{min}^{-1}$ $k_r^R = 0.048 \,\mathrm{min}^{-1}, k_r^T = 0.001 \,\mathrm{min}^{-1}$
$\dfrac{d[C]}{dt} = k_f^R[L][R] - k_r^R[C] - k_c[C][P]$ $- k_c[C]^2 + 2k_{uc}[C_2] - k_{int}[C]$	$k_f^R = 2.5 \times 10^8 \,\mathrm{M}^{-1}\,\mathrm{min}^{-1}$ $k_r^R = 0.048 \,\mathrm{min}^{-1}, k_c = 0.001 \,\mathrm{min}^{-1}(\#/\mathrm{cell})^{-1}$ $k_{uc} = 1 \,\mathrm{min}^{-1}, k_{int} = 0.005 \,\mathrm{min}^{-1}$
$\dfrac{d[C_2]}{dt} = \dfrac{k_c}{2}[C]^2 - k_{uc}[C_2] - k_{int}^D[C_2]$	$k_c = 0.001 \,\mathrm{min}^{-1}(\#/\mathrm{cell})^{-1}, k_{uc} = 1 \,\mathrm{min}^{-1}$ $k_{int}^D = 0.078 \,\mathrm{min}^{-1}$
$\dfrac{d[G]}{dt} = -k_f^R[L][R] + k_r^R[C] + k_r^T[T]$ $- k_c[R][G] - k_{int}[R]$	$k_f^R = 2.5 \times 10^8 \,\mathrm{M}^{-1}\,\mathrm{min}^{-1}$ $k_r^R = 0.048 \,\mathrm{min}^{-1}, k_r^T = 0.001 \,\mathrm{min}^{-1}$ $k_c = 0.001 \,\mathrm{min}^{-1}(\#/\mathrm{cell})^{-1}, k_{int} = 0.005 \,\mathrm{min}^{-1}$
$\dfrac{d[G_2]}{dt} = \dfrac{k_c}{2}[G]^2 - k_{uc}[G_2] - k_{int}[G_2]$	$k_c = 0.001 \,\mathrm{min}^{-1}(\#/\mathrm{cell})^{-1}, k_{uc} = 1 \,\mathrm{min}^{-1}$ $k_{int}^D = 0.078 \,\mathrm{min}^{-1}$
$\dfrac{d[T]}{dt} = k_c[R][G] + k_c[C][P] - k_r^T[T]$ $- k_c[T]^2 + 2k_{uc}[T_2] - k_{int}[T]$	$k_c = 0.001 \,\mathrm{min}^{-1}(\#/\mathrm{cell})^{-1}, k_r^T = 0.001 \,\mathrm{min}^{-1}$ $k_{uc} = 1 \,\mathrm{min}^{-1}, k_{int} = 0.005 \,\mathrm{min}^{-1}$
$\dfrac{d[T_2]}{dt} = \dfrac{k_c}{2}[T]^2 - k_{uc}[T_2] - k_{int}^D[T_2]$	$k_c = 0.001 \,\mathrm{min}^{-1}(\#/\mathrm{cell})^{-1}, k_{uc} = 1 \,\mathrm{min}^{-1}$ $k_{int}^D = 0.078 \,\mathrm{min}^{-1}$

A comparison was then made between the single-grid and multigrid computations, shown in Fig. 5(b), where the number of iterations means the number of outer iterations on the finest grid (160×20). Only the residual from the pressure equation is shown as it had the largest value for the single-grid (Fig. 5(a)), but similar differences were found for the other residuals (data not shown). In single-grid computation, the pressure residual can only be reduced to the order of 10^{-3} while in the case of the three-level multigrid computation, it can be reduced to the order of 10^{-5} within less than 100 outer iterations. Table 2 lists the number of outer iterations, CPU-time, and speed-up ratio for the capillary flow with various mesh size and different linear system solvers. In Table 2, the recorded CPU-time for multigrid is the total computing time for all levels of grids that are involved. For all three solvers studied (SIP, BiCGSTAB, and GMRES), a great amount of savings in

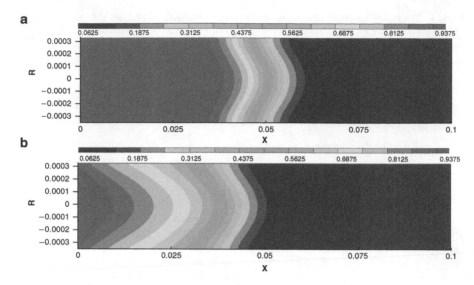

Fig. 4 A snapshot of ligand concentration distribution at $t = 10$ min in the capillary at the temperature of 37°C and the uniform inlet velocity of $u = 5.2$ mm/min. (**a**) Surface binding is not considered. (**b**) Ligands bind to receptors and HSPGs on the capillary surface

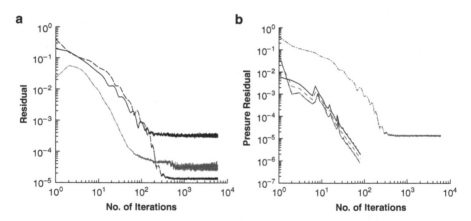

Fig. 5 Comparison of convergence history between single-grid and multigrid computations. (**a**) Residuals of pressure and velocities in single-grid computation, where dashed line is the U residual, dashed-dotted-dotted line the V residual, solide line the P residual. (**b**) Comparison of pressure residuals between single-grid and multigrid computation with different linear system solvers, where dashed-dotted line is for single grid, solid line for multigrid with SIP, dashed line for multigrid with BiCGSTAB, and dashed-dotted-dotted line for multigrid with GMRES

CPU time has been observed by using multigrid. For example, the CPU time needed for a 160×20 grid using the SIP method was 176.45 seconds for the single-grid and 5.39 for the multigrid. This difference in time was due to the significant difference

Table 2 Performance comparison of multigrid method with different linear system solvers

Solver	Mesh	No. outer iterations		CPU-time (s)		Speed-up ratio
		Single-grid	Multigrid	Single-grid	Multigrid	
SIP	40×5	234	234	0.41	0.41	1.0
	80×10	898	28	6.87	0.64	10.7
	160×20	6,000	54	176.45	5.39	32.7
BiCGSTAB	40×5	114	114	0.35	0.35	1.0
	80×10	432	26	5.24	0.56	9.36
	160×20	6,000	50	174.66	5.43	32.16
GMRES	40×5	101	101	0.22	0.22	1.0
	80×10	389	23	5.43	0.55	9.87
	160×20	6,000	52	186.75	5.82	32.08

in the number of outer iterations required which was 6,000 for the single-grid and only 54 for the multigrid. This can result in a speed-up ratio for the capillary flow of over 30 for the finest grid. Three iterative solvers, SIP [19], BiCGSTAB [20] and GMRES [17] are used to solve the inner linear system and their performance are compared. In this particular case with a five diagonal coefficient matrix, BiCGSTAB and GMRES solvers do not have obvious advantages over SIP in reducing the number of outer iterations of Navier–Stokes equations in the multigrid computation.

4 Discussions

The binding kinetics and signaling pathways of FGF-2 are very complicated, and many mathematical models have been proposed to predict the behavior of FGF-2 [3, 7, 9–11]. These models include pure reaction models [7, 10, 11], where a system of ordinary differential equations is provided and it is assumed that the fluid movement of FGF-2 does not impact the solution, and reaction-diffusion models [3, 9], which are relatively more complex and the movement of FGF-2 molecules from fluid to cell surface is modeled by diffusion. The diffusion model is valid only if the fluid is quiescent, which is not consistent with the actual biological environment of moving bio-fluids. Our model is unique, in which a coupled convection-diffusion-reaction model is applied, and the motion of bio-fluids is fully considered.

We model the process of growth-factor binding and dissociation under flow as fluid flow in capillary by incompressible Navier–Stokes equation, protein transport by convection-diffusion, and local bioreaction on capillary surface. The flow is assumed to be laminar, and both the fluid density and viscosity are taken as constants. This is due to the fact that the fluid in the bioreactor fibers is mainly water with some additives (FiberCell Systems, Inc., Frederick, MD, USA). The impact of fluid flow on FGF-2 binding is through the transport equations in the coupled non-linear convection-diffusion-reaction model, where the flow velocity affects the distribution of FGF-2 in the solution, and that of FGFR, HSPG, and their compounds on capillary surface.

The solution of Navier–Stokes equations consists of two loops. The inner iteration handles each of the individual equations of momentum, energy, and turbulent kinetics if necessary, and the outer iteration deals with the coupling and nonlinearity. For unsteady flow using implicit discretization, the linear equations need not be solved very accurately at each outer iteration. Usually a few iterations of a linear system solver is enough. More accurate solution will not reduce the number of outer iterations but may increase the computing time [21].

A colocated grid system is used to discretize the Navier–Stokes and transport equations. The naive way of colocated arrangement makes the solution of Navier–Stokes equations not very pleasant due to the checkerboard pressure distribution. Therefore staggered grid was once considered as a standard way for the calculation of incompressible flow [13]. A remedy was proposed to deal with the pressure-velocity coupling on colocated grids by eliminating the oscillations in the pressure field [8]. Our work uses a colocated grid and adopts the remedy to filter out the oscillations. A colocated arrangement is attractive when using a non-orthogonal grid, complex geometry, and multigrid methods.

Developing efficient solvers for nonlinear system of equations arising from fluid flow and mass transfer problems is of practical interest. When nonlinear partial differential equations are discretized in a particular spatial mesh, the result is a set of ordinary differential equations, with time as the independent variable. The temporal terms may be further discretized using either an explicit or an implicit scheme. In an explicit method, a large number of independent equations, one for each control volume, have to be solved. In the case of an implicit discretization, we have to solve a large set of simultaneous equations.

$$F(\vec{u}) = \frac{f(\vec{u}) - f(\vec{u})^o}{\Delta t} + \Phi(\vec{u}) = 0 \tag{13}$$

where $f(\vec{u})^o$ is the value at the previous time step. If the number of components is m and the number of control volumes is n, the number of independent variables becomes mn. Such large sets of nonlinear equations are generally solved by a variant of Newton's method, where a sequence of linear systems are solved [8]. The great advantage of the Newton iteration is its quadratic convergence. However, Newton iteration converges only if the initial guess is close to the solution. Instead, in this paper, the method of Picard iteration is applied, in which the nonlinear convective term and source term are linearized by using values from the previous outer iterations. This kind of linearization requires many more iterations than a coupled technique such as Newton-like linearization, but an initial guess close to the solution is not critical for the convergence. The number of outer iterations, however, can be substantially reduced by using multigrid techniques, as shown previously.

5 Concluding Remarks

In conclusion, this paper presents a multigrid computation strategy for the numerical solution of FGF-2 transport and cellular binding within a capillary (i.e., cylindrical geometry). The capillary flow is predicted using the incompressible Navier–Stokes equations by the finite volume method with collocated mesh arrangement with the assumption that the capillary is a circular pipe. To reduce the computation cost, a multigrid V-cycle technique, which includes restriction and prolongation to restrict fine grid solution to coarse grid, and to interpolate coarse grid solution to fine grid, is applied for the nonlinear Navier–Stokes equations. The multigrid method is extended to solve the mass transport equation. Computational results indicate that the multigrid method can reduce CPU time substantially. The computed profile of FGF-2 distribution in the capillary is presented for a set of conditions and no advantage with regard to CPU time or the magnitude of residue has been observed among the three linear system solvers being used.

References

1. P. Brown, G. Byrne, A. Hindmarsh, VODE: a variable coefficient ODE solver. SIAM J. Sci. Stat. Comput. **10**, 1038–1051 (1989)
2. K. Chang, D. Hammer, The forward rate of binding of surface-tethered reactants: effect of relative motion between two surfaces. Biophys. J. **76**, 1280–1292 (1999)
3. C.J. Dowd, C.L. Cooney, M.A. Nugent, Heparan sulfate mediates bFGF transport through basement membrane by diffusion with rapid reversible binding. J. Biol. Chem. **274**, 5236–5244 (1999)
4. S. Elhadj, R.M. Akers, K. Forsten-Williams, Chronic pulsatile shear stress alters insulin-like growth factor-I (IGF-I) bonding protein release *in vitro*. Annal. Biomed. Eng. **31**, 163–170 (2003)
5. A. Ezekowitz, J. Mulliken, J. Folkman, Interferon alpha therapy of haemangiomas in newborns and infants. Br. J. Haematol. **79**(Suppl 1), 67–68 (1991)
6. M. Fannon, M.A. Nugent, Basic fibroblast growth factor binds its receptors, is internalized, and stimulates DNA synthesis in Balb/c3T3 cells in the absence of heparan sulfate. J. Biol. Chem. **271**, 17949–17956 (1996)
7. M. Fannon, K.E. Forsten, M.A. Nugent, Potentiation and inhibition of bFGF binding by heparin: a model for regulation of cellular response. Biochemistry, **39**, 1434–1444 (2000)
8. J.H. Ferziger, M. Peri\acute{c}, in *Computational Methods for Fluid Dynamics* (Springer, Berlin, 1999)
9. R.J. Filion, A.S. Popel, A reaction-diffusion model of basic fibroblast growth factor interactions with cell surface receptors. Annal. Biomed. Eng. **32**, 645–663 (2004)
10. K. Forsten-Williams, C. Chua, M. Nugent, The kinetics of FGF-2 binding to heparan sulfate proteoglycans and MAP kinase signaling. J. Theor. Biol. **233**, 483–499 (2005)
11. F.M. Gabhann, M.T. Yang, A.S. Popel, Monte Carlo simulations of VEGF binding to cell surface receptors in vitro. Biochim. Biophys. Acta **1746**, 95–107 (2005)
12. W. Glaser, Antigen-antibody binding and mass transport by convection and diffusion to a surface: a two-dimensional computer model of binding and dissociation kinetics. Analyt. Biochem. **213**, 152–161 (1993)

13. F. Harlow, J. Welsh, Numerical calculation of time dependent viscous incompressible flow with free surface. Phys. Fluid **8**, 2182–2189 (1965)
14. N. Itoh, The Fgf families in humans, mice, and zebrafish: their evolutional processes and roles in development, metabolism, and disease. Biol. Pharm. Bull. **30**, 1819–1825 (2007)
15. D. Myszka, X. He, et al., Extending the range of rate constants available from BIACORE: interpreting mass transport-influenced binding data. Biophys. J. **75**, 583–594 (1998)
16. E. Schreck, M. Peri\acute{c}, Computation of fluid flow with a parallel multigrid solver. Int. J. Numer. Meth. Fluid **16**, 303–327 (1993)
17. Y. Saad, M. Schultz, GMRES: a generalized minimal residual algorithm for solving nonsymmetric linear systems. SIAM J. Sci. Stat. Comput. **7**, 856–869 (1986)
18. W. Shen, C. Zhang, M. Fannon et al., A computational model of FGF-2 binding and HSPG regulation under flow condition. IEEE Trans. Biom. Eng. **56**, 2147–2155 (2009)
19. H. Stone, Iterative solution of implicit approximations of multidimensional partial differential equations. SIAM J. Numer. Anal. **5**, 530–558 (1968)
20. A. Van der Vorst, Bi-CGSTAB: a fast and smoothly converging variant of Bi-CG for the solution of nonsymmetric linear systems. SIAM J. Sci. Stat. Comput. **13**, 631–644 (1992)
21. J. Zhang, H. Sun, J.J. Zhao, High order compact scheme with multigrid local mesh refinement procedure for convection diffusion problems. Comput. Meth. Appl. Mech. Engi. **191**, 4661–4674 (2002)

Chapter 24
Integrated Mining Fuzzy Association Rules For Mineral Processing State Identification

Seraphin C. Abou and Thien-My Dao

Abstract Mineral processes are multi-variable, power-intensive and strongly coupled with large delay and nonlinearities. The properties of controllability, observability and theory of minimal realization for linear systems are well understood and have been very useful in analyzing such systems. This paper deals with analogous questions for nonlinear systems with application to mineral processing. A method that can control and provide accurate prediction of optimum milling condition and power consumption, water and chemical additive requirement is developed for mineral plants operation. A fuzzy mining algorithm is proposed for extracting implicit generalized knowledge on grading process performance as qualitative values. It integrates fuzzy-set concepts and generalized data mining technologies to achieve this purpose. Using a generalized similarity transformation for the error dynamics, simulation results show that under boundedness condition the proposed approach guarantees the global exponential convergence of the error estimation. Although the nominal performance of the process is improved, the robust stability still is not guaranteed to fully avoid the mill plugging.

1 Introduction

Data mining is the process for automatic extraction of high level knowledge from the information provided by sensors. A control-engineering method for fuzzy control was provided in [1]. In mineral processing, understanding how to modify the rheological characteristics of fine particle systems is a key for the process performance. These characteristics include particle settling, pH, bulk/carrier fluid viscosity, particulate flocculation or dispersion, attrition, pipe/fitting/impeller

S.C. Abou (✉)
Mechanical and Industrial Engineering Department, University of Minnesota Duluth, 1305 Ordean Ct, Duluth, MN 55812, USA
e-mail: sabou@d.umn.edu

S.-I. Ao et al. (eds.), *Machine Learning and Systems Engineering*,
Lecture Notes in Electrical Engineering 68,
DOI 10.1007/978-90-481-9419-3_24, © Springer Science+Business Media B.V. 2010

wear, degradation of flocculated or friable solids and the pumpability of the slurry. Moreover, fine particle systems exhibit a range of rheological properties that influence processing and handling. The properties, settling behavior and the rheology of fine particulate slurries are determined by both the physical properties and surface chemistry of the particles. Figure 1 depicts two stages structure considered for developing both useful fuzzy association rules and suitable membership functions from quantitative values: (1) adaptive learning procedure to learn the membership functions, and (2) a method to mine fuzzy association rules.

A wet grinding plant shown in Fig. 2 has been analyzed with the objective of evaluating the effects of many variables on particle size reduction in continuous grinding processes. Detailed phenomenological model that describes the charge behaviour has been developed and validated against real data [2]. Indeed, mineral

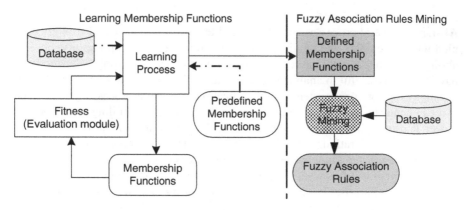

Fig. 1 Scheme for fuzzy association rules and membership functions

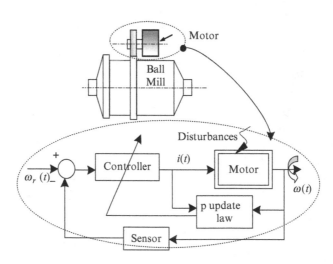

Fig. 2 Motor servo-system of the ball mill

processes present non-linear/chaotic dynamic behaviour. Considerable efforts have been developed in controlling such systems [3, 4]. In [2], a comprehensive model integrating physical mechanisms and fundamental understanding of the charge behaviour was developed. The poor quality of fine grinding is due to lacks of an appropriate control of the power draw of the mill. This causes increase of energy consumption, and production cost [5]. A classical PID controller is worthless to efficiently monitor the mineral plant. Through the past few decades, advanced control methods are being developed to address the system as a whole, dealing with all the problems that arise out of the complexity of the breakage process, its distributed nature, the networked interactions of the components [6–8]. Nevertheless, several approaches mainly depend on the model of the controlled plant. Fuzzy control does not need the model of the controlled plant. In general, the fuzzy logic rules are obtained from the knowledge of experts and operators [3, 9, 10]. Since its introduction in 1993 [11], the task association rule mining finds interesting relationships among a large set of data items [4]. In this paper the association rules mining algorithm is adopted to design a fuzzy logic controller as a control law.

2   Grinding Process Modelling

Besides in *batch mode operation*, grinding circuit can operate in *continuous or fed-batch mode*. As shown in Fig. 1, the motor load is strongly influenced by the filling percentage, the speed, the mill geometry and other relevant material properties such as stiffness and the coefficient of friction, etc.

The theoretical position of the charge at different rotation speeds was first derived by Davis [5] based on the total force balance. We are interested in the constitutive characteristics of the charge motion defined by a function $f(x, u)$ that better describes continuous grinding phenomena:

$$\frac{\partial}{\partial t}[x(.)] = \frac{\partial}{\partial z}\left[\Psi_n(.)\frac{\partial m_i}{\partial z} - \Psi_n(.)m_i\right] \tag{1}$$

where, m_i (kg), is particle mass of size i

The left side term of Eq. (1) expresses the rate of mineral production. Without lacking the physical sense for the process, we can write:

$$\Psi_n(.) = \Psi_n(x, u) \tag{2}$$

Based on volumetric analysis, the fraction of the total mass broken within a tiny volume dV of the charge is assumed to be $\sigma(t)$ that is determined as follows:

$$\sigma(t) = \iiint_V a\rho_c dV \tag{3}$$

Where ρ_c is the charge bulk density, a is defined as a mass volume of material of classes i; Therefore, the flow rate of particle through the mill is:

$$\frac{d\sigma}{dt} = \iiint\limits_{V} \frac{\partial(a\rho_c)}{\partial t} dV + \iiint\limits_{V} a\rho_c \frac{d(dV)}{dt} \qquad (4)$$

However, as the mass could not be transferred by conduction phenomena, the mass flux vanishes, so that we could write:

$$\frac{d\sigma}{dt} = -\iint\limits_{F} \vec{J}_i \cdot d\vec{F} + \iiint\limits_{V} \vartheta\sigma_p\, dV \qquad (5)$$

where, $\vec{J}_i$ = longitudinal diffusion flux of the mass in class I; ϑ = piecewise parameter; σ_p = local fine particle.

3 The Controller Design

Wet communition (grinding) and particle size classification can be effected by viscosity, particle size distribution, fines concentration etc. For example, hydro-cyclone classification is effected by the presence of a yield stress (the minimum force required for slurry to flow). An interlaced approach, called platform-based control design, can be developed using fuzzy logic controller based on association rules mining [11]. The main problem of the grinding process lies in developing a sufficiently simple and reliable method for measuring the load in the Ball mill during operation. Although the fineness distribution capability of the particle can not be measured directly, it is obviously related to ball mill load, the stability of the grinding and the floatation processes.

Figure 3 depicts the characteristics of ball mill in function of the ball mill load, l. Functions $m(l)$ and $p(l)$ represent the grinding capability and the driving motor

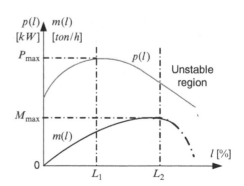

Fig. 3 Static characteristics of ball mill

power, respectively. Our motivation is to design a fuzzy system to produce alarms and reconfigure the control variables if critical conditions occur in the process (pH, rheology, degradation of flocculated solids, slurry pumpability, etc.). Assume the fuzzy system uses $x_1(t)$, and $x_2(t)$ as inputs, and its output is an indication of what type of warning condition occurred along with the certainty that this warning condition has occurred.

The alarm signals represent certain warnings characterized by the decision regions shown in Fig. 4. For instance the density of the slurry is greater than the upper threshold limit; this occurs if: $x_1(t) > \alpha_1$ and $x_1(t) \geq x_2(t) + \alpha_3$. Small changes in chemical properties (in terms of lapping) may have large effects on the mineral quality and the grinding system dynamics. In the next section, we consider the system as multi-state system. The grinding process is complex ill-defined process consisting of n elements, any element j, $1 \leq j \leq n$ can have k_j different states with corresponding performance rates (levels), which can be represented by the ordering set as follows [3]:

$$g_j = \{g_{j1}, \cdots g_{ji_j}, \cdots , \ g_{jk_j}\} \tag{6}$$

where g_{ji_j} is the performance rate (level) of the element j in the state i_j; $i_j \in \{1, 2, \ldots, k_j\}$. The performance rate $G_j(t)$ of element j at any instant $t \geq 0$ is a random variable that takes its values from $g_j : G_j(t) \in g_j$. Thus, the probabilities associated with different states for the element j can be represented by a set:

$$p_j = \{p_{j1}, \ldots, p_{jk_j}\}; p_j \to 1 \leq j \leq n$$

The mapping $g_{ji_j} \to p_{ji_j}$ is usually called the probability mass function as defined for multi-state system [3]. However, for some multi-state systems, evaluating precisely the state probability and performance rate of an element is difficult. As above pointed out for the cause of the deterioration of the grinding quality, let define the error e_r and change of error e_c at sampled times k as follows:

$$e_r(k) = \frac{p(k) - p(k-1)}{V_m(k) - V_m(k-1)} \ \text{and} \ e_c(k) = e_r(k) - e_r(k-1) \tag{7}$$

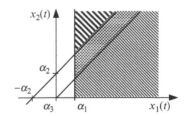

Fig. 4 Decision regions for the rheology state

where p is the power and can be measured by means of measuring the electric current of the ball mill.

The noise and modeling errors will cause different distortions to the probability assignments in the different models. Assume the probability distribution σ_d of performance rates for all of the system elements at any instant $t \geq 0$ and system structure function as follows:

$$\phi(G_1(t) \ldots G_n) \tag{8}$$

Accordingly, the total number of possible states or performance rates of the system are:

$$\pi_p = \prod_{j=1}^{n} k_j \tag{9}$$

Let $L^n = \{g_{11}, \ldots, g_{1k_1}\} \times \cdots \times \{g_{n1}, \ldots, g_{nk_n}\}$ be the space of possible combinations of performance rates for all system elements and $M = \{g_1, \ldots, g_{\pi_p}\}$ be the space of possible values of entire system performance levels. The transform $\phi(G_1(t) \ldots G_n(t)) : L^n \to M$ which maps the space of performance rates of system elements into the space of system's performance rates, is the system structure function [10]. The probability of the system state is given as: $\sigma_i = \prod_{j=1}^{n} \sigma_{ij}$; the performance rate for state i is:

$$g_i = \phi(g_{ni_1}, \ldots, g_{ni_n}) \tag{10}$$

For the system under consideration, the estimation of a single number for the probabilities and performance levels is very difficult. To avoid the "*dimension damnation*", the model is reduced to decrease the computational burden. Fuzzy logic controllers (FLC) have the advantage to be robust and relatively simple to design since they do not require the knowledge of the exact model. The fuzzy system that serves the implementation of general decision-making in tuning the driving motor power of the mill is presented in Fig. 5.

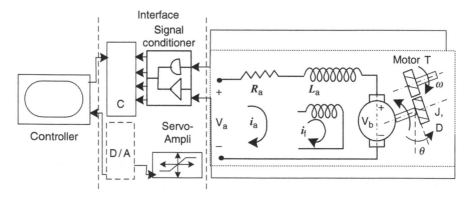

Fig. 5 Hardware setup

In this study, the proposed fuzzy logic controller has three inputs and outputs. The tuned control surface is nonlinear corresponding to the property of controlled plant l, ω_r and n_p are the measured value of the ball mill load, the rotation speed and the inlet negative pressure, respectively. In addition, p is the measured value of the ball mill driving motor power. e_l, e_ω and e_{np} which are input variables of the fuzzy logic controller, represent the error of l, ω_r and n_p respectively. u_l, u_w and u_{np} are the output variables of the fuzzy logic controller, which are usually used to control the raw ore feeder, the driving motor speed and the recycle air damper, respectively. Therefore, the probability distribution σ_d of the system is:

$$\phi(g_{1i_1}, \, \cdots \, , \, g_{3i_3}), \sigma_d = \prod_{j=1}^{3} \sigma_{ji_j} \tag{11}$$

Furthermore, the max-min algorithm is used in fuzzy logic inference, and the defuzzification is accomplished by the largest of maximum method.

3.1   Fuzzy Logic Controller

In the fuzzy control design methodology, we made use of heuristic information as from an operator who has acted as a "human-in-the-loop" controller for the process. To develop a set of rules applicable to mineral processing control, the practical expertise is drawn on our knowledge performed through extensive mathematical modeling, analysis, and development of control algorithms for diverse processes. Then we incorporate these into a fuzzy controller that emulates the decision-making process of the human, Fig. 6.

The function $\phi(\cdot)$, (11) is strictly defined by the type of connection between elements in the reliability logic-diagram sense, i.e. on the structure of the logic-diagram representing the system/subsystem. Despite the fact that the universal generating function resembles a polynomial, it is not a polynomial because: (i) its exponents are not necessary scalar variables, but can be arbitrary mathematical objects (e.g. vectors); (ii) operators defined over the universal generating function

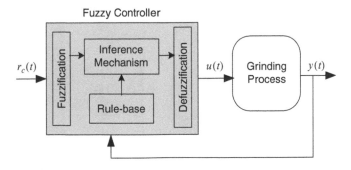

Fig. 6 Fuzzy controller architecture

can differ from the operator of the polynomial product (unlike the ordinary generating function technique in which only the product of polynomials is defined) [12]. Figure 7 idealizes a general flow transmission through out the system (e.g., ore, particle size, fluid flow, energy). For instance, consider a flow transmission system shown in Fig. 7 which consists of three elements. As a result, the system performance rate which is defined by its transmission capacity can have several discrete values depending on the state of control equipments.

Assume the element 1, (the rheology of the slurry) has three states with the performance rates: $g_{11} = 1.5$, $g_{12} = 1$, and $g_{13} = 0$; the corresponding probabilities are respectively: $\sigma_{p11} = 0.8$, $\sigma_{p12} = 0.1$ and $\sigma_{p13} = 0.1$. The element 2, (the pH) has three states with the performance rates $g_{21} = 2$, $g_{22} = 1.5$, $g_{23} = 0$ and the corresponding probabilities $\sigma_{p21} = 0.7$, $\sigma_{p22} = 0.22$ and $\sigma_{p23} = 0.08$. The element 3, (the density) has two states with the performance rates $g_{31} = 4$, $g_{32} = 0$ and the corresponding probabilities $\sigma_{p31} = 0.98$ and $\sigma_{p32} = 0.02$. According to (9) the total number of possible combinations of the states of elements are $\pi_p = 3 \times 3 \times 2 = 18$.

In order to obtain the output σ_p for entire system with the arbitrary structure function $\phi(.)$ [9] used a general composition operator ∂_ϕ over individual universal z-transform representations of n system elements:

$$
\begin{cases}
U(z) = \partial_\phi(u_1(z), \ldots, u_n(z)) \quad \text{and} \quad u(z) = \sum_{i=1}^{k_j} \sigma_{d\,ji_j} . Z^{g_{ji_j}} \\[2ex]
U(z) = \sum_{i_1}^{k_1} \sum_{i_2}^{k_2} \cdots \sum_{i_n}^{k_n} \left(\prod_{j=1}^{n} \sigma_{d\,j} . z^{\phi(g_{ji_j}, \ldots, g_{ni_n})} \right)
\end{cases}
\tag{12}
$$

where $U(z)$ is z-transform representation of output performance distribution for the entire system.

Figure 6 illustrates the basic control structure. The scheme includes a classical PID control structure together with fuzzy corrector. The fuzzy corrector uses the command input $r_c(t)$ and the plant output y to generate a command signal $u_c(t)$, described by the following equations:

$$
\begin{cases}
e(t) = r_c(t) - y(t) \\
\Delta e(k) = e(k) - e(k-1) \\
\mu(k) = F[e(k), \Delta e(k)]
\end{cases}
\tag{13}
$$

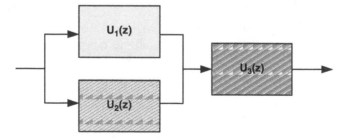

Fig. 7 A flow transmission structure

Fig. 8 The fuzzy sets

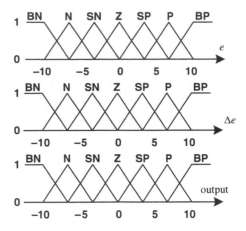

In the above, $e(k)$ is the position error between the command input $r_c(t)$ and the process output $y(k)$; $\Delta e(k)$ is the change in position error. The term $F[e(k), \Delta e(k)]$ is a nonlinear mapping of $e(k)$ and $\Delta e(k)$ based on fuzzy logic. The term $\mu(k)$ represents a correction term. The control $u(k)$ is applied to the input of the grinding circuit. The purpose of the fuzzy corrector is to modify the command signal to compensate for the overshoots and undershoots present in the output response when the load dynamics has unknown nonlinearities.

Let x_1, x_2, x_3, y_1, y_2 and y_3 represent l, ω_r, n_p, u_l, u_ω and u_{np} respectively. The expertise and knowledge method used to build a rule base and membership functions provide the description of $e(k)$ and $\Delta e(k)$ as inputs, and $\mu(k)$ as the output. The unified fuzzy inverse is given as $[-0.1, 0.1]$, Fig. 8. The fuzzy states of the inputs and the output, all are chosen to be equal in number and use the same linguistic descriptors: *Big Negative* (BN), *Negative* (N), *Small Negative* (SN), *Zero* (Z), *Small Positive* (SP), *Positive* (P) and *Big Positive* (BP). Figure 8 illustrates the membership functions. Despite the operator expertise and knowledge at the level of the inference rules and the membership functions, some defects may appear. The degree of membership of each value of attribute i_k in any of its fuzzy sets is directly based on the evaluation of the membership function of the particular fuzzy set with the value of i_k as input.

To improve the conventional FLC, association rules mining algorithms are used to find the optimal membership functions. This is achieved based on the membership functions depicted in Fig. 9.

3.2   Association Rules Miming Algorithm

Specified fuzzy linguistic terms in fuzzy association rules can be given only when the properties of the attributes are estimated. In real life, contents of columns (i.e., values of attributes) may be unknown and meaningful intervals are usually

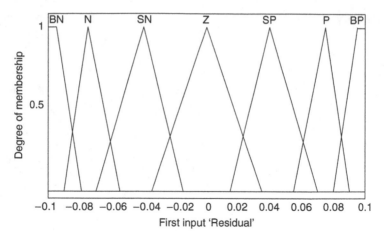

Fig. 9 Membership functions

not concise and crisp enough. In this paper, the target is to find out some interesting and potentially useful regularities, i.e., fuzzy association rules with enough support and high confidence. We use the following form for fuzzy association rules.

Let be $\{x_i^1, \ldots, x_i^k\}$ and $\{y_i^1, \ldots, y_i^k\}$ the antecedent set and the consequence set respectively, in a database. A fuzzy association rule mining is expressed as:

If $X = \{x_i^1, \ldots, x_i^k\}$ is $A = \{\alpha_i^1, \ldots, \alpha_i^k\} \rightarrow$ Then
$Y = \{y_i^1, \ldots, y_i^k\}$ is $B = \{\delta_i^1, \ldots, \delta_i^k\}$. Here, X and Y are disjoint sets of attributes called item-sets, i.e., $X \subset I; Y \subset I$ and $X \cap Y = \phi$. A and B contain the fuzzy sets associated with corresponding attributes in X and Y, respectively.

Let $\mu(.)$ represent the membership value of each element of the antecedent and the consequence set. Under fuzzy taxonomies, using the measurements could result in some mistakes. Consider for instance the following conditions:

1. $\mu(x_i^k) \geq \mu(y_i^k)$ and $\mu(x_i^k) \geq \mu(y_i^m)$
2. $\mu(x_i^k) < \mu(y_i^k)$ and $\mu(x_i^k) < \mu(y_i^m)$

In the first condition, the confidence under fuzzy taxonomies of the two rules is equal, while in the second, the coverage of the two rules is equal. This situation gives rise to the following question: *which rule can be judged as best evidence rule?*

In the proposed method the algorithm iterations alternate between the generation of the candidate and frequent item-sets until large item-sets are identified. The fuzzy support value of item-set Z is calculated as:

$$S(Z, F) = \frac{\sum_{t_i \in T} \prod_{z_j \in Z} \mu(\alpha_j \in F, t_i(z_j))}{n_T} \tag{14}$$

where n_T is the number of transactions in the database. Also, to ensure the accuracy of rules base, the consequent strength measure, ψ is used to estimate the mined rules as:

$$\psi = \sum_{i=1}^{n} \left(\prod_{k=1}^{k_X} \mu(y_i^k) \wedge \prod_{m=1}^{k_Y} \mu(y_i^m) \right) \cdot \sum_{i=1}^{n} \prod_{j=1}^{k_Y} \mu(y_i^j) \qquad (15)$$

Based on the literature review, the fuzzy-PID control is implemented in such way that the fuzzy system has $\dot{\psi}$ (derivative of the PID output) as input and returns an output $\dot{g}$ that has a nonlinear dependence upon the input. The dependence upon $\dot{\psi}$ ensures that changes in the steady value of ψ do not change the closed-loop response characteristics of the fuzzy-PID controller. Hence the simplest form of fuzzy-PID controller is:

$$\frac{dg}{d\tau} = \frac{1}{\beta} f\left(\alpha \frac{d\psi}{d\tau} \right) \qquad (16)$$

where α is an input scaling factor and β is an output scaling factor. The input magnitude is scaled to unity by:

$$\alpha = \left(\max\left(\left| \frac{d\psi}{d\tau} \right| \right) \right)^{-1} \qquad (17)$$

The evaluation of $\max\left(\left| \frac{d\psi}{d\tau} \right| \right)$ is normally estimated from one or more PID runs. The final system implemented is presented as:

$$f(\varsigma) = \begin{cases} \varphi_\varsigma(|\varsigma|(\varphi - 1) + 1)^{-1} & \text{for } |\varsigma| \leq 1 \\ -1 & \text{for } \varsigma \leq -1 \\ 1 & \text{for } \varsigma \geq 1 \end{cases} \qquad (18)$$

The graph of the function $f(\varsigma)$ is presented in Fig. 10. It depicts sigmoidal dependence on ζ. As a result, it may corrupt all slowly varying signals of the process. In order to suppress the noise and get the nonlinear control surface, the parameters of input membership functions and output membership functions of the fuzzy rule base are tuned by a nonlinear optimization. In this study a sequential quadratic programming (SQP) algorithm as presented in Matlab is used.

This approach represents state-of-the-art in non-linear programming methods, because a non-linearly constrained problem can often be solved in fewer iterations using SQP than an unconstrained problem. One of the reasons is that due to the limits on the feasible area, the optimizer can make well informed decisions regarding directions of search and step length. Note that widely overlapping membership functions give good numerical results. However, fuzzy membership functions

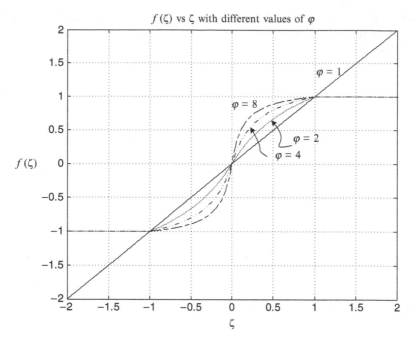

Fig. 10 $f(\zeta)$ with varying ϕ

lacking any physical interpretation and loosing locality are possible. To avoid this, different kinds of constraints may be put on the optimization: kinds of constraints, inequality constraints and parameter bounds. SQP efficiently solve this constrained nonlinear optimization problem in which the objective function and constraints may be nonlinear functions of variables.

4 Simulation Results

Through available actuators it is possible to adjust the speed of the grinding circuit feeders, the fluid added to the mill and sump, the pump and the driving motor speed. The level of the sump is controlled by adjusting the set-point of pump velocity. The water flow to the flotation circuit is kept proportional to the load reference. As mentioned, the above process is subject to many disturbances, being the hardness of the raw feeding ore the most significant one. This exerts a strong influence on the quality of the grinding product. Modeling of the process is further complicated by the fact that different operation points are defined by changes in mineral hardness. Based on a reasonably simple transfer function, the adjustable input ports of the driving motor and entrance negative pressure of the breakage circuit are initialized with step signal. The strongly coupling characteristics of variables implies that if regulating one controlled variable is attempted with one manipulated variable, the

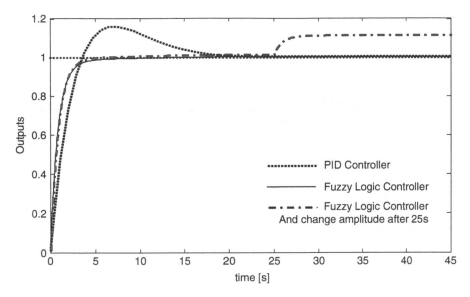

Fig. 11 Fuzzy logic controller step response

other controlled variables will, more likely than not, be influenced in an undesired fashion. Using the aforementioned fuzzy logic rules the simulated step response of the grinding circuit is shown in Fig. 11. The amplitude changes are followed smoothly to reach the steady state. To ensure both controller stability and performance, although guidelines exist, the control procedure will require a number of iterations until a solution is found, satisfying all requirements.

The existence of large time delays, time-varying parameters and nonlinearities are some of the other difficulties generally encountered in mineral processing control.

As illustrated in Fig. 12, one of the major attractions of the proposed methodology is that it can be readily used to design robust multivariable controllers that take account of all control loop interactions while still guaranteeing all users specified stability and performance requirements. As shown in Fig. 12, in the presence of disturbances the fuzzy logic controller model presents a response better than the conventional PID model.

5   Conclusion

Understanding how to modify rheological characteristics of fine particle is a key for mineral process performance. Constrained optimization methods with integrated association rules mining based on a nonlinear, fuzzy dynamic control scheme were designed to improve the grinding process. The proposed structure of fuzzy logic controller combines the advantages of the fuzzy networks approach and the association mining rules. The proposed algorithms can control the process in a better way

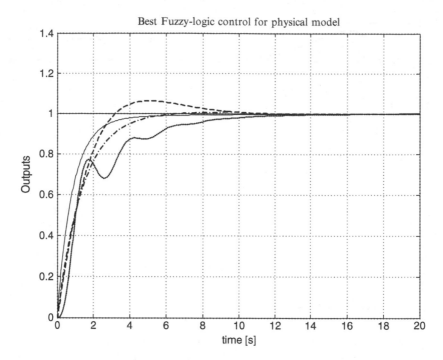

Fig. 12 Tuned fuzzy logic controller using physical model

than the conventional PID approach. Based on simulation results it can be deduced that the fuzzy controller is faster than the conventional controller in the transitional state, and also presents a much smoother signal with less fluctuation at steady state. The proposed method has a strong adaptability and can overcome nonlinear and strong coupling features of mineral processing in a wide range.

References

1. S.C. Abou, M.D. Thien, Fuzzy logic controller based on association rules mining: application to mineral processing. International Conference on Modeling, Simulation and Control (ICMSC'09), San Francisco, USA, 20–22 Oct 2009
2. S.C. Abou, Contribution to ball mill modeling. Doctorate Thesis, Laval University, Quebec, Canada, 1998
3. G. Levitin, *Universal Generating Function and Its Applications* (Springer, Berlin, 2005)
4. S. Morell, The prediction of powder draws in wet tumbling mills. Doctorate Thesis, University of Queensland, 1993
5. E.W. Davis, Fine crushing in ball mills. AIME Trans. **61**, 250–296 (1919)
6. A. Desbiens, K. Najim, A. Pomerleau, D. Hodouin, Adaptive control-practical aspects and application to a grinding circuit. Optim. Control Appl. Methods **18**, 29–47 (1997)
7. R.K. Rajamani, J.A. Herbst, Optimal control of a ball mill grinding circuit: II. Feedback and optimal control. Chem. Eng. Sci. **46**(3), 871–879, 1991

8. L. Zhai, T. Chai, Nonlinear decoupling PID control using neural networks and multiple models. J. Control Theory Appl. **4**(1), 62–69, 2006
9. W. Kuo, R. Wan, Recent advances in optimal reliability allocation. IEEE Trans. Syst. Man Cybern. Part A Syst. Hum. **37**, 143–156, 2007
10. H.J. Zimmermann, *Fuzzy Set Theory and its Application*, 2nd edn. Kluwer Academic, Dordrecht, 1991
11. R. Agrawal, T. Imielinski, A. Swami, Mining association rules between sets of items in large databases. International conference on management of data (ACM SIGMOD'93), Washington, DC, 1993
12. J. Guan, Y. Wu, Repairable consecutive-k-out-of-n: F system with fuzzy states. Fuzzy Sets Syst. **157**, 121–142 (2006)
13. J. Huang, M.J. Zuo, Y. Wu, Generalized multi-state k-out-of-n: G systems. IEEE Trans. Reliab. **49**(1), 105–111 (2000)

Chapter 25
A Combined Cycle Power Plant Simulator: A Powerful, Competitive, and Useful Tool for Operator's Training

Eric Zabre, Edgardo J. Roldán-Villasana, and Guillermo Romero-Jiménez

Abstract In this chapter we present the development of a combined cycle power plant simulator for operator's training. This simulator has been designed and developed by the Simulation Department of the Instituto de Investigaciones Eléctricas. This is one of the several technological developments carried out by the Simulation Department, and it is part of a full scale simulators group that belongs to the Comisión Federal de Electricidad that offers the electrical service in the whole country. The simulator is currently under testing and evaluation by the final user, before entering on service at the National Center for Operator's Training and Qualification. Tendencies of these development and impact within the operators' scope as well as some results and future works are also presented.

1 Introduction

The Instituto de Investigaciones Eléctricas (IIE, Mexican Electric Research Institute, founded in 1975) has been the right hand as research and development institution of the Comisión Federal de Electricidad (CFE, Mexican electrical utility company) offering technical innovations and developments. The Simulation Department (SD), one of the technical areas of the IIE, has developed a series of software programs and applications to generate full scale simulators for operators' training. In this chapter, a full scope simulator of a combined cycle power plant (CCPP) is presented and its different parts are described within training purposes.

E. Zabre (✉)
Instituto de Investigaciones Eléctricas, Simulation Department, Reforma 113, Col. Palmira, Cuernavaca, Morelos, Mexico 62490
e-mail: ezabre@iie.org.mx

S.-I. Ao et al. (eds.), *Machine Learning and Systems Engineering*,
Lecture Notes in Electrical Engineering 68,
DOI 10.1007/978-90-481-9419-3_25, © Springer Science+Business Media B.V. 2010

The development of simulators' technology involves many areas and, to reach the goals, different specialists were required.

Engineers of software, processes modeling, control, communications network, maintenance and operators have shared the same commitment.

2 Antecedent

According to Webopedia's Online Dictionary, a combined cycle power plant is defined as "the coexistence of two thermodynamic cycles in the same system", and simulation as "the process of imitating a real phenomenon with a set of mathematical formulas; in addition to imitating processes to see how they behave under different conditions, simulations are also used to test new theories".

Generally speaking, fluids of work: the gas and the steam water, product of combustion and a heat recovery steam generator, respectively, produce energy in the so called combined cycle. A combined cycle power plant may be very large, typically rated in the hundreds of mega watts. If any combined cycle power plant operates only the gas turbine to generate electricity, and diverts the exhaust gases, there is a substantial loss of efficiency. A typical large gas turbine is in the low 30% efficiency range, but combined cycle plants can exceed 60% of global efficiency, because the waste heat of the gas turbine is used to make steam to generate additional electricity through a steam turbine, getting a high efficiency of electricity generation over a wide range of loads [1, 2]. The most modern combined cycle power plants implement innovative strategies in order to reduce emissions.

In 2006, the IIE's SD developed a gas turbine full scope simulator (GTS) [3] based on the combined cycle power plant "El Sauz", for the Centro Nacional de Adiestramiento y Capacitación (CENAC, National Center for Operator's Training and Qualification), using proprietary technology of the IIE. The GTS is formed by a computer platform, an instructor console (IC), a set of interactive process diagrams (IPD), and mathematical models of the plant systems.

In this chapter, the development of a combined cycle power plant simulator (CCPPS) is presented. It is based on 1, 2, and 3 unit's package from the Chihuahua II CCPP [1, 2, 4] (also known as "El Encino", located at the north of the Mexican territory, generates 450 MW and went into operation on May, 2001). The CCPPS has a configuration of two gas turbines (GT1 and GT2), two heat recovery steam generators (HRSG1 and HRSG2), and one steam turbine. One of the gas turbines derives from the actual CENAC0's GTS; the second one is a simplified gas turbine with a reduced scope, this means that it has the values of the main processes variables, in such a way to simulate the necessary variables to interact with the steam turbine models.

Nowadays, combined cycle power plants (CCPP) represent the most frequent form of energy production in México [5, 6]. Figure 1 shows a typical diagram of the main components of a CCPP (with only one HRSG and without auxiliary systems, for simplicity). An advantage of this type of plant is the possibility of constructing

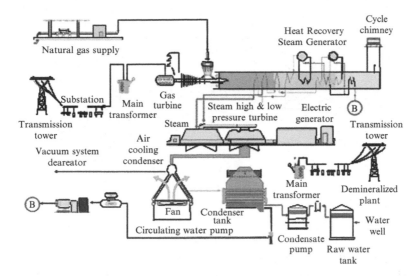

Fig. 1 Combined cycle power plant diagram

them in two stages. The first one corresponds to the gas unit, which can be finished in a brief term and immediately initiate its operation; later it is possible to finish the construction of the steam unit, and then the combined cycle may be completed.

3   Architecture Configuration

3.1   Software Architecture

In the CCPPS several configuration modules are identified, as shown in Fig. 2. All modules are strongly related to a real time system (RTS) that control and sequence the different tasks running in the simulator. The modules are the following:

- Instructor Console® (IC). This is an important man machine interface (MMI) of the instructor, which controls and coordinates the scenarios seen by the operator under training. These scenarios could be presented to the operator as a normal plant operation or under any malfunction of a particular system or equipment. A more detailed functional description of this interface is given below.
- CONINS. A module to retrieve all the static information during simulation session, for example: malfunctions, remote functions, local instrumentation, external parameters, etc.
- Operator Console (OC). This system is a replica of the real operation station and allows the operator to monitor and control the plant parameters from three different sets of graphical interfaces (GI): gas turbine, steam turbine and

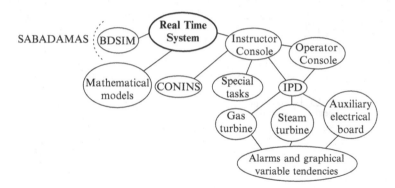

Fig. 2 Conceptual software architecture of the simulator

auxiliary emulated electrical external boards. These three GI have a special display for the alarms and graphical variable tendencies.

- IPD. This module executes all interactive processes diagrams or graphical interfaces, accessed from the IC or from the OC.
- Mathematical models. This system has a set of mathematical models that represent all the simulated systems of the plant. The RTS coordinates the execution sequence of each mathematical model, which runs sequentially in two processors within a 1/10 s execution frame.
- The special task module which coordinates other applications like environmental plant noises such as spinning turbine, start-up or shutdown pumps, etc.
- BDSIM. A module to store the simulator information in a data base using Microsoft Access™ interfaces and programs written on SQL language. The data base can be updated by means of external (off-line) tool as data base administrator system, named SABADAMAS®. It is dedicated to receive/update any information required by the executive system. The CCPPS uses and manages about 27,000 global variables.

3.1.1 Instructor Console

The IC is the MMI, which is constituted by functions as shown in Fig. 3. The simulation session is controlled since the IC. The instructor has a main menu of functions. The main functions of the IC are the following:

1. Run/Freeze. The instructor may start or freeze a dynamic simulation session. The mathematical models of the control and process respond to any action of the trainee in a very similar way as it occurs in the real plant.
2. Initial conditions. The instructor may select an initial condition to initiate the simulation session, either from the general pre-selected list of initial conditions or from their own catalogue (each instructor has access up to 100 initial conditions). Also in this function, it is possible to create an initial condition or to erase

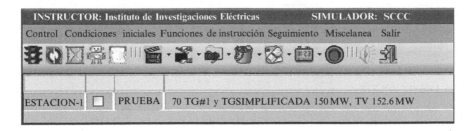

Fig. 3 Part of the main display of the Instructor Console

an old one. The simulators gets an automatic snapshoot every 15 s and may be configured up to a frequency of 10 min.

3. Simulation speed. Since the beginning of the simulation session the simulator is executed in real time, but the instructor may execute the simulator up to two or three times faster than real time. For example, slow thermal processes such as the turbine iron-heating can be simulated to occur faster. To slow the simulation speed (up to ten times) may be useful to study some fast events with detail.

4. Malfunctions. It is used to introduce, modify, or remove a simulated failure of plant equipment, for example: pumps trips, heater exchanger tube breaking, electrical switch opening, valve obstruction, etc. The instructor has the option to define the malfunction initial time and its permanence time as well as the evolution time.

5. External parameter. The external conditions such as atmospheric pressure and temperature, voltage and frequency of the external system, among others can be modified by the instructor.

6. Repetition. Simulation sessions may be repeated as many times as the instructor considers, including the trainees actions.

7. Actions register. This is an automatic registration of the actions carried out by the trainee in order to be re-played exactly in the same sequence and time with the purpose of analyzing what the trainee did during the simulation session.

8. Development tools. The simulator has implemented some others helpful tools to use during simulation session development, for example: to monitor and change on line any selected list of global variable, tabulate any selected list of variable and plot them.

3.1.2   Operator Console

The OC is a replica of the real operation station, and it is formed by the interfaces communication system and it consists of three main displaying sets of plant interfaces: gas turbine, combined cycle and auxiliary electrical external board interface. From there it is possible to display any logical or analogical plant parameter and to operate and handle some manual or automatic controls related to the plant operation. Its objective is to maintain the optimal functional conditions, to monitor or graphic tendencies values of diverse parameters, and to re-establish the normal conditions, among others functions.

Gas Turbine Interface

The gas turbine interface (GTI), is a reproduction of the dynamic screens on which it is possible to monitor and control the gas turbine parameters, including: ready to start/trips; vibration analysis; combustion flashback; lube oil system; trend overview; turbine cooling system; emissions; synchronization; simplified gas turbine (SGT) model; etc. [7].

Combined Cycle Interface

The combined cycle interface (CCI) is a translation of the real operator console, known as engineering console. An example is shown in Fig. 4.

Auxiliary Electrical External Board Interface

The auxiliary electrical external board interface represents a couple of screens that lead all the necessary external instrumentation to synchronize the electrical unit.

3.2 Software Platform

The CCPPS has Windows XP™ as operating system, it was programmed using MS Visual Studio 2005; Fortran Intel™ for the mathematical models; Flash and

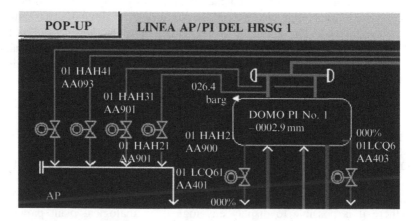

Fig. 4 CCI High and intermediate pressure of the HRSG1 (detail)

VisSim™ for the gas turbine screens; the steam turbine screens were translated from the CCPP control, and C# was used to program the modules of the simulation environment. The simulation environment (called MAS®, proprietary software of the IIE [8]) has three main parts: the real time executive, the operator module, and the console module. Each module runs in a different personal computer (PC), and all of them are communicated by means of a TCP/IP protocol. The modules of the MAS are programmed on C# under MS Visual Studio™ software development platform.

3.3   Hardware Architecture

As Fig. 5 shows, the CCPPS is constituted by four PC interconnected through a fast Ethernet local area network. Each PC has a mini-tower PentiumD™ processor with 3.6 GHz, 1 GB of RAM memory, 40 GB HD, and Windows XP™ as operating system. Figure 6 shows a schematic of this architecture.

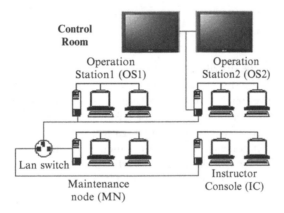

Fig. 5 Hardware architecture

Fig. 6 View of the Combined Cycle Power Plant Simulator on site.

The Instructor Console PC, also named simulation node (SN) has two 20″ flat panel monitors, two operator stations (OS1 and OS2) PC, each one has also 20″ flat panel monitors, but the OS1 has two additional 42″ flat panel monitors as auxiliary monitors.

The maintenance node (MN) is used to make any modifications to the software or process or control models and are tested and validated by an instructor before to install it in the simulator. Figure 6 shows the real and final simulator architecture as described above.

4 Modeled Systems

The control models were translated into C# by means of a graphical software tool, while the system processes were programmed on Fortran Intel™.

4.1 Control System

Generally speaking, the distributed control system (DCS) is a set of PLC in where are allocated the algorithms to control, in an automatic or semi-automatic way, all the systems of a power plant. These control algorithms are organized in components with specific function or task, for example: PID controllers, high/low detectors, timers, memories set/reset, etc. This organization is represented by means of a network of these components, which communicates information through connections, see Fig. 7. These networks are organized in a hierarchical way, in the bottom levels there are the basic *elements* like AND, OR, NOT gates, in the middle level

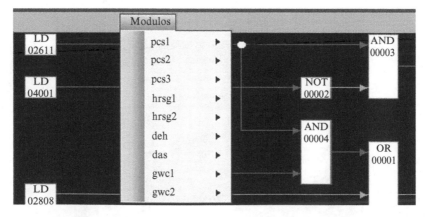

Fig. 7 Control diagram (detail) for speed control on the digital electro-hydraulic control (EHC) system

are *diagrams*, and finally in the top level are *modules*. Then, with a set of *modules*, a DCS was built [9].

4.2   DCS Model for Real-Time Simulation

For the gas turbine, the control models were developed by drawing the diagrams (considering the original ones provided by the designer) into VisSim which generates the code of the drawing into standard C language routines (one per each diagram).

For the control of the steam processes the DCS model was developed with C# programming language, and taking into account the modularization made in the power plant of reference. The main control modules modeled were:

- **pcs1 and pcs2**, auxiliary control system part 1 & 2
- **pcs3**, electric network control
- **hrsg1 and hrsg2**, control of heat recovery system generator, part 1 & 2
- **deh**, digital electro-hydraulic control system
- **das**, digital acquisition system

4.3   The Graphic Visualization Tool

The graphic visualization tool (GVT) is a software application developed to visualize components in diagrams of the Simulator Control Model. This tool was very useful during the simulator development and adjustments because it allows to verify and to visualize signals, states, inputs, outputs and parameters of components on line.

The GVT allows to disable diagrams, modules or components in such a way that the components may be isolated to verify its behavior without the influence of the entire control model.

4.4   Processes System

There are 14 different system processes modeled of the previous gas part plus 26 systems processes modeled of the steam part which together are the mathematical representation of the behavior of the CCPP. All models were developed using Fortran Intel™ in a methodology proposed by the IIE [3], which is summarized in a few steps: (a) to obtain and classify process information directly from the power plant; (b) to analyze information and state a conceptual model to be simulated; (c) to simplify the analyzed information to show only the simulated equipments

Table 1 Classified common groups of the plant systems

Group	Systems included
Water	Condensed water, including air-condenser; HRSG water supply
Steam	HRSG1 and HRSG2 water; HP, IP, and LP water and steam
Turbine	HP, IP, and LP steam turbine; turbine metals
Electric and generator	Electric (generator gas unit1, unit2, and steam unit); primary and secondary regulation for the three generators
Auxiliaries	Auxiliary steam; cooling water; control oil; unit efficiencies; generator cooling air; turbine seal steam; lubricating oil
Auxiliaries minimized	Chemical's dosage to the cycle water-steam; instrument and service air; HRSG1 and HRSG2 drain; steam drain; chemical analysis; potable water; services water; water of sanitary garbage; demineralized water; common drain; and chemical drain. This group contains a set of simplified models with a reduced scope, because it is not necessary to use the complete physical representation
Gas	Fuel gas; gas turbine; compressed air; combustor; air for instruments; lubrication oil; control oil; water-ethylene glycol cooling; electrical network; generator; generator cooling with hydrogen; turbine (metals temperatures and vibrations); performance calculations (heat rate and efficiencies) and a simplified gas turbine

and its nomenclature; (d) to justify all possible assumptions; and (e) to obtain the flow and pressure network with an operational point, namely at 100% of load. The plant systems were classified on groups as listed on Table 1.

5 Project Control

The Project was divided in 10 main phases and 50 tasks. Time, work, and resources were loaded and controlled with Microsoft Project Manager™ Some overwork was needed in order to accomplish the programmed dates. The main sources of delays were some discrepancies that needed to be solved when the local tests were performed and the delay in the delivering of the plant information.

The project lasted seventeen months. The CFE participated actively in getting the real plant data and defining with the SD the data not available. In particular CFE created the control diagrams that were no part of the DCS and helped in performing the plant operation on the simulator to determine and adjust the models behavior.

There was a head of project, one responsible of the software activities, one for the hardware, one for the process modelling and one for the control models. A total of seventeen researchers participated during peak stages of the development of the project. All activities were closely followed in order to assure a punctual and precise ending. Weekly revisions were performed with the heads of area. One summarized report was delivered to the costumer monthly and a full revision was performed every two months. The quality system implanted in the SD allowed the opportune correction actions when needed.

6   Results

The simulator validation was carried out proving its response against the very detailed operation acceptance simulator test procedures elaborated by specialized personnel from CFE.

As example of the models behavior, some results of the GTS are presented. The customer provided plant data for both, an automatic start up procedure and the down load operation. The simulator results were compared with these data. No data for other transients were available. Figure 8 presents a comparison between the expected and simulator speed.

In Fig. 9, the x-axis is the sum of the turbine speed (rad/s) and then power produced (MW). Pressure in the combustor, its temperature and the temperature

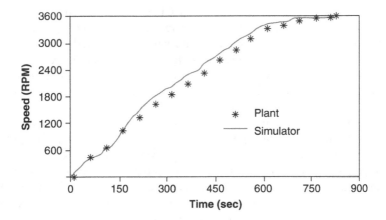

Fig. 8 Turbine speed plant and simulator comparison

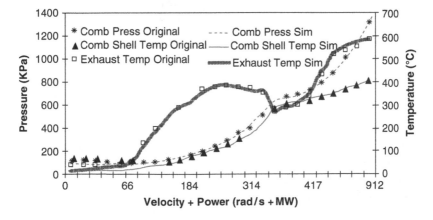

Fig. 9 Plant and simulator variables comparison

in the exhaust were chosen to exemplify the results. While the turbine is increasing its speed the power is zero and when the plant produces power (up to 150 MW) the speed is constant at 376.8 rad/s. In the graphic (with a non-linear x-axis) may be observed that the behavior of the simulator variables is similar than the real variables values.

Besides, the simulator was probed in all the normal operation range, from cold start conditions to full charge, including the response under malfunctions and abnormal operation procedures. In all cases the response satisfied the ISA-S77.20-1993 Fossil-Fuel Power Plant Simulators Functional Requirements norm.

7 Conclusions

The CCPPS validation is been carried out by CENAC specialized personnel under rigorous acceptance simulator testing procedures, and with a lot of experience in the use of simulators, to ensure that this simulator will fulfill the performance specified by the end user in order to have one more useful tool to train the future generation of operators of the combined cycle power plants. The results demonstrate than the simulator is a replica, high-fidelity, plant specific simulator for operator training. Realism is provided by the use of DCS screens emulation.

This simulator had several adjustments during its development, to assure that its behavior and dynamics will operate similarly to those obtained in the real plant.

An additional aspect that gives a lot of certitude and robustness to the simulator behavior is that the simulator has the same DCS that the Chihuahua CCPP. Only minimum adjustments and changes in the controller parameters of regulation loops were required, which means that the process models that were obtained by the SD engineers reproduce correctly the real behavior of the reference plant. The combined cycle power plants have powerful features that include high thermal efficiency, low installed cost, wide range of burning fuels, short installation cycle, compared to conventional thermal, nuclear, and steam plants, as well as low operation costs, it is expected that this simulators will be so strongly claimed for training purposes in a near future.

8 Future Works

As future works, we have considered some projects that might improve and robustness this development, with the purpose of giving a better approach to the control room of the power plant. Among these are the following: (1) to standardize the MMI's simulator to original CCPP interface view, because it is the real equipment the Chihuahua combined cycle power plant has, so it will be very valuable as well as useful to this power plant trainee operators; (2) to expand the simplified gas turbine model; (3) to include thermal regimen of a combined cycle power plant; and (4) to incorporate a touch screen keyboard or replace the virtual keyboard by a physical one with the same functionality that the used in the control room.

Acknowledgment The simulators developed by the SD of the IIE have been projects carried out thanks to the effort, enthusiastic and very professional hard work of researchers and technicians of the IIE as well as the co-operation of operators, trainers and support personnel of CFE training center. This project was carried out thanks to the financial support of CFE; R. Cruz was the technical manager as client counterpart; A. Matías took part as specialized operator during the simulator tests.
About the SD personnel's labor, it is necessary to mention that all researchers participants contributed with their valuable experience during design steps, implementation and final tests concerning to this simulator.

References

1. L. Bo, M. Shahidehpour, Short-term scheduling of combined cycle units. IEEE Trans. Power Syst. **19**(3), 1616–1625 (Aug. 2004)
2. J.Y. Shin et al., Analysis of the dynamic characteristics of a combined-cycle power plant. Energy Int. J. **27**(12), 1085–1098 (Dec 2002). ISNN 0360-5442
3. J. Roldán-Villasana Edgardo, M.J. Cardoso, Y. Mendoza-Alegría, Modeling methodology for operator's training full scope simulators applied in models of a gas-turbine power plant, *in Memorias del 9o. Congreso Interamericano de Computación Aplicada a la Industria de Procesos, 25 al 28 de agosto*, Montevideo, Uruguay, 2009
4. Mitsubishi Heavy Industries, Chihuahua Combined Cycle Power Plant Begins Commercial Operation in Mexico, http://www.mhi.co.jp/en/ power/news/sec1/2001_sep_08.html, Sept 2001
5. CFE Comisión Federal de Electricidad, Una empresa de clase mundial, Centrales, http://www. cfe.gob.mx/QuienesSomos/queEsCFE/ estadisticas/Paginas/Indicadoresdegeneración.aspx, 2009
6. D.L. Chase, GE power systems, *Combined-Cycle Development Evolution and Future*, Schenectady, NY, 1998 http://www.gepower.com/prod_serv/products/tech_docs/en/downloads/ ger4206.pdf
7. E.J. Roldán-Villasana, Y. Mendoza-Alegría, J. Zorrilla-Arena, Ma.J. Cardoso, EMS 2008, Development of a Gas Turbine Full Scope Simulator for Operator's Training, in *Second UKSIM European Symposium on Computer Modeling and Simulation*, Liverpool, UK, 8–10 Sept 2008
8. L.A. Jiménez-Fraustro, Desarrollo del Medio Ambiente de Simulación para Simuladores en Tiempo Real basado en sistema operativo Windows XP, Reporte Interno, Gerencia de Simulación, IIE, México, 2005
9. G. Romero-Jiménez, V. Jiménez-Sánchez, E.J. Roldán-Villasana, WASET 2008, in *Proceedings of World Academy of Science, Engineering and Technology*, vol. 30, July 2008, ISSN 1307-6884, Graphical Environment for Modeling Control Systems in Full Scope Training Simulators, Paris, France, 5–7 July 2008, pp. 792–797

Chapter 26
Texture Features Extraction in Mammograms Using Non-Shannon Entropies

Amar Partap Singh and Baljit Singh

Abstract This paper deals with the problem of texture-features-extraction in digital mammograms using non-Shannon measures of entropy. Texture-features-extraction is normally achieved using statistical texture-analysis method based on gray-level histogram moments. Entropy is important texture feature to measure the randomness of intensity distribution in a digital image. Generally, Shannon's measure of entropy is employed in various feature-descriptors implemented so far. These feature-descriptors are used for the purpose of making a distinction between normal and abnormal regions in mammograms. As non-Shannon entropies have a higher dynamic range than Shannon's entropy covering much wider range of scattering conditions, they are more useful in estimating scatter density and regularity. Based on these considerations, an attempt is made to develop a new type of feature-descriptor using non-Shannon's measures of entropy for classifying normal and abnormal mammograms. Experiments are conducted on images of mini-MIAS (Mammogram Image Analysis Society) database to examine its effectiveness. The results of this study are quite promising for extending the work towards the development of a complete Computer Aided Diagnosis (CAD) system for early detection of breast cancer.

1 Introduction

Cancer remains one of the frequently occurring fatal diseases affecting men and women throughout the world. Among the cancer diseases, breast cancer is especially a concern in women. According to the statistics, breast cancer is one of the

A.P. Singh (✉)
Department of Electronics & Communication Engineering, SLIET, Longowal, 148106, Sangrur, Punjab, India
e-mail: amarpartapsingh@yahoo.com

S.-I. Ao et al. (eds.), *Machine Learning and Systems Engineering*,
Lecture Notes in Electrical Engineering 68,
DOI 10.1007/978-90-481-9419-3_26, © Springer Science+Business Media B.V. 2010

major causes for the increase in mortality among middle-aged women in both the developed and developing countries. However, the etiologies of breast cancer are unknown and no single dominant cause has emerged. Still, there is no known way of preventing breast cancer but early detection allows treatment before it is spread to other parts of the body. However, it is evident that the earlier breast cancer is found, the better chance a women gets for a full recovery. Moreover, the early detection of breast cancer can play a very important role in reducing the morbidity and mortality rates.

Mammography [1, 2] is the single most effective, reliable, low cost and highly sensitive method for early detection of breast cancer. Mammography offers high-quality images at low radiation doses and is the only widely accepted imaging method for routine breast cancer screening. It is recommended that women at the ages of 40 or above should have a mammogram every 1–2 years. Although mammography is widely used around the world for breast cancer detection, there are some difficulties when mammograms are searched for signs of abnormality by expert radiologists. Such difficulties are that mammograms generally have low contrast compared with normal breast structure and sign of early disease are often small or subtle. This is the main cause of many missed diagnoses that can be mainly attributed to human factors such as subjective or varying decision criteria, distraction by other image features, or simple oversight. As the consequences of errors in detection or classification are costly and since mammography alone can not prove that a suspicious area is tumorous, malignant or benign the tissue has to be removed for closer examination using breast biopsy techniques. Nevertheless, a false-positive detection causes unnecessary biopsy. On the other hand, in false-negative detection an actual tumor remains undetected. Thus, there is a significant necessity for developing methods for automatic classification of suspicious areas in mammograms, as a means of aiding radiologists to improve the efficacy of screening programs and avoid unnecessary biopsies.

A typical mammogram contains a vast amount of heterogeneous information that depicts different tissues, vessels, ducts, chest skin, breast edge, the film and the x-ray machine characteristics [3–5]. In order to build a robust diagnostic system towards correctly classifying abnormal and normal regions of mammograms, present all the available information that exits in mammograms to the diagnostic system so that it can easily discriminate between the abnormal and normal tissues. However, the use of all the heterogeneous information, results to high dimensioned features vectors that degrade the diagnostic accuracy of utilized systems significantly as well as increase their computational complexity, Therefore, reliable feature vectors should be considered the reduce the amount of irrelevant information thus producing robust mammographic descriptors of compact size [6–8].

Texture is one of the important characteristics used in classifying abnormal and normal regions in mammograms. The texture of images refers to the appearance, structural and arrangement of the parts of an object within the image [9, 10]. Images used for diagnostic purposes in clinical practice are digital. A two dimensional digital image is made up of little rectangular blocks or pixels (picture elements). Each is represented by a set of coordinates in space and each has a value representing the gray-level intensity of that picture element in space. A feature value is a real number, which encodes some discriminatory information about a property of an object. Generally speaking, texture feature extraction methods can be classified into three major categories, namely, statistical, structural and spectral. In statistical approaches, texture statistics such as the moments of the gray level histogram, or statistics based on gray level co-occurrence matrix are computed to discriminate different textures [11]. For structural approaches, texture primitive, the basic element of texture, is used to form more complex texture pattern by grammar rules which specify the generation of texture pattern. Finally, in spectral approaches, the textured image is transformed into frequency domain. Then, the extraction of texture features can be done by analyzing the power spectrum. Various texture descriptors have been proposed in the past. In addition to the aforementioned methods, Law's texture energy measures, Markov random field models, texture spectrum etc. are some other texture descriptors [12–14].

This paper deals with the problem of statistical approaches to extract texture features in digital mammogram. Gray level histogram moments method is normally used for this purpose. Entropy [15–17] is an important texture feature, which is computed based on this method, to build a robust descriptor towards correctly classifying abnormal and normal regions of mammograms. Entropy measures the randomness of intensity distribution. In most feature descriptors, Shannon's measure is used to measure entropy. However, in this paper, non-Shannon measures are used to measure entropy. Non-Shannon entropies have a higher dynamic range than Shannon entropy over a range of scattering conditions, and are therefore more useful in estimating scatter density and regularity [18].

2 Gray Level Histogram Moments

Regarding the statistical approach for describing texture, one of the simplest computational approaches is to use statistical moments of the gray level histogram of the image. The image histogram carries important information about the content of an image and can be used for discriminating the abnormal tissue from the local healthy background. Considering the gray level histogram $\{h_i, i = 0, 1, 2, \ldots, N_{g-1}\}$, where N_g is the number of distinct gray levels in the ROI (region of interest). If n is the total number of pixels in the region, then the normalized histogram of the ROI is the set $\{H_i, i = 0, 1, 2, \ldots, N_{g-1}\}$, where $H_i = h_i/n$. Generally, Shannon's measure is used to estimate entropy in most of the descriptors. Shannon's entropy (S) is a degree of randomness and is defined as

$$S = - \sum_{i=0}^{N_g-1} H_i \, log_2(H_i)$$

However, non-Shannon's entropies have a higher dynamic range than Shannon entropy covering a wider range of scattering conditions. Important non-Shannon measures of entropies include Renyi's entropy (R), Havrda and Charvat's Entropy (HC) and Kapur's entropy (K). These entropies are defined as

$$R = \frac{1}{1-\alpha} \log_2 \left(\sum_{i=0}^{N_g-1} H_i^{\alpha} \right)$$

$$\alpha \neq 1, \alpha > 0$$

$$HC = \frac{1}{1-\alpha} \left(\sum_{i=0}^{N_g-1} H_i^{\alpha} - 1 \right)$$

$$\alpha \neq 1, \alpha > 0$$

$$K_{\alpha,\beta} = \frac{1}{\beta - \alpha} \log_2 \frac{\sum_{i=0}^{N_g-1} H_i^{\alpha}}{\sum_{i=0}^{N_g-1} H_i^{\beta}}$$

$$\alpha \neq \beta, \alpha > 0, \beta > 0$$

3   Experimental Results

Experiments were conducted on images of mini-MIAS database (Mammogram Image Analysis Society) database, UK. For implementation, MATLAB 7.0 was used. The mammograms of about 200 images were considered for simulating the proposed work. For the completeness of description several experimental results are depicted in Tables 1 and 2 at different values of α and β. The exact values of parameters are of less significant and hence the range values have been considered for true or best classification

The basic classification of normal, benign and malignant mammogram images based on the values of the texture parameters are shown in Tables 3 and 4. The original images and their corresponding histograms for basic classification are shown in Figs. 1–3. Graphical representation of these classifications is shown in Figs. 4 and 5. In order to validate the results, uncompressed fatty, fatty and non-uniform breast mammogram images are considered to examine the versatility of the proposed feature descriptor. These images and their corresponding histograms are shown in Figs. 6–8. These classification results are shown in Tables 5 and 6. Graphical representation of these classifications is shown in Figs. 9 and 10.

Table 1 Entropy measure of mammograms based on gray level histogram moments at $\alpha = 0.5$ for Renyi, $\alpha = 0.7$ for Havrda & Charvat and for Kapur $\alpha = 0.5$, $\beta = 0.7$

Image samples	S	R $\alpha = 0.5$	HC $\alpha = 0.7$	$K_{\alpha, \beta}$ $\alpha = 0.5$ $\beta = 0.7$
Mam1	4.7658	6.9242	8.7940	7.9945
Mam2	5.1761	6.9060	9.0540	7.7959
Mam3	5.2400	7.0073	9.3439	7.8821
Mam4	5.1644	6.7166	8.6331	7.5718
Mam5	4.8436	6.7857	8.5198	7.8131
Mam6	5.1044	6.7977	8.7193	7.7227
Mam7	5.5805	7.0348	9.7094	7.7459
Mam8	4.2316	6.6505	7.8027	7.9252
Mam9	2.9738	6.1737	6.0264	7.9867
Mam10	4.1991	6.4753	7.3892	7.7603
Mam11	4.5951	6.7681	8.2881	7.9115
Mam12	4.9022	6.8990	8.8035	7.9257
Mam13	4.9559	6.8047	8.6226	7.7983
Mam14	4.4958	6.7285	8.1625	7.8909
Mam15	4.9878	6.9086	8.9161	7.8832

Table 2 Entropy measure of mammograms based on gray level histogram moments at $\alpha = 0.1$ for Renyi, $\alpha = 0.5$ for Havrda & Charvat and for Kapur $\alpha = 0.1$, $\beta = 0.9$

Image samples	S	R $\alpha = 0.1$	HC $\alpha = 0.5$	$K_{\alpha, \beta}$ $\alpha = 0.1$ $\beta = 0.9$
Mam1	4.7658	7.7796	20.0410	8.0925
Mam2	5.1761	7.7505	19.9023	8.0219
Mam3	5.2400	7.7346	20.6845	7.9934
Mam4	5.1644	7.6608	18.5109	7.9306
Mam5	4.8436	7.7167	19.0077	8.0217
Mam6	5.1044	7.7035	19.0954	7.9825
Mam7	5.5805	7.6587	20.9020	7.8736
Mam8	4.2316	7.7460	18.0460	8.1189
Mam9	2.9738	7.6251	14.9927	8.1298
Mam10	4.1991	7.7053	16.8654	8.0857
Mam11	4.5951	7.7552	18.800	8.0900
Mam12	4.9022	7.7795	19.8491	8.0820
Mam13	4.9559	7.7529	19.1466	8.0516
Mam14	4.4958	7.7473	18.5954	8.0910
Mam15	4.9878	7.7747	19.9222	8.0668

Table 3 Classification mammograms based on Shannon entropy, Renyi's entropy at $\alpha = 0.5$, Havrda & Charvat's entropy at $\alpha = 0.7$ and Kapur's entropy at $\alpha = 0.5$, $\beta = 0.7$

Mammogram classification	Average S	Average R $\alpha = 0.5$	Average HC $\alpha = 0.7$	Average $K_{\alpha, \beta}$ $\alpha = 0.5$ $\beta = 0.7$
Normal	4.2315	6.4321	7.6275	7.6789
Benign	4.4236	6.7543	8.235	7.7678
Malignant	4.9875	7.0024	8.9678	7.8123

Table 4 Classification mammograms based on Shannon entropy, Renyi's entropy at $\alpha = 0.1$, Havrda & Charvat's entropy at $\alpha = 0.5$ and Kapur's entropy at $\alpha = 0.1$, $\beta = 0.9$

Mammogram classification	Average S	Average R $\alpha = 0.1$	Average HC $\alpha = 0.5$	Average K $_{\alpha, \beta}$ $\alpha = 0.1$ $\beta = 0.9$
Normal	4.2315	7.6975	18.2013	7.8020
Benign	4.4236	7.7204	19.5750	7.9652
Malignant	4.9875	7.7824	20.2542	8.0125

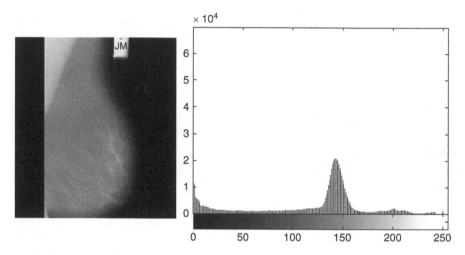

Fig. 1 Normal Mammogram:-mdb006 and its Histogram

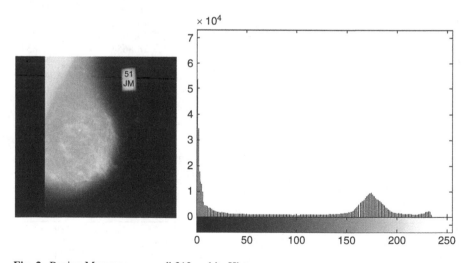

Fig. 2 Benign Mammogram:-mdb312 and its Histogram

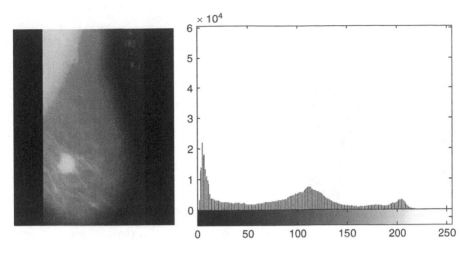

Fig. 3 Malignant Mammogram:-mdb028 and its Histogram

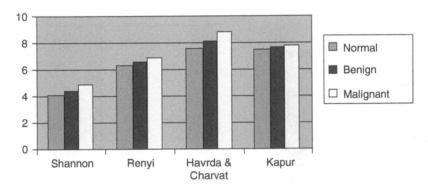

Fig. 4 Graphical representation of classification of mammogram images based on Shannon entropy, Renyi's entropy at $\alpha = 0.5$, Havrda & Char vat's entropy at $\alpha = 0.7$ and Kapur's entropy at $\alpha = 0.5$, $\beta = 0.7$

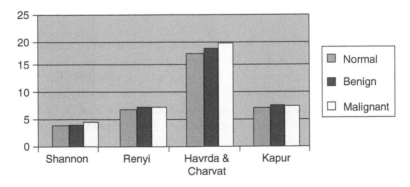

Fig. 5 Graphical representation of classification of mammograms based on Shannon entropy, Renyi's entropy at $\alpha = 0.1$, Havrda & Charvat's entropy at $\alpha = 0.5$ and Kapur's entropy at $\alpha = 0.1$, $\beta = 0.9$

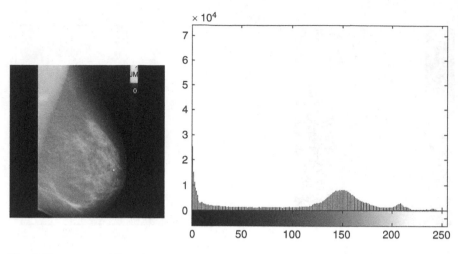

Fig. 6 Uncompressed Fatty image:-mdb008 and its Histogram

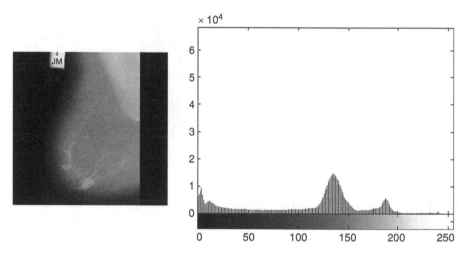

Fig. 7 Fatty image:-mdb005 and its Histogram

From the above results, it can be inferred that non-Shannon's entropies (Renyi, Havrda and Charvat, and Kapur) are useful parameters for the classification of normal and abnormal mammogram images. Specially, Havrda and Charvat is most suitable for this purpose. Experimental results indicate that, in non-Shannon entropies, the values of constants α, β play important role in classification. So, selection of suitable values of these constants is necessary. Havrda and Charvat entropy at $\alpha = 0.5$ gives the best suitable results for classification. This in turn also helps in the diagnosis of palpable or microcalcifications or tumors in the breast. This work may be employed to develop a Computer Aided Decision (CAD) System for early detection of breast cancer.

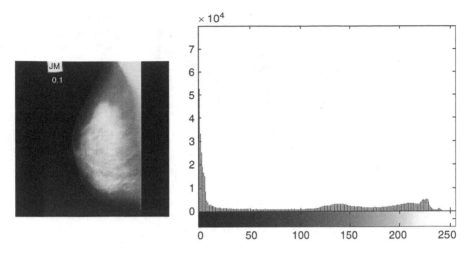

Fig. 8 Non-uniform Fatty image:-mdb003 and its Histogram

Table 5 Classification of uncompressed fatty, fatty and non-uniform fatty breasts based on Shannon entropy, Renyi's entropy at $\alpha = 0.5$, Havrda & Charvat's entropy at $\alpha = 0.7$ and Kapur's entropy at $\alpha = 0.5$, $\beta = 0.7$

Mammogram classification	Average S	Average R $\alpha = 0.5$	Average HC $\alpha = 0.7$	Average K $_{\alpha,\,\beta}$ $\alpha = 0.5$ $\beta = 0.7$
Uncompressed fatty	4.9878	6.9086	8.9161	7.8832
Fatty breast	5.1761	6.9060	9.2540	7.7959
Non-uniform fatty breast	4.5951	6.7681	8.2881	7.9115

Table 6 Classification of uncompressed fatty, fatty and non-uniform fatty breasts based on Shannon entropy, Renyi's entropy at $\alpha = 0.1$, Havrda & Charvat's entropy at $\alpha = 0.5$ and Kapur's entropy at $\alpha = 0.1$, $\beta = 0.9$

Mammogram classification	Average S	Average R $\alpha = 0.1$	Average HC $\alpha = 0.5$	Average K $_{\alpha,\,\beta}$ $\alpha = 0.1$ $\beta = 0.9$
Uncompressed fatty	4.9878	7.7747	19.9222	8.0668
Fatty breast	5.1761	7.7505	20.2023	8.0219
Non-uniform fatty breast	4.5951	7.7552	18.8800	8.0900

4   Conclusions and Future

In this paper, an attempt is made to develop a new features descriptor based on non-Shannon's entropies (Renyi, Havrda and Charvat, and Kapur) for classifying normal and abnormal mammogram images. Experiment results have demonstrated that Havrda and Charvat entropy based feature works satisfactorily for classifying normal, benign and malignant digital mammograms including uncompressed-fatty, fatty and non-uniform fatty breast mammograms.

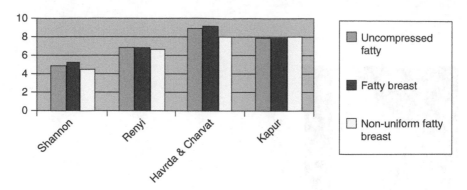

Fig. 9 Graphical representation of classification of uncompressed fatty, fatty breast and non-uniform fatty breast based on Shannon entropy, Renyi's entropy at α =0.5, Havrda & Charvat's entropy at α =0.7 and Kapur's entropy at α =0.5, β =0.7

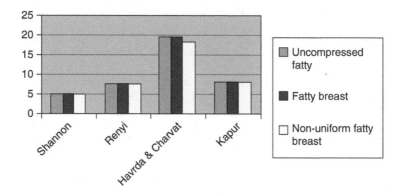

Fig. 10 Graphical representation of classification of uncompressed fatty, fatty and non-uniform fatty breasts based on Shannon entropy, Renyi's entropy at α = 0.1, Havrda & Charvat's entropy at α = 0.5 and Kapur's entropy at α =0.1, β =0.9

In future work, Havrda and Charvat entropy based feature will be included with other existing statistical features to build a robust features descriptor for developing a complete Computer Aided Diagnosis (CAD) for early detection of breast cancer. The work is under further progress in this direction.

References

1. H.S. Sheshadri, A. Kandaswamy, Experimental investigation on breast tissue classification based on statistical feature extraction of mammogram, computerized. Med. Imaging Graph. **31**, 46–48 (2007)
2. I. Christoyianni, A. Koutras, E. Dermatas, G. Kokkinakis, Computer aided diagnosis of breast cancer in digitized mammograms. Comput. Med. Imaging Graph. **26**, 309–319 (2002)

3. B. Verma, P. Zhang, A novel neural-genetic algorithm to find the most significant combination of features in digital mammograms. Appl. Soft Comput. **1**(7), 612–625 (2007)
4. S.-K. Lee, C.-S. Lo, C.-M. Wang, P.-C. Chung, C.-I. Chang, C.-W. Yang, P.-C. Hsu, A computer-aided design mammography screening system for detection and classification of microcalcifications. Int. J. Med. Inf. **60**, 29–57 (2000)
5. H.D. Cheng, X. Cai, X. Chen, L. Hu, X. Lou, Computer-aided detection and classification of microcalcifications in mammograms: a survey. Pattern Recogn. **36**, 2967–2991 (2003)
6. H.D. Cheng, M. Cui, Mass lesion detection with a fuzzy neural network. Pattern Recogn. **37**, 1189–1200 (2004)
7. H.D. Cheng, X.J. Shi, R. Min, L.M. Hu, X. P. Cai, H.N. Du, Approaches for automated detection and classification of masses in mammograms. Pattern Recogn. **39**, 646–668 (2006)
8. M.E. Mavroforakis, H.V. Georgiou, N. Dimitropoulos, D. Cavouras, S. Theodoridis, Mammographic masses characterization based on localized texture and dataset fractal analysis using linear, neural and support vector machine classifiers. Artif. Int. Med. **37**, 145–162 (2006)
9. B. Verma, J. Zakos, A computer-aided diagnosis system for digital mammograms based on fuzzy-neural and feature extraction techniques. IEEE Trans. Inf. Technol. Biomed. **5**(1), (Mar 2001)
10. J.K. Kim, H.W. Park, Statistical textural features for detection of microcalcifications in digitized mammograms. IEEE Trans. Med. Imaging **18**(3), (Mar 1999)
11. J.K. Kim, J. Mi Park, K.S. Song, H.W. Park, Texture analysis and artificial neural network for detection of clustered microcalcifications on mammograms, in *Neural Networks for Signal Processing [1997] VII, Proceedings of the 1997 IEEE Workshop*, Amelia Island, Florida, USA, pp. 199–206
12. A Karahaliou, S Skiadopoulos, I Boniatis, P Sakellaropoulos, E Likaki, G Panayiotakis, L Costaridou, Texture analysis microcalcifications on mammograms for breast cancer diagnosis. Brit. J. Radiol. **80**, 648–656 (2007)
13. Y.-Y. Wan, J.-X. Du, D.-S. Huang, Z. Chi, Y.-M. Cheung, X.-F. Wang, G.-J. Zhang, Bark texture feature extraction based on statistical texture analysis, in *Proceedings of 2004 International Symposium on Intelligent Multimedia, Video and Speech Processing*, Hong Kong, Oct 20–22, 2004
14. M.R. Chandraratne, Comparison of three statistical texture measures for lamb grading, in *1st International Conference on Industrial and Information Systems,ICIIS2006* Sri Lanka, 8–11 Aug 2006
15. J.N. Kapur, *Measure of Information and Their Applications* (Wiley Eastern Limited, New Delhi, 1994)
16. L.I. Yan, F. Xiaoping, L. Gang, An application of Tsallis entropy minimum difference on image segmentation, in *Proceeding of the 6th World Congress on Intelligent Control and Automation*, Dalian, China, 21–23 June 2006
17. M. Portes de Albuquerque, I.A. Esquef, A.R. Gesualdi Mello, Image thresholding using Tsallis entropy. Pattern Recogn. Lett. **25**(2004), 1059–1065 (2004)
18. R. Smolikova, M.P. Wachowiak, G.D. Tourassi, A. Elmaghraby, J.M. Zurada, Characterization of ultrasonic back scatter based on generalized entropy, in *Proceeding of the 2nd Joint EMBS/BMES Conference*, Houston, TX, USA, 23–26 Oct 2002

Chapter 27
A Wideband DOA Estimation Method Based on Arbitrary Group Delay

Xinggan Zhang, Yechao Bai, and Wei Zhang

Abstract Direction-of-arrival (DOA) estimation is an important algorithm in array processing. Most traditional DOA estimation methods focus on narrow-band sources. A new wideband DOA estimation approach based on arbitrary group delay is proposed in this paper. In the proposed algorithm, echo gain at each direction is calculated based on digital group delay compensation. Digital group delay can realize time delay precisely, and has no restriction on delay step. The proposed method is able to operate arbitrary waveform and is suitable for linear frequency modulation signals, nonlinear frequency modulation signals, and even carrier free signals. Simulations with Gaussian function are carried out to show the effectiveness of the proposed algorithm.

1 Introduction

DOA estimation is an important algorithm in array signal processing systems [1, 2]. Various estimation methods such as Maximum Likelihood (ML), Multiple Signal Classification (MUSIC) [3], and Estimation of Signal Parameters via Rotational Invariance Technique (ESPRIT) [4] were proposed in recent two decades. However, most of these methods are based on narrow band models. Several extensions to the wideband sources are also proposed [5], for example, the Coherent Signal-Subspace Processing (CSSP) [6], Wideband Cyclic MUSIC [7] and Wideband ESPRIT [8]. A frequency domain algorithm based on ML criterion called AML method has been proposed for source localization and DOA estimation [9, 10]. A method proposed by Ward [11] performs beam-space processing using time-domain frequency-invariant beamformers (TDFIB). In this paper, we propose a

X. Zhang (✉)
School of Electronic Science and Engineering, Nanjing University, Nanjing, P. R. China
e-mail: zhxg@nju.edu.cn

S.-I. Ao et al. (eds.), *Machine Learning and Systems Engineering*,
Lecture Notes in Electrical Engineering 68,
DOI 10.1007/978-90-481-9419-3_27, © Springer Science+Business Media B.V. 2010

new wideband DOA estimation approach based on arbitrary group delay. This
method is able to operate arbitrary waveform and is suitable for linear frequency
modulation signals, nonlinear frequency modulation signals, and even carrier free
signals.

2   Method of Digital Group Delay

When the object direction is far from the normal direction, the delay between array
elements is so large that signals need to be aligned to get a better distance resolution
and a higher power gain. The time delay compensation is usually operated in two
steps: the integral sampling period part and fractional sampling period part. The
integral part can be simply carried out by register shifting. The subsample time
delay is compensated by interpolation commonly, and the compensation effect lies
on the interpolation function. Dense sampling, in which sampling frequency is
much higher than the Nyquist frequency is also used in some situation to achieve a
higher accuracy. Unfortunately, this method brings too much data redundancy and
raises a higher demand to the sampling device. Especially, when the bandwidth of
ultra-wideband signals is very high, for example 1 GHz, it is hard to sample with
frequency several times higher, and the expense is also too huge.

Suppose that the wideband signal is $s(t)$, and the delayed signal is $s(t-\tau)$.
According to the time shifting property of continuous time Fourier transform, we
can get:

$$s(t - \tau) = F^{-1}\{F\{s(t)\}e^{-j\omega\tau}\} \tag{1}$$

If the sampling frequency is so high that the spectral aliasing can be ignored,
the digital signal delay can be implemented by

$$s(nT_s - \tau) = IDFT\{DFT\{s(nT_s)\}e^{-\frac{j2\pi k\tau}{NT_s}}\} \tag{2}$$

where T_s is the sampling period, and $s(nT_s)$ is the digital signal. The time delay τ in
Eq. (2) need not to be integral multiple of the sampling period. N is the length of the
operated signal, consisting of original signal and $INT(\tau/f_s)$ padded zeroes, where
INT denotes rounding operation. Zeros are padded at the tail of the original signal
to avoid cyclic shifting of the original signal.

Gaussian modulated sine signal is used to demonstrate the method of digital
group delay. The analytical expression of the signal is

$$s(t) = e^{-\pi\left(\frac{t-T_0}{T_g}\right)^2} \cos[2\pi f(t - T_0)] \tag{3}$$

where T_g, T_0 and f are the time width parameter of Gaussian window, the time
center and the frequency of sine signal, respectively. T_g and f were chosen to be 50

and 0.1. T_0 was chosen to be 0 before signal is delayed. The results of simulation of digital group delay with MATLAB are shown in Figs. 1 and 2. In simulations, the signal delayed for 3.7 samples by the method of digital group delay is compared with the theoretical waveform obtained by setting T_0 to 3.7 in Eq. (3).

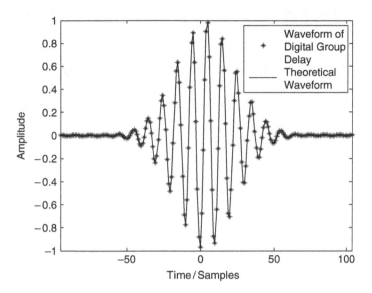

Fig. 1 Comparison between waveform of digital group delay and theoretical waveform

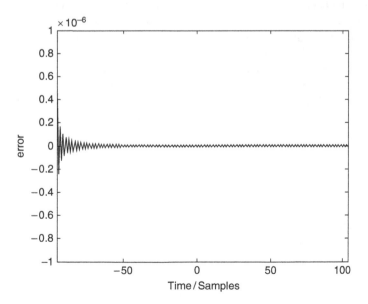

Fig. 2 Error between waveform of digital group delay and theoretical waveform

It can be seen that digital group delay can realize time delay precisely for approximate band limit signals from Figs. 1 and 2. In addition, digital group delay operates the integral sampling period part and fractional sampling period part together, and has no restriction on delay step. Furthermore, this method has no requirement for the signal waveform, and is suitable for linear frequency modulation signals, nonlinear frequency modulation signals, and even carrier free signals.

3   DOA Estimation Based on Digital Group Delay

A simple time delay manifests itself as the slope of an additive linear phase in the phase spectrum, which is an additional group delay

$$\tau_g(f) = -\frac{1}{2\pi}\frac{d\varphi}{df} = -\tau \tag{4}$$

where τ_g is a constant.

Consider an equally spaced linear array composed of N_e identical elements. The distance between consecutive sensors is d. The relative delay of the ith element is

$$\tau_i = \frac{id\sin\alpha}{c} \tag{5}$$

where α is the beam direction, and c is the velocity of electromagnetic wave.

The target echoes of array elements are aligned according to each direction. At the direction α, signals are aligned completely after group delay compensation, and the gain after match filter reaches the maximum. In contrast, at other directions, echo signals can not be totally aligned, and the gain decreases.

DOA estimation based on arbitrary group delay compensates the echo signals according to different beam directions, and finds the target direction by locating the maximum of the match filter gain. Considering that the searching directions are discrete and the accurate is limited, we fit a convex parabola around the peak and obtain the required direction by calculating the symmetry axis of the parabola.

4   Simulation

Gaussian function [12] (GGP, generalized Gaussian pulse) was employed in the simulation (using MATLAB) to demonstrate the effectiveness of the proposed algorithm. One of forms based on experimentation with UWB impulse radiators and sensors is:

Table 1 The RMSE in the DOA estimation versus SNR

SNR (dB)	−10	0	10	20	30
RMSE (°)	1.4457	0.4584	0.1627	0.0579	0.0187

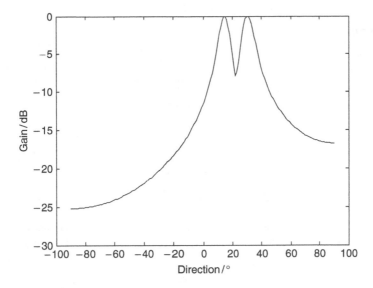

Fig. 3 DOA estimation for two targets at 15° and 30°, SNR = 20 dB

$$f(t) = \frac{E_0}{1-\alpha}\left\{e^{-4\pi\left(\frac{t-t_0}{\Delta T}\right)^2} - \alpha e^{-4\pi\alpha^2\left(\frac{t-t_0}{\Delta T}\right)^2}\right\} \tag{6}$$

where E_0 is peak amplitude at time $t = t_0$, α is scale parameter. In the simulation, $E_0 = 1$, $\Delta T = 2$ μs and $\alpha = 10$. In addition, element amount N_e, distance between consecutive sensors d and the sampling frequency were chosen to be 100, 0.4 m and 100 MHz, respectively.

The root mean square error (RMSE) in the DOA estimation versus signal to noise ratio (SNR) is shown in Table 1. The target is fixed at direction 45.5°. The simulation for each SNR was performed 1,000 times.

From Table 1, it is easily seen that the proposed algorithm has a very small RMSE, and the proposed algorithm also performs well at low SNR.

Figure 3 shows the resolution performance of the proposed algorithm. In the simulation, E_0, ΔT, α and the sampling frequency were chosen to be 1, 2 μs, 200 and 1 GHz, respectively. The echo contains two targets located at 15° and 30°, and SNR is 20 dB.

5 Conclusion

The digital group delay can realize time delay precisely for approximate band limit signals. In this paper, we apply the digital group delay method to DOA estimation. The simulations show the effectiveness of the proposed algorithm. The proposed algorithm has no requirement for the signal waveform, and is suitable for wideband and ultra-wideband radars, sonar systems, and microphone arrays.

References

1. M.R. Azimi-Sadjadi, A. Pezeshki, N. Roseveare, Wideband DOA estimation algorithms for multiple moving sources using unattended acoustic sensors. IEEE Trans. Aerosp. Electron. Syst. **44**, 1585–1599 (Oct 2008)
2. Y.-S. Yoon, L.M. Kaplan, J.H. McClellan, TOPS: new DOA estimator for wideband signals. IEEE Trans. Signal Process. **54**, 1977–1989 (June 2006)
3. R. Schmidt, Multiple emitter location and signal parameter estimation. IEEE Trans. Antennas Propag. **34**, 276–280 (Mar 1986)
4. R. Roy, A. Paulraj, T. Kailath, Direction-of-arrival estimation by subspace rotation methods - ESPRIT, IEEE International Conference on ICASSP, Apr 1986, pp. 2495–2498
5. L. Yip, C.E. Chen, R.E. Hudson, K. Yao, DOA estimation method for wideband color signals based on least-squares Joint Approximate Diagonalization, IEEE Sensor Array and Multi-channel Signal Processing Workshop, July 2008, pp. 104–107
6. H. Wang, M. Kaveh, Coherent signal-subspace processing for the detection and estimation of angles of arrival of multiple wide-band sources. IEEE Trans. Acoust. Speech Signal Process. **33**, 823–831 (Aug 1985)
7. K.D. Mauck, Wideband cyclic MUSIC, ICASSP **4**, 288–291 (Apr 1993)
8. B. Ottersten, T. Kailath, Direction-of-Arrival estimation for wideband signals using the ESPRIT algorithm, IEEE Trans. Acoust. Speech Signal Process. **38**, 317–327 (Feb 1990)
9. J.C. Chen, R.E. Hudson, K. Yao, Maximum-likelihood source localization and unknown sensor location estimation for wideband signals in the near Field, IEEE. Trans. Signal Process. **50**, 1843–1854 (Aug. 2002)
10. J.C. Chen, L. Yip, J. Elson, H. Wang, D. Maniezzo, R.E. Hudson, K. Yao, D. Estrin, Coherent acoustic array processing and localization on wireless sensor networks, Proc. IEEE **91**, 1154–1162 (Aug. 2003)
11. D.B. Ward, Z. Ding, R.A. Kennedy, Broadband DOA estimation using frequency invariant beamforming, IEEE Trans. Signal Process. **46**, 1463–1469 (May 1998)
12. W. Min, W. Shunjun A time domain beamforming method of UWB pulse array, IEEE Int. Radar Conf. 697–702 (May 2005)

Chapter 28
Spatial Speaker Spatial Positioning of Synthesized Speech in Java

Jaka Sodnik and Sašo Tomažič

Abstract In this paper, we propose a "Spatial speaker", an enhancement of a Java FreeTTS speech synthesizer with the additional function of spatial positioning of both the speaker and the listener. Our module enables the reading of an arbitrary text from a file or webpage to the user from a fixed or changing position in space through normal stereo headphones. Our solution combines the following modules: FreeTTS speech synthesizer, a custom made speech processing unit, MIT Media Lab HRTF library, JOAL positioning library and Creative X-Fi sound card. The paper gives an overview of the design of the "Spatial Speaker" and proposes three different practical applications of such a system for visually impaired and blind computer users. Some preliminary results of user studies confirmed the system's usability and showed its great potential also in other types of applications and auditory interfaces. The entire system is developed as a single Java class which can be imported and used in any Java application.

1 Introduction

The most important qualities of modern speech synthesis systems are their naturalness and intelligibility [1]. By naturalness we mean how closely the synthesized speech resembles real human speech. Intelligibility, on the other hand, describes the ease with which the speech is understood. The maximization of these two criteria is the main research and development goal in the TTS field.

At the moment, numerous examples of TTS software can be found on the market, for example stand-alone speech programs that convert all types of text inserted through an input interface into high quality realistic speech output

J. Sodnik (✉)
University of Ljubljana, Faculty of Electrical Engineering, Tržaška 25, 1000 Ljubljana, Slovenia, Europe
e-mail: jaka.sodnik@fe.uni-lj.si

S.-I. Ao et al. (eds.), *Machine Learning and Systems Engineering*,
Lecture Notes in Electrical Engineering 68,
DOI 10.1007/978-90-481-9419-3_28, © Springer Science+Business Media B.V. 2010

(TextAloud [2], VoiceMX STUDIO [3], etc.). Many of them can save the readings as a media file that can later be played on demand. On the other hand, there are also several speech synthesis libraries that can be included and used in various program languages, for example FreeTTS, a TTS library written in Java programming language [4].

In this article, we propose an extension to the FreeTTS software that enables text-to-speech conversion in a 3D audio environment. Our solution enables the positioning of the voice (i.e. the text reader) at any arbitrary spatial position in relation to the listener. The relative spatial positions of the speaker and the listener can be updated dynamically in real time while performing text-to-speech conversion. For example, the text can be read from a specific spatial position in relation to the listener or the position can be dynamically changed and updated.

We believe there are numerous potential applications of such technology, from screen readers for visually impaired computer users to advanced auditory interfaces for desktop and mobile devices. In the next chapter, we summarize some related work which uses both spatial audio and TTS.

2 Related Work

In the past, spatial auditory interfaces were mostly used to enrich audio-only applications for visually impaired computer users.

Mynatt proposed a general methodology for transforming the graphical interfaces into non-visual interfaces [5] by using 3D audio output techniques. Various salient components of graphical interfaces were transformed into auditory components with the so-called auditory icons. The final goal was to present object hierarchies with sound.

The transformation of MS-Windows interface into spatial audio was proposed by Crispien [6] and Sodnik [7], both transforming the hierarchical navigation scheme into a "ring" metaphor. Sodnik used speech output for explaining various commands to the user.

Nowadays, one of the most commonly used computer applications is a web browser. Web browsers are used to represent extensive and complex web content comprising of text, pictures, movies, animations, etc. Visually impaired computer users use web browsers together with screen readers which process pages sequentially and read the content. Several researchers proposed different and more advanced auditory versions of web browsers.

The HearSay solution, for example, enabled flexible navigation and form-filling and also context-directed browsing through extensible VoiceXML dialog interface [8]. The Csurf application was intended to help with the identification of relevant information on web pages [9]. When the user followed a web link, CSurf captured the content and preprocessed it in order to identify the relevant information based on statistical machine-learning models. The content of the web page was then read to the user from the most to the least relevant section.

Roth et al. proposed AB-Web, an audio-haptic internet browser for visually impaired users [10]. AB-Web created an immersive 3D sound environment, into which HTML objects were mapped. Web elements such as text, link or image were represented by corresponding auditory elements whose simulated 3D locations depend on their original locations displayed on the web browser. According to our understanding, only non-speech items were positioned in space, while TTS mechanisms were used without spatial positioning. The audio rendering was based on Intel library RSX using HRTF technology.

Goose and Moller proposed a 3D audio-only interactive web browser with a new conceptual model of the HTML document and its mapping into 3D audio space [11]. The audio rendering techniques included the use of various non-speech auditory elements as well as a TTS synthesizer with multiple voices, intonation, etc. The rendering was performed by regular PC audio hardware and commercially available 3D audio software toolkit (MS DirectSound, Intel RSX and Aureal). The authors' initial idea was to simulate the act of reading a physical page from top to bottom. Due to very poor localization of up-down direction (elevation) with the audio equipment selected, the authors decided to project the moving source to the left–right direction (azimuth). The HTML document was therefore read sequentially from left to right. Different synthesized voices were used to indicate different portions of the document: header, content and links.

Synthesized speech seems extremely suitable also for various interfaces of mobile devices, as these are often used in situations when the visual interaction is limited or even impossible: when walking down the street, riding a bicycle, driving a car, etc.

An example of an auditory interface for a communication device in a vehicle was proposed and evaluated by Sodnik et al. [12]. The interface was based on speech commands positioned in space and manipulated through a custom interaction device. The audio rendering was preformed locally on the mobile device itself by commercially available hardware and software (Creative X-Fi and OpenAL).

A server-based 3D audio rendering for mobile devices was proposed by Goose et al. [13]. In this case, a server received a request and dynamically created multiple simultaneous audio objects. It spatialized the audio objects in 3D space, multiplexed them into a single stereo audio and sent over an audio stream to a mobile device. A high quality 3D audio could therefore be played and experienced on various mobile devices. The same authors also proposed a system called Conferencing3, which offers virtual conferencing based on 3D sound on commercially available wireless devices [14].

One of the earliest solutions in this field, which is closely related to our work, was proposed by Crispien et al. [15]. It also combined a screen reader and spatial sound. It was based on a professional audio processing system called Beachtron, which was used for spatial positioning of synthesized speech. The system enabled the user to identify the position of spoken text parts in relation to their visual representation on the screen.

2.1 Our Research Contribution

The aim of our research was the development of a spatial speaker (i.e. synthesized voice) that can be positioned in arbitrary positions in space and manipulated in real time. The solution can be used in any of the abovementioned applications. Since the majority of auditory interfaces is designed to be used by a large group of users (e.g. visually impaired computer users or users of various mobile devices), generally available hardware and software should preferably be used for audio rendering. Our solution is based on Java platform, standard Creative X-Fi soundcard [16] and some additional components which are also available off the shelf or even for free. The major drawback of all general spatial audio rendering equipment is its poor localization accuracy due to the use of simplified HRTF libraries. In our experiment, we tried to improve the localization accuracy with some additional preprocessing of the sound signals.

Our work is different from previous audio rendering techniques in a number of ways:

- It combines software and hardware spatial audio rendering techniques.
- It improves the accuracy of elevation localization by adding additional HRTF processing.
- Due to its simplicity, the system can be used in various desktop or mobile platforms (hardware processing is optional and can be ignored if 3D sound hardware is not available).
- We propose some innovative ideas for the practical use of such a system and we already give some results of preliminary evaluation studies and user feedback.

In the following chapter, we describe the system design. All major components of the system are listed here along with the most important design problems and solutions. We also propose some future work and additional improvements in this field.

3 System Design and Architecture

The system architecture consists of five major modules: four pieces of software (libraries and Java classes) and a standard sound card with hardware support for 3D sound. The software part is based on Java programming language with some additional external plug-ins. The five modules are:

- FreeTTS: speech synthesis system written entirely in Java
- JOAL: implementation of the Java bindings for OpenAL API [17, 18]
- HRTF library from MIT Media Lab (measurements of KEMAR dummy head) [19]
- A custom made signal processing module
- Creative Sound Blaster X-Fi ExtremeGamer sound card

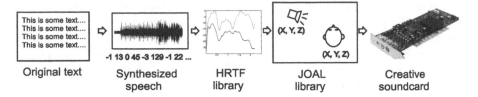

Fig. 1 System design

Figure 1 shows the basic concept of the system.

The FreeTTS module reads an arbitrary text from the file and synthesizes the samples of the spoken sound signal. The samples are then reformatted and convoluted with the samples of external HRTF function in order to add the elevation information to the signal. The correct HRTF function has to be determined according to the relative positions of the sound source and the listener. The result of convolution is then reformatted again to fit the JOAL input buffer. JOAL is a positioning library which determines the positions of the source and the listener and adds the information on the azimuth of the source. All positions are defined in the Cartesian coordinate system. The output of JOAL library is then sent directly to the sound card with hardware support for 3D sound. The final positioning is done by the hardware itself and the sound signal is then played through normal stereo headphones.

In the following subchapters, we describe the individual modules and present the development necessary for the entire system to function.

3.1   FreeTTS

FreeTTS is a speech synthesis program written in Java. It is based on Flite, a speech synthesis engine developed at Carnegie Mellon University. Flite is further based on two additional systems: Festival (University of Edinburgh) and FestVox (also Carnegie Mellon University). FreeTTS comes with extensive API documentation. The system includes three basic voices. Additional voices can be imported from FestVox or MBROLA systems. In our system, we use kevin16, a 16 kHz diphone male US voice, and three additional MBROLA US voices (two male and one female). All voices enable the setting of different pitch or speaking rate and therefore changing the voice characteristics.

With FreeTTS included in any Java application, one has to define the instance of the class voice. The selected voice is determined through the class called voiceManager. The speech synthesis can then be done simply by calling a method speak (a method of voice class) with the text to be spoken as parameter. We wrote a wrapper function which accepts additional parameters on pitch and speaking rate. By default, Java uses its default audio player for playback, and the player then sends the synthesized sound directly to the default sound card. In our case, we wanted the output to be a buffer of sound samples for further processing. We had to develop a custom audio player by implementing Java class AudioPlayer. Our audio player named BufferPlayer outputs synthesized speech as an array of bytes. The output is an array of 16-bit samples with little endian byte order. Each sound sample is presented as signed integer with two's complement binary form (amplitude value between 2^{15} and -2^{15}).

3.2   MIT Media Lab HRTF Library

HRTFs (Head Related Transfer Functions) are transfer functions of head-related impulse responses (HRIRs) that describe how a sound wave is filtered by diffraction and reflection of torso, head and pinna, when it travels from the sound source to the human eardrum [20]. These impulse responses are usually measured by inserting microphones into human ears or by using dummy heads. The measurements for different spatial positions are gathered in various databases or libraries and can be used as Finite Impulse Response (FIR) filters for creating and playing spatial sounds through the headphones. A separate function has to be used for each individual ear and each spatial position.

MIT Media Lab library, which is used in our system, contains HRIR measurements for 710 spatial positions: azimuths from $-180°$ to $180°$ and elevations from $-40°$ to $90°$. The MIT library is available online in various formats and sizes and it contains individualized as well as generalized functions (measured with a dummy head) [19]. We use the library in order to improve the elevation positioning of the sound sources (Fig. 2).

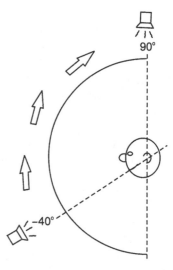

Fig. 2 The elevation area of the filters used from HRIR library

The elevation localization depends strongly on individualized human factors: torso, shoulder and pinna shape [21]. Ideally, individualized HRIR should be used for each user, but this is virtually impossible when the system is intended to be used by a large number of users. We therefore use the generalized compact measurements in wav format. Only 14 functions are used in our application: elevations from $-40°$ to $90°$ at azimuth $0°$. At azimuth $0°$, the same function can be used for both ears. The azimuth positioning is done by JOAL library.

3.3 Signal Processing Module

The MIT HRTF library contains functions in PCM format consisting of 16-bit samples at 44.1 kHz sampling frequency. In order to be used with the output of FreeTTS, the samples had to be down-sampled to 16 kHz. In Java, the HRIRs are defined as arrays of bytes.

The filtering of synthesized samples *inp* with $HRIR_{el}$ is done by calculating the convolution between the two arrays.

$$res_{el}[n] = \sum_{m=1}^{N-1} inp[n-m] \cdot HRIR_{el}[m] \tag{1}$$

In order to calculate the correct convolution, both arrays have to be converted to arrays of float numbers. After convolution, the resulting array res_{el} is converted to a 16-bit array of unsigned samples with big endian byte order, appropriate for input to JOAL library functions.

3.4 JOAL Library

JOAL is a reference implementation of the Java bindings for OpenAL API. OpenAL is a cross-platform 3D audio API appropriate for use in various audio applications. The library models a collection of audio sources moving in 3D space that are heard by a single listener located somewhere in that space. The positioning of the listener or the sources is done by simply defining their coordinates in the Cartesian coordinate system, for example: $Source_1(x_1,y_1,z_1)$, $Source_2(x_2,y_2,z_2)$ and $Listener(x_0,y_0,z_0)$.

The actual processing of input sounds is done by software (the library itself) or hardware (if provided within the soundcard). At the beginning of the experiment, we expected we would be able to use JOAL for all spatial positioning (i.e. azimuth and elevation), but the elevation perception proved to be too poor to be used in our application (similar findings were reported also by Goose et al. [9]). The addition of the afore-mentioned MIT Media Lab HRTF library was an attempt to improve the elevation coding in the signals.

JOAL provides the functions for reading external wav files which are then positioned and played. The samples from wav files are then written to special buffers and the buffers are attached to the sources with specific spatial characteristics. In our case, the input to JOAL is an array of samples (the output of FreeTTS convoluted with HRIR), which is then written directly to buffers.

By using only JOAL library for spatial positioning, the spatial arrangement of the sources and the listener can be changed at any time. However, in our case only the changes of the horizontal position can be preformed dynamically through JOAL. The vertical changes, on the other hand, require preprocessing with new HRIR.

3.5 Soundcard

Creative Sound Blaster X-Fi Extreme Gamer was used within the system. The soundcard has a special DSP unit called CMSS-3D, which offers hardware support for spatial sound generation. CMSS is another type of a generalized HRTF library used for filtering input sounds. In general, the soundcard can be configured for output to various speaker configurations (i.e. headphones, desktop speakers, 5.1 surround, etc.), but in our case the use of an additional HRTF library required the playback through stereo headphones.

Creative Sound Card works well with JOAL (OpenAL) positioning library. However, if no hardware support can be found for spatial sound, the latter is performed by the library itself (with a certain degradation of quality).

4 Prototype Applications and Preliminary User Studies

The "Spatial speaker" is meant to be a part of various auditory interfaces and systems used mostly by visually impaired users or the so called users-on-the-move: drivers, runners, bikers, etc. In this paper, we describe the prototypes of three different applications which are meant to be used primarily by visually impaired computer users:

- Spatial audio representation of a text file
- Spatial story reader
- Multiple simultaneous files reader

We preformed some preliminary user studies with the three applications. Ten users participated in the experiments. All of them reported normal sight and hearing. The tests were preformed in a semi-noise environment with Sennheiser HD 270 headphones. The headphones enable approximately 10–15 dB attenuation of ambient noise. Since the primary focus of this paper is the design of the "Spatial speaker" system, only a short summary of the results and user comments is provided.

4.1   Spatial Audio Representation of a Text File

A typical screen reader, the most important tool for visually impaired users to access and read web content or word processing files, reads the content of the file sequentially from top left to bottom right corner. Some screen readers describe the positions of the elements on the page with the aid of the coordinate system. Our prototype application reads the content of the page as if coming from different spatial positions relative to the listener. The spatial positions are defined according to the original position of the element on the page. The reader starts reading the page at the left top corner and then it moves from left to right and follows the positions of the text on the page. It finishes at the bottom right corner of the page. For example, a text appearing on the left side of the page can actually be heard on the left, the text on the right can be heard on the right, etc. In this way, the application enables the perception of the actual physical structure of the web page (Fig. 3).

In a short experiment, the users were asked to describe the approximate structure of the page by just listening to its content. The majority reported that the positions of the individual texts could be perceived clearly, but the perception of the entire structure remained blurry, since no information on images, banners and other web page elements was given.

4.2   Spatial Story Reader

The speech synthesizer can also be used to read various messages and notes to mobile users or to read fiction to visually impaired people. The normal output of various speech synthesizers is usually very monotonous and dull. We developed a story reader which separates different characters in the story by assigning them a location in space. Each specific character in the story as well as the narrator speaks from a different spatial position according to the listener. The characters with their corresponding texts can therefore be easily separated from one another. In addition, the pitch of synthesized speech is changed for each character, which makes the separation even better.

The input for the speech synthesizer is a normal txt file. All characters or voices which are to be manipulated and positioned in space need to be tagged with appropriate coordinates and information about the pitch in order to be recognized by the system (Fig. 4).

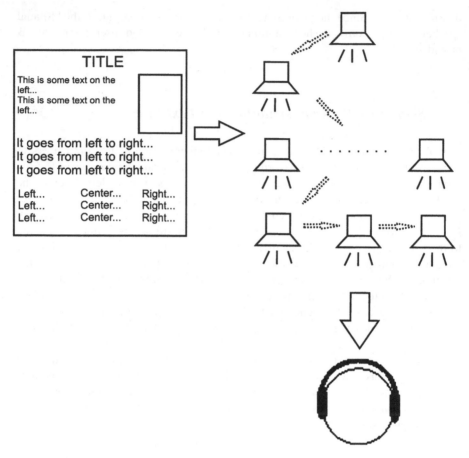

Fig. 3 The basic concept of auditory representation of a text file

The users listened to five different stories. They were asked to subjectively evaluate the system without focusing on the content of the stories. The majority especially liked the additional separation of the characters by pitch. Several users commented that the accompanying text explaining which of the characters said something was in some cases almost redundant, since the designated spatial positions and the corresponding pitch of the speech clearly identified the individual characters.

4.3 Multiple Simultaneous Files Reader

A human listener is unable to perceive and understand more than one speech source at a time, unless the sources are coming from different spatial positions [22]. General text-to-speech converters can therefore be used with just one text input

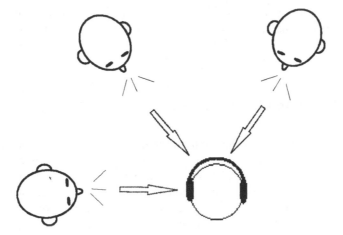

Fig. 4 Multiple voices at different spatial positions

Fig. 5 Two simultaneous
speech sources

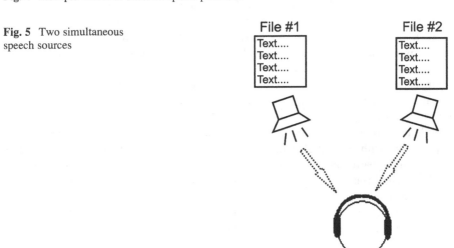

and one spoken output. Our "Spatial speaker" can, however, produce multiple outputs that are separated by their locations in space. We developed a test application that reads an arbitrary number of text files and synthesizes speech flows at different spatial positions relative to the speaker. We evaluated the intelligibility with two and three sources. The users were positioned at the origin of the coordinate system (i.e. 0°,0°,0°) and the speech sources were positioned as follows (Fig. 5):

- Two sources: (−30°,0°,0°) and (30°,0°,0°)
- Three sources: (−30°,0°,0°), (0°,0°,0°) and (30°,0°,0°)

The preliminary results show that a maximum of two simultaneous speech sources can be perceived and understood simultaneously. The users reported that with three sources it was still possible to concentrate on one individual source and understand the content, but it was impossible to understand all of the speech sources simultaneously.

5   Conclusion and Future Work

This paper describes the development and system design of a 3D speech synthesizer. The system is based on Java platform and comprises of standard programming libraries, custom developed modules and a sound card, all of which are reprogrammed and modified to fit together. An external MIT Media Lab HRTF library is used in order to improve the localization accuracy of the system. A personalized HRTF library can also be measured and used for a specific user, thus enabling an extremely accurate localization performance and spatial awareness if required. The current system is realized as a single Java class with some additional libraries and files and can therefore be imported and used in any Java application. The system works with any soundcard, since hardware support for 3D sound is optional and is automatically replaced by the software library if not present.

In addition to its use as a typical speech synthesizer or TTS converter, the "Spatial speaker" could also be used as a key component of various auditory interfaces. It can, for example, be used effectively by visually impaired people or mobile users-on-the-go. Three different prototype applications are provided in this paper along with some preliminary results of evaluation studies. Only users with normal sight participated in the preliminary experiments in order to get some feedback and ideas for improvements. All three prototypes proved to be interesting and innovative and potentially useful for visually impaired people. Our next step will be an extensive user study with visually impaired computer users. We intend to establish some major advantages of a 3D speech synthesizer over the non-spatial one.

Acknowledgements This work has been supported by the Slovenian Research Agency within the research project: "Assisting visually impaired people in using the computer with the aid of spatial sound".

References

1. R.A. Cole, Survey of the State of the Art in Human Language Technology (1996)
2. TextAloud (5 Feb 2010). http://www.nextuptech.com
3. VoiceMX STUDIO (5 Feb 2010). http://www.tanseon.com/products/voicemx.htm
4. FreeTTS (5 Feb 2010). http://freetts.sourceforge.net/docs/index.php
5. E.D. Mynatt, Transforming graphical interfaces into auditory interfaces. Doctoral Dissertation, Georgia Institute of Technology, 1995

6. K. Crispien, K. Fellbaum, A 3D-auditory environment for hierarchical navigation in non-visual interaction, in *Proceedings of ICAD96*, 1996
7. J. Sodnik, S. Tomažič, Spatial auditory interface for word processing application, in *Proceedings of IEEE ACHI09* 2009, pp. 271–276
8. Y. Borodin, J. Mahmud, I.V. Ramakrishnan, A. Stent, The HearSay non-visual web browser. Proc. Int. Cross Discip. Conf. Web Accessibility, **225**, 128–129 (2007)
9. J.U. Mahmud, Y. Borodin, I.V. Ramakrishnan, Csurf: a context-driven non-visual web-browser, in *Proceedings of 16th International Conference on World Wide Web*, pp. 31–40, 2007
10. P. Roth, L.S. Petrucci, A. Assimacopoulos, T. Pun, Audio-haptic internet browser and associated tools for blind and visually impaired computer users, in *Workshop on Friendly Exchanging Through the Net*, pp. 57–62, 2000
11. S. Goose, C. Moller, A 3D audio only interactive Web browser: using spatialization to convey hypermedia document structure, in *Proc. of Seventh ACM International Conference on Multimedia*, pp. 363–371, 1999
12. J. Sodnik, C. Dicke, S. Tomažič, M. Billinghurst, A user study of auditory versus visual interfaces for use while driving. Int. J. Human Comput. Stud. **66**(5), 318–332 (2008)
13. S. Goose, S. Kodlahalli, W. Pechter, R. Hjelsvold, Streaming speech3: a framework for generating and streaming 3D text-to-speech and audio presentations to wireless PDAs as specified using extensions to SMIL, in *Proceedings of the International Conference on World Wide Web*, pp. 37–44, 2002
14. S. Goose, J. Riedlinger, S. Kodlahalli, Conferencing3: 3D audio conferencing and archiving services for handheld wireless devices. Int. J. Wireless Mobile Comput. 1(1), 5–13 (2005)
15. K. Crispien, W. Wurz, G. Weber, Using spatial audio for the enhanced presentation of synthesised speech within screen-readers for blind computer users, *Computers for Handicapped Persons*, vol. 860. (Springer, Berlin, Heidelberg, 1994), pp. 144–153
16. Creative (5 Feb 2010). http://www.soundblaster.com
17. JOAL (5 Feb 2010). https://joal.dev.java.net
18. OpenAL (5 Feb 2010). http://connect.creativelabs.com/openal/default.aspx
19. HRTF Measurements of a KEMAR Dummy-Head Microphone (5 Feb 2010). http://sound.media.mit.edu/resources/KEMAR.html
20. C.I. Cheng, G.H. Wakefield, Introduction to head-related transfer functions (HRTF's): representations of HRTF's in time, frequency, and space (invited tutorial). J. Audio Eng. Soc. **49**(4), 231–249 (2001)
21. J. Blauert, *Spatial Hearing: The Psychophysics of Human Sound Localization*, Revised edn. (MIT Press, Cambridge, MA, 1997)
22. B. Arons, A review of the cocktail party effect, J. Am. Voice I/O Soc. **12**, 35–50 (1992)

Chapter 29
Commercial Break Detection and Content Based Video Retrieval

N. Venkatesh and M. Girish Chandra

Abstract This chapter presents a novel approach for automatic annotation and content based video retrieval by making use of the features extracted during the process of detecting commercial boundaries in a recorded Television (TV) program. In this approach, commercial boundaries are primarily detected using audio and the detected boundaries are validated and enhanced using splash screen of a program in the video domain. Detected splash screen of a program at the commercial boundaries is used for automatic annotation of recorded video which helps in fast content based video retrieval. The performance and validity of our approach is demonstrated using the videos recorded from different Indian Television broadcasts.

1 Introduction

The Television (TV) remains and always has been one of the most powerful medium of communication since its inception. Today, we are in the next generation of television where, the additional hardware and software have completely changed the way the television is watched. The recent technologies such as set-top box coupled with the reduction of costs for digital media storage have led to the introduction of personal video recorders (PVR) which are now fairly established and fast growing part of the TV landscape.

In this context, audiovisual analysis tools that help the user to manage the huge amount of data are very important to introduce the novel recording devices in the highly competitive market. Among other analysis tools, detection of TV advertisements is a topic with many practical applications. For instance, from the point of

---

N. Venkatesh (✉)
Innovation Lab, Tata Consultancy Services, Plot no 96, Abhilash building, EPIP Industrial area, white field main road, Bangalore 560066, India
e-mail: venk.n@tcs.com

S.-I. Ao et al. (eds.), *Machine Learning and Systems Engineering*,
Lecture Notes in Electrical Engineering 68,
DOI 10.1007/978-90-481-9419-3_29, © Springer Science+Business Media B.V. 2010

view of a TV end user, it could be useful to avoid commercials in personal recordings. Also of great utility is to appropriately index and retrieve the large number of recorded videos automatically.

Existing commercial detection approaches can be generally divided into two categories: feature based and matching-based approaches. While the feature-based approaches use some inherent characteristics of TV commercials to distinguish commercials and other types of videos, the matching based methods attempt to identify commercials by searching a database that contains known commercials. The challenges faced by both approaches are the same: how to accurately detect commercial breaks, each of which consits of a series of commercials and how to automatically perform fast commercial recognition in real time.

Similarly, the video retrieval approaches can also be categorized into two different approaches [1]. First category focuses mainly on visual features, such as colour histograms, shapes, textures, or motion, which characterize the low-level visual content. Although these approaches use automatically extracted features representing the video content, they do not provide semantics that describe high-level video information (concepts), which is much more appropriate for users when retrieving video segments. The second one focuses on annotation based approaches [2], which makes use of keyword annotation or text attribute to represent high level concepts of the video content. Performance of this approach is affected by several factors, including the fact that the search process depends solely on the predefined attributed information. Furthermore, manual annotation process is very tedious and time consuming. The challenges associated with the management of large database of recorded programs involve: (1) knowledge-based methods for interpreting raw data into semantic contents, (2) better query representations, and (3) incorporation of relevant spatiotemporal relations.

Commercial break detection [3, 4] have based their strategies in studying the relation between audio silences and black frames as an indicator of commercials boundaries. The analysis is performed in either compressed [3] or uncompressed [4] domain. In [5], specific country regulations about commercials broadcast is used as a further clue. Another interesting approach is presented in [6], where the overlaid text trajectories are used to detect commercial breaks. The idea here is that overlaid text (if any) usually remains more stable during the program time than in the case of commercials. In [7], Hidden Markov Model (HMM) based commercial detection is presented with two different observations taken for each video shot: logo presence and shot duration.

In many of the existing works, black frame [8] is used as the crucial information for detecting the commercial in video processing. But, in Indian TV stations black frames are absent and our methodology works without using black frames as video clues. Our simple and accurate approach for commercial detection uses audiovisual features. Initially, we find the commercial boundaries using a combination of different audio features and the confidence level of the boundary detected will be further enhanced and validated using splash screen (program name) video frames. This detection is carried out around the commercial boundaries detected by audio.

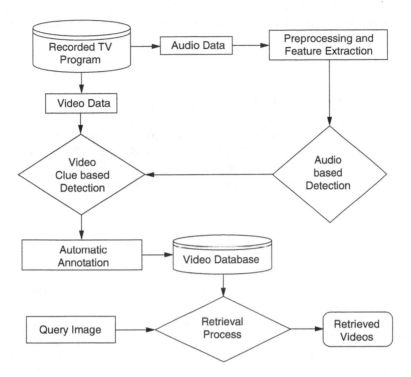

Fig. 1 System overview

Image matching of splash screen frame appearing at the start and end of the commercial is performed using simple pixel intensity based matching.

Further, we make use of the information extracted during the commercial detection process for automatic annotation and retrieval system which tries to fill the gap between the low-level media features and high-level concepts for yielding better performance using computationally simple procedures.

The overall scheme is depicted in the block diagram shown in Fig. 1. In the remaining sections of this chapter the requisite details of the proposed method and the relevant performance analysis results are provided.

2   Preprocessing and Feature Extraction

As discussed in the previous section the methodology adopted in this chapter is based on two step approach combining both audio and video based approach to accurately detect the commercial break boundaries in recorded TV program. The rationale for this two-step approach is to reduce the search space for splash screen detection during the video processing step, hence significantly reducing the computational time.

In the following two subsections audio and video feature extraction issues are provided.

2.1 Audio Feature Extraction

As a pre-processing step, the audio signal is extracted from the recorded TV program and is segmented into audio frames of 23 ms duration with 50% overlap.

By banking on the general observation that the energy of the audio signal would be comparatively high during the commercial break, we have opted for an energy based feature. A simple energy feature is the usual short time energy given by

$$E_i = \left(\frac{1}{N} \sum_{n=0}^{N-1} x_i^2(n) \right) \tag{1}$$

where, $i = 1, 2, \ldots, F$ and N is the frame length (i.e., the number of samples in the frame).

As evident from Eq. (1), the short term energy is computed per frame as the sum of squares of the signal samples normalized by the frame length. But, using only short time energy would result in large number of falsely detected commercial boundaries and therefore we have come out with a new feature termed as Energy Peak Rate (EPR), which is defined as the ratio of total number of energy peaks exceeding the threshold T to the total number of frames (F) in the evaluation time interval.

The requisite threshold is heuristically fixed as

$$T = 0.5 \times \max(E_i) \tag{2}$$

The EPR is computed using the following expression

$$EPR = \frac{1}{F} \sum_{i=1}^{F} E_p(E_i) \tag{3}$$

where, $E_p(x)$ is the binary-valued function, defined as

$$E_p(x) = \begin{cases} 1 & \text{if } x > T \\ 0 & \text{else} \end{cases} \tag{4}$$

Based on the elaborate experimentation, the evaluation time interval is chosen as 5 s.

Using the suggested audio feature we will only be able to find the approximate commercial break boundaries. As mentioned earlier, it is possible to enhance the boundary detected by the audio features to arrive at the exact commercials boundaries using video processing. The methodology we have adopted is based on the extraction of Harris corner points which is presented in the next section.

2.2   Video Feature Extraction

As a preprocessing step, we convert colour image to gray scale image on which the subsequent processing is carried out. For detecting the exact boundaries, splash screen (program name) frames are matched at the start and end of a commercial break. A typical splash screen, including the program name (from one of the TV recorded programs) is shown in Fig. 2. The matching is carried out using Harris corner points, which are nothing but the points on the image where there is significant change in the brightness in orthogonal directions [2].

2.2.1   Harris Corner Points

The Harris corner detector is a popular interest point detector because of its strong invariance to rotation, scale, illumination variation and image noise [9]. The Harris corner detector essentially finds a small number of points useful for matching or finding correspondence and tracking [10]. This detector is based on the local auto-correlation function of the image, which measures the local changes of the image. $c(x, y; \Delta x, \Delta y)$ quantifies how similar is the image function $I(x, y)$ at point (x, y) to itself, when shifted by $(\Delta x, \Delta y)$[10]:

Fig. 2 Splash screen
(Program name)

$$c(x, y; \Delta x, \Delta y) = \sum_{(u,v) \in W(x,y)} w(u, v)[I(u, v) - I(u + \Delta x, v + \Delta y)]^2 \tag{5}$$

where, $W(x, y)$ is a window centered at point (x,y), and $w(u, v)$ is either Gaussian. Approximating the shifted function $I(u + \Delta x, v + \Delta y)$ by the first-order Taylor expansion, the autocorrelation can be in turn approximated as

$$c(x, y; \Delta x, \Delta y) \cong [\Delta x, \Delta y] \, Q(x, y) \begin{bmatrix} \Delta x \\ \Delta y \end{bmatrix} \tag{6}$$

where $Q(x, y)$ is a 2×2 matrix computed from the image partial derivatives I_x and I_y:

$$Q(x, y) = \sum_W \begin{bmatrix} I_x^2 & I_x I_y \\ I_x I_y & I_y^2 \end{bmatrix} \tag{7}$$

where, $\sum_{(u,v) \in W(x,y)} w(u, v)$ is represented with its short-hand notation $\sum_W$. The requisite corner points are obtained as the local maxima of the corner response function $H(x, y)$ given by

$$H(x, y) = Det \, Q - k \, (Trace \, Q)^2 \tag{8}$$

where, Det and $Trace$ denote the determinant and trace of a matrix respectively and k is a constant, usually set to the value 0.04 [10]. The local maximum in an image is defined as a point greater than its neighbors in 3×3 or even 5×5 neighborhood. Figure 3 shows the Harris corner points of the image depicted in Fig. 2.

Fig. 3 Harris corner points of splash screen (program name)

3   Commercial Detection Scheme

The approximate commercial boundaries based on the audio feature are detected by applying heuristics found by elaborate experimentation.

3.1   Audio Feature Based Detection

From Eq. (3) if Energy Peak Rate (*EPR*) exceeds the experimentally found threshold θ then we declare that frames of particular time interval lies in the commercial break segment and hence we will be able to detect the approximate commercial boundaries. The exact commercial boundaries will be detected by video processing algorithm as outlined below.

3.2   Video Feature Based Detection

After detecting the approximate commercial boundaries using audio features, we validate and further look for exact commercial boundaries using the splash screen (program name) video frames matching at the start and end of commercial break. The video processing is carried out to detect splash screen (program name) around the boundaries detected by audio. Figure 2 shows an example of splash screen (program name) appearing in one of the TV program video.

The image matching using the extracted features is carried out by simple pixel intensity matching algorithm as explained below:

Step-1: $M_P = 0$ where, M_P is the number of pixels matched.

Step-2: $\sum_{x,y}$ if $(R(x, y) = 1 \,\&\, T(x, y) = 1)$ then $M_P = M_P + 1$

where, $R(x, y)$ denotes the reference image and $T(x, y)$ is the test image.

Step-3: $P_M = \frac{M_P}{W_R} \times 100$

Where, W_R is the number of white pixels in reference template image R.

Step-4: If P_M is greater than 80% then one can say that two images are similar. It is to be noted that the step 2 is carried out using the Harris corner points of the respective images for computationally efficient matching.

4   Mechanism for Automatic Annotation and Retrieval

4.1   Automatic Annotation

Once the commercial boundaries have been detected, the next step is to remove the commercial part by retaining only the recorded TV program and during this process, the splash screen (program name) frame detected in video based boundary detection process is inserted at the beginning of the recorded TV program video (without commercial) which is not going to affect the visual display because we are inserting only one key frame at the beginning and normally video quality will be 25 frames per second. Additionally the key frame can be inserted with the useful information like recording date, time and length of the video, which can be useful during the retrieval process, for short listing the closest matching videos in response to the query.

4.2   Content Based Video Retrieval

There will be a set of query images where each image is a splash screen (program name) representing the recorded TV programs. It is to be noted that these set of query images can be incorporated into the modern day set-top box and the associated remote can be provided with a dedicated key to facilitate the display of query images. Later, the query image will be compared with all the first frame of all the videos and the best match is obtained through correlation matching as explained below.

The correlation matching technique is simple and computationally fast and hence it has been used to find the best match between the query image and the set of images in the database. Query image (A) and the extracted key frame (B) from the video are converted to binary images and then correlation coefficient is computed between these two images using Eq. (9).

$$r_i = \frac{\sum_m \sum_n \left(A_i(m,n) - \overline{A}_i\right)\left(B(m,n) - \overline{B}\right)}{\sqrt{\left(\sum_m \sum_n \left(A_i(m,n) - \overline{A}_i\right)^2\right)\left(\sum_m \sum_n \left(B(m,n) - \overline{B}\right)^2\right)}} \qquad (9)$$

where, $I = 1,\ldots, N$, N is number of stored key frames in the database.
A_i is the key frame belonging to ith class of a database
$\overline{A}_i$ is the mean of A_i image
B is the query/test image
$\overline{B}$ is the mean of an image
Largest r_i value indicates us the closest match of key frame to the query image.

5   Results and Discussion

In order to examine the proposed methodology and to carry out the relevant experimentation, we have recorded eleven videos from seven different Indian TV stations.

Table 1 Details of data

Video	Video length (min:s)	Commercial break duration (min:s)	Class label
1	17:30	i.12:25–17:12	1
2	21:00	i. 08:10–11:45	1
		ii. 20:10–21:00	
3	07:09	i. 00:40–4:12	1
4	23:19	i. 04:52–52	2
		ii. 17:20–22:28	
5	21:17	i. 03:0–06:45	2
		ii. 16:15–20:58	
6	18:54	i. 06:42–13:38	3
7	18:26	i. 02:08–05:25	4
		ii. 11:30–15:20	
8	20:41	i. 03:00–06:35	5
		ii. 14:20–17:30	
9	19:55	i. 07:40–12:45	6
10	34:14	i. 03:46–09:10	7
		ii. 14:45–19:25	
		iii. 27:53–31:10	
11	33:24	i. 07:30–11:35	7
		ii. 17:28–21:25	
		iii. 30:10–33:24	

Videos are captured at 25 frames per second with 288×352 resolution and the audio was extracted from these videos and stored in uncompressed wav format at 44,100 kHz with 16 bits per sample for further processing. Commercial details for the recorded data are provided in Table 1. We have performed our experiments using MATLAB® programming in Windows operating system. Videos with same class label indicate that they are different episodes of same TV program which is helpful during retrieval process.

From Table 3 we observe that using our simple video clues of matching splash screen video frames at commercial boundaries for refining the boundaries detected by audio we are able to increase the confidence level of the commercial boundary detection and able to find the exact commercial boundaries.

One important observation to be made from Tables 2 and 3 is that false events detected by audio processing is eliminated by the information obtained through video processing step as the splash screen will not be present at the start and the end of falsely detected commercial boundaries using audio. We can have a desired trade off between number of falsely detected events and the number of missed events by varying a threshold in audio based feature given by Eq. (2).

Once the commercial detection process is completed, the video will be automatically annotated using the key frame with the additional information (see Section 4.1) appended at the beginning of the video. Currently, the experimentation has been carried out for seven classes (seven types of recorded TV programs). The content based video retrieval is checked for all seven classes (query images) using correlation classifier to match between query and reference images to retrieve the closest matching automatically annotated videos. Retrieval performance results are

Table 2 Results using only audio feature

Video	Total no of events detected	No of false events detected	No of missed event detected	No of true events detected
1	1	0	0	1
2	2	0	0	2
3	1	0	0	1
4	2	0	0	2
5	3	1	0	2
6	1	0	0	1
7	2	0	0	2
8	2	0	0	2
9	1	0	0	1
10	3	0	0	3
11	4	1	0	3

Table 3 Results using both audio and video feature

Video	Total no of events detected	No of false events detected	No of missed event detected	No of true events detected
1	1	0	0	1
2	2	0	0	2
3	1	0	0	1
4	2	0	0	2
5	2	0	0	2
6	1	0	0	1
7	2	0	0	2
8	2	0	0	2
9	1	0	0	1
10	3	0	0	3
11	3	0	0	3

comparatively fast and accurate, with no error detected in the limited data set considered.

6 Conclusion and Future Work

This chapter describes the commercial detection for recorded TV program using audiovisual features. Further, automatic indexing and content based retrieval algorithm which makes use of the information extracted during commercial detection is proposed. The system provides the user with complete solution of recording TV program without advertisement and more importantly, database management of the recorded TV programs is achieved through automatic annotation and fast content based video retrieval. The proposed scheme is worked out keeping in mind the simplicity and time efficiency of the used algorithms.

Our future work will be focused on testing the robustness of the methodology by applying to wide variety of TV programs like news, serials, reality shows etc, encompassing both Indian and international TV broadcasts. Further, in order to demonstrate the efficacy of the scheme for content based retrieval, it is essential to experiment with large video archives. Also of interest is the study of amenability of the scheme and any requisite modifications for easy implementation on real-time embedded systems (like TV set-top box).

References

1. C.A. Yoshitaka, T. Ichikawa, A survey on content-based retrieval for multimedia databases. IEEE Trans. Knowl. Data Eng. **11**(1), 81–93 (1999)
2. M.S. Kankanhalli, et al, Video modeling using strata-based annotation. IEEE Multimedia **7** (1), 68–73 (2000)
3. D.A. Sadlier, et al., Automatic TV advertisement detection from mpeg bitstream. J. Patt. Rec. Soc. **35**(12), 2–15 (Dec 2002)
4. Y. Li, C.-C. Jay Kuo, Detecting commercial breaks in real TV program based on audiovisual information, in *SPIE Proceedings on IMMS*, vol. 421, Nov. 2000
5. R. Lienhart, C. Kuhmnch, W. Effelsberg, On the detection and recognition of television commercials, in *Proceedings of the IEEE Conference on MCS*, Otawa, Canada, 1996
6. N. Dimitrova, Multimedia content analysis: the next wave, in *Proceedings of the 2nd CIVR*, Illinois, USA, Aug 2003
7. A. Albiol, M.J. Fullà, A. Albiol, L. Torres, Commercials Detection using HMMs, 2004
8. A. Albiol, M.J. Fullà, A. Albiol, Detection of TV Commercials, in *Proceedings of the International Conference on Acoustics, Speech and Signal Processing*, 2004, Rocky Mountain Research Lab., Boulder, CO, private communication, May 1995, ed. C.J. Kaufman
9. C. Harris, M. Stephens, A combined corner and edge detector in *Proceedings of the 4th Alvey Vision Conference*, 1988, pp. 147—151
10. Tomas Werner, (2 Oct 2008). http://cmp.felk.cvut.cz/cmp/courses/X33PVR/2009-2010ZS/ Docs/harris.pdf
11. N. Venkatesh, B. Rajeev, M. Girish Chandra, Novel TV commercial detection in cookery program videos, in *Proceedings of the World Congress on Engineering and Computer Science 2009 Vol II, WCECS 2009*, San Francisco, USA, 20–22 Oct 2009

Chapter 30
ClusterDAM: Clustering Mechanism for Delivery of Adaptive Multimedia Content in Two-Hop Wireless Networks

Hrishikesh Venkataraman and Gabriel-Miro Muntean

Abstract In the recent years, there has been an increasing demand for streaming high quality multimedia content over wireless networks. Demand for triple play services (voice, video, data) by a number of simultaneously communicating users often results in lack of acceptable multimedia quality. In addition, the large transmission distance and the limited battery power of the hand-held wireless device serves as a major bottleneck over transmitting video directly from the network operator to the end-users. There has been no successful mechanism developed till date that would provide live video streaming over wireless networks. In this paper, *clusterDAM* – a novel cluster-based architecture is proposed for *delivering adaptive multimedia* content over two-hop wireless networks. An intermediate relay node serves as a proxy-client-server between the base station and the mobile users which ensures that the multimedia delivery is adapted in real-time according to the channel conditions and the quality of the multimedia content. An extensive simulation-based analysis with different kinds of network traffic and protocols indicate that the cluster-DAM architecture is significantly superior to the traditional single-hop design, not only in terms of the perceived video quality, but also in terms of the loss rate and the average bit rate.

1 Introduction

With the advent of third generation (3G) mobile systems, video conferencing and downloading of movies/documentaries that requires huge file transfers of the order of several tens of megabytes per second (Mbps) have been in great demand [1]. Multimedia transmission and VoD streaming requires high data rates which can be

H. Venkataraman (✉)
S220, Performance Engineering Laboratory, School of Electronic Engineering, Dublin City University, Glasnevin, Dublin 9, Ireland
e-mail: hrishikesh@eeng.dcu.ie

S.-I. Ao et al. (eds.), *Machine Learning and Systems Engineering*,
Lecture Notes in Electrical Engineering 68,
DOI 10.1007/978-90-481-9419-3_30, © Springer Science+Business Media B.V. 2010

achieved either by increasing the bandwidth or by increasing the signal power at the receiver. However, the total bandwidth of the network is controlled by the Government and cannot be increased arbitrarily. Similarly, the wireless devices are energy-constrained units; and hence, the transmission power of these devices cannot be increased indiscriminately. Significantly however, the power required at the hand-held device for multimedia decoding and playing is quite high; and this itself drains the battery power considerably. In this scenario, an efficient mechanism to achieve high-data rate communication is to use multihop transmission between the source and destination node [2].

A hierarchical multihop design with a single central entity and several smaller devices that would serve as intermediate relays is shown in Fig. 1. The relays reduce the distance between a transmission-reception (Tx-Rx) pair which in turn reduces the power requirement. At the same time, this increases the achievable maximum data rate of a communicating link [3]. In addition, with the existence of different kinds of wireless networks in today's scenario (GPRS, CDMA2000, UMTS, WiMAX, WLAN, etc.), the relays could serve as a switching point between the different networks. In this context, the multihop design can efficiently model the multiplicity in wireless networks. It has been shown [4] that time division multihop cellular networks (MCN) offers the potential to integrate various multihop infrastructure-based wireless networks. In fact, it has been shown in [4] that in a multihop design with up to *five hops*, the spectral efficiency is significantly increased as compared to the single-hop transmission.

A novel cluster-based design for two-hop hierarchical networks has been recently proposed in [5], wherein, it has been shown that the cluster-based two-hop design shows a significantly superior performance as compared with the state-of-the-art resource algorithms. In this paper, *Cluster-DAM* - a novel cluster-based architecture is proposed for delivery of adaptive multimedia content in two-hop cellular networks. A double dumbbell topology that fits the two-hop cellular design

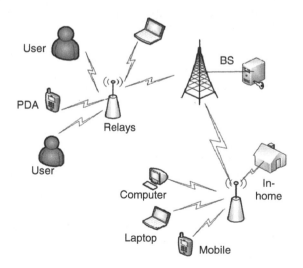

Fig. 1 Next generation two-hop hierarchical wireless architecture

is considered; and analyzed for efficient multimedia transmission. Different adaptive multimedia techniques, viz., LDA+ (enhanced loss delay adaptation protocol), TFRCP (TCP friendly rate control protocol), RBAR (receiver based auto-rate protocol) and QOAS (quality oriented adaptive design) are considered in the network design, and their performance is analyzed for both single-hop (dumbbell topology) and cluster-based two-hop design (double dumbbell topology). It is shown that the QOAS-based wireless network with double dumbbell topology performs significantly better than any other adaptive solutions and topology, not only with regard to the video perceived quality, but also in terms of the bit rate and the average loss rate.

The paper is organized as follows: 2 describes the cluster-DAM architecture over two-hop wireless network model in detail, along with QOAS and other adaptive solutions. 3 describes the simulation model, the set up, and the simulation scenarios. The simulation results are explained in 4, and the conclusions are written in 5.

2   Cluster-Dam Architecture

2.1   Cluster-based Two-Hop Design for WiMAX Networks

The basic architecture of *cluster-DAM* is a cluster-based hierarchical two-hop cellular network based on IEEE 802.16j standards. A multihop relay networks is established between the server/base station (BS) and the end-user mobile stations (MSs), as shown in Fig. 2. As per this design, there are *six* clusters in a coverage area. The circular coverage area has a radius, r, and is divided into two layers. The wireless nodes in the inner-layer communicate directly with the server; whereas the wireless terminals in the outer-layer are grouped into several clusters (six clusters in Fig. 2) [6]. In each cluster, a wireless terminal located at the boundary of the inner

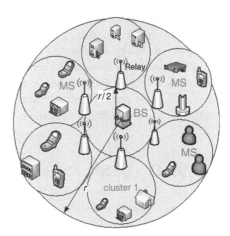

Fig. 2 Cluster-based
hierarchical two-hop cellular
networks

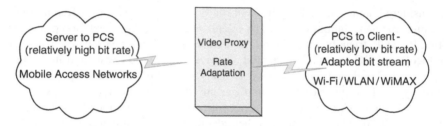

Fig. 3 Proxy-client-server as video proxy be'tween base station and end-users

and outer layer of the network region is selected as a cluster-head node, alternately known as *relay*. The server always communicates with the users in the outer-layer through the relay. Hence, the maximum transmission distance of a communicating pair in the network is $r/2$ [7].

The relay node serves as a client during its communication with the BS (server); and as a server while communicating with the end-users. Hence, the relay node can be described as a proxy-client-server (PCS). In the adaptive multimedia delivery, there would be several instances where there are multiple clients requesting for the same video. In case of a single-hop design, this will require multiple encoding of a single video stream, thereby creating unnecessary load on the BS. However, in the two-hop design, the video stream could be stored in the PCS. After that, the adapted stream could be transmitted individually to different clients. Hence, in this regard, the PCS can also be considered as the database server. A two-hop design between the BS and the end-users is shown in Fig. 3, wherein the PCS serves as the video proxy. An adaptation framework incorporated at the PCS divides the incoming multimedia stream from the server to several chunks of multimedia packets. The size and length of each chunk depends upon the total size, length and the type of video to be streamed.

In order to reduce the additional interference arising from the simultaneous communication of multiple communicating pairs, a *Protocol Model* is considered [8] for interference avoidance in the two-hop design. The reusability of the available spectrum resource (time slot in a time division multiple access system and a frequency band in a frequency division multiple access system) is increased in the cluster-based design by allowing two multihop clusters in the network to utilize the same spectrum resource. It should be noted that the number of clusters in the cluster-based design need not be always six [9]. But it should be an *'even'* number due to the basic principle of simultaneous transmission of communication pairs located in the diametrically opposite clusters.

2.2 QOAS - Quality Oriented Adaptive Scheme

The primary aim of integrating QOAS with the cluster-based design in the IEEE 802.16j wireless model is to maintain a high end-user perceived quality even with an increase in the number of wireless devices in the network. QOAS relies on the fact that the impact on the end-user perceived quality is greater in case of random losses than that of controlled reduction in quality [10]. The system architecture of the feedback-based QOAS includes multiple instances of the end-to-end adaptive client and server applications [11]. Following ITU-T R. P.910 standard [12], a five state model is defined for the multimedia streaming process. The QOAS client continuously monitors the transmission parameters and estimates the end-user perceived quality. The Quality of Delivery Grading Scheme (QoDGS) regularly computes Quality of Delivery (QoD) scores that reflect the multimedia streaming quality in current delivery conditions. These grades are then sent as feedback to the server arbitration scheme (SAS). The SAS assesses the values of a number of consecutive QoD scores received as feedback in order to reduce the effect of noise in the adaptive decision process [13]. Based on these scores SAS suggests adjustments in the data rate and other parameters.

2.3   Other Adaptive Solutions

With an increase in the demand for multimedia streaming in wireless networks, there have been significant approaches researched in the recent past. TFRCP is a unicast transport layer protocol, designed for multimedia streaming, and provides nearly the same amount of throughput as that of TCP on wired networks. The TFRCP controls rate based on network conditions expressed in terms of RTT and packet loss probability [14]. Similar to TFRCP, LDA+ (enhanced loss delay adaptation) also aims to regulate the transmission behavior of multimedia transmitters in accordance with the network congestion state [15]. LDA+ uses RTP protocol for calculating loss and delay and uses them for regulating transmission rates of the senders. LDA+ adapts the streams in a manner similar to that of TCP connections. In comparison, RBAR is a receiver based auto-rate mechanism. It is a MAC layer protocol and is based on RTS/CTS mechanism [16]. The main feature of RBAR is that both channel quality estimation and rate selection mechanism are on the receiver side. This allows the channel quality estimation mechanism to directly access all of the information made available to it by the receiver (number of multipath components, symbol error rate, received signal strength, etc.) for more accurate rate selection.

3   Simulation Model and Testing

3.1   Dumbbell and Double Dumbbell Topology

In *cluster-DAM*, the delivery of adaptive multimedia content is achieved by using a multimedia/web server. The web server acts as the multimedia source, which transmits the multimedia content to all the wireless devices in its coverage area.

The end-users are the web-clients which receive the multimedia information. Figure 4 shows a dumbbell topology for achieving the single-hop communication, wherein, B1-B2 forms the bottleneck link. B1 is the multimedia source and transmits information to the n clients C1, C2 $\sim$ L $\sim$ Cn. In case of a two-hop communication, there is an intermediate relay between the web source and the end-user client. This can be represented by a double dumbbell topology, as shown in Fig. 5. The multimedia server is represented by B0, whereas the diametrically opposite relays are represented by B1 and B2. The end-users S1, S2, ..., Sn on one end and C1, C2, ..., Cn on the diametrically opposite cluster are the multimedia clients. The major advantage of the double dumbbell topology (cluster-based two-hop design) over the dumbbell topology (single-hop wireless network) is the hierarchical formation of the network, and the provision for peer-to-peer communication among the wireless nodes.

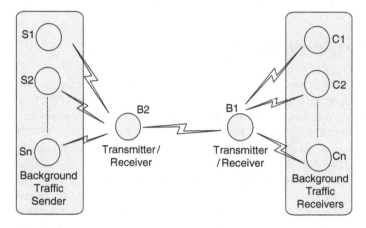

Fig. 4 Dumbbell network topology for single-hop client-server wireless networks

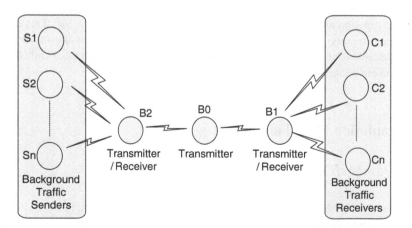

Fig. 5 Double dumbbell network topology for two-hop client-server networks

In the simulation environment, B1-B2 in the dumbbell topology and B0-B1, B0-B2 links in the double dumbbell topology are the bottleneck links, with 2 Mbps bandwidth and 100 ms latency. The other communicating links connected to the end-users in the network are over-provisioned. Hence, the congestion in the traffic, packet loss and delays occur mainly because of the bottleneck link. The buffering at the ends of the bottle-neck link uses a drop-tail queue of size proportional to the product of round trip time (RTT) and bottleneck link bandwidth.

3.2   Simulation Setup

The simulation setup consists of a number of mobile nodes distributed in the given coverage area. There is a centrally located server which covers the whole area. In order to increase the coverage area and the data rate provided to each node, two fixed access points (APs) located equidistantly from the server and diametrically opposite to each other are considered in the system design. The nodes located between the server and the AP communicates with the server directly. However, the mobile nodes that are located farther from the server and relatively closer to the AP communicate with the server through the AP. The APs therefore act as proxy-client-server between the actual client and server, as shown in Fig. 4. Hence, a hierarchical structure exists between the server, the relays and the mobile nodes.

The system is simulated using the server and client instances inbuilt in network simulator, NS-2. The length of all NS-2 simulations is 250s. The video streams are modeled using Transform Expand Sample (TES) [17] and then encoded using MPEG4. The primary reason for using MPEG4 is that it supports media streaming and is suitable for home networking applications with its low bit rate as well as its interoperability of audio and video signals [18]. The topology proposed assumes a bandwidth of 5 MHz and 100 ms latency. In the above topology, B0-B1 and B0-B2 links are the bottleneck links. The other links connected to the mobile nodes are over provisioned. Hence, the congestion in the traffic, packet loss and delays occur mainly because of the bottleneck link. The buffering at the ends of the bottle-neck link uses a drop-tail queue of size proportional to the product of round trip time (RTT) and bottleneck link bandwidth. In each of the envisaged scenarios 95–99% of the total available bandwidth is used for video and multimedia communication and the remaining 1–5 is reserved for feedback purpose. A constant PLR of 10^{-7} is assumed throughout the analysis. Similarly, a constant transmission delay of 10 ns is assumed between the PCS and the mobile nodes.

A binary phase shift keying (BPSK) modulation technique is used at the physical layer. A slow varying flat fading channel is assumed throughout the simulations. In addition, a two-ray model with a lognormal shadowing of 4 dB standard deviation is considered [19]. At the MAC layer, an IEEE 802.11 g based distributed coordination function (DCF) is used. The simulation is done at the packet level and the performance is evaluated in terms of average bit rate, loss rate and estimated user perceived quality. The end-user perceived quality is measured by developing a relationship between coding bitrate, packet loss ratio and user-level quality. Traffic with different sizes and shapes are considered as in [20], so as to emulate the real life scenario of varieties of traffic sources with different average bit rates. The network performance is analyzed in terms of the perceived quality, the loss rate and the average bit rate, not only for QOAS technique, but also compared with other protocols, i.e., LDA+, TFRCP and RBAR.

4 Results

Tables 1 and 2 present the performance results for UDP-CBR periodic traffic case, using 'dumbbell' and 'double dumbbell' topology respectively. The traffic patterns are kept identical for both single-hop and clustered two-hop scenario. It can be

Table 1 Performance of dumbbell topology – UDP CBR periodic traffic

Chars	Traffic (Mbps)	1 × 0.6 20 s on– 40 s off	1 × 0.6 30 s on– 60 s off	1 × 0.8 20 s on– 40 s off	1 × 0.8 30 s on– 60 s off	1 × 1.0 20 s on– 40 s off	1 × 1.0 30 s on– 60 s off
Perc. video quality (1–5)	QOAS	3.908	3.809	3.942	3.608	3.809	3.565
	LDA+	3.722	3.624	3.754	3.586	3.609	3.405
	TFRCP	3.608	3.493	3.571	3.216	3.282	3.165
	RBAR	3.621	3.467	3.279	3.011	2.945	2.829
Loss rate (%)	QOAS	0.18	0.18	0.14	0.14	0.10	0.10
	LDA+	0.42	0.42	0.40	0.38	0.24	0.24
	TFRCP	0.20	0.18	0.16	0.16	0.14	0.14
	RBAR	0.22	0.20	0.18	0.16	0.14	0.14
Avg. bit-rate (Mbps)	QOAS	0.726	0.734	0.776	0.763	0.802	0,802
	LDA+	0.713	0.726	0.736	0.734	0.772	0.772
	TFRCP	0.708	0.718	0.741	0.740	0.762	0.772
	RBAR	0.702	0.721	0.731	0.745	0.758	0.762

Table 2 Performance of double dumbbell topology – UDP CBR periodic traffic

Chars	Traffic (Mbps)	1 × 0.6 20 s on– 40 s off	1 × 0.6 30 s on– 60 s off	1 × 0.8 20 s on– 40 s off	1 × 0.8 30 s on– 60 s off	1 × 1.0 20 s on– 40 s off	1 × 1.0 30 s on– 60 s off
Perc. video quality (1–5)	QOAS	4.209	4.032	4.112	3.879	3.989	3.691
	LDA+	3.814	3.724	3.855	3.586	3.809	3.505
	TFRCP	3.809	3.692	3.674	3.315	3.308	3.261
	RBAR	3.693	3.423	3.404	3.016	3.124	3.083
Loss rate (%)	QOAS	0.08	0.08	0.04	0.04	0.01	0.01
	LDA+	0.38	0.38	0.34	0.34	0.21	0.21
	TFRCP	0.12	0.12	0.16	0.16	0.10	0.10
	RBAR	0.18	0.18	0.16	0.16	0.12	0.12
Avg. bit-rate (Mbps)	QOAS	0.732	0.732	0.756	0.756	0.812	0.813
	LDA+	0.756	0.756	0.788	0.787	0.793	0.793
	TFRCP	0.726	0.738	0.757	0.766	0.782	0.792
	RBAR	0.704	0.718	0.735	0.758	0.779	0.786

observed that for all three kinds of periodic traffic considered, the perceived quality obtained for the cluster-based two-hop design (double dumbbell topology) is significantly superior to that obtained using the single-hop design (dumbbell topology).

For example, in case of 1×0.6 Mbps traffic with 20 s on -40 s off, the perceived quality obtained with QOAS in the single-hop model is 3.908, whereas that obtained using the two-hop model is 4.209, an increase of 7.71%. In addition, for the same traffic model, the perceived quality of QOAS in the single-hop design is better than any other scheme (LDA+) by 5.25% (QOAS has a Q of 3.908 whereas LDA+ has a Q of 3.7025). However, in case of a cluster-based two-hop design, the improvement in the perceived quality between QOAS and LDA+ is at least 9.97%, almost twice the benefit obtained from 'dumbbell' topology. As seen from Tables 1 and 2, the improvement of double dumbbell topology over dumbbell topology is observed consistently across different UDP-CBR traffics.

In a similar result, in case of UDP-CBR staircase traffic, the perceived quality obtained from the 'double dumbbell' topology scores significantly over the 'dumbbell' scheme, as can be seen from Tables 3 and 4. For example, in case of 4×0.4 Mbps (Up 40 s steps), the perceived quality of QOAS using 'dumbbell' scheme is 3.808, whereas the same using 'double dumbbell' scheme is 4.298, an increase of 12.87%. It can be observed from Tables 3 and 4 that the improvement in the average bit rate and the loss rate is also notably high in case of the double dumbbell topology, as compared to the dumbbell scheme. Similarly, the improvement in loss rate for QOAS is over *six* times (0.04 using 'double dumbbell' and 0.24 using *dumbbell*). Importantly, this performance improvement remains consistent over different scenarios of staircase traffic, as shown in Tables 3 and 4.

In the Internet environment, traffic is generally bursty, as in case of FTP traffic. Tables 5 and 6 shows the performance for different FTP and WWW traffic

Table 3 Performance of UDP-CBR staircase traffic in dumbbell topology

Char.	Traffic (Mbps)	1×0.6 20 s on– 40 s off	1×0.6 30 s on– 60s off	1×0.8 20 s on– 40 s off	1×0.8 30 s on– 60 s off	1×1.0 20 s on– 40 s off	1×1.0 30 s on– 60 s off
Perc. video quality (1–5)	QOAS	3.808	3.862	3.726	3.887	3.727	3.581
	LDA+	3.683	3.677	3.496	3.579	3.561	3.321
	TFRCP	3.585	3.481	3.176	3.202	2.871	2.687
	RBAR	3.481	3.179	3.006	2.954	2.784	2.677
Loss rate (%)	QOAS	0.24	0.18	0.18	0.16	0.14	0.14
	LDA+	0.46	0.42	0.40	0.28	0.24	0.24
	TFRCP	0.26	0.18	0.18	0.18	0.14	0.13
	RBAR	0.24	0.20	0.18	0.18	0.16	0.16
Avg. bit-rate (Mbps)	QOAS	0.716	0.763	0.753	0.802	0.846	0.858
	LDA+	0.706	0.746	0.748	0.752	0.814	0.805
	TFRCP	0.686	0.721	0.734	0.742	0.801	0.808
	RBAR	0.671	0.711	0.725	0.738	0.762	0.801

Table 4 Performance of UDP-CBR staircase traffic-double dumbbell topology

Char.	Traffic (Mbps)	1 × 0.6 20 s on– 40 s off	1 × 0.6 30 s on– 60 s off	1 × 0.8 20 s on– 40 s off	1 × 0.8 30 s on– 60 s off	1 × 1.0 20 s on– 40 s off	1 × 1.0 30 s on– 60 s off
Perc. video quality (1–5)	QOAS	4.298	4.124	4.061	4.098	3.961	3.835
	LDA+	4.096	3.929	3.861	3.803	3.662	3.635
	TFRCP	3.834	3.675	3.636	3.597	3.441	3.335
	RBAR	3.721	3.485	3.241	3.144	3.004	2.955
Loss rate (%)	QOAS	0.04	0.00	0.00	0.06	0.01	0.00
	LDA+	0.24	0.18	0.18	0.16	0.11	0.09
	TFRCP	0.08	0.03	0.04	0.10	0.04	0.03
	RBAR	0.08	0.14	0.10	0.12	0.12	0.12
Avg. bit-rate (Mbps)	QOAS	0.736	0.809	0.818	0.712	0.797	0.817
	LDA+	0.725	0.789	0.768	0.682	0.767	0.807
	TFRCP	0.705	0.787	0.798	0.682	0.767	0.784
	RBAR	0.702	0.768	0.756	0.678	0.702	0.725

Table 5 Performance of TCP-FTP traffic using dumbbell topology

Char.	Traffic (Mbps)	50 x FTP	54 x FTP	58 x FTP	40 x WWW	50 x WWW	60 x WWW
Perc. video quality (1–5)	QOAS	4.093	3.722	3.701	4.794	4.451	4.002
	LDA+	3.901	3.725	3.612	4.393	4.160	3.922
	TFRCP	3.609	3.473	3.305	4.203	4.065	3.682
	RBAR	3.512	3.307	3.288	2.754	2.659	2.804
Loss rate rate (%)	QOAS	0.10	0.08	0.08	0.10	0.08	0.15
	LDA+	0.20	0.20	0.18	0.14	0.14	0.18
	TFRCP	0.14	0.14	0.14	0.14	0.14	0.19
	RBAR	0.18	0.22	0.22	0.24	0.24	0.18
Avg. bit-rate (Mbps)	QOAS	0.743	0.774	0.775	0.701	0.701	0.758
	LDA+	0.714	0.723	0.739	0.684	0.684	0.748
	TFRCP	0.714	0.737	0.752	0.691	0.691	0.716
	RBAR	0.693	0.716	0.739	0.752	0.752	0.781

Table 6 Performance of TCP-FTP traffic using double dumbbell topology

Char.	Traffic (Mbps)	50 x FTP	54 x FTP	58 x FTP	40 x WWW	50 x WWW	60 x WWW
Perc. video quality (1–5)	QOAS	4.109	3.972	3.801	4.809	4.565	4.162
	LDA+	4.003	3.902	3.712	4.599	4.361	4.062
	TFRCP	3.809	3.672	3.501	4.409	4.265	3.862
	RBAR	3.712	3.506	3.387	2.924	2.829	2.904
Loss rate (%)	QOAS	0.00	0.01	0.04	0.01	0.01	0.05
	LDA+	0.10	0.10	0.14	0.11	0.12	0.15
	TFRCP	0.10	0.10	0.12	0.04	0.04	0.09
	RBAR	0.10	0.12	0.12	0.14	0.14	0.12
Avg. bit-rate (Mbps)	QOAS	0.756	0.787	0.797	0.712	0.721	0.776
	LDA+	0.724	0.741	0.759	0.701	0.719	0.753
	TFRCP	0.736	0.757	0.767	0.702	0.711	0.737
	RBAR	0.702	0.722	0.754	0.769	0.769	0.791

models for the duration of 250 s. Similar to the results obtained in case of UDP transmission, the performance of the 'double dumbbell' topology is superior to that obtained from the 'dumbbell' topology for both FTP and WWW traffic.

5 Conclusions

This paper proposes *cluster-DAM*, a resource-efficient cluster-based design for delivery of adaptive multimedia content in IEEE 801.16j enabled two-hop cellular network. A 'double dumbbell' topology is considered in the cluster-based two-hop design so that the clients that are located diametrically opposite to the web server could communicate simultaneously. The 'state-of-the-art' cluster-DAM solution - *QOAS combined with the cluster-based design* not only results in superior multimedia transmission as compared to single-hop network, but also outperforms other protocols like LDA+, TFRCP and RBAR. In addition, the loss rate of the video frames is reduced, *even up to a factor of 10*, when the two-hop cluster-based design is used. This is a significant result for next generation data-centric wireless transmission systems. It demonstrates the feasibility of multimedia streaming over distantly located wireless users. This would boost the network operator to incorporate both the aspects of the *cluster-DAM* model in designing the next generation wireless networks.

Acknowledgements The authors would like to acknowledge Dr. Sinan Sinanovic and Prof. Harald Haas from the University of Edinburgh for their input during the initiatory stages. Also, the authors would like to thank Irish Research Council for Science Engineering and Technology (IRCSET) for supporting this research work.

References

1. B. Li, H. Yin, Peer-to-peer Live Video Streaming on the Internet: Issues, Existing Approaches and Challenges. IEEE Commun. Mag. **45**(6), 94–99 (2007)
2. S. Toumpis, A.J. Goldsmith, Capacity regions for wireless ad hoc networks. IEEE Trans. Wireless Commun. **2**(4), 736–748 (2003)
3. H. Venkataraman, G.M. Muntean, Analysis of random data hopping in distributed multihop wireless networks, in *Proceedings of IEEE Region Ten Conference (TENCON)*, Hyderabad, India, 18–21 Nov 2008
4. Y. Liu, R. Hoshyar, X. Yang, R. Tafazolli, Integrated radio resource allocation for multihop cellular networks with fixed relay stations. IEEE J. Selected Areas Commun. **24**(11), 2137–2146 (Nov 2006)
5. H. Venkataraman, S. Sinanovic, H. Haas, Cluster-based design for two-hop cellular networks. Int. J. Commun. Netw. Syst. (IJCNS) **1**(4), Nov 2008
6. H. Venkataraman, P. Kalyampudi, A. Krishnamurthy, J. McManis G.M. Muntean, Clustered architecture for adaptive multimedia streaming in WiMAX based cellular networks, in *Proceedings of World Congress on Engineering and Computer Science (WCECS)*, Berkeley, San Francisco, CA, Oct 2009

7. H. Venkataraman, P.K. Jaini, S. Revanth, Optimum location of gateways for cluster-based two-hop cellular networks. Int. J. Commun. Netw. Syst. (IJCNS) **2**(5) (Aug 2009)
8. P. Gupta, P.R. Kumar, The capacity of wireless networks. IEEE Trans. Inf. Theory **46**(2) 388–404 (Feb 2000)
9. H. Venkataraman, S. Nainwal, P. Srivastava, Optimum number of gateways in cluster-based two-hop cellular networks. AEUe J. Electron. Commun. Elsevier **64**(4), 310–321 (April 2010)
10. G.M. Muntean, N. Cranley, Resource efficient quality-oriented wireless broadcasting of adaptive multimedia content. IEEE Trans. Broadcasting **53**(1), 362–368 (Mar 2007)
11. G.M. Muntean, P. Perry, L. Murphy, A New Adaptive Multimedia Streaming System for all-IP Multi-Service Networks. IEEE Trans. Broadcasting **50**(1), 1–10 (2004)
12. ITU-T Recommendation P.910, Subjective Video Quality Assessment Methods for Multimedia Applications, Sept 1999
13. G.M. Muntean, P. Perry, L. Murphy, A comparison-based study of quality-oriented video on demand. IEEE Trans. Broadcasting **53**(1), 92–102 (Mar 2007)
14. M. Miyabayashi, N. Wakamiya, M. Murata, H. Miyahara, MPEG-TFRCP: video transfer with TCP-friendly rate control protocol, in *Proceedings of IEEE International Conference on Communications (ICC' 01)*, vol. 1, Helsinki, Finland, June 2001, pp. 137–141
15. D. Sisalem, A. Wolisz, LDA+ TCP Friendly Adaptation: A Measurement and Comparison Study, in *ACM International Workshop on Network and Operating Systems Support for Digital Audio and Video (NOSSDAV)*, Chapel Hill, NC, USA, June 2000
16. G. Holland, N. Vaidy, V. Bahl, A rate-adaptive MAC protocol for multihop wireless networks, Rome, Italy, in *Proceedings of ACM Mobile Computing and Networking (MOBICOM' 01)*, 2001, pp. 236–251
17. A. Matrawy, I. Lambadaris, C. Huang, MPEG4 traffic,modeling using the transform expand sample methodology, in *Proceedings of 4th IEEE International Workshop on Networked Appliances*, 2002, pp. 249–256
18. H. Venkataraman, S. Agrawal, A. Maheshwari, G.M. Muntean, Quality-controlled prioritized adaptive multimedia streaming in two-hop wireless networks, in *Proceedings of World Congress on Engineering and Computer Science (WCECS)*, Berkeley, San Francisco, CA, Oct 2009
19. 3rd Generation Partnership Project (3GPP), Technical Specification Group Radio Access Network, *Selection procedures for the choice of radio transmission technologies of the UMTS*, 3GPP TR 30.03U, May 1998
20. Z. Yuan, H. Venkataraman, G.M. Muntean, iPAS: an user perceived quality based intelligent prioritized adaptive scheme for IPTV in wireless home networks, in *Proceedings of IEEE Broadband Multimedia Systems and Broadcasting (BMSB) Conference*, Shanghai, China, 24–26 Mar 2010

Chapter 31
Ranking Intervals in Complex Stochastic Boolean Systems Using Intrinsic Ordering

Luis González

Abstract Many different phenomena, arising from scientific, technical or social areas, can be modeled by a system depending on a certain number n of random Boolean variables. The so-called complex stochastic Boolean systems (CSBSs) are characterized by the ordering between the occurrence probabilities $\Pr\{u\}$ of the 2^n associated binary strings of length n, i.e., $u=(u_1,\ldots,u_n) \in \{0,1\}^n$. The intrinsic order defined on $\{0,1\}^n$ provides us with a simple positional criterion for ordering the binary n-tuple probabilities without computing them, simply looking at the relative positions of their 0s and 1s. For every given binary n-tuple u, this paper presents two simple formulas – based on the positions of the 1-bits (0-bits, respectively) in u – for counting (and also for rapidly generating, if desired) all the binary n-tuples v whose occurrence probabilities $\Pr\{v\}$ are always less than or equal to (greater than or equal to, respectively) $\Pr\{u\}$. Then, from these formulas, we determine the closed interval covering all possible values of the rank (position) of u in the list of all binary n-tuples arranged by decreasing order of their occurrence probabilities. Further, the length of this so-called ranking interval for u, also provides the number of binary n-tuples v incomparable by intrinsic order with u. Results are illustrated with the intrinsic order graph, i.e., the Hasse diagram of the partial intrinsic order.

1 Introduction

In many different scientific, technical or social areas, one can find phenomena depending on an arbitrarily large number n of basic random Boolean variables. In other words, the n basic variables of the system are assumed to be stochastic and they only take two possible values: either 0 or 1. We call such a system: a *complex*

L. González

Research Institute IUSIANI, Department of Mathematics, University of Las Palmas de Gran Canaria, Campus Universitario de Tafira, 35017 Las Palmas de Gran Canaria, Spain

e-mail: luisglez@dma.ulpgc.es

S.-I. Ao et al. (eds.), *Machine Learning and Systems Engineering*,
Lecture Notes in Electrical Engineering 68,
DOI 10.1007/978-90-481-9419-3_31, © Springer Science+Business Media B.V. 2010

stochastic Boolean system (CSBS). Each one of the 2^n possible elementary states associated to a CSBS is given by a binary n-tuple $u=(u_1,\ldots,u_n) \in \{0,1\}^n$ of 0s and 1s, and it has its own occurrence probability $\Pr\{(u_1,\ldots,u_n)\}$.

Using the statistical terminology, a CSBS on n variables $x_1,\ldots,x_n$ can be modeled by the n-dimensional Bernoulli distribution with parameters $p_1,\ldots,p_n$ defined by

$$\Pr\{x_i = 1\} = p_i,\ \Pr\{x_i = 0\} = 1 - p_i,$$

Throughout this paper we assume that the n Bernoulli variables x_i are mutually statistically independent, so that the occurrence probability of a given binary string $u=(u_1,\ldots,u_n) \in \{0,1\}^n$ of length n can be easily computed as

$$\Pr\{u\} = \prod_{i=1}^{n} p_i^{u_i}(1 - p_i)^{1-u_i}, \tag{1}$$

that is, $\Pr\{u\}$ is the product of factors p_i if $u_i=1$, $1-p_i$ if $u_i=0$ [6].

Example 1. Let $n=4$ and $u=(0,1,0,1) \in \{0,1\}^4$. Let $p_1=0.1$, $p_2=0.2$, $p_3=0.3$, $p_4=0.4$. Then, using Eq. 1, we have

$$\Pr\{(0, 1, 0, 1)\} = (1 - p_1)p_2(1 - p_3)p_4 = 0.0504.$$

The behavior of a CSBS is determined by the ordering between the current values of the 2^n associated binary n-tuple probabilities $\Pr\{u\}$. For this purpose, in [1] we have established a simple positional criterion that allows one to compare two given binary n-tuple probabilities, $\Pr\{u\},\Pr\{v\}$, without computing them, simply looking at the positions of the 0s and 1s in the n-tuples u,v. We have called it the *intrinsic order criterion*, because it is independent of the basic probabilities p_i and it *intrinsically* depends on the positions of the 0s and 1s in the binary strings.

In this context, the main goal of this paper is to determine, in a simple way, the ranking interval of u for every fixed $n \geq 1$ and for every fixed binary n-tuple u. This is defined as the closed interval that covers all possible ranks (i.e., positions) of u in the list of all binary n-tuples when they are arranged by decreasing order of their occurrence probabilities. Of course the position of each n-tuple u in such lists depends on the way in which the current values $\{p_i\}^n_{i=1}$ of the Bernoulli parameters affect the ordering between the occurrence probability of u and the occurrence probability of each binary n-tuple v incomparable with u.

For this purpose, this paper has been organized as follows. In Section 2, we present some previous results on the intrinsic order relation, the intrinsic order graph and other related aspects, enabling non-specialists to follow the paper without difficulty and making the presentation self-contained. Section 3 is devoted to provide new formulas for counting (and generating) all the binary n-tuples intrinsically greater than, less than, or incomparable with u. In Section 4, we present our main result: We exactly obtain the ranking interval for each given binary string u, and we illustrate our propositions with a simple, academic example. Finally, in Section 5, we present our conclusions.

2 The Intrinsic Ordering

2.1 Intrinsic Order Relation on $\{0,1\}^n$

Throughout this paper, the decimal numbering of a binary string u is denoted by the symbol $u_{(10}$, while the Hamming weight of a binary n-tuple u (i.e., the number of 1-bits in u) will be denoted, as usual by $w_H\{u\}$, i.e.,

$$(u_1,\ldots,u_n) \equiv u_{(10} = \sum_{i=1}^n 2^{n-i} u_i, \quad w_H(u) = \sum_{i=1}^n u_i,$$

e.g., for $n=5$ we have

$$(1,0,1,1,1) \equiv 2^0 + 2^1 + 2^2 + 2^4 = 23, \ w_H(1,0,1,1,1) = 4.$$

According to Eq. 1, the ordering between two given binary string probabilities $\Pr\{u\}$ and $\Pr\{v\}$ depends, in general, on the parameters p_i, as the following simple example shows.

Example 2. Let $n = 3$, $u = (0,1,1)$ and $v=(1,0,0)$. Using Eq. 1 we have

$$p_1 = 0.1, \ p_2 = 0.2, \ p_3 = 0.3 : \ \Pr\{u\} = 0.054 < \Pr\{v\} = 0.056,$$

$$p_1 = 0.2, \ p_2 = 0.3, \ p_3 = 0.4 : \ \Pr\{u\} = 0.096 > \Pr\{v\} = 0.084.$$

As mentioned in Section 1, to overcome the exponential complexity inherent to the task of computing and sorting the 2^n binary string probabilities (associated to a CSBS with n Boolean variables), we have introduced the following intrinsic order criterion [1], denoted from now on by the acronym IOC (see[5] for the proof).

Theorem 1 (The intrinsic order theorem). *Let $n \geq 1$. Suppose that $x_1, \ldots, x_n$ are n mutually independent Bernoulli variables whose parameters $p_i = \Pr\{x_i = 1\}$ satisfy*

$$0 < p_1 \leq p_2 \leq \cdots \leq p_n \leq 0.5. \tag{2}$$

Then the probability of the binary n-tuple $v = (v_1,\ldots,v_n)$ is intrinsically less than or equal to the probability of the binary n-tuple $u = (u_1,\ldots,u_n)$ (that is, for all set $\{p_i\}^n_{i=1}$ satisfying (2)) if and only if the matrix

$$M_v^u := \begin{pmatrix} u_1 \ldots u_n \\ v_1 \ldots v_n \end{pmatrix}$$

either has no $\binom{1}{0}$ columns, or for each $\binom{1}{0}$ column in M_v^u there exists (at least) one corresponding preceding $\binom{0}{1}$ column (IOC).

Remark 1. In the following, we assume that the parameters p_i always satisfy condition (2). Note that this hypothesis is not restrictive for practical applications because, if for some $i : p_i > \frac{1}{2}$, then we only need to consider the variable $\overline{x_i} = 1 - x_i$, instead of x_i. Next, we order the n Bernoulli variables by increasing order of their probabilities.

Remark 2. The $\binom{0}{1}$ column preceding to each $\binom{1}{0}$ column is not required to be necessarily placed at the immediately previous position, but just at previous position.

Remark 3. The term *corresponding*, used in Theorem 1, has the following meaning: For each two $\binom{1}{0}$ columns in matrix M_v^u, there must exist (at least) two *different* $\binom{0}{1}$ columns preceding to each other. In other words: For each $\binom{1}{0}$ column in matrix M_v^u, the number of preceding $\binom{0}{1}$ columns must be strictly greater than the number of preceding $\binom{1}{0}$ columns.

The matrix condition IOC, stated by Theorem 1, is called the *intrinsic order criterion*, because it is independent of the basic probabilities p_i and it *intrinsically* depends on the relative positions of the 0s and 1s in the binary n-tuples u,v. Theorem 1 naturally leads to the following partial order relation on the set $\{0,1\}^n$ [1]. The so-called intrinsic order will be denoted by "$\preceq$", and we shall write $v \preceq u$ ($u \succeq v$) to indicate that v (u) is intrinsically less (greater) than or equal to u (v).

Definition 1. *For all* $u,v \in \{0,1\}^n$

$$v \preceq u \quad \text{iff} \quad \Pr\{v\} \leq \Pr\{u\} \quad \text{for all set } \{p_i\}_{i=1}^n \text{ s.t. (2)}$$

$$\text{iff} \quad M_v^u \text{ satisfies IOC.}$$

From now on, the partially ordered set (poset, for short) $(\{0,1\}^n, \preceq)$ will be denoted by I_n.

Example 3. For $n=3$, we have $3 \equiv (0,1,1) \preceq 4 \equiv (1,0,0)$ and also $4 \equiv (1,0,0) \preceq 3 \equiv (0,1,1)$ because the matrices

$$\begin{pmatrix} 1 & 0 & 0 \\ 0 & 1 & 1 \end{pmatrix} \text{ and } \begin{pmatrix} 0 & 1 & 1 \\ 1 & 0 & 0 \end{pmatrix}$$

do not satisfy IOC (Remark 3). Thus, (0,1,1) and (1,0,0) are incomparable by intrinsic order, i.e., the ordering between $\Pr\{(0,1,1)\}$ and $\Pr\{(1,0,0)\}$ depends on the parameters $\{p_i\}_{i=1}^3$, as Example 2 has shown.

Example 4. For $n=5$, we have $24\equiv(1,1,0,0,0)\preceq 5\equiv(0,0,1,0,1)$ because matrix

$$\begin{pmatrix} 0 & 0 & 1 & 0 & 1 \\ 1 & 1 & 0 & 0 & 0 \end{pmatrix}$$

satisfies IOC (Remark 2).Thus, for all $\{p_i\}_{i=1}^5$ s.t. (2)

$$\Pr\{(1,1,0,0,0)\} \le \Pr\{(0,0,1,0,1)\}.$$

Example 5. For all $n\ge 1$, the binary n-tuples

$$\left(0,\overset{n}{\dots},0\right) \equiv 0 \quad \text{and} \quad \left(1,\overset{n}{\dots},1\right) \equiv 2^n - 1$$

are the maximum and minimum elements, respectively, in the poset I_n. Indeed, both matrices

$$\begin{pmatrix} 0 & \dots & 0 \\ u_1 & \dots & u_n \end{pmatrix} \quad \text{and} \quad \begin{pmatrix} u_1 & \dots & u_n \\ 1 & \dots & 1 \end{pmatrix}$$

satisfy the intrinsic order criterion, since they have no $\begin{pmatrix} 1 \\ 0 \end{pmatrix}$ columns!.Thus, for all $u \in \{0,1\}^n$ and for all $\{p_i\}_{i=1}^n$ s.t. (2)

$$\Pr\left\{\left(1,\overset{n}{\dots},1\right)\right\} \le \Pr\{(u_1,\dots,u_n)\} \le \Pr\left\{\left(0,\overset{n}{\dots},0\right)\right\}.$$

Many different properties of the intrinsic order relation can be derived from its simple matrix description IOC (see, e.g., [1–3]). For the purpose of this paper, we need recall here the following necessary (but not sufficient) condition for intrinsic order; see [2] for the proof.

Corollary 1. *For* all $u,v \in \{0,1\}^n$

$$u \succeq v \quad \Rightarrow \quad w_H(u) \le w_H(v).$$

2.2   The Intrinsic Order Graph

Now, we present, in Fig. 1, the graphical representation of the poset I_n. The usual representation of a poset is its Hasse diagram (see, e.g., [7] for more details about posets and Hasse diagrams). This is a directed graph (digraph, for short) whose vertices are the binary n-tuples of 0s and 1s, and whose edges go downward from u to v whenever u covers v (denoted by $u\,B\,v$), that is, whenever u is intrinsically greater than v with no other elements between them, i.e.,

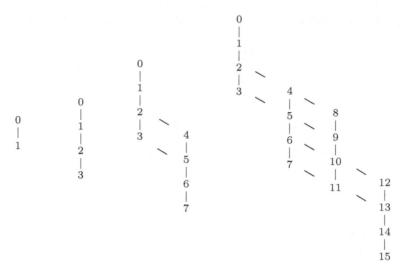

Fig. 1 The intrinsic order graph for n=1,2,3,4

$$uBv \quad \text{iff} \quad u \succ v \text{ and there is no } w \in \{0,1\}^n \text{ s.t. } u \succ w \succ v.$$

The Hasse diagram of the poset I_n will be also called the intrinsic order graph for n variables. For all $n \geq 2$, in [3] we have developed an algorithm for iteratively building up the digraph of I_n from the digraph of I_1. Basically, I_n is obtained by adding to I_{n-1} its isomorphic copy $2^{n-1} + I_{n-1}$: A nice fractal property of I_n!

In Fig. 1, the intrinsic order graphs of I_1, I_2, I_3 and I_4 are shown, using the decimal numbering instead of the binary representation of their 2^n nodes, for a more comfortable and simpler notation. Each pair (u,v) of vertices connected in the digraph of I_n either by one edge or by a longer descending path (consisting of more than one edge) from u to v, means that u is intrinsically greater than v, i.e., $u \succ v$. On the contrary, each pair (u,v) of non-connected vertices in the digraph of I_n either by one edge or by a longer descending path, means that u and v are incomparable by intrinsic order, i.e., $u \nsucc v$ and $v \nsucc u$.

Looking at any of the four graphs in Fig. 1, we can confirm the properties of the intrinsic order stated by both Example 5 and Corollary 1.

2.3   Three Sets of Bitstrings Related to a Binary n-tuple

In the following two definitions, we present three subsets of $\{0,1\}^n$ related to each binary n-tuple $u \in \{0,1\}^n$.

Definition 2. *For every binary n-tuple* $u \in \{0,1\}^n$, *the set* C^u *(the set* C_u, *respectively) is the set of all binary n-tuples* v *whose occurrence probabilities* $Pr\{v\}$ *are always less (greater, respectively) than or equal to* $Pr\{u\}$, *i.e., according to Definition 1, those n-tuples* v *intrinsically less (greater, respectively) than or equal to* u, *i.e.,*

$$C^u = \{v \in \{0,1\}^n \mid u \succeq v\}, \tag{3}$$

$$C_u = \{v \in \{0,1\}^n \mid u \preceq v\}. \tag{4}$$

Definition 3. *For every binary n-tuple* $u \in \{0,1\}^n$, $Inc\{u\}$ *is the set of all binary n-tuples* v *intrinsically incomparable with* u, *i.e.,*

$$Inc(u) = \{v \in \{0,1\}^n \mid u \succeq v, \; u \preceq v\} = \{0,1\}^n - (C^u \cup C_u). \tag{5}$$

For instance, due to Example 5, we have, for all $n \geq 1$,

$$C^{\left(0,\overset{n}{\ldots},0\right)} = \{0,1\}^n = C_{\left(1,\overset{n}{\ldots},1\right)}, Inc\left(\left(0,\overset{n}{\ldots},0\right)\right) = \{\} = Inc\left(\left(1,\overset{n}{\ldots},1\right)\right)$$

as Fig. 1 shows, for $n=1,2,3,4$.

Remark 4. Note that, because of the reflexive and antisymmetric properties of the intrinsic order relation $\preceq$, we obviously have, for all $n \geq 1$ and for all $u \in \{0,1\}^n$

$$C^u \cap C_u = \{u\}. \tag{6}$$

The sets C^u and C_u are closely related via complementary n-tuples, as described with precision by the following definition and theorem (see [4] for the proof).

Definition 4. *The complementary n-tuple of a given binary n-tuple* $u \in \{0,1\}^n$ *is obtained by changing its 0s by 1s and its 1s by 0s*

$$u^c = (u_1, \ldots, u_n)^c = (1 - u_1, \ldots, 1 - u_n).$$

The complementary set of a given subset $S \subseteq \{0,1\}^n$ *of binary n-tuples is the set of the complementary n-tuples of all the n-tuples of* S

$$S^c = \{u^c \mid u \in S\}.$$

Theorem 2. *For all* $n \geq 1$ *and for all* $u \in \{0,1\}^n$

$$C_u = \left[C^{u^c}\right]^c, \qquad C^u = \left[C_{u^c}\right]^c. \tag{7}$$

3   Generating and counting the elements of C^u and C_u

First, we need to set the following nomenclature and notation.

Definition 5. *Let n≥1 and let* u ∈ *{0,1}ⁿ with Hamming weight* $w_H(u)$=m. *Then*

(i) *The vector of positions of 1s of* u *is the vector of positions of its m 1-bits, displayed in increasing order from the left-most position to the right-most position, and it will be denoted by*

$$u = [i_1^1, \ldots, i_m^1]_n^1, \ 1 \le i_1^1 < \cdots < i_m^1 \le n, \ 0 < m \le n.$$

(ii) *The vector of positions of 0s of* u *is the vector of positions of its* (n−m) *0-bits, displayed in increasing order from the left-most position to the right-most position, and it will be denoted by*

$$u = [i_1^0, \ldots, i_{n-m}^0]_n^0, \ 1 \le i_1^0 < \cdots < i_{n-m}^0 \le n, \ 0 \le m < n.$$

Example 6. Let *n*=6 and *u* = 27 ≡ (0, 1, 1, 0, 1, 1). Then we have $m = w_H(u) = 4$, $n - m = 2$, and

$$u = [i_1^1, i_2^1, i_3^1, i_4^1]_6^1 = [2, 3, 5, 6]_6^1, \quad u = [i_1^0, i_2^0]_6^0 = [1, 4]_6^0.$$

In [4], the author presents an algorithm for obtaining the set C^u, for each given binary *n*-tuple *u*. Basically, this algorithm determines C^u by expressing the set difference {0,1}ⁿ−C^u as a set union of certain half-closed intervals of natural numbers (decimal representation of bitstrings). The next two theorems present new, more efficient algorithms for rapidly determining C^u and C_u, using the binary representation of their elements. Moreover, our algorithms allow us not only to generate, but also to count the number of elements of C^u and C_u.

Theorem 3. *Let u ∈ {0,1}ⁿ, u≠0, with (nonzero) Hamming weight* $w_H\{u\} = m$ *(0 < m ≤ n). Let u = $\{i_1^1, \ldots, i_m^1\}_n^1$ be the vector of positions of 1s of u. Then C^u is the set of binary n-tuples v generated by the following algorithm:*

   The Algorithm for C^u
   (i) Step 1 (Generation of the sequences $\{j_1^1, \ldots j_m^1\}$) :
       For $j_1^1 = 1$ to i_1^1 Do:
       For $j_2^1 = j_1^1 + 1$ to i_2^1 Do:
       ·

           For $j_m^1 = j_{m-1}^1 + 1$ to i_m^1 Do:
               Write $\{j_1^1 j_2^1, \ldots j_m^1\}$
           EndDo
       ·

       EndDo
   EndDo

(ii) Step 2 (Generation of the n-tuples $v \in C^u$) :
   For every sequence $\{j_1{}^1,\ldots,j_m{}^1\}$ generated by Step 1, define the binary
   n-tuple $v = \{v_1,\ldots,v_n\}$ as follows

$$v_i = \begin{cases} 1 & \text{if } i \in \{j_1^1,\ldots,j_m^1\}, \\ 0 & \text{if } i \notin \{j_1^1,\ldots,j_m^1\}, \ i < j_m^1, \\ 0,1 & \text{if } i \notin \{j_1^1,\ldots,j_m^1\}, \ i > j_m^1. \end{cases} \tag{8}$$

Moreover, the cardinality of the set C^u is given by the multiple sum

$$|C^u| = \sum_{j_1^1=1}^{i_1^1} \sum_{j_2^1=j_1^1+1}^{i_2^1} \cdots \sum_{j_m^1=j_{m-1}^1+1}^{i_m^1} 2^{n-j_m^1}. \tag{9}$$

Proof. According to (3) and Corollary 1, we have

$$v \in C^u \Leftrightarrow u \succeq v \Rightarrow m = w_H(u) \leq w_H(v),$$

that is, the Hamming weight of every n-tuple $v \in C^u$ is necessarily greater than or
equal to the Hamming weight m of u. So, all n-tuples v of C^u must contain at least
m 1-bits.

First, note that, according to Definition 1, $u \succeq v$ if and only if matrix M_v^u satisfies
IOC. That is, either M_v^u has no $\binom{1}{0}$ columns, or for each $\binom{1}{0}$ column in M_v^u there exists
(at least) one corresponding preceding $\binom{0}{1}$ column. But, obviously, IOC can be
equivalently reformulated as follows: For each 1-bit in u there exists at least one
corresponding 1-bit in v, placed at the same or at a previous position. This reduces
the characterization of the elements $v \in C^u$ to the vectors of positions of 1-bits in u
and v and thus, according to the notation used in Definition 5-(i), we derive that $v \in C^u$ with $w_H(v) = m = w_H(u)$ if and only if

$$1 \leq j_1^1 \leq i_1^1, \ j_1^1 + 1 \leq j_2^1 \leq i_2^1, \ldots, j_{m-1}^1 + 1 \leq j_m^1 \leq i_m^1, \tag{10}$$

but these are the sequences $\{j_1{}^1 j_2{}^1,\ldots,j_m{}^1\}$ generated by Step 1.

Second, note that the substitution of 0s by 1s in any n-tuple v such that $u \succeq v$ does
not avoid the IOC condition, because such a substitution changes the $\binom{0}{0}$ and $\binom{1}{0}$
columns of matrix M_v^u into $\binom{0}{1}$ and $\binom{1}{1}$ columns, respectively. Hence, to obtain all
the binary strings of the set C^u, it is enough to assign, in all possible ways, both
values 0, 1, to any of the $n-j_m{}^1$ (if $j_m{}^1 < n$) null right-most components $v_{j_m^1+1},\ldots,v_n$
of all the binary strings $v=[j_1{}^1,\ldots,j_m{}^1]_n{}^1$ with weight m, generated by (10). In other
words, we define v as described by (8) in Step 2.

Finally, since there are exactly $2^{n-j_m^1}$ different ways of assigning the values
$v_i=0,1$ for all $j_m^1 < i \leq n$, and since by this procedure we generate all elements of
C^u without repetitions, then the cardinality of C^u is given by (9). $\qquad \square$

Theorem 4. *Let* $u \in \{0,1\}^n$, $u \neq 2^n - 1$, *with Hamming weight* $w_H(u) = m$ $(0 \leq m < n)$. *Let* $u = \left[i_1^0, \ldots, i_{n-m}^0\right]_n$ *be the vector of positions of 0s of* u. *Then* C_u *is the set of binary n-tuples* v *generated by the following algorithm:*

The Algorithm for C_u

(i) *Step 1 (Generation of the sequences* $\{j_1^0, \ldots, j_{n-m}^0\}$ *) :*

   For $j_1^0 = 1$ to i_1^0 Do:

   For $j_2^0 = j_1^0 + 1$ to i_2^0 Do:

   . . . .

   For $j_{n-m}^0 = j_{n-m-1}^0 + 1$ to i_{n-m}^0 Do:

   Write $\{j_1^0 j_2^0 \ldots j_{n-m}^0\}$

   EndDo

   . . . .

   EndDo

EndDo

(ii) *Step 2 (Generation of the n-tuples* $v \in C_u$*):*

For every sequence $\{j_1^0, \ldots, j_{n-m}^0\}$ *generated by Step 1, define the binary n-tuple* $v = (v_1, \ldots, v_n)$ *as follows*

$$
v_i = \begin{cases} 0 & \text{if } i \in \{j_1^0, \ldots, j_{n-m}^0\}, \\ 1 & \text{if } i \notin \{j_1^0, \ldots, j_{n-m}^0\}, \ i < j_{n-m}^0, \\ 0,1 & \text{if } i \notin \{j_1^0, \ldots, j_{n-m}^0\}, \ i > j_{n-m}^0. \end{cases} \tag{11}
$$

Moreover, the cardinality of the set C_u *is given by the multiple sum*

$$
|C_u| = \sum_{j_1^0=1}^{i_1^0} \sum_{j_2^0=j_1^0+1}^{i_2^0} \cdots \sum_{j_{n-m}^0=j_{n-m-1}^0+1}^{i_m^0} 2^{n-j_{n-m}^0}. \tag{12}
$$

Proof. For proving this theorem, it is enough to use Theorems 2 and 3. More precisely, due to the duality property stated by Theorem 2, we get that $v \in C_u$ iff $v \in \left[C^{u^c}\right]^c$ iff $v^c \in C^{u^c}$. Then the proof is concluded using the obvious fact that the 0-bits and 1-bits in u and v become the 1-bits and 0-bits, respectively, in u^c and v^c. □

Remark 5. Note the strong duality relation between Theorems 3 and 4. The statement of each theorem is exactly the statement of the other one after interchanging 1s and 0s, C^u and C_u. Indeed, due to Theorem 2, one can determine the set C^u (C_u, respectively) by determining the set $\left[C_{u^c}\right]^c$ ($\left[C^{u^c}\right]^c$, respectively) using Theorem 4 (Theorem 3, respectively).

4 Ranking Intervals

For each given binary n-tuple $u \in \{0,1\}^n$, the next theorem provides us with the closed interval covering all possible values of the rank (position) of u in the list of all binary n-tuples arranged by decreasing order of their occurrence probabilities.

We call this interval "the ranking interval of u". The following is the main result of this paper.

Theorem 5 (The ranking interval). *Let $n \geq 1$ and $u \in \{0,1\}^n$. Then the ranking interval of u is*

$$[\, |C_u| \,,\, 2^n + 1 - |C^u| \,], \tag{13}$$

where $|C_u|$ and $|C^u|$ are respectively given by Eqs. 12 and 9. Moreover, the length of the ranking interval of u coincides with the number of binary n-tuples incomparable by intrinsic order with u, i.e.,

$$|Inc(u)| = l([\, |C_u| \,,\, 2^n + 1 - |C^u| \,]) = 2^n + 1 - |C^u| - |C_u|. \tag{14}$$

Proof. On one hand, using Eq. 6 we have that $u \in C^u$. Then, in the list of all binary n-tuples arranged by decreasing order of their occurrence probabilities, at least $|C^u| - 1$ binary strings are always below u. Thus, in that list, u is always among the positions 1 (the first possible position) and $2^n - (|C^u| - 1)$ (the last possible position), i.e.,

$$u \in [\, 1 \,,\, 2^n + 1 - |C^u| \,]. \tag{15}$$

On the other hand, using again Eq. 6 we have that $u \in C_u$. Then, in the list of all binary n-tuples arranged by decreasing order of their occurrence probabilities, at least $|C_u| - 1$ binary strings are always above u. Thus, in that list, u is always among the positions $|C_u|$ (the first possible position) and 2^n (the last possible position), i.e.,

$$u \in [\, |C_u| \,,\, 2^n \,]. \tag{16}$$

Hence, from Eqs. 15 and 16, we get

$$u \in [\, 1 \,,\, 2^n + 1 - |C^u| \,] \cap [\, |C_u| \,,\, 2^n \,]. \tag{17}$$

Now, we prove that

$$1 \leq |C_u| \leq 2^n + 1 - |C^u| \leq 2^n. \tag{18}$$

First, the left-most inequality in Eq. 18, i.e., $1 \leq |C_u|$ is obvious taking into account that $u \in C_u$ (see Eq. 6).

Second, the right-most inequality in Eq. 18, i.e., $2^n + 1 - |C^u| \leq 2^n$, that is, $1 \leq |C^u|$ is obvious taking into account that $u \in C^u$ (see Eq. 6).

Third, the central inequality in Eq. 18, i.e., $|C_u| \leq 2^n + 1 - |C^u|$, that is, $|C_u| + |C^u| \leq 2^n + 1$ immediately follows from Eq. (6). That is, using the inclusion-exclusion formula, we have

$$|C_u| + |C^u| = |C_u \cup C^u| + |C_u \cap C^u| \leq |\{0,1\}^n| + |\{u\}| = 2^n + 1.$$

Hence, from Eqs. 17 and 18, we get

$$u \in [\,1\,,\, 2^n + 1 - |C^u|\,] \cap [\,|C_u|\,,\, 2^n\,] = [\,|C_u|\,,\, 2^n + 1 - |C^u|\,].$$

Finally, the last assertion of the theorem is an immediate consequence of Eq. 5 and the inclusion-exclusion formula

$$\begin{aligned}
|Inc(u)| &= |\{0,1\}^n - (C^u \cup C_u)| = |\{0,1\}^n| - |C^u \cup C_u|\\
&= 2^n - (|C^u| + |C_u| - |C_u \cap C^u|) = 2^n + 1 - |C^u| - |C_u|,
\end{aligned}$$

as was to be shown.   □

The following example illustrates Theorems 3, 4 and 5.

Example 7. Let $n=5$ and $u = 21 \equiv (1,0,1,0,1)$. Then, we have $m = w_H(u) = 3$, $n - m = 2$, and

$$u = \left[i_1^1, i_2^1, i_3^1\right]_5^1 = [1,3,5]_5^1, \quad u = \left[i_1^0, i_2^0\right]_5^0 = [2,4]_5^0.$$

Using Theorems 3 and 4 we can generate all the 5-tuples $v \in C^u$ and all the 5-tuples $v \in C_u$, respectively. Results are depicted in Tables 1 and 2, respectively. The first column of Tables 1 and 2, show the auxiliary sequences generated by Step 1 in Theorems 3 and 4, respectively. The second column of Tables 1 and 2, show all the binary 5-tuples $v \in C^u$ and $v \in C_u$ generated by Step 2 in Theorems 3 and 4, respectively. The symbol "$*$" means that we must choose both binary digits: 0 and 1, as described by the last line of (8) and (11). According to (9) and (12), the sums of all quantities in the third column of Tables 1 and 2, give the number of elements of C^u

Table 1 The set $C^{(1,0,1,0,1)}$ and its cardinality

$\{j_1^1 j_2^1 j_3^1\}$	$v \in C^u$	$2^{n-j_m^1} = 2^{5-j_3^1}$
$\{1,2,3\}$	$(1,1,1,*,*) \equiv 28,29,30,31$	$2^{5-3} = 4$
$\{1,2,4\}$	$(1,1,0,1,*) \equiv 26,27$	$2^{5-4} = 2$
$\{1,2,5\}$	$(1,1,0,0,1) \equiv 25$	$2^{5-5} = 1$
$\{1,3,4\}$	$(1,0,1,1,*) \equiv 22,23$	$2^{5-4} = 2$
$\{1,3,5\}$	$(1,0,1,0,1) \equiv 21$	$2^{5-5} = 1$
		$|C^u| = \sum 2^{5-j_3^1} = 10$

Table 2 The set $C_{(1,0,1,0,1)}$ and its cardinality

$\{j_1^0 j_2^0\}$	$v \in C_u$	$2^{n-j_{n-m}^0} = 2^{5-j_2^0}$
$\{1,2\}$	$(0,0,*,*,*) \equiv 0,1,2,3,4,5,6,7$	$2^{5-2} = 8$
$\{1,3\}$	$(0,1,0,*,*) \equiv 8,9,10,11$	$2^{5-3} = 4$
$\{1,4\}$	$(0,1,1,0,*) \equiv 12,13$	$2^{5-4} = 2$
$\{2,3\}$	$(1,0,0,*,*) \equiv 16,17,18,19$	$2^{5-3} = 4$
$\{2,4\}$	$(1,0,1,0,*) \equiv 20,21$	$2^{5-4} = 2$
		$|C_u| = \sum 2^{5-j_2^0} = 20$

and C_u, respectively, using just the first columns, so that we do not need to obtain these elements by the second columns.

So, on one hand, regarding the bitstrings intrinsically comparable with u, from Tables 1 and 2 we have

$$C^u = \{21, 22, 23, 25, 26, 27, 28, 29, 30, 31\}, \quad |C^u| = 10,$$

$$C_u = \{0, 1, 2, 3, 4, 5, 6, 7, 8, 9, 10, 11, 12, 13, 16, 17, 18, 19, 20, 21\}, |C_u| = 20$$

and then, using Eq. 13, the ranking interval of $u{=}21$ is

$$[\,|C_u|\,,\,2^n + 1 - |C^u|\,] = [\,20\,,\,23\,].$$

On the other hand, regarding the bitstrings intrinsically incomparable with u, using Eq. 5, we have

$$Inc(u) = \{0, 1\}^5 - (C^u \cup C_u) = \{14, 15, 24\},$$

so that $Inc(u)$ is a 3-set, in accordance with Eq. 14, i.e.,

$$|Inc(u)| = l([\,|C_u|\,,\,2^n + 1 - |C^u|\,]) = l([\,20\,,\,23\,]) = 3.$$

5   Conclusions

For any fixed binary n-tuple u, two dual algorithms for rapidly generating the set C^u (C_u, respectively) of binary n-tuples that are always (i.e., intrinsically) less (more, respectively) probable than u have been presented. These algorithms, exclusively based on the positions of the 1-bits (0-bits, respectively) of u, also allow one to count the elements of C^u (C_u, respectively) with no need to previously obtain them. From the cardinalities of the sets C^u and C_u, we have exactly obtained the ranking interval for the binary n-tuple u, i.e., the interval of all possible positions (ranks) of u in the list of all the 2^n binary n-tuples arranged by decreasing order of their occurrence probabilities. By the way, the length of the ranking interval for u, gives us the exact number of binary n-tuples v intrinsically incomparable with u. Applications can be found in many different scientific or engineering areas modeled by a CSBS.

Acknowledgements This work was partially supported by Ministerio de Ciencia e Innovación (Spain) and FEDER through Grant contract: CGL2008-06003-C03-01/CLI.

References

1. L. González, A new method for ordering binary states probabilities in reliability and risk analysis. Lect. Notes Comp. Sci. 2329, **137–146**, (2002)
2. L. González, N-tuples of 0s and 1s: necessary and sufficient conditions for intrinsic order. Lect. Notes Comp. Sci. **2667**, 937–946 (2003)
3. L. González, A picture for complex stochastic boolean systems: the intrinsic order graph. Lect. Notes Comput. Sci. **3993**, 305–312(2006)
4. L. González, Algorithm comparing binary string probabilities in complex stochastic Boolean systems using intrinsic order graph. Adv. Complex Syst. **10**(Suppl. 1), 111–143 (2007)
5. L. González, D. García, B. Galván. An intrinsic order criterion to evaluate large, complex fault trees. IEEE Trans. Reliab. **53**, 297–305 (2004)
6. W.G. Schneeweiss, *Boolean Functions with Engineering Applications and Computer Programs* (Springer, New York, 1989)
7. R.P. Stanley, *Enumerative Combinatorics*, vol. 1 (Cambridge University Press, Cambridge, 1997)

Chapter 32
Predicting Memory Phases

Michael Zwick

Abstract The recent years, phase classification has frequently been discussed as a method to guide scheduling, compiler optimizations and program simulations. In this chapter, I introduce a new classification method called *Setvectors*. I show that the new method outperforms classification accuracy of state-of-the-art methods by approximately 6–25%, while it has about the same computational complexity as the fastest known methods. Additionally, I introduce a new method called *PoEC* (Percentage of Equal Clustering) to objectively compare phase classification techniques.

1 Introduction

It is well known that the execution of computer programs shows cyclic, reoccuring behavior over time. In one time interval, a program may get stuck waiting for I/O, in another time interval, it may likely stall on branch misprediction or wait for the ALU (Arithmetic Logic Unit) to complete its calculations. These intervals of specific behavior are called *phases*. *Phase classification* is the method that analyzes programs and groups program intervals of similar behavior to equal *phases* [5, 7, 8, 12].

To identify phases, programs are split into intervals of a fixed amount of instructions. Then, these intervals are analyzed by some method to predict their behavior. Intervals showing similar behavior are grouped together to form a specific class of a *phase*. Therefore, a phase is a set of intervals that show similar behavior while discarding temporal adjacency. The phase information of a program can be used to guide scheduling, compiler optimizations, program simulations, etc.

Several phase classification techniques have been proposed the recent years, many of which rely on code metrics such as *Basic Block Vectors* [12] and *Dynamic*

M. Zwick
Lehrstuhl für Datenverarbeitung, Technische Universität München, Arcisstr. 21, 80333 München, Germany
e-mail: zwick@tum.de

S.-I. Ao et al. (eds.), *Machine Learning and Systems Engineering*,
Lecture Notes in Electrical Engineering 68,
DOI 10.1007/978-90-481-9419-3_32, © Springer Science+Business Media B.V. 2010

Branch Counts [3]. Since code related methods only have low correlation with memory hierarchy behavior, several memory related phase classification methods have been proposed to predict L1 and L2 processor cache misses, such as *wavelet based phase classification* [5], *Activity Vectors* [11] and *Stack Reuse Distances* [4].

In this chapter, I introduce a new phase classification technique called *Setvector* method to predict L2 cache performance. On the basis of ten SPEC2006 benchmarks, I show that the mean accuracy of the *Setvector* method outperforms the *Activity Vector* method by about 6%, the *Stack Reuse Distance* method by about 18% and the *wavelet* based method by about 25%. Further, I introduce a new metric called *PoEC* (Percentage of Equal Clustering) to objectively evaluate different phase classification methods and make them comparable to one another.

The remainder of this chapter is organized as follows: Section 2 describes state-of-the-art phase classification techniques, Section 3 presents the *Setvector* method, Section 4 introduces *PoEC* as a methodology to evaluate phase classification accuracy, Section 5 presents the results and Section 6 concludes this chapter.

2 Phase Classification Techniques

In this section, I describe state-of-the-art techniques I compare the *Setvector* method to. As I focus on phase classification for L2 cache performance prediction, I exclusively consider memory based techniques.

All the methods I apply were evaluated using the same tracefiles comprised of memory references that were gathered using the *Pin tool* described in [9], as it has been done in [5]. For cross validation, I used the MCCCSim (Multi Core Cache Contention Simulator) simulator described in [14] to obtain the hitrates for each interval. As I initially started the evaluation of the *Setvector* method in comparison to the *wavelet* technique presented in [5], I chose the same interval size of 1 million instructions as they did to make the results better comparable. All methods were evaluated on ten SPEC2006 test benchmarks, each comprised of 512 million instructions.

2.1 Wavelet Based Phase Classification

Huffmire and Sherwood [5] use Haar wavelets [13] to perform phase classification.

First, they create *16 × 16 matrices* for each interval of 1 million instructions. Therefore, they split each set of 10^6 instructions into 20 subsets of $10^6/20 = 50,000$ instructions forming 20 column vectors. They determine the elements of every such column vector by calculating $m = ((address\%M) \cdot (400/M))$ for each address in the corresponding subset, where '%' is the modulo operator and $M = 16k$ the modulo size that has been matched to the L1 cache size. By iterating over each address of the $50k$ instructions, Huffmire and Sherwood fill up the rows of the column vectors by summing up the occurences of each m ($0 \leq m < 400$) in a histogram manner. After having calculated each of the 20 400 × 1 column vectors, they scale the resulting 400 × 20 matrix to a 16 × 16 matrix using methods from image scaling.

In a second step, they calculate the *Haar wavelet transform* [13] for each 16×16 matrix and weight the coefficients according to [6].

In a third step, Huffmire and Sherwood apply the k-means clustering algorithm [10] on the scaled wavelet coefficients and compare the clusters with the hitrates of the corresponding intervals, gained by a cache simulator.

In my implementation of the wavelet technique, I followed Huffmire's and Sherwood's description except for the following: I split each 10^6 instructions in 16 intervals of $64k$ instructions each to omit the scaling from 20 to 16 columns. Additionally, I utilized the MCCCSim Simulator [14] that is based on Huffmire's and Sherwood's cache simulator anyway. I implemented everything else as it as been presented by [5].

Since the top-left element of the coefficient matrix corresponds to the mean value of the original matrix, I also clustered all top-left elements of the coefficient matrices and compared the results of the clustering process to the L2 hitrates of the corresponding intervals.

As the wavelet transform seems not an obvious choice of algorithm for this problem, yet it achieved good results shown by Huffmire and Sherwood, I decided to replace the wavelet transformation by a SVD (Singular Value Decomposition) $M = U\Sigma V^T$ in another experiment to probe if a more general method could find more information in the data matrix. I clustered both columns and rows of U and V respectively but could not find any impressive results, as I will show in Section 5.

2.2 Activity Vectors

Settle et al. [11] propose to use *Activity Vectors* for enhanced SMT (simultaneous multi-threaded) job scheduling. The *Activity Vectors* are used like a phase classification technique to classify program behavior with respect to memory hierarchy performance. The *Activity Vector* method has been proposed as an *online* classification method that relies on a set of on-chip event counters that count memory accesses to so-called *Super Sets*. Note that Settle et al. include both access count and cache miss information in their Activity Vectors. In my evaluation however, I exclusively incorporate access count information.

I implemented the Activity Vector method in software and applied the same tracefile data that I applied to all the other simulations mentioned in this chapter. To use the Activity Vector method as a phase classification method, I clustered both the *vectors* and the *length of each vector* and compared the clustering results with the L2 cache performance of the corresponding intervals.

In Section 5, I show that the Activity Vector method on average achieves better results than the wavelet based method.

2.3 Stack Reuse Distances

Beyls and D'Hollander [1] propose to use the so-called *Stack Reuse Distance* as a metric for cache behavior. They define the Stack Reuse Distance of a memory access as the number of unique memory addresses that have been referenced since the last access to the requested data. In Section 5, I show that on average, the classification performance of the Stack Reuse Distance method lies between the wavelet and the Activity Vector method, whereas its calculation takes a huge amount of time.

2.4 Other Techniques

Although many other techniques for phase classification have been proposed such as *Basic Block Vectors* [12], Local/Global Stride [8], Working Set Signatures [2] etc., I omitted to compare the *Setvector* technique to those methods as it has been shown that they are outperformed by other methods, for example the *wavelet* based method [5].

3 Setvector Based Phase Classification

In this section, I describe the *Setvector* based phase classification method. *Setvectors* are as easily derived as they are effective: For all addresses of an interval and an n-way set-associative cache, determine the number of addresses with *different key* that are mapped to the same cache set. That means: Given a L2 cache with 32 bit address length that uses b bits to code the byte selection, s bits to code the selection of the cache set and $k = 32 - s - b$ bits to code the key that has to be compared to the tags stored in the tag RAM, do the following:

- Extract the set number from the address, e.g. by shifting the address k bits to the left and then unsigned-shifting the result $k+b$ bits to the right.
- Extract the key from the address, e.g. by unsigned-shifting the address $s+b$ bits to the right.
- In the list for the given set, determine whether the given key is already present.
- If the key is already present, do nothing and process the next address.
- If the key is not in the list yet, add the key and increase the counter that corresponds to that set.

More formally written: Starting with a tracefile

$$\mathbf{T} = \{a_i | 1 \le i \le t\} \tag{1}$$

that contains t memory addresses a_i, $1 \le i \le t$ that are grouped into intervals of a fixed amount of instructions, split the tracefile into a set of *access vectors* $\mathbf{a}_i$, each representing an interval of several a_i:

$$\mathbf{T} = [\mathbf{a}_1 \cdots \mathbf{a}_c] \tag{2}$$

Now, for each $\mathbf{a}_i$ and caches with a row size of r byte and a way size of w byte, derive the set vector

$$\mathbf{s}_i = [s_1 \cdots s_S]^T \tag{3}$$

by

$$S_i \leftarrow \left\{ \frac{a}{r} \Big| a \in \mathbf{a}_i, \frac{(a\%w) - ((a\%w)\%r)}{r} = i, \frac{a}{r} \notin S_i \right\} \tag{4}$$

$$s_i = min(2^n - 1, \sum_{a/r \in S_i} 1) \tag{5}$$

with '%' representing the modulo operator and n the maximum number of bits allowed to code the set vector elements, if there should be such a size constraint. This way, each element of a set vector contains for each corresponding interval the number of addresses that belong to the same cache set, but have a different cache key – saturated by n, the number of bits at most to be spent for each vector element. Due to this composition, the Setvectors directly represent cache set saturation, a measure that is highly correlated with cache misses.

In Section 5, I show that on average, the *Setvector* method outperforms all methods mentioned in Section 2.

4 Metrics to Compare Phase Classification Techniques

Lau et al. [8] define the *Coefficient of Variation* (CoV) as a metric to measure the effectiveness of phase classification techniques. CoV measures the standard deviation as percentage of the average and can be calculated by

$$CoV = \sum_{i=1}^{phases} \frac{\frac{\sigma_i}{average_i} \cdot intervals_i}{total\ intervals}. \tag{6}$$

Huffmire and Sherwood adapt this metric by omitting the division by the average, resulting in the *weighted standard deviation*

$$\sigma_{weighted} = \sum_{i=1}^{phases} \frac{\sigma_i \cdot intervals_i}{total\ intervals}. \tag{7}$$

Being derived from standard deviation, both CoV and $\sigma_{weighted}$ denote better clustering performance by smaller values. However, the CoV metric (as well as $\sigma_{weighted}$) may describe the standard deviation of the L2 cache performance in each phase, but not the correlation between L2 cache performance and the different phases, what should be the key evaluation metric for phase classification. Therefore, I developed the *PoEC* (Percentage of Equal Clustering) metric that can be calculated as follows:

Consider the cluster vector γ as a vector that holds, for each index, the phase of the corresponding 16×16 scaled matrix. In a first step, sort the elements of the cluster vector γ according to its phases, such that $\forall_{i \in 1..indeces} : \gamma_i \leq \gamma_{i+1}$. Then, in a second step, calculate the percentage of equal clustering by

$$PoEC = 2 \cdot \left[min \left(\frac{\sum_{i=1}^{indeces} (\gamma_{h,i} == \gamma_{x,i})}{indeces}, 0.5 \right) - 0.5 \right] \tag{8}$$

This way, high correlation between L2 cache performance and the cluster vectors result in PoEC values near 1 and low correlation corresponds to values near 0, with $0 \leq PoEC \leq 1$.

Figures 1 (CoV) and 2 (PoEC) show the difference between those metrics by clustering some SPEC2006 benchmarks in two phases ("good ones" and "bad ones") using the wavelet method and plotting the phases (each ring corresponds to one interval) against the L2 hitrate of the corresponding intervals. As L2 cache hitrates of the same magnitude should be classified into the same class, a good clustering is achieved if one class contains higher hitrates and the other contains lower hitrates, as it is the case for the "milc", "soplex", "lbm" and "bzip2" benchmarks.

Figure 1 shows the clustering of the SPEC2006 benchmark programs sorted by their *CoV* value (calculated according to formula 6). While analyzing Figure 1, one can observe the following: There are benchmarks that achieve good clustering, such as "soplex", "milc", "bizp2" and "lbm". And there are benchmarks that do not cluster well at all, such as "hmmer", "libquantum", "gobmk". But the point is: The clustering quality *does not* fit the *CoV* value the plots are arranged by. Although not plotted in this place, the $\sigma_{weighted}$ metric shows similar behavior.

In Figure 2, the programs are sorted according to their *PoEC* value (calculated according to formula 8). Although the clustering is the same, this time the clustering quality *does fit* the PoEC value the plots are arranged by.

As the *PoEC* metric is obviously more applicable to evaluate clustering performance, I decided to omit both $\sigma_{weighted}$ and *CoV* and to perform the evaluation of the phase classification techniques using the *PoEC* metric.

5   Results

In this section, I present the results gathered from my simulations.

5.1   Classification Accuracy

Figure 3 shows the *PoEC* values for each of the mentioned methods for ten SPEC2006 benchmarks. For the *Activity Vectors*, I clustered the *vector* as well as the *magnitude of the vector* |Activity Vector|. For the *Haar wavelet* method, I

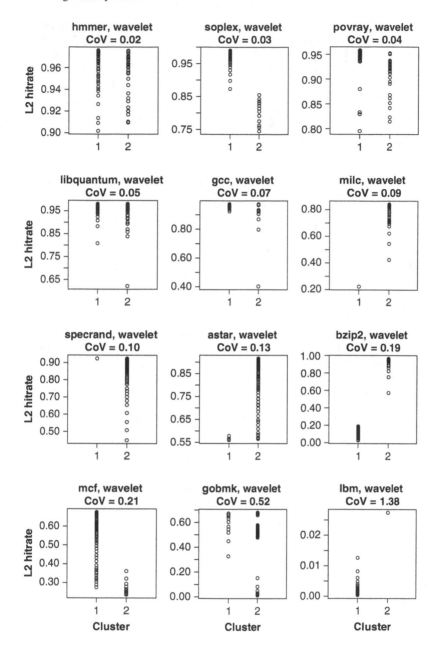

Fig. 1 CoV metric

clustered both the scaled matrix and the left-top matrix element (Haar wavelet[0] [0]). For the *Setvectors*, I clustered the magnitude of the Setvectors; for the *Stack Reuse Distances*, I clustered the Stack Distance Vector. The results shown for the

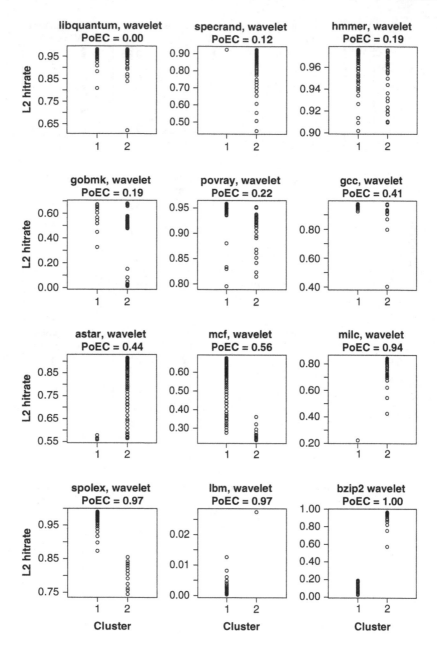

Fig. 2 PoEC metric

SVD originate from the column space of **U**, that has been calculated from the scaled matrices described above. Column/Rowspace of **V** didn't achieve any better results. *PoEC* values near 1 indicate good classification performance, *PoEC* values near

0 indicate poor classification performance. The benchmark *mcf* for example shows superior performance for the *Activity Vector* method, the *wavelet* method and the *Setvector* approach and poor classification performance for the *Stack Distance* and *SVD* method.

Figure 4 depicts the *mean* classification performance of each method averaged over the ten SPEC2006 benchmarks. The *Setvector* approach outperforms all other methods and achieves about 6% better classification performance than the next best method, the *Activity Vectors* method, while the *Stack Reuse Distance* method shows about 18% inferior performance than the Setvector method. While the *wavelet* based method still shows fair results (about 75% of the performance of the Setvector method), the *Singular Value Decomposition* of the scaled matrix has apparently not been a good idea at all.

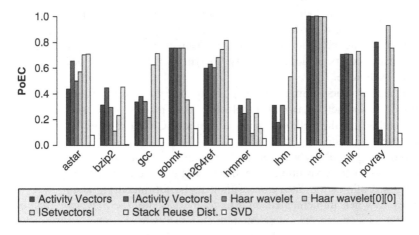

Fig. 3 Accuracy of phase classification techniques measured by the PoEC metric

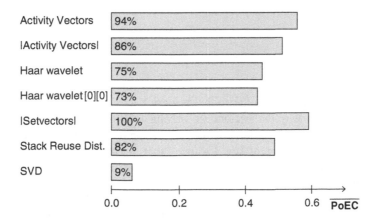

Fig. 4 Mean PoEC classification performance

5.2   Computational Performance

The computational performance of the mentioned methods cannot easily be compared in this chapter, because the methods have been implemented in different
programming languages. The *Setvector* and *Activity Vector* methods have completely been implemented in slow Ruby 1.8.6, the *wavelet* and *SVD* methods have
been implemented in a mixture of Ruby and R and the *Stack Reuse Distances*
method has completely been implemented in C for performance reasons.

The calculation of the scaled matrices used by the *wavelet* method has been
timed for 14 μs per simulated instruction. The calculation of wavelet coefficients in
Ruby was about 15 ns, the calculation of the k-means algorithm in R was about
2.5 ns, including data exchange over the Ruby-R-bridge in both cases.

The *Setvector* method suffered – as the *wavelet* method did – from the slow
Ruby interpreter and took about 3.97 μs per instruction. The magnitude calculation
and the clustering algorithm for the Setvectors finished in less than 0.5 ns. The
better clustering performance originates from using scalars (vector magnitues)
rather than vectors of dimensionality $16 \cdot 16 = 256$, as it has been done in the
wavelet technique.

The by far worst computational performance was demonstrated by the *Stack
Reuse Distance* method, that also took about 3.97 μs of time per simulated instruction, but this time the method was implemented in C-code to finish the simulation in
just some days instead of months.

Comparing the Ruby wavelet implementation (about 14.0175 μs per simulated
instruction) to the execution time measued by [5] (about 10 ns per simulated
instruction) allows the *Setvector* method to be estimated for about 5–10 ns per
instruction when implemented on a C code basis; i.e. the *Setvector* method has
about the same magnitude of calculation complexity as the other mentioned methods, except the *Stack Reuse Distance method*. All values have been averaged over
several executions or files.

6   Conclusion

Within this chapter, I introduced a new method for phase classification called
Setvectors. The method is similar to the *Activity Vector* method proposed in [11],
but it differs in the way the vectors are obtained. While Settle et al. just count
accesses to *Super Sets*, the *Setvector* method calculates the number of accesses to
a set that reference a different key. I showed that the proposed *Setvector* method
outperforms state-of-the-art methods with respect to classification accuracy by
approximately 6–25%, having about the same computational complexity. As a
second contribution, I introduced the *PoEC* metric that can be used to objectively
evaluate phase classification methods in a more intuitive way than the known
metrics *CoV* and $\sigma_{weighted}$. Although I proved the better performance of the *PoEC*

method compared to the $CoV/\sigma_{weighted}$ method just qualitatively "by inspection", it obviously is a more reasonable approach.

References

1. K. Beyls, E.H. D'Hollander, Reuse distance as a metric for cache behavior, in *Proceedings of the IASTED Conference on Parallel and Distributed Computing and Systems*, 2001
2. A.S. Dhodapkar, J.E. Smith, Managing multi-configuration hardware via dynamic working set analysis, in *International Symposium on Computer Architecture (ISCA'02)*, 2002
3. E. Duesterwald, C. Cascaval, S. Dwarkadas, Characterizing and predicting program behavior and its variability, in *12th International Conference on Parallel Architectures and Compilation Techniques (PACT'03)*, 2003
4. C. Fang, S. Carr, S. Önder, Z. Wang, Reuse-distance-based miss-rate prediction on a per instruction basis, in *Proceedings of the 2004 Workshop on Memory System Performance*, 2004
5. T. Huffmire, T. Sherwood, Wavelet-based phase classification, in *Parallel Architectures and Compilation Techniques (PACT'06)*, 2006
6. C. Jacobs, A. Finkelstein, D. Salesin, Fast multiresolution image querying, in *Proceedings of the 22nd Annual Conference on Computer Graphics and Interactive Techniques*, 1995
7. J. Lau, J. Sampson, E. Perelman, G. Hamerly, B. Calder, The strong correlation between code signatures and performance. *International Symposium on Performance Analysis of Systems and Software*, 2005
8. J. Lau, S. Schoenmackers, B. Calder, Structures for phase classification. *International Symposium on Performance Analysis of Systems and Software*, 2004
9. C.-K. Luk, R. Cohn, R. Muth, H. Patil, A. Klausner, G. Lowney, S. Wallace, V.J. Reddi, K. Hazelwood, Pin: building customized program analysis tools with dynamic instrumentation. *Programming Language Design and Implementation*, 2005
10. J. MacQueen, Some methods for classification and analysis of multivariate obervations, in *Proceedings of the Fifth Berkeley Symposium on Mathematical Statistics and Probability*, vol. 1, pp. 281–297, 1967
11. Settle, A., Kihm, J.L., Janiszewski, A., and Connors, D. A. (2004). Architectural Support for Enhanced SMT Job Scheduling, in *Proceedings of the 13th International Conference of Parallel Architectures and Compilation Techniques*
12. T. Sherwood, E. Perelman, G. Hamerly, B. Calder, Automatically characterizing large scale program behavior, in *ASPLOS-X: Proceedings of the 10th International Conference on Architectural Support for Programming Languages and Operating Systems*, 2002
13. E. Stollnitz, T. DeRose, D. Salesin, Wavelets for computer graphics: a primer. *IEEE Computer Graphics and Applications*, 1995
14. M. Zwick, M. Durkovic, F. Obermeier, W. Bamberger, K. Diepold, MCCCSim – a highly configurable multi core cache contention simulator. *Technical Report – Technische Universität München*, https://mediatum2.ub.tum.de/doc/802638/802638.pdf, 2009

Chapter 33
Information Security Enhancement to Public–Key Cryptosystem Through Magic Squares

Gopinath Ganapathy and K. Mani

Abstract The efficiency of a cryptographic algorithm is based on its time taken for encryption and decryption and the way it produces different cipher text from a plain text. The RSA, the widely used public key algorithm and other public key algorithms may not guarantee that the cipher text is fully secured. As an alternative approach to handling ASCII value of the text in the cryptosystem, a magic square of order 16 equivalent to ASCII set generated from two different approaches is thought of in this work. It attempts to enhance the efficiency by providing add-on security to the cryptosystem. Here, encryption/decryption is based on numerals generated by magic square rather than ASCII values. This proposed work provides another layer of security to any public key algorithms. Since, this model is acting as a wrapper to a public-key algorithm, it ensures that the security is enhanced. Further, this approach is experimented in a simulated environment with 2, 4, and 8 processor model using Maui scheduler which is based on back filling philosophy.

1 Introduction

Today security is an important thing that we need to send data from one location to another safely. As there is strong security, there is a great hacker and spire at the other side. Therefore, many models and systems are looking for optimization of the electronic connected-world. Cryptography accepts the challenges and plays the vital role of the modern secure communication world. It is the study of mathematical techniques related to aspects of information security such as confidentiality,

G. Ganapathy (✉)
Department of Computer Science, University of Bharthidasan, 620 024, India
e-mail: gganapathy@gmail.com

S.-I. Ao et al. (eds.), *Machine Learning and Systems Engineering*,
Lecture Notes in Electrical Engineering 68,
DOI 10.1007/978-90-481-9419-3_33, © Springer Science+Business Media B.V. 2010

data integrity, entity authentication, and data origin authentication. Cryptographic algorithms are divided into public-key and secret-key algorithms. In public-key algorithms both public and private keys are used, with the private key computed from the public key. Secret-key algorithms rely on secure distribution and management of the session key, which is used for encrypting and decrypting all messages. Though, public-key encryption is slower than symmetric-key encryption, it is used for bulk-data encryption. Hence, in practice cryptosystems are a mixture of both [1, 2].

There are two basic approaches used to speed up the cryptographic transformations. The first approach is to design faster (symmetric or asymmetric) cryptographic algorithms. This approach is not available most of the time. The speed of cryptographic algorithm is typically determined by the number of rounds (in private-key) or by the size of messages (in public-key case). The second approach is the parallel cryptographic system. The main idea is to take a large message block, divide it into blocks of equal sizes and each block can be assigned to one processor [3]. To perform the operations in parallel, effective scheduling is very important so that it can reduce the waiting time of message for processing. The back filling is one such approach.

The security of many cryptographic systems depends upon the generation of unpredictable components such as the key stream in the one-time pad, the secret key in the DES algorithms, the prime p, and q in the RSA encryption etc. In all these cases, the quantities generated must be sufficient in size and the random in the sense that the probability of any particular value being selected must be sufficiently small. However, RSA is not semantically secure or secure against chosen cipher text attacks even if all parameters are chosen in such a way that it is infeasible to compute the secret key d from the public key (n, e), choosing p, q are very large etc. Even if the above said parameters are taken carefully, none of the computational problems are fully secured enough [4]. Because to encrypt the plaintext characters, their ASCII values are taken and if a character occurs in several places in a plaintext there is a possibility of same the cipher text is produced. To overcome the problem, this paper attempts to develop a method *with different doubly even magic squares of order 16* by two different approaches and each magic square is considered as one ASCII table. Thus, instead of taking ASCII values for the characters to encrypt, preferably different numerals representing the position of ASCII values are taken from magic square and encryption is performed using RSA cryptosystem.

2 Methodology

The working methodology of the proposed add-on security model is discussed stepwise.

- Construct different doubly even magic square of order 16 as far as possible and each magic square corresponds to one ASCII set.

- To encrypt the character, use the ASCII value of the character to determine the *numeral in the magic square* by considering the position in it. Let N_P and N_C denote the numeral of the plaintext and cipher text respectively. Based on N_P and N_C values, all plaintext and cipher text characters are encrypted and decrypted respectively using RSA algorithm.
- To speed up the operations, perform them in parallel in a simulated environment. For that, use Maui scheduler with back filling philosophy.

The methodology of Add-on Security Model is shown in Fig. 1

2.1 Magic Squares and Their Construction

Definition 1: A magic square of order n is an arrangement of integers in an n * n magic square such that the sums of all the elements in every row, column and along the two main diagonals are equal.

A normal magic square contains the integers from 1 to n^2. It exists for all orders $n \geq 1$ except $n = 2$, although the case $n = 1$ is trivial as it consists of a single cell containing the number. The constant sum in every row, column and diagonal is

a

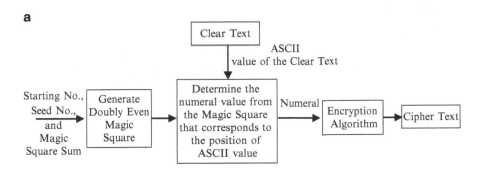

b

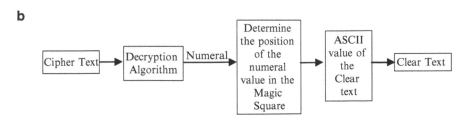

Fig. 1 Add-on wrapper Model (**a**) encryption process (**b**) decryption process

called the magic constant or magic sum, M [5]. The magic constant of a normal magic square depends only on n and has the value

$$M(n) = \frac{n(n^2 + 1)}{2} \tag{1}$$

For normal magic squares of order $n = 3, 4, 5, 6, 7, 8, \ldots$ the magic constants are $15, 34, 65, 111, 175, 265, \ldots$ respectively. Though there is only one magic square of order 3 apart from trivial notations and reflections, the number of squares per order quickly sky rockets. There are 880 distinct order 4 squares, and 275, 305, 234 distinct order 5 magic squares [6]. Magic squares can be classified into three types: odd, doubly even (n divisible by four) and singly even [7] (n even but not divisible by four). This paper focuses only on doubly even magic square implementation and their usefulness for public-key cryptosystem.

While constructing the doubly even magic squares of order 16 based on different views and fundamental properties of magic square the following *notations* are used.

MS: Magic Square

n: Order of MS where n = 4m, where

m = 1, 2, 3 and 4

MSn: MS of order n

MSB4: Base MS of order 4

MS$_{start}$: Starting number of MS

MSTn$_{sum}$: Total sum of MS of order n

MSDn$_{sum}$: Diagonal sum of MS of order n

T_No_MS: Total Number of MS

S_1 S_2 S_3 S_4: First, Second, Third and Fourth digit of 4 digits seed number. Each digit has a value from 0 to 7. If $S_1 = S_2 = S_3 = S_4 = 0$, i.e. 0000 then it is MSB4. The values in the MSB4 are filled as shown in Table 1. Call it as MS4_fill_order (Min$_{start}$, Max$_{start}$)

where the numerals with positive sign represent value to be added with MS$_{start}$ and the numerals with negative sign represent value to be subtracted from MST2$_{sum}$.

MS4_min_start_value: Minimum starting value of MS

MS4_fill_order (MS$_{start}$, MST4$_{sum}$): function used to fill the values in MS the order is shown in Table 1.

MSn_base i: ith nth order base MS

MS4_sub i: ith 4th order MS obtained from MSB4 by using suitable transformation based on S_i, where i= 1, 2, 3, and 4

Table 1 Magic square filling order

−4	MS$_{start}$	−8	+12
−10	+14	−6	+2
+8	−12	+4	MST2$_{sum}$
+6	−2	+10	−14

MS4_prop_sub i: *i*th *4th order property based MS obtained from MSB4 by using the fundamental property of MS4magic square based on Sp$_i$, where p$_i$= 1, 2, 3, and 4*
    Prop: properties of fundamentalMS4 with four character values
    ||: concatenation

2.2 Construction of Doubly Even Magic Square based on different Views of fundamental Magic Square

In this work, to generate the doubly even magic square, any seed number, starting number, and magic sum may be used and the numbers generated will not be in consecutive order. The algorithm 2.2.1 proposed by Gopinath Ganapathy, and Mani [8] starts with building 4 × 4 magic square. Incrementally 8 × 8 and 16 × 16 magic squares are built using 4 × 4 magic squares as building blocks.

2.2.1 Algorithm

Input: 4 digit Seed number, Starting Number and
        Magic Square Sum
Output: Doubly Even Magic Square of order 16.
1. **Read** *MS$_{start}$; S$_i$, i = 1, 2, 3, 4 ;MST16$_{sum}$ and T_No_MS*
2. MST4$_{sum}$← ⌈ *MST16$_{sum}$/ 4*⌉*; 3. MST8$_{sum}$←* ⌈*MST16$_{sum}$/ 2*⌉
4. Count ← 1
5. **While** *Count ≤ T_No_MS do // to generate MS16*
    5. a. prev← 1;
    5. b. **for** *j ←* 1 to 4 *// to generate four MS8*
        5.b.1 **for** *i ←* 1 to 4 *// to generate four MS4*
            5.b.1.1 **if** *((j=2)&&(i = 3)) || ((j = 3) &&(i =2))*
                MS4_min_start_value
                    ← MS4$_{start}$*+ 16 *(iprev-1) +1*
                else
                MS4_min_start_value
                    ← MS$_{start}$*+ 16* (prev-1)*
                end if
            5.b.1.2 MS4_max_start_value
                ← MST4$_{sum}$*– 16* (prev- 1) - MS$_{start}$*
            5.b.1.3 MS4_base i ← call MS4_fill_order
            (MS4_min_start_value, MS4_max_ start_value)
            5.b.1.4 **Case** *S$_i$ in 0,1,2,3,4,5,6,7*
                0: MS4_sub i ← rotate MS4_base i by 0°
                1: MS4_sub i ← rotate MS4 _base i by 90°
                2: MS4_sub i ← rotate MS4_base i by 180°

3: *MS4_sub i ← rotate MS4_base i by 270°*

4: *MS4_sub i ← rotate MS4_base i by 180° about a vertical axis*

5: *MS4_sub i ← rotate MS4_base i by 90° through anticlock wise*

6: *MS4_sub i ← rotate MS4_base i by 180° about horizontal axis*

7: *MS4_sub i ← rotate MS4_base i by 180° about the main diagonal*

end case S_i

5.b.1.5 prev← prev+1

end for i

5.b.2 MS8_sub j ←MS4_sub 1|| MS4_sub 2 || MS4_sub 3 || MS4_sub 4

5.b.3 **case** *j in 2, 3, 4*

    2: S_1← S_2; S_2← S_3; S_3← S_4; S_4← S_1

    3: S_1← S_3; S_2← S_4; S_3← S_1; S_4← S_2

    4: S_1← S_4; S_2← S_1; S_3← S_2; S_4← S_3

end case j

5.b.4 S_i← S_1|| S_2|| S_3|| S_4, i ← 1, 2, 3, 4

    end for j

5.c MS16_count ← MS8_sub1|| MS8_sub2 ||MS8_sub3||MS8_sub4

5.d S_i←(S_i+ rnd(10)) mod 8, i = 1,2,3,4

5.e count ← count + 1

end while count

2.2.2   Example

Based on the Algorithm 2.2.1 the first four MS4_sub$_i$, i = 1, 2, 3 and 4 are generated and they are shown in Tables 2 – 5.

Other magic squares MS8_sub2, MS8_sub3 and MS4_sub4 are generated in this manner and they are concatenated so that they form MS16 which is shown in Table 6.

Similarly other MS16 magic squares are generated by using suitable transformations of the seed number. In this paper, only four MS16 magic squares are generated.

2.3   Construction of Doubly Even Magic Square of Order 16 based on the Properties of 4 × 4 Magic Square

In this paper, Section 2.1focuses on construction of a 16 × 16 magic square from sixteen 4 × 4 different views of fundamental magic squares. To obtain the different views of magic squares, each one can be rotated or reflected so that it can generate eight different views of a 4 × 4 fundamental magic square. Thus, any

Table 2 MS4_sub1 with (4,1539)

1,535	4	1,531	16
1,529	18	1,533	6
12	1,527	8	1,539
10	1,537	14	1,525

Table 3 MS4_sub2 with (20, 1,523)

(a) MS4_sub2 with (20, 1,523)			
1,519	20	1,515	32
1,513	34	1,517	22
28	1,511	24	1,523
26	1,521	30	1,509
(b) 90° Rotation of Table 3a			
26	28	1,513	1,519
1,521	1,511	34	20
30	24	1,517	1,515
1,509	1,523	22	32

Table 4 MS4_sub3 with (36, 1,507)

1,503	36	1,499	48
1,497	50	1,501	38
44	1,495	40	1,507
42	1,505	46	1,493

Table 5 MS4_sub4 with (52, 1,491)

1,487	52	1,483	64
1,481	66	1,485	54
60	1,479	56	1,491
58	1,489	62	1,477

square is one of a set of eight that are equivalent under rotation or reflection. In order to generate different views of a magic square, the general format shown in Table 7 is used.

2.3.1 Sub-types of 4 × 4 Magic Square

• Normal

A magic square is said to be normal in which every row, column, and the main and secondary diagonals sum to the magic sum. That is $B + C + N + O = E + H + I + L = F + G + J + K = A + D + M + P = MST4_{sum}$

Table 6 First MS16 using starting no. 4, seed no. 0100, and magic sum 12,345

1535	4	1531	16	26	28	1513	1519	74	76	1465	1471	1455	85	1451	96
1529	18	1533	6	1521	1511	34	20	1473	1463	82	68	1449	98	1454	86
12	1527	8	1539	30	24	1517	1515	78	72	1469	1467	93	1447	88	1459
10	1537	14	1525	1509	1523	22	32	1461	1475	70	80	90	1457	94	1446
1503	36	1499	48	1487	52	1483	64	1439	101	1435	112	1423	116	1419	128
1497	50	1501	38	1481	66	1485	54	1433	114	1438	102	1417	130	1421	118
44	1495	40	1507	60	1479	56	1491	109	1431	104	1443	124	1415	120	1427
42	1505	46	1493	58	1489	62	1477	106	1441	110	1430	122	1425	126	1413
1407	132	1403	144	1391	149	1387	160	1343	196	1339	208	1327	212	1323	224
1401	146	1405	134	1385	162	1390	150	1337	210	1341	198	1321	226	1325	214
140	1399	136	1411	157	1383	152	1395	204	1335	200	1347	220	1319	216	1331
138	1409	142	1397	154	1393	158	1382	202	1345	206	1333	218	1329	222	1317
1375	165	1371	176	186	188	1353	1359	1311	228	1307	240	250	252	1289	1295
1369	178	1374	166	1361	1351	194	180	1305	242	1309	230	1297	1287	258	244
173	1367	168	1379	190	184	1357	1355	236	1303	232	1315	254	248	1293	1291
170	1377	174	1366	1349	1363	182	192	234	1313	238	1301	1285	1299	246	256

Table 7 General format of a fundamental 4 × 4 magic square

A	B	C	D
E	F	G	H
I	J	K	L
M	N	O	P

- Pan-Magic

It has the additional property that the broken diagonals also sum to the magic sum. That is $B + G + L + M = D + E + J + O = C + F + I + P = A + H + K + N = C + H + I + N = B + E + L + O = MST4_{sum}$. The magic squares depicted shown in Tables 2 through 5 are normal and Pan Magic.

- Associative

A magic square is an associative, it has the additional property that the eight symmetrically opposite the centre of the square shown in Table 2 are AP, BO, CN, DM, EL, FK, GJ and HI having the magic sum magic constant $MST4_{sum}$.

- Quadrant Associative

It has the property that all diagonally opposite pairs of numbers within the four main quadrants sum to $MST4_{sum}$. That is, the diagonally opposite pairs of numbers within the four quadrants AF, BE, CH, DG, IN, JM, KP and LO are having the magic sum $MST4_{sum}$.

- Distributive

It has the property that each of the four integers in the set of numbers $(1, 2, 3, 4)$, and $(5, 6, 7, 8)$ and $(9, 10, 11, 12)$, and $(13, 14, 15, 16)$ are located in a row and column where none of the other three numbers in the set are located [9, 10].

2.3.2 Construction of Doubly Even Magic Square of Order 16 Based on the Properties

In this work, to generate the doubly even magic square based on the properties the algorithm 2.3.3 is used similar to algorithm 2.2.1 with a slight modification. Apart from any starting number, and magic sum, the fundamental magic square seed number 0000, and the four character value for the Prop variable are given as input. The algorithm starts with by constructing 4 × 4 fundamental magic square and there are sixteen 4 × 4 fundamental magic squares are generated. In each magic square, the properties of fundamental magic square are applied based on the cyclic left shift operation in the value of Prop variable. The algorithm is shown in 2.3.3.

2.3.3   Algorithm

Input: 4 digit seed number, starting Number, magic square property, and magic squares sum

   Output: Doubly Even Magic Square of order 16.

To generate 4×4 fundamental magic squares, the algorithm uses the steps from 1 to 5.b.1.3 illustrated in Algorithm 2.2.1. To generate further, the following steps are used.

>   *5.b.1.4 Case* Sp_i *in P. A, Q, D*
>   *'P': MS4_prop_sub i ← Pan-magic of MS4_base i*
> •   *'A': MS4_prop_sub i ← Associativec of MS4_base i*
>   *'Q': MS4_prop_sub i ← Quadrant Associative of*
>   *MS4_base i*
>   *'D': MS4_prop_sub i ← Distributive of MS4_base i*
>   **end Case**
>   *5.b.1.5 prev←* prev+1
> **end for i**
> *5.b.2 MS8_prop_sub j ←MS4_prop_sub 1|| MS4_prop_sub 2*
>   *|| MS4_prop_sub 3 || MS4_prop_sub 4*
> *5.b.3 case j in 2, 3, 4*
>   *2: Sp_1← Sp_2; Sp_2← Sp_3; Sp_3← Sp_4; Sp_4← Sp_1*
>   *3: Sp_1← Sp_3; Sp_2← Sp_4; Sp_3← Sp_1; Sp_4← Sp_2*
>   *4: Sp_1← Sp_4; Sp_2← Sp_1; Sp_3← Sp_2; Sp_4← Sp_3*
>   **end case j**
> *5.b.4 Sp_i← Sp_1|| Sp_2|| Sp_3|| Sp_4, i ← 1, 2, 3, 4*
>   **end for j**
> *5.c MSP16_count ←*
>   *MS8_prop_sub1||MS8_prop_sub2||MS8_prop_sub3||MS8_prop_sub4*
> *5.d S_i←(S_i+ rnd(10)) mod 8, i = 1, 2, 3, 4*
> *5.e count ← count + 1*
> **end while** *count*

2.3.4   Magic Square Construction Based on Properties-Example

To generate a 16×16 doubly even magic square equivalent to ASCII set based on the properties of 4×4 magic square, already generated the fundamental magic squares MS4_sub1, MS4_sub2, MS4_sub3 and MS4_sub4 are considered. Let, the value for seed number, starting number, magic sum and Prop are given 0000, 4, 12345 and 'PAQD' respectively as input. The Tables 2 – 5 are considered and the 4×4 magic squares based on the properties are generated as follows.

Since the first value of 'Prop' variable is P i.e. pan-magic, apply this property to the MS4_sub1. The resulting MS4_prop_sub1 is shown in Table 8. Based on the second value 'A' i.e associative, the MS4_prop_sub2 is generated from MS4_sub2 and it is shown in Table 9.

Since the third character is 'Q' i.e. quadrant associative, the MS4_prop_sub3 is generated from MS4_sub3 and it is shown in Table 10.

Now, the fourth letter is 'D' i.e. distributive the MS4_prop_sub4 is generated from MS4_sub4 and it is shown in Table 11. Thus, to form MS8_prop_sub1 concatenate all four MS4_prop_sub$_i$, I = 1, 2, 3 and 4 (shown in Tables 8 through 11) and the first quadrant 8×8 magic square is shown in Table 12.

The second quadrant MS8_prop_sub2 magic square is generated by transforming the Seed No. as $Sp_1 \leftarrow Sp_2$; $Sp_2 \leftarrow Sp_3$; $Sp_3 \leftarrow Sp_4$; $Sp_4 \leftarrow Sp_1$ and the 'Prop' value is changed to 'AQDP', the second quadrant 8×8 magic square is generated from the next four fundamental 4×4 magic squares shown in Table 13.

Table 8 MS4_prop_sub1 with (4, 1,539)

1,535	4	1,531	16
1,529	18	1,533	6
12	1,527	8	1,539
10	1,537	14	1,525

Table 9 MS4_prop_sub2 with (20, 1,523)

1,523	1,503	22	32
428	26	1,513	1,519
24	30	1,517	1,515
1,511	1,521	34	20

Table 10 MS4_sub3 with (36, 1,507)

1,507	42	1,493	44
1,501	36	1,499	50
38	1,503	48	1,497
40	1,505	40	1,495

Table 11 MS4_sub4 with (52, 1,491)

52	1,489	62	1,483
66	1,479	56	1,485
1,481	60	1,491	54
1,487	58	1,477	64

Table 12 MS8_prop_sub1 with (4, 1,539)

MS4_prop_sub1	MS4_prop_sub2
MS4_prop_sub3	MS4_prop_sub4

The third quadrant MS8_prop_sub3 is obtained by transforming the seed number as $Sp_1 \leftarrow Sp_3$; $Sp_2 \leftarrow Sp_4$; $Sp_3 \leftarrow Sp_1$; $Sp_4 \leftarrow Sp_2$ and the 'Prop' value is changed to 'QDPA'. By applying the properties, the resultant magic square is shown is Table 14.

Similarly, the fourth quadrant MS8_prop_sub4 is generated in this manner by shifting 'QDPA' one position towards left, the resultant magic square is shown in Table 15.

In order to obtain first MS16_prop magic square, concatenate all four MS8_prop_sub$_i$(i = 1, 2, 3, and 4) magic squares and it is shown Table 16.

Similarly, the other MS16_prop magic squares are generated by using suitable transformations of the seed number.

Table 13 MS8_prop_sub2 with (68, 1,475)

1471	68	1467	80	1456	84	1451	96
1465	82	1469	70	1449	98	1454	86
76	1463	72	1475	92	1448	88	1459
74	1473	78	1461	90	1457	94	1446
1439	101	1435	112	1423	116	1419	128
1433	114	1438	102	1417	130	1421	118
109	1431	104	1443	124	1415	120	1427
106	1441	110	1430	122	1425	126	1413

Table 14 MS8_prop_sub3 with (132, 1,411)

1411	138	1397	140	149	1393	158	1387
1405	132	1403	146	162	1383	152	1390
134	1407	144	1401	1385	157	1395	150
136	1409	142	1399	1391	154	1382	160
1375	165	1371	176	1363	1349	182	192
1369	178	1374	166	188	186	1353	1359
173	1367	168	1379	184	190	1357	1355
170	1377	174	1366	1351	1361	194	180

Table 15 MS8_prop_sub4 with (196, 1,347)

196	1345	206	1339	1327	212	1323	224
210	1335	200	1341	1321	226	1325	214
1337	204	1347	198	220	1319	216	1331
1343	202	1333	208	218	1329	222	1317
1315	1301	230	240	1291	1285	256	254
236	234	1305	1311	258	252	1289	1287
232	238	1309	1307	244	250	1295	1297
1303	1313	242	228	1293	1299	246	248

Table 16 First MS16_prop magic square using starting no. 4, seed no. 0100, 'PAQD', and magic sum 12345

MS8_prop_sub1	MS8_prop_sub2
MS8_prop_sub3	MS8_prop_sub4

3   Encryption/Decryption of Plain Text Using RSA Cryptosystem with Magic Square

To show the relevance of this work to the security of public-key encryption schemes, a public-key cryptosystem RSA is taken. RSA was proposed by Rivest et al. To encrypt a plaintext M the user computes $C = M^e$ mod n and decryption is done by calculating $M = C^d$ mod n. In order to thwart currently known attacks, the modulus n and thus M and C should have a length of 512–1,024 bits in this paper.

3.1   Wrapper Implementation-Example

In order to get a proper understanding of the subject matter of this paper, let p = 11, q = 17 and e = 7, then n = 11(17) = 187, (p − 1)(q − 1) = 10(16) = 160. Now d = 23. To encrypt, $C = M^7$mod 187 and to decrypt $M = C^{23}$mod 187. Suppose the message is to be encrypted is "BABA". The ASCII values for A and B are 65 and 66 respectively. To encrypt B, the numerals which occur at 66th position in first (Table 6) and third MS16 (not shown here) are taken because B occurs in first and third position in the clear text. Similarly, to encrypt A, the numerals at 65th position in second and fourth MS16 (not shown here) are taken. Thus $N_P(A) = 42$ and 48, $N_P(B) = 36$ and 44. Hence $N_C(B) = 36^7$mod 187 = 9, $N_C(A) = 42^7$mod 187 = 15, $N_C(B) = 44^7$mod 187 = 22, $N_C(A) = 48^7$mod 187 = 157. Thus, for same A and B which occur more than once, different cipher texts are produced. Similar encryption and decryption operations can be performed by considering the numerals occur at the positions of 66 and 65 for the characters B and A respectively from MS16_prop magic square.

4   Parallel Cryptography

To speed up cryptographic transformation, the parallel cryptography is used. Effective scheduling is important to improve the performance of crypto system in parallel. The scheduling algorithms are divided into two classes: time sharing and space sharing. Backfilling is the space sharing optimization technique. Maui is a job scheduler specifically designed to optimize system utilization in policy driver, heterogeneous high performance cluster (HPC) environment. The philosophy behind it is essentially schedule jobs in priority order, and then backfill in the holes. Maui has a two-phase scheduling algorithm. In the first phase, the high priority jobs are scheduled using advanced reservation. A backfill algorithm is used in the second phase to schedule low-priority jobs between previously selected jobs [11]. In our study the Maui Scheduler with back filling scheduling technique is used to calculate the encryption/decryption time of given message in simulated environment.

Table 17 Encryption and decryption time using RSA on a pentium processor based on properties of magic squares

File Size (MB)	No. of processors											
	1			2			4			8		
	E (ms)	D (ms)	T (ms)	E (ms)	D (ms)	T (ms)	E (ms)	D (ms)	T (ms)	E (ms)	D (ms)	T (ms)
1	362	405	767	560	607	1,167	502	515	1,017	450	447	897
2	700	707	1,407	737	756	1,493	590	599	1,189	514	530	1,044
3	1,332	1,370	2,702	1,073	1,097	2,170	753	672	1,425	610	605	1,215
4	2,601	2,618	5,219	1,700	1,741	3,441	1,084	1,100	2,184	765	776	1,541

E - encryption time, D - decryption time, ms - milliseconds

5   Experimental Result

The methodology proposed is implemented in visual C++ version 6.0. The time taken for encryption and decryption of various file sized message in simulated parallel environment using RSA public-key crypto system based on the properties of magic squares is shown in Table 17. The simulation scenario configured in our implementation consisting of 2, 4, and 8 processors.

From Table 17, we observe that as the file size is increased in double, encryption and decryption time is also increased in double for single processor. Moreover, the time taken for encryption and decryption is almost same. Parallel encryption and decryption have more effect if the file size is increased.

6   Conclusion

An alternative approach to existing ASCII based cryptosystem a number based on magic squares generated by two different schemes is thought of and implemented. It is noted that that the numerals generated using different views of magic square for encryption may be obtained by the eavesdropper (but it is very difficult) because of some kind of relationship exists among the numerals. But, in the case of magic squares generated using the properties, it is very difficult to obtain the numerals by him because of the numbers do not have any kind of relationship and hence it is very difficult to obtain such numerals for encryption. Thus, the second approach is better than the first one. This methodology will add-on one more layer of security, it adds numerals for the text even before feeding into public-key algorithm for encryption. This add-on security is built using magic square position equivalent to the character to be encrypted.

Further efficiency of cryptographic operation depends on performing them in parallel. Simulation using different number of processors for encryption/decryption has shown that the time taken for decryption is approximately 1.2 times larger than the corresponding time for encryption.

References

1. A. Hamalainenn, M. Tommiska, J. Skytta, *6.78 Gigabits per Second Implementation of the IDEA Cryptographic Algorithm LNCS 2438* (Springer-Verlag, Berlin, 2002), pp. 760–769
2. T. Wollinger, J. Guajardo, C. Paar, Cryptography in embedded systems: an overview, in *Proceedings of the Embedded World 2003 Exhibition and Conference*, 18–20 Feb, 2003, pp. 735–744
3. J. Piepryk, D. Pointcheval, Parallel authentication and public-key encryption, The Eigth Australasian Conference on Information Security and Privacy (Springer-Verlag, July 2003), pp. 383–401
4. A.J. Menezes, P.C. Van Oorschot, S. Vanstone, *Handbook of Applied Cryptography*, (CRC Press, Boca Ration, FL, 1997)
5. Wikipedia (Feb 2007) http://en.wikipedia.org/wiki/Magicsquare, pp. 1–3
6. A. Rogers, P. Loly, *The Inertial Properties of Magic Squares and Cube*, Nov 2004, pp. 1–3.
7. C.A. Pickover, *The Zen of Magic Squares, Circles and Stars* (Universities Press, Hyderabad, India, 2002)
8. G. Ganapathy, K. Mani, *Add-On Security Model for Public-KeyCryptosystem Based on Magic Square Implementation*, WCECS, (20–22 Oct 2009, San Francisco, CA), p. 317–321.
9. G.A. Foster, St. Joseph's Healthcare Hamilton, Hamilton, et al. (Global Forum 2009) *SAS®- Magic Squares*
10. J.P. Robertson, *Theory of 4 × 4 Magic Squares* (17 Jan 2009)
11. Y. Zhang, H. Franke, J.E. Moreira, A. Sivasubramanian, *An Integrated Approach to Parallel Scheduling using Gang-Scheduling, Backfilling and Migration* (Springer-Verlag, JSSPP, UK, 2006), pp. 133–158

Chapter 34
Resource Allocation for Grid Applications: An Economy Model

G. Murugesan and C. Chellappan

Abstract With the emergence of grid environments featuring dynamic resources and varying user profiles, there is an increasing need to develop reliable tools that can effectively coordinate the requirements of an application with available computing resources. The ability to predict the behavior of complex aggregated systems under dynamically changing workloads is particularly desirable, leading to effective resource usage and optimization of networked systems. To explore these issues, some economic/market based models have been introduced in the literature, where users, external schedulers, and local schedulers negotiate to optimize their objectives. In this chapter, we have proposed a mathematical model for Grid resource allocation with aims to minimize the cost of the grid users when multiple grid users are trying to utilize the grid system at the same time.

1 Introduction

Grid computing is the sharing, selection and aggregation of resources such as supercomputers, mainframes, storage systems, data sources which are geographically dispersed and the individual resources can vary widely in capacity, capability and availability. In Grid computing the user (customer) can rent the resource residing on the grid for executing their computationally intensive applications instead of purchasing their own special expensive resources. The definition of Grid is taken from the electrical power Grid that provides pervasive access to power without caring where and how the power was produced. In the same way, the computational Grid should provide pervasive access to high and computing resources, without consideration for where and how the jobs were executed.

G. Murugesan (✉)
Department of Computer Science and Engineering, Anna University, Chennai, 600 025, Tamil Nadu, India
e-mail: murugesh02@gmail.com

S.-I. Ao et al. (eds.), *Machine Learning and Systems Engineering*,
Lecture Notes in Electrical Engineering 68,
DOI 10.1007/978-90-481-9419-3_34, © Springer Science+Business Media B.V. 2010

The major classifications of Grids are computational grid, scavenged grid and data grid. The first one is mainly concerned with computing power, the second classification focuses on how the unused resources can be utilized and the last one is concerned with data access capability.

The most challenging task in grid computing is the allocation of resources with respect to the user application requirement; i.e. mapping of jobs to various resources. This is a known NP-complete problem. For example mapping of 10 jobs on to 10 resources produce 10^{10} possible mappings. This is because every job can be mapped to any of the resources. Another complexity in the resource allocation process is the lack of availability of information about the status of resources. Much of the work was done on finding an optimal allocation of resources. Some of the work used conventional strategies that were concerned with the overall system performance but did not consider economics (prices) for allocating jobs to the resources. On the other hand, a large number of projects use economic strategies for allocating jobs. This work also follows the economic strategies for allocating jobs to the resources. In our scenario, there are three classes of entities: Users, Scheduler, and Resources. There are n users (also called as sources) $S_1, S_2, \ldots, S_n$, a Scheduler and m number of Resources (also called as Processors) $R_1, R_2, \ldots, R_m$. Each user i has W_i jobs and all the jobs are divisible jobs.

The users have jobs that need to be executed. The users send the jobs to the scheduler which in turn identifies the resources with respect to the job requirement. The resources represent the computing power of the grid and are where the users' jobs get executed. Resource providers make a profit by selling computing power. Users are the entities who can only send the job to the scheduler. Scheduler acts as a mediator between the user and the resources. Finally resources are the entities that just receive the jobs and process them.

2 Grid Economy Model

While comparing the grid system with the system around us, it is found that the Grid system has high similarity with the market. The market is a decentralized, competitive and dynamic system where the consumer and the producer have their own objectives. Based on this observation, economic models are introduced to the Grid in order to optimize the resource management and scheduling problems. The basic components of the market model are the producers, consumers and commodities, and which is analogous to the resource owners, resource users and various computing resources in the Grid environment. The decision making parameter in the market models are prices and quality of commodities. For example, a consumer usually wants to get better services (e.g., a smaller makespan) with a budget as small as possible, while a provider usually wants to get a higher resource utilization to raise its profits.

The use of economic methods in Grid scheduling involves interacting processes between resource providers and users, analogous to various market behaviors, such as bargain, bid, auction and so on. Buyya et al. discusses some economic models that can be applied to the Grid world, including the Commodity Market Model, Tender/contract-net Model, and Auction Model.

Due to the introduction of economic model in Grid computing, new research opportunities have arisen. Because the economic cost and profit are considered by Grid users and resource providers respectively, new objective functions and scheduling algorithms optimizing them are proposed. There are several algorithms called deadline and budget constrained (DBC) scheduling algorithms are presented which consider the cost and makespan of a job simultaneously. These algorithms implement different strategies. For example, guarantee the deadline and minimize the cost or guarantee the budget and minimize the completion time. The difficulties to optimize these two parameters in an algorithm lie in the fact that the units for cost and time are different, and these two goals usually have conflicts (for example, high performance resources are usually expensive). Economic models are not only useful when economic cost/profit is explicitly involved, but also to construct new scheduling methods with traditional objectives.

Economic methods for scheduling problems are very interesting because of their successes in our daily lives. Most of the models can only support relatively simple Grid scheduling problems such as independent tasks. For more complex applications, such as those consisting of dependent tasks and requiring cross-site cooperation, more sophisticated economic models might be needed. With our knowledge most of the economic models are focused on feasible solutions with respect to the deadline and budget constraints. In our work we initially focuses on independent workload model and later focusing on dependent workload model to achieve both feasible and optimal allocation with linear programming approach.

3 Resource Management Challenges

The Grid environment contains heterogeneous resources, local management systems (single system image OS, queuing systems, etc.) and policies, and applications (scientific, engineering, and commercial) with varied requirements (CPU, I/O, memory, and/or network intensive). The producer (also called resource owners) and consumers (who are the users) have different goals, objectives, strategies, and demand patterns. Most importantly, both resources and end-users are geographically distributed with different time zones. A number of approaches for resource management architectures have been proposed and the prominent ones are: centralized, decentralized, and hierarchical.

In managing the complexities present in large-scale Grid-like systems, traditional approaches are not suitable as they attempt to optimize system-wide measures of performance. Traditional approaches use centralized policies that need complete state information and a common resource management policy, or decentralized consensus based policy. Due to the complexity in constructing successful Grid environments, it is impossible to define an acceptable system-wide performance matrix and common fabric management policy. Therefore, hierarchical and decentralized approaches are suitable for Grid resource and operational management. Within these approaches, there exist different economic models for management and regulation of supply-and-demand for resources.

4 Resource Allocation Model

A generic grid computing system infrastructure considered here comprises a network of supercomputers and/or a cluster of computers connected by local area networks, as shown in Fig. 1. We consider the problem of scheduling large-volume loads (divisible loads) within a cluster system, which is a part of grid infrastructure. We envisage this cluster system as a cluster node comprising a set of computing nodes. Communication is assumed to be predominant between such cluster nodes and is assumed to be negligible within a cluster node. The underlying computing system within a cluster can be modeled as a fully connected bipartite graph comprising sources, which have computationally intensive loads to be processed (very many operations are performed on them) and computing elements called sinks, for processing loads (data). To organize and coordinate distributed resources participating in the grid computing environment, we utilize the resource model which contains more than one source and a resource, based on which the uniform and scalable and allocating computational resources are identified.

In a dynamic Grid environment, nodes may join or leave frequently, and nodes status may change dynamically. In our model, we are considering the environment as static; i.e. once the node joins the scheduling process, it has to be active throughout the processing schedule. But in divisible load theory all the resources that we propose unequal division of load from a resource with respect to the processor capacity and the availability and the divisible load is assign to a set of processors participate in the scheduling process.

We assume a star like structure with a scheduler as the central component; acts as a controller to the model. The right of the scheduler; a set of processors

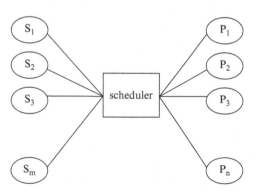

Fig. 1 Resource allocation model

(processing elements) to execute the jobs assigned by the scheduler. The other side of the scheduler are a set of sources; grid users. The scheduler collects the load from all the sources and distribute to a set of processors p_1, p_2, ..., p_n. Initially all the loads are held by the scheduler. All the communications are starts from the scheduler and there is no direct communication between the processors. For simplicity here we are assuming that the scheduler is only communicating with processors to distribute the loads and get back the result.

Here we are considering the communication delay (transmitting time) to send a portion of load to a processor and the result returning time is negligible. The processor starts processing only after receiving entire workload which is assigned to them. Also we assume that processors having independent communication hardware which allows for simultaneous communications and computations on the previously received loads. Additional constraints may be imposed on the processors to the existence of other more urgent computations or maintenance periods the availability of processor p_i may be restricted to some interval $[r_i, d_i]$. Where r_i be the release time of the processor p_i and d_i be the deadline of the processor p_i. By such a restriction we mean that computations may take place only in the interval $[r_i, d_i]$. A message with the load may arrive or start arriving before r_i. We assume that computations start immediately after the later of the two events: r_i or the load arrival. The computation time must be fit between the later of the above two events. Also the loads are received from different sources with different capacity and each load become divisible without any restriction. Each load received from different sources is divided into a set of different tasks (portion of workload). Maximum of one task may be assigned or allotted to a processor. The load portion depends upon the capacity of the processor. Suppose, if there are m number of sources $\{S_1, S_2, ..., S_m\}$ and n number of processors $\{P_1, P_2, ..., P_n\}$ and the workloads from each sources become $L = \{L_1, L_2, ..., L_m\}$. Where L_1 be the total workload received from the source S_1 and so on. Each workload L_i may be divided into T_1, T_2, ... The load L can be reordered by the scheduler to achieve good performance of the computation. The scheduler splits the workload into tasks and sends them to processors to perform the specified process. Only a set of processors may be used to perform the workload of a source.

We denote that α_{ij} be the size of the task assigned to the processor p_j. It is expressed in load units (e.g. in bytes). $\alpha_{ij} = 0$ implies that p_j is not in the set of processor selected to perform the process of ith source workload. The total workload from a source is the sum of the sizes of the workload parts. i.e. $\Sigma\alpha_{ij} = L_i$. Not only p_j is selected by the scheduler, but also the sequence of activating the processors in p_j and the division of load L_i into chunks α_{ij}. Here we consider the processors as unrelated processors; its communication links and start-up time are specific for the task. Similarly the processor computing rates depend on the processor and the task.

Let z_j be the communication rate of the link to processor p_j perceived by task T_j. Transferring the portion of workload unit α_{ij} to p_j takes $z_j\alpha_{ij}$ time units. Let t_j be the processing rate of processor p_j perceived by task T_j. To process α_{ij} portion of load by processor p_j takes $t_j\alpha_{ij}$ time units. Let c_j the processing cost to process the potion of workload T_j. The total processing cost to process a portion of workload α_{ij} by the processor p_j becomes $c_j\alpha_{ij}$ cost units.

In this work we analyze the complexity of scheduling the divisible loads L_1, $L_2, \dots, L_m$ of sizes $T_1, T_2, \dots$ on n parallel processors $P_1, P_2, \dots, P_n$ which are interconnected together. We assume that the processors have sufficient memory buffers to store the received portion of the workloads and computations. All the processor will start processing immediately after receiving their entire portion of workloads. One processor can process more than one source's portion of workload.

By constructing a schedule the scheduler decides on the sequence of the tasks, the set of processors assigned to each portion of workloads, the sequence of processor activation and the size of the load parts. Our objective is to minimize the usage of the grid user cost. We assumed that there is no separate start-up time for individual processors and there is no fixed cost to utilize the processors. All the processors are dedicated processors. But practically it is not possible, to simplify our model as well as reduce the number of variables and constraints. The following notations are used to formulate the mathematical model.

5 Design of Economy Model

To design the mathematical model we have used the following notations/variables are used.

c_j – Amount to spend to utilize jth processor
α_{ij} – Portion of workload from ith source to jth processor
x_{ij} – Binary variable
d_i – Deadline to complete the ith source job
b_i – Budget allotted for the ith source job
s_j – Schedule period of jth processor
t_j – Time required to perform operation on one unit of job by jth processor
ω_i – Total workload of ith source
z_j – Time taken to transfer a unit of workload to jth processor
s_j – Scheduled time ith source

Minimize
$$\sum_i \sum_j c_j\alpha_{ij}x_{ij}$$

Subject to

1. $\sum_i \sum_j z_j\alpha_{ij}x_{ij} + \sum_i \sum_j t_j\alpha_{ij}x_{ij} \leq d_i$
2. $\sum_i \sum_j z_j\alpha_{ij}x_{ij} + \sum_i \sum_j t_j\alpha_{ij}x_{ij} \leq s_i$
3. $\sum_i \sum_j c_j\alpha_{ij}x_{ij} \leq b_i$
4. $\sum_j \alpha_{ij} = \omega_i$

5. $\sum_i x_{ij} = 1$
6. $\alpha_{ij} \geq 0$
7. $x_{ij} = \{0, 1\}$
8. $s_j \geq 0$
9. $z_j = 0$

The objective is to minimize the total cost of the grid user those who are assigning job to the grid system. The Eq. (1) represent the cost of all jobs that are being assigned to a resource is the objective function. The Eqs. (1) through (7) specifies the constraints used the mathematical model. The constraints (1) represent the deadline associated with each sources. Constraints (2) match the workload within the availability of resources. Constraints (3) are the budget for each source's workload. Constraint (4) represents the total workload of each sources involved in the scheduling process. Constraints (5) makes sure that a portion of workload is assigned to only one resource; i.e., there is no overlap between processing of workloads. Constraint (6) makes sure that a portion of workload divided from the total workload becomes a whole number. Constraint (7) is to set the binary value either 0 or 1. The constraint (8) is a non-negativity constraints and the constraint (9) we are assuming that there is no communication delay. But practically it is not possible, for the simplicity of the model we assumed $z_j = 0$.

6 Experimental Results

Let us assume that the Grid system consists of five processors (resources) namely P_1, P_2, P_3, P_4, and P_5 with three sources S_1, S_2, and S_3 are trying to utilize the grid system to execute their workloads.

Table 1 shows the processors involved in the Grid system, the processing capacity of each processor per unit workloads, the processing cost of each processor

Table 1 Processor capacity

Processors	Processing time per unit Workload (h)	Processing cost ($)	Available time (hrs)
P1	2	4	30
P2	3	3	50
P3	4	2	50
P4	3	3	50
P5	2	5	50

Table 2 Details of workloads

Sources	Work loads (kb)	Budget ($)	Deadline (hrs)
S1	25	60	60
S2	30	100	70
S3	40	120	130

Table 3 Processor allotments

Source	Allotted processor	Workload allotted (kb)	Time taken to complete (hrs)
S1	P1	15	57
	P2	7	
	P5	3	
S2	P3	3	68
	P4	2	
	P5	25	
S3	P2	16	124
	P3	6	
	P4	16	
	P5	2	

to execute a unit workload and the available time of each processor. Table 2 shows the details of the different sources which are trying to utilize the grid system, total workload of each source and the budget allotted to each of the sources to complete their workloads and the expected time to complete the process of each workload. Using the details given in the table we have formed the mathematical model and solved the equation using LINGO package. After execution of the mathematical model, the maximum cost for processing all the three source workloads are Rs. 351.

Table 3 shows the details of workload allotment to the processors involved in the process. From the table it is clear that the total workload of S_1 is divided into three parts, the total workload of S_2 is divided into three parts and the total workload of S_3 is divided into four parts and allotted into processors. Also it shows that the completion time of each source's workloads.

7 Related Works

Till date several grid scheduling algorithms have been proposed to optimize the overall grid system performance. The study of managing resources in the Grid environment started from 1960s. The economic problem [1] results from having different ways for using the available resource, so as how to decide what the best way to use them is. Utilization of Grid must be cheaper for the Grid user than purchasing their own resources [2] and must satisfy their requirements. On the other hand, resource providers must know if it is worth to provide their resources for usage by the Grid users. Also the pricing of resources should not be per time unit or slot (e.g. cost per minutes) [3]. Because it leads to big difference in speeds of processors, so the price per unit time of a processor might be cheaper, but the user must pay large amount of money due to slow processing resources. Moreover the users have to know how many time units they need to finish their jobs. Thus the cost of Grid user must be determined based on the tasks the resource is processing.

Deadline scheduling algorithm [4] is one of the algorithms which follow the economic strategy. The aim of this algorithm is to decrease the number of jobs that do not meet their deadlines. The resources are priced according to their performance. This algorithm also has a facility of fallback mechanism; which mean that it can inform the grid user to resubmit the jobs again and the jobs which do not met the deadline of the available resources. Nimrod/G [5, 6] includes four scheduling algorithms which are cost, time, conservative time and cost-time. Cost scheduling algorithm try to decrease the amount of money paid for executing the jobs with respect to the deadline. Time scheduling algorithms attempt to minimize the time required to complete the jobs with respect to their budget allotment. The conservative time scheduling algorithm aims to execute the jobs within the stipulated budget and the deadlines. The cost-time scheduling algorithm works similar to cost scheduling algorithm except that when there are two or more resources with the same price, it employs time scheduling algorithm. It does not deal with co-allocation.

The effective workload allocation model with single source has been proposed [7] for data grid system. Market-based resource allocation for grid computing (Gomaluch and Schroeder) supports time and space shared allocations. Furthermore it supports co-allocation. It is supposed that resources have background load which change with time and that has the highest priority for execution, so they are not fully dedicated to the grid. Gridway [8] is an agent based scheduling system. It aims to minimize the execution time, total cost and the performance cost ratio of the submitted job. Market economy based resource allocation in Grids [9] is an auction based user scheduling policies for selecting resources were proposed. GridIs [10] a P2P decentralized framework for economic scheduling using tender model. The author tries to perform the process without considering the deadline and the algorithm is implemented with the help of a variable called conservative degree, its value between 0 and 1.

The time and cost trade-off has been proposed [11] with two meta-scheduling heuristics algorithms, that minimize and manage the execution cost and time of user applications. Also they have presented a cost metric to manage the trade-off between the execution cost and time. Compute power market [12] is architecture responsible for managing grid resources, and mapping jobs to suitable resources according to the utility functions used. Parallel virtual machine [13, 14] enables the computational resources to be used as if they are a single high performance machine. It supports both execution on single and multiple resources by splitting the task into subtasks. Grid scheduling by using mathematical model was proposed [15] with equal portion of load to all the processors; i.e., the entire workload received from a source is equally divided and a portion of load is assigned to a processor with the help of random numbers to divide the entire workload from a source. The work closest to ours is [6] where the authors proposed the algorithm that claims to meet the budget and deadline constraints of jobs. However, the algorithm proposed is ad-hoc and does not have an explicit objective.

8 Conclusion

In this chapter, we have proposed scheduling strategies for processing multiple divisible loads on grid systems. In real life the loads will come dynamically and the resources are also in dynamic in nature. But here we have considered both loads and the resources are in static manner. This work can be extended for dynamic nature also. We are in the process of dynamic arrival of loads and the resources. Also we have assumed that there is sufficient buffer space available in all the resources. The experimental result demonstrates the usefulness of this strategy. We need a system to find out the execution time of a task and also the cost of usage of a processor/resource. The execution time entirely depends upon the processor speed.

References

1. J. Nakai, Pricing computing resources: reading between the lines and beyond. Technical Report, National Aeronautics and Space Administration, 2002
2. L. He, X. Sun, G.V. Laszewski, A qos guided scheduling algorithm for grid scheduling, 2003
3. L. He, Designing economic-based distributed resource and task allocation mechanism for self-interested agents in computational grids, in *Proceeding of GRACEHOPPER*, 2004
4. A. Takefusa, S. Matsuoka, H. Casanova, F. Berman, A study of deadline scheduling for client-server system on the computational grid, in *HPDC'01: Proceedings of the 10th IEEE International Symposium on High Performance Distributed Computing*, pp. 406–415, 2001
5. R. Buyya, Economic-based distributed resource management and scheduling for grid computing. PhD Thesis, Monash University, Melbourne, Australia, 2002
6. R. Buyya, D. Abramson, J. Giddy, Nimrod/g: An architecture for a resource management and scheduling system in a global computational grid, *HPC ASIA'2000*, 2000
7. M. Abdullah, M. Othman, H. Ibrahim, S. Subramaniam, Load allocation model for scheduling divisible data grid applications. J. Comput. Sci. 5(10), 760–763 (2009)
8. R.A. Moreno, A.B. Alonso-Conde, Job scheduling and resource management technique in economic grid environments. *European Across Grids Conference*, pp. 25–32, 2003
9. S.R. Reddy Market economy based resource allocation in grids. Master's Thesis, Indian Institute of technology, Kharagpur, India. 2006
10. L. Xiao, Y. Zhu, L.M. Ni and Z. Xu, Gridis: An incentive-based grid scheduling, in IPDPS'05: *Proceedings of the 19th IEEE International Parallel and Distributed Processing Symposium*, Washington, DC, 2005, IEEE Computer Society
11. S.K. Garg, R. Buyya, H.J. Siegel, Scheduling parallel applications on utility grids: time and cost trade-off management, in *Proceedings of the Thirty-Second Australasian Computer Science Conference (ACSC2009)*, Wellington, Australia, *Conferences in Research and Practice in Information Technology* (CRPIT), vol. 91
12. R. Buyya, S. Vazhkudai, Computer power market: towards a market oriented grid. *CCGRID*, pp. 574–581, 2001
13. A. Beguelin, J. Dongarra, A. Geist, V. Sunderam, Visualization and debugging in a heterogeneous environment. Computer 26(6), 88–95 (1993)
14. V.P.V.M. Sunderam, A framework for parallel distributed computing. Concurrency Prac. Exp. 2(4), 315–340 (1990)
15. G. Murugesan, C. Chellappan, An optimal allocation of loads from multiple sources for Grid Scheduling, *ICETIC'2009, International Conference on Emerging Trends in Computing*, 2009
16. R. Buyya, M. Murshed, D. Abramson, A deadline and budget constrained cost-time optimization algorithm for scheduling task farming applications on global grids, *Technical Report CSSE-2002/109*, Monash University, Melbourne, Australia, 2002
17. J. Gomaluch, M. Schroeder, Market-based resource allocation for grid computing: *A model and simulation, Middleware workshops*. pp. 211–218
18. G. Murugesan, C. Chellappan, An economic allocation of resources for multiple grid applications, in *WCECS2009, Proceedings of the World Congress on Engineering and Computer Science*, vol I, San Francisco, CA, 2009

19. M. Othman, M. Abdullah, H. Ibrahim, S. Subramaniam, Adaptive divisible load model for scheduling data-intensive grid applications: computational science, LNCS, **4487**, 246–253 (2007)
20. S. Viswanathan, B. Veeravalli, T.G. Robertazzi, Resource-aware distributed scheduling strategies for large-scale computational cluster/grid systems. IEEE Trans. Parallel Distributed Syst.18: 1450–1461, 2007
21. Q. Xiao, Design and analysis of a load balancing strategy in data grids, Future Generation Comput. Syst. **16**, 132–137, 2007

Chapter 35
A Free and Didactic Implementation of the SEND Protocol for IPv6

Say Chiu and Eric Gamess

Abstract IPv6 adds many improvements to IPv4 in areas such as address space, built-in security, quality of service, routing and network auto-configuration. IPv6 nodes use the Neighbor Discovery (ND) protocol to discover other nodes on the link, to determine their link-layer addresses, to find routers, to detect duplicate address, and to maintain reachability information about the paths to active neighbors. ND is vulnerable to various attacks when it is not secured. The original specifications of ND called for the use of IPsec as a security mechanism to protect ND messages. However, its use is impractical due to the very large number of manually configured security associations needed for protecting ND. For this reason, the Secure Neighbor Discovery Protocol (SEND) was proposed. In this work, we present Easy-SEND, an open source implementation of SEND that can be used in production environment or as a didactic application for the teaching and learning of the SEND protocol. Easy-SEND is easy to install and use, and it has an event logger that can help network administrators to troubleshoot problems or students in their studies. It also includes a tool to generate and verify Cryptographically Generated Addresses (CGA) that are used with SEND.

1 Introduction

IPv6 [1–3] (Internet Protocol version 6) is a solution to the problem of the shortage of public IPv4 addresses that faces Internet. IPv6 adds many improvements to IPv4 in areas such as quality of service, routing and network auto-configuration. Even if IPv6 has been around for more than 10 years now, there is a lack of IPv6 network

S. Chiu (✉)
Universidad Central de Venezuela, Facultad de Ciencias, Escuela de Computación, Paseo Los Ilustres, Urb. Valle Abajo, Caracas 1020, Venezuela
e-mail: saychiu@gmail.com

S.-I. Ao et al. (eds.), *Machine Learning and Systems Engineering*,
Lecture Notes in Electrical Engineering 68,
DOI 10.1007/978-90-481-9419-3_35, © Springer Science+Business Media B.V. 2010

specialists in Venezuela and around the world. Therefore, the training of IPv6 specialists has become an important issue. In the undergraduate program of Computer Science at *Central University of Venezuela* (in Spanish: Universidad Central de Venezuela), some courses have been upgraded or added to the curriculum to face the problem. For example, *Advanced Network Protocols* (in Spanish: Protocolos Avanzados en Redes) is a new course that was introduced to the curriculum of the undergraduate program of Computer Science in 2005. Its objectives include the understanding of IPv6 standards, such as the ND [4] (Neighbor Discovery) protocol and the SEND [5] (Secure Neighbor Discovery) protocol.

Since ND [4] is supported by all the actual modern operating systems, a variety of laboratories are done in the course (*Advanced Network Protocols*) to strengthen student's knowledge about this important protocol. However, we have been facing the lack of support for SEND [4] (as stated is Section 5) from manufacturers and it has been almost impossible for us to do laboratories to clarify the complex procedures involved in SEND. Therefore, we decided to develop a new application (*Easy-SEND*) from scratch that implements the SEND protocol, with good support for the teaching and learning process. Its main goal is to be used as a didactic application in advanced courses related to networks at *Central University of Venezuela*. Additionally, *Easy-SEND* can be used in production networks where security is important as a replacement of the ND protocol.

The rest of this research is organized as follows: An overview of ND is presented in Section 2. Vulnerability issues for ND are discussed in Section 3. SEND is presented in Section 4. Related works are viewed in Section 5. Easy-SEND is introduced and justified in Section 6. Conclusions and future work are discussed in Section 7.

2 Neighbor Discovery Protocol Overview

The ND [4] (Neighbor Discovery) protocol solves a set of problems related to the interaction between nodes attached to the same link. It defines mechanisms to solution each of the following problems:

- Router discovery: During router discovery, a host discovers the local routers on an attached link. This process is equivalent to ICMPv4 router discovery [6].
- Prefix discovery: How hosts discover the set of network prefixes that define which destinations are on-link for an attached link.
- Parameter discovery: Through this process, nodes learn some operating parameters such as the link MTU (Maximum Transmission Unit) or the default hop-limit value to place in outgoing packets.
- Address auto-configuration: How nodes automatically configure addresses for an interface in either the presence or absence of an address configuration server, such as a DHCPv6 [7] (Dynamic Host Configuration Protocol for IPv6) server.

- Address resolution: Is a process by which nodes determine the link-layer address of an on-link destination (neighbor) given only the destination's IPv6 address. It is equivalent to ARP (Address Resolution Protocol) in IPv4.
- Next-hop determination: How nodes determine the IPv6 address of the neighbor to which a packet must be forwarded, based on the destination address. The next-hop address is either the destination address or the address of an on-link router.
- Neighbor Unreachability Detection (NUD): Is a process by which nodes determine that a neighbor is no longer reachable. For neighbors used as routers, alternate default routers can be tried. For both routers and hosts, address resolution can be performed again.
- Duplicate Address Detection (DAD): During duplicate address detection, a node determines that an address it wishes to use is not already in use by another node. This process is equivalent to using gratuitous ARP frames in IPv4.
- Redirect: How a router informs a host of a better first-hop node to reach a particular destination. It is equivalent to the use of the ICMPv4 Redirect message.

The ND protocol has five messages or PDUs (Protocol Data Units) provided by ICMPv6 [8] (Internet Control Message Protocol version 6), an updated version of ICMPv4 [9] (Internet Control Message Protocol version 4). These messages are: RS (Router Solicitation), RA (Router Advertisement), NS (Neighbor Solicitation), NA (Neighbor Advertisement), and Redirect. RS are sent by IPv6 hosts to discover neighboring routers on an attached link. RA are sent by IPv6 routers periodically (unsolicited multicast router advertisements) or in response to a RS message (solicited router advertisements). NS are sent by IPv6 nodes to resolve a neighbor's IPv6 address to its link-layer address (MAC address) or to verify if an IPv6 node is still reachable (NUD). NA are sent by IPv6 nodes in response to a NS message or to propagate a link-layer address change. Redirect messages are sent by IPv6 routers to inform hosts of a better first-hop for a destination.

ND messages consist of an ND message header, composed of an ICMPv6 header, some ND message-specific data, and zero or more ND options. Fig. 1 shows the format of an ND message.

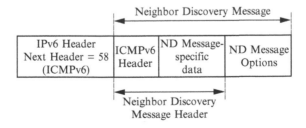

Fig. 1 Format of an ND message

Table 1 Options for Neighbor Discovery

Type	Option name	Description
1	Source link-layer address	Included in NS, RS and RA messages to indicate the link-layer address of the sender of the packet
2	Target link-layer address	It indicates the link-layer address of the target. It is used in NA and Redirect messages
3	Prefix information	Appear in RA messages to provide hosts with on-link prefixes and prefixes for address auto-configuration.
4	Redirected header	It is used in Redirect messages and contains all or part of the packet that is being redirected
5	Maximum Transmission Unit (MTU)	This option is used in RA messages to ensure that all nodes on a link use the same MTU value in those cases where the link MTU is not well known

ND message options provide additional information, such as link-layer addresses, on-link network prefixes, on-link MTU information, redirection data, mobility information, and specific routes. These options are formatted in type-length-value (TLV) format. Table 1 shows the main options of the ND protocol with a brief description.

3 Vulnerabilities of the Neighbor Discovery Protocol

ND can suffer different attacks that can be classified in three types as follow:

- Impersonation/Spoofing: class of attacks in which a malicious node successfully masquerades as another by falsifying data and thereby gaining an illegitimate advantage. This type of attacks is easily made since there is no control over MAC addresses and they can be change with little effort.
- DoS (Denial of Service): type of attacks in which a malicious node prevents communication between the node under attack and all other nodes.
- Redirection: class of attacks in which a malicious node redirects packets away from the legitimate receiver to another node on the link.

According to [10], the ND protocol suffers frequent attacks that include, but are not limited to: Neighbor Solicitation Spoofing, Neighbor Advertisement Spoofing, Neighbor Unreachability Detection Failure, Duplicate Address Detection DoS Attack, Malicious Last Hop Router, Spoofed Redirect Message, Bogus On-Link Prefix, Bogus Address Configuration Prefix, Parameter Spoofing, Replay Attacks, and Neighbor Discovery DoS Attack.

4 Secure Neighbor Discovery Protocol

As stated in Section 3, ND is insecure and susceptible to malicious interference. To counter the threats of the ND protocol, IPsec [11, 12] (Internet Protocol Security) can be used. One of the main objectives of ND is the auto-configuration of hosts with no human intervention. With IPsec, the key exchange can be done automatically when hosts already have a valid IPv6 address. Without an initial IPv6 address, the usage of IPsec can be quite tedious and impracticable. For these reasons, the IETF (Internet Engineering Task Force) developed the SEND [5] (Secure Neighbor Discovery) protocol.

SEND is a security extension of the ND protocol. It introduces new messages (CPS and CPA) and options (CGA, Timestamp, Nonce, RSA Signature, Trust Anchor and Certificate) to the ND protocol, as well as defense mechanisms against attacks on integrity and identity. The main components used in the SEND protocol are:

- Certification paths: anchored on trusted parties certify the authority of routers. A host must be configured with a trust anchor to which the router has a certification path before the host can adopt the router as its default router. CPS (Certification Path Solicitation) and CPA (Certification Path Advertisement) messages are used to discover a certification path to the trust anchor.
- Cryptographically Generated Addresses (CGA): are used to make sure that the sender of a Neighbor Discovery message is the owner of the claimed address. A public-private key pair is generated by all nodes before they can claim an address. A new ND option, the CGA option, is used to carry the public key and associated parameters.
- RSA Signature option: is used to protect all messages relating to Neighbor and Router Discovery.
- Timestamp option: Provides replay protection against unsolicited advertisements and redirects.
- Nonce option: Ensures that an advertisement is a fresh response to a solicitation sent earlier by the node.

Similarly to the ND protocol, SEND uses ICMPv6 [8] messages. Two new messages were added for the ADD (Authorization Delegation Discovery) process. Table 2 shows the value of the Type field from the ICMPv6 message and a brief description.

SEND defines two mechanisms (CGA and ADD) presented in the following sections for securing the ND protocol.

Table 2 ICMPv6 messages for SEND

ICMPv6 type	SEND message	Description
148	Certification Path Solicitation (CPS)	Sent by hosts during the ADD process to request a certification path between a router and one of the host's trust anchors
149	Certification Path Advertisement (CPA)	The CPA message is sent in reply to the CPS message and contains the router certificate

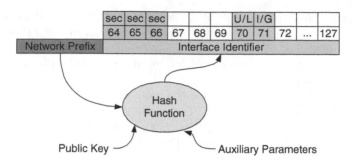

Fig. 2 Generation process of a CGA

4.1 Cryptographically Generated Address

CGAs [13] (Cryptographically Generated Addresses) are IPv6 addresses generated with a hash function, from a public key and auxiliary parameters. This provides a method for securely associating a cryptographic public key with an IPv6 address in the SEND protocol. The node generating a CGA address must first obtain a RSA (Rivest, Shamir, and Adelman) key pair (SEND uses an RSA public/private key pair). The node then computes the interface identifier part (which is the rightmost 64 bits) and appends the result to the prefix to form the CGA address (see Fig. 2).

4.2 Authorization Delegation Discovery

The ADD (Authorization Delegation Discovery) is used to certify the authority of routers via a trust anchor. A "trust anchor" is an authoritative source that can be trusted by hosts, and which is used to certify the authority of routers: "certification paths" can be established from the trust anchors to enable a certified router to perform certification of another router; before any host can adopt a router as its default router, the host must be configured with a trust anchor that can certify the router via a certification path. Hence, the certificate format mandated by the SEND protocol enables the hosts to learn a certification path of a router, and to validate the authenticity of a router.

5 Related Works

At the present time, manufacturers of network devices and developers of operating systems do not include functionalities of the SEND protocol in their products. The unique implementation of SEND had been developed by members of the working group that did its specification [5], especially James Kempf who works at DoCoMo

USA Labs. DoCoMo's Open Source SEND Project is an implementation of SEND developed by DoCoMo USA Labs that works in the operating system user-space. It can be installed in Linux (kernel 2.6) or FreeBSD 5.4.

The SEND implementation from DoCoMo USA Labs has a limited debugging mode where users can see information related to the internal states and the operations executed by the application. It also includes two tools (*cgatool* and *ipexttool*) to generate keys, CGA addresses and the extension for IPv6 addresses required in X.509 certificates [14]. The application allows two configuration modes: basic and advanced modes. In the basic mode, users can create a SEND environment for hosts only, without on-link router discovery. In the advanced mode, hosts can discover on-link routers by using certification path and their correspondent trust anchor.

The SEND application from DoCoMo USA Labs has some drawbacks. It is not able to manage address collisions during the DAD (Duplicate Address Detection) process, due to the difficulties carried by its integration to the kernel. Additionally, users must have good knowledge in the specification of firewall rules (*ip6tables*), since they are essential for the configuration of the application. Also, DoCoMo USA Labs has announced a few weeks ago that it will no longer maintain the SEND project[1]. Hence, the source code is no longer available to download in the web site and support has been canceled.

JSend[2] is announced in SourceForge as a free and open source implementation of the SEND protocol in Java. Its main objective is to be used for learning and testing purpose. For now, this project has not released any files.

6   A Didactic Implementation of the SEND Protocol

Easy-SEND is a free and open source application developed in Java that implements the SEND [5] (Secure Neighbor Discovery) protocol. It also includes a tool (*CGA-Gen*) to generate and verify CGAs [13] (Cryptographically Generated Addresses).

We developed the application in the Linux user-space that does not require modification at the kernel level. *Easy-SEND* works as a firewall between the network interface card and the IPv6 stack, by using the *ip6tables* filtering rules. The actual version is limited to the creation of a secure environment for IPv6 hosts; that is, hosts are not able to participate in the Router Discovery process.

The architecture and the functionality of *Easy-SEND* are summarized in Fig. 3. *libipq*[3] provides a mechanism for passing packets out of the stack for queuing to user-space, then receiving these packets back into the kernel with a verdict

---

[1]http://www.docomolabs-usa.com/lab_opensource.html

[2]http://sourceforge.net/projects/jsend

[3]http://www.netfilter.org/projects

specifying what to do with the packets (such as acceptation or rejection). These packets may also be modified in the user-space prior to reinjection back into the kernel.

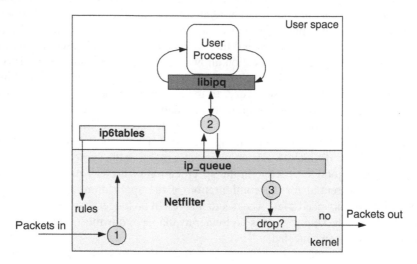

Fig. 3 Easy-SEND architecture

Initially, ND messages are selected based on *ip6tables* rules (see Fig. 4) and sent to *ip_queue*, a queue managed by *libipq*. From the first to the fifth line of Fig. 4, *ip6tables* rules capture RS (type 133), RA (type 134), NS (type 135), NA (type 136), and Redirect (type 137) messages, respectively. Library *libipq* allows the manipulation of ND packets (aggregation and verification of options) based on the SEND specifications before the determination of the verdict (acceptation or rejection of the packets). Messages marked for rejection will be dropped. Message marked for acceptation will be further processed.

```
ip6tables -A INPUT -p icmpv6 -j QUEUE --icmpv6-type "133"
ip6tables -A INPUT -p icmpv6 -j QUEUE --icmpv6-type "134"
ip6tables -A INPUT -p icmpv6 -j QUEUE --icmpv6-type "135"
ip6tables -A INPUT -p icmpv6 -j QUEUE --icmpv6-type "136"
ip6tables -A INPUT -p icmpv6 -j QUEUE --icmpv6-type "137"
```

Fig. 4 Capture rules for ND messages with *ip6tables*

Since library *libipq* was written in C, we used a wrapper, called *Virtual Services IPQ*[4] or *VServ IPQ* for short, which allows Java users to call the *libipq* native functions from Java applications.

---

[4]http://www.savarese.org/software/vserv-ipq

To develop the application, we used the waterfall methodology [15, 16]. It is a sequential methodology, in which development is seen as flowing steadily downwards (like a waterfall) through the phases of Conception, Requirement, Analysis, Design, Coding, and Testing. The application's general structure consists of a set of classes that interacts with each others. Fig. 5 shows the principal classes of *Easy-SEND*.

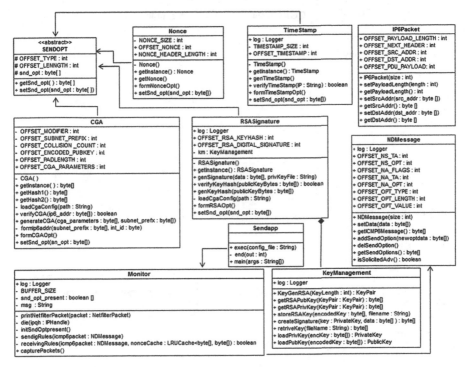

Fig. 5 Diagram of principal classes

Class *SENDOPT* is an abstract class where get and set methods are defined to access and modify SEND options. Class *CGA* allows the generation and manipulation of CGA addresses, as well as the validation of CGA options included in ND messages. Methods of the *RSASignature* class permit the creation and verification of digital signatures. *TimeStamp* and *Nonce* are classes with methods to produce and handle timestamps for the protection against replay attacks. Class *Monitor* is for the capture and manipulation of packets that arrive to the *ip6tables* queue. The rules for the reception and transmission of packets are also defined in class *Monitor*. *NDMessage* is a class for the aggregation (outgoing packets) and removal (incoming packets) of SEND options in messages. Its methods automatically update the checksum field. Class *KeyManagement* allows users to generate, load and save RSA keys.

To validate our implementation of the SEND protocol, we make several test experiments. For reasons of space, in this work we will only briefly describe two of our experiments. In the first experiment, we setup a test-bed (see Fig. 6) where we

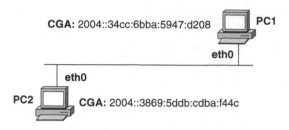

Fig. 6 Test-bed topology

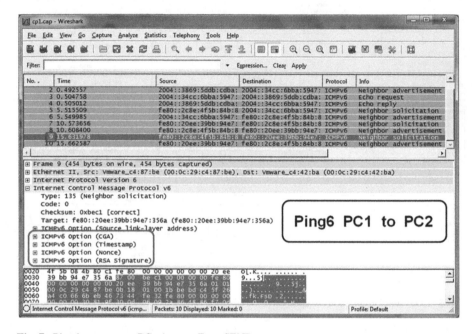

Fig. 7 Ping between two PCs that run Easy-SEND

connected two PCs in an Ethernet LAN. Each PC had 512 MB of RAM, and we installed Ubuntu 8.04 LTS (a community developed, Linux-based operating system), Java 2SE 1.6.0_07 and *Easy-SEND*.

Fig. 7 shows the ND messages sent by PC1 and PC2 during the execution of a ping command from PC1 to PC2. We can observe that *Easy-SEND* modifies the original messages (NS and NA) to add SEND options (CGA, Timestamp, Nonce, and RSA Signature). We used Wireshark [17] (a popular network analyzer formerly known as Ethereal) to analyze the SEND protocol.

In the second experiment, we made interoperability tests between Easy-SEND and the SEND implementation of DoCoMo USA labs. We made a test-bed similar to the one shown in Fig. 6, but we installed the SEND implementation of DoCoMo USA Labs in PC1 and Easy-SEND in PC2. We made different communication tests (FTP, SSH, PING, TFTP, etc.) between the two PCs and had a perfect interoperability.

Easy-SEND has a powerful event logger to record all the SEND events for troubleshooting and learning purposes. It is based on *Apache log4j*[5], a Java-based logging framework. Our event logger has three messages levels (INFO, DEBUG, and ERROR) that can be activated. INFO messages show information related to the states and processes that occur in the application. DEBUG messages allow users to get a hexadecimal dump of the messages sent or received by *Easy-SEND*. ERROR messages are shown when abnormal situations occur.

Typically, logs can be displayed in real-time in a Unix shell. It is also possible to redirect the logs to text files for later analysis. Our event logger can show many information to users helping them in the troubleshooting or learning process. Fig. 8

```
[06:53:07,577] INFO  Monitor   - NetfilterPacket Hook Local Input
[06:53:07,578] INFO  Monitor   - Incoming Interface: eth0
[06:53:07,579] INFO  Monitor   - Verifying SEND options
[06:53:07,579] INFO  Monitor   - ICMPv6 Type: <- Neighbor Solicitation
[06:53:07,579] INFO  CGA   - Verifying CGA
[06:53:07,580] INFO  CGA   - Verifying address: fe80:0:0:0:2c8e:4f5b:84b:80c1

[06:53:07,580] DEBUG CGA   - CGA Parameters:

1b be bd c4 5f 26 a4 c0 66 6b eb 46 73 44 fe 32
fe 80 00 00 00 00 00 00 00 30 81 9f 30 0d 06 09
2a 86 48 86 f7 0d 01 01 01 05 00 03 81 8d 00 30
81 89 02 81 81 00 b0 89 3b 51 88 d3 f2 3c 9b a0
98 4f 99 01 b4 5f de 81 89 b3 d6 ba 96 ff a7 05
b3 49 4c 28 ce 60 2e ba e7 f6 76 f4 c2 87 f9 51
36 be 08 12 ab c2 e5 00 50 8a b5 be fb 04 79 99
79 b4 ee 88 5b 45 b7 f1 88 a9 e2 e4 52 3a 75 9b
ea 94 87 38 56 49 0d 63 66 b6 78 1e d0 41 4a f5
25 e9 cf 69 50 9b f2 b9 68 91 a9 8c e6 8a 6f 61
8f 45 8d 40 6a fc a4 26 ae 1e a8 7d 8b 48 88 c2
20 b3 c3 e3 93 2b 02 03 01 00 01

[06:53:07,581] INFO  CGA   - Verification of CGA was successful!
[06:53:07,581] INFO  TimeStamp   - Verifying Timestamp
[06:53:07,581] INFO  TimestampCache   - 1.223292216E9 > 1.223292209E9 ?
[06:53:07,581] INFO  RSASignature   - Verifying KeyHash
[06:53:07,582] INFO  RSASignature   - Keyhash was successfully verified
[06:53:07,582] INFO  RSASignature   - Verifying digital signature
[06:53:07,585] INFO  RSASignature   - Digital signature successfully verified

[06:53:07,585] DEBUG Monitor   - Modified Packet

60 00 00 00 00 20 3a ff fe 80 00 00 00 00 00 00
2c 8e 4f 5b 08 4b 80 c1 fe 80 00 00 00 00 00 00
20 ee 39 bb 94 e7 35 6a 87 00 7b a6 00 00 00 00
fe 80 00 00 00 00 00 20 ee 39 bb 94 e7 35 6a
01 01 00 0c 29 c4 87 be

[06:53:07,585] INFO  Monitor   - Reinjected!!!
```

Fig. 8 Example of logs

---

[5]http://logging.apache.org/log4j/1.2/index.html

shows the trace generated by packet 9 (a Neighbor Solicitation message) for the capture of Fig. 7.

At the present time, *Easy-SEND* is the unique free and open source implementation of the SEND protocol that is active, since DoCoMo USA Labs is no longer supporting its implementation. The source code of *Easy-SEND* can be downloaded from SourceForge (a famous web-based source code repository) at http://sourceforge. net/projects/easy-send To facilitate the learning and teaching process, a virtual appliance is also available. A virtual appliance is a minimalist virtual machine image designed to run under some sort of virtualization technology and aimed to eliminate the installation, configuration and maintenance costs. Our virtual appliance can be run with *VMWare Player*.[6]

7   Conclusions and Future Work

Security aspects of IPv6 networks are not fundamentally different from IPv4 networks, therefore, many of the possible attacks to IPv4 protocols can also be done in IPv6 protocols, and the ND protocol is not the exception. SEND has been developed to minimize the attacks that can be done to IPv6 nodes during their auto-configuration process. Unlike IPsec, SEND is more easy and practicable to be implemented. Note that [18] proposes a new protocol called TRDP (Trusted Router Discovery Protocol) to secure the router discovery process. Compared with the ADD (Authorization Delegation Discovery) process introduced in the SEND protocol, TRDP obviates the burdensome work for a host to parse the lengthy certification path, improves efficiency of network communication between the router and hosts during the router authentication process, and also reduces the exposure to attacks on both hosts and the access router.

At *Central University of Venezuela*, SEND is an important topic in the syllabus of the *Advanced Network Protocols* course. Instructors had a hard time teaching SEND, so some laboratories were introduced to clarify the complex processes associated to SEND. Initially, the SEND implementation of DoCoMo USA Labs was used, but this implementation has been discarded since its debugging system has some weaknesses and DoCoMo USA Labs announced that it will no longer maintain the project. *Easy-SEND* was developed to face this situation.

Easy-SEND is a free and open source implementation of the SEND protocol developed under the GNU General Public License in Java. It has a powerful event logger with three levels of messages. It main goal is to be used for teaching and learning purpose, but can also be used in a production environment. Since it is open source, its source code can be enhanced or adapted by other researchers for specific needs.

---

[6]http://www.vmware.com/products/player

At the present time, *Easy-SEND* is the unique open source project that implements the SEND protocol that is active. We plan to further develop it. Our goal is to offer the application to other national and international universities to be used in their curriculum for teaching the SEND protocol. As future work, we plan to integrate to our application the process of Router Discovery with the ADD trust mechanism, and extension fields to the CGA parameter data structure to support different hash functions [19]. Also, the implementation of *Easy-SEND* in the Linux user-space can not be considered as a final solution, so we will adapt our application to the kernel to optimize the aggregation and the verification of SEND message options. Finally, we want to port our application to Microsoft Windows operating systems.

References

1. J. Davies, *Understanding IPv6*, 2nd edn. (Microsoft Press, Jan 2008)
2. S. Deering, R. Hinden, *Internet Protocol, Version 6 (IPv6) Specification*. RFC 2460. Dec 1998
3. S. Hagen, *IPv6 Essentials*, 2nd edn. (O'Reilly, Jan 2008)
4. T. Narten, E. Nordmark, W. Simpson, H. Soliman, *Neighbor Discovery for IP Version 6 (IPv6)*. RFC 4861. Sep 2007
5. J. Arkko, J. Kempf, B. Zill, P. Nikander, *Secure Neighbor Discovery (SEND)*. RFC 3971. Mar 2005
6. S. Deering, *ICMP Router Discovery Messages*. RFC 1256. Sep 1991
7. R. Droms, J. Bound, B. Volz, T. Lemon, C. Perkins, M. Carney, *Dynamic Host Configuration Protocol for IPv6 (DHCPv6)*. RFC 3315. Jul 2003
8. A. Conta, S. Deering, M. Gupta, *Internet Control Message Protocol (ICMPv6) for the Internet Protocol Version 6 (IPv6) Specification*. RFC 4443. Mar 2006
9. A. Farrel, *The Internet and Its Protocols: A Comparative Approach*, 1st edn. (Morgan Kaufmann, May 2004)
10. P. Nikander, J. Kempf, E. Nordmark, *IPv6 Neighbor Discovery (ND) Trust Models and Threats*. RFC 3756. May 2004
11. S. Kent, *IP Authentication Header*. RFC 4302. Dec 2005
12. S. Kent, *IP Encapsulating Security Payload (ESP)*. RFC 4303. Dec 2005
13. T. Aura, *Cryptographically Generated Addresses (CGA)*. RFC 3972. Mar 2005
14. R. Housley, W. Polk, W. Ford, D. Solo, *Internet X.509 Public Key Infrastructure Certificate and Certificate Revocation List (CRL) Profile*. RFC 3280. Apr 2002
15. J. Charvat, *Project Management Methodologies: Selecting, Implementing, and Supporting Methodologies and Processes for Projects* (John Wiley & Sons. Feb 2003)
16. W. Royce, Managing the development of large software systems: concepts and techniques, in *Proceedings of the 9th International Conference on Software Engineering*, Monterey, California, United States. 30 Mar–2 Apr 1987
17. A. Orebaugh, G. Ramirez, J. Beale, *Wireshark & Ethereal Network Protocol Analyzer Toolkit*. (Syngress, Feb 2007)
18. J. Zhang, J. Liu, Z. Xu, J. Li, X. Ye, TRDP: a trusted router discovery protocol, in *Proceeding of the 7th International Symposium on Communications and Information Technologies*, Sydney, Australia. 17–19 Oct 2007
19. M. Bagnulo, J. Arkko, *Cryptographically Generated Addresses (CGA) Extension Field Format*. RFC 4581. Oct 2006

Chapter 36
A Survey of Network Benchmark Tools

Karima Velásquez and Eric Gamess

Abstract Nowadays, there are a wide variety of network benchmark tools, giving researchers and network administrators many options to work with. However, this variety tends to hinder the selection process of the appropriate tool. Furthermore, sometimes users are forced to try several tools in order to find one that calculates a desired gauge, so they have to learn how to manipulate different tools and how to interpret the obtained results. This research offers a compilation of network benchmark tools currently used, with the purpose of guiding the selection of one tool over the others, by outlining their main features, strengths and weaknesses.

1 Introduction

It is common that researchers and/or network administrators have to measure different network indicators. Several tools are available for this purpose, but not all of them have the same characteristics and evaluate the same performance parameters. Sometimes, researchers are obliged to use a number of these tools and complete several experiments in order to find significant results. According to the test to be performed, and to the network or device under test, one benchmark tool can be more appropriated than another. This can complicate and delay the evaluation process, since researchers would have to learn to use different tools. For example, Gamess and Velásquez [1] had to use two different performance tools and write their own benchmarks in order to evaluate the forwarding performance on different operating systems.

K. Velásquez (✉)
Universidad Central de Venezuela, Facultad de Ciencias, Escuela de Computación, Paseo Los Ilustres, Urb. Valle Abajo, Caracas 1020, Venezuela
e-mail: karima.velasquez@ciens.ucv.ve

S.-I. Ao et al. (eds.), *Machine Learning and Systems Engineering*,
Lecture Notes in Electrical Engineering 68,
DOI 10.1007/978-90-481-9419-3_36, © Springer Science+Business Media B.V. 2010

This research presents an up-to-date survey of network benchmark tools currently used, to help researchers and network administrators to make a right decision about the more appropriate tool for their needs. Additionally, this work outlines desirable features for benchmark tools to be developed in the future.

This work is organized as follow. In Section 2, we discuss previous works related to our study. Section 3 contains a deep study and description of seven popular benchmark tools. Section 4 shows a comparative analysis of the seven tools. Conclusions and future work are discussed in Section 5.

2   Related Works

The interest of network benchmark tools has been very important in the Internet community, to the point that Parker and Schmechel [2] wrote an RFC (Request For Comment) to present some of them, when RFCs are usually used to specify protocols. Botta et al. [3] present an exhaustive state-of-the-art survey on available bandwidth estimation tools. They divided the tools in three categories: (1) end-to-end capacity estimation tools, (2) available bandwidth estimation tools, and (3) TCP throughput and bulk transfer capacity measurement tools. They also introduce a tool called *BET* (Bandwidth Estimation Tool) and they compare it with other performance tools, in terms of accuracy and total time used to complete the measurement process. Ubik and Král [4] summarize their experience with bandwidth estimation tools; they focus on finding the size and location of bottlenecks. They present a classification of end-to-end bandwidth estimation tools based on a number of properties, which include determining bandwidth (installed bandwidth) vs. throughput (available bandwidth), sender only software vs. sender and receiver software, etc. They also describe properties for a few selected tools, present the results of measurement with one tool (*pathload*) on a high-speed scenario and results of combined measurements with several tools over a real fast long-distance network. Gupta et al. [5] perform an experimental comparison study of both passive and active bandwidth estimation tools for 802.11-based wireless networks, and conclude that for wireless networks a passive technique provides greater accuracy. Strauss et al. [6] describe *Spruce*, a simple tool for measuring available bandwidth, and then compare it with other existing tools over many different Internet paths. Their study is based on accuracy, failure patterns, probe overhead, and implementation issues. Montesino [7] presents a comparative analysis of state-of-the-art active bandwidth estimation techniques and tools. He offers a short overview of a set of tests performed over different conditions and scenarios, which were done to assess the performance of active bandwidth estimation tools. Mingzhe et al. [8] present *WBest*, a wireless bandwidth estimation tool designed to accurately estimate the available bandwidth in IEEE 802.11 networks, explaining that most of the existing tools were not designed for wireless networks. They define the algorithm employed by *WBest* and present a set of experiments and analysis of the results. They also compare their tool with other tools like *pathchirp* [9] and *pathload* [10].

3 Network Benchmark Tools

For our study, we selected seven different throughput and bulk capacity measurement tools: *Netperf, D-ITG, NetStress, MGEN, LANforge, WlanTV*, and *TTCP*. We chose these tools for their popularity and also because they are active projects. Many performance tools available in Internet have no updated versions and seem to be dead. These seven tools also offer different features, so they provided us with a much wider range for our analysis.

Our goal is to present a clear description of the available tools that can help users choose one that is best suited for a specific experiment. Additionally, we gathered some desirable features that a throughput and bulk transfer capacity measurement tool should have since our goal is to develop a new network benchmark tool from scratch.

During our analysis, each necessary step, questions that arose, the efforts it took to find the answers to them, features that were helpful or confusing, and the required time were carefully recorded. Progress and experiences were discussed among the authors.

According to [11], most of the tools included in our study belong to the group of Throughput and Bulk Transfer Capacity (BTC) measurement tools, which are benchmark tools that use large TCP and/or UDP transfers to measure the available throughput in an end-to-end path. The experiences with each of the tools are described below. For tools that require a Unix environment, tests were conducted under Linux Debian 5.0 (Lenny) with kernel version 2.6.26. For Windows, tests were performed under Windows XP Professional SP3.

3.1 Netperf

Netperf[1] is an open source benchmark tool that can be used to measure various aspects of networking performance. Its primary focus is on bulk data transfer and request/response performance using either TCP or UDP and the BSD socket interface.

Netperf is designed around the basic client-server model. There are two executables (*netperf* and *netserver*). The *netserver* program can be invoked by *inetd* (the system daemon), or can be run as a standalone daemon. In the first method, users must have administrator privileges; the second method implies that users must remember to run the program explicitly. Unless users want to change the default port (option *-p*) or want to work with IPv6 (option *-6*) instead of IPv4, the server is invoked without options.

---

[1]http://www.netperf.org

3.1.1 Installation and Documentation

We installed *Netperf* using Aptitude, a Debian package management utility. In order to do so, we had to edit the */etc/apt/sources.list* file (root privileges are required), to add the non-free branch indicator to the listed mirrors, so the package could be found. Installation from source files is also possible when the binary distribution is not available. For our Debian version, the compilation of the source code was quite easy.

Documentation can be found at the official website. It includes detailed information on how to install the tool using the source files and a general description of the tool. It also describes how different tests can be done, including bulk data transfer and request/response performance experiments. Finally, it has a section with examples and some troubleshooting recommendations.

The information reported in the man pages was not accurate. In fact, most options described were inconsistent to their actual use in the tool. We suspect that the man pages have not been upgraded with the new releases of the tool and correspond to an old release. So, we recommend to follow the indications from the tool's help (invoked with the *-h* option) instead of using the man pages.

3.1.2 Testing and Usability

Netperf can be executed from a console using the command *netperf -H <hostName|IPAddress>*, specifying the remote host name or IP address. When *netperf* is executed, a control connection is established to the remote system (*netserver*) to pass test configuration parameters and results to and from the remote system. The control connection is a TCP connection using BSD sockets.

The duration of the test can be configured. Final reports include sender and receiver buffer socket size in bytes, sent message size in bytes, duration of the test in seconds, and throughput. Results are reported as a table. It is possible to change units for the output, but they have to be set at the beginning of the test.

3.1.3 Supported Platforms and Protocols

Netperf is only supported by Unix platforms. It is possible to use IPv4 or IPv6 as the network layer protocol, just by specifying the option *-4* or *-6* respectively. It allows the usage of both TCP and UDP as the transport layer protocol.

3.2 D-itg

D-ITG[2] (Distributed Internet Traffic Generator) is an open source packet level traffic generation platform that produces IPv4/IPv6 traffic for variable

---

[2]http://www.grid.unina.it/software/ITG/index.php

inter-departure time and packet size. It was conceived to be used as a distributed performance tool, able to measure one-way-delay (OWD), round-trip-time (RTT), packet loss rate, jitter and throughput [12].

D-ITG follows the client-server model. There are four basic executables that form the platform components: *ITGSend*, *ITGRecv*, *ITGLog*, and *ITGDec*. *ITGSend* acts as the client, and can generate several data flows simultaneously as specified by the input file (configuration file). *ITGRecv* acts as the server and can receive several data flows from different clients simultaneously. *ITGLog* is the log server of the platform, and receives information from *ITGSend* and *ITGRecv*. *ITGDec* is a utility that analyzes the results of the experiments. Additionally there are two more executables, *ITGPlot* and *ITGapi*. *ITGPlot* is an *Octave*[3] based tool to plot the data contained in the log files obtained with *ITGDec* (such as *delay.dat*, *bitrate.dat*, *jitter.dat* and *packetloss.dat*), that contain an average delay, bit rate, jitter and packet loss rate, respectively, which are calculated each millisecond and reported in their respective files. *ITGapi* is a C++ API that enables the remote control of traffic generation.

When *D-ITG* is executed, a control connection is established to the remote system, to pass test configuration parameters such as network protocol to use.

3.2.1   Installation and Documentation

We downloaded the *D-ITG-2.3-Familiar.zip* file for our Linux version, which contains the precompiled binaries to execute this tool. This version did not work on our system. Then we tried the *D-ITG-2.6.1d.zip* file, which contains the source files for compilation. Following the compilation instructions included in this distribution, we used the *make* command but were not able to compile the source files. We presume incompatibility issues between the tool and our Linux version. Finally, we tried on an older version of Debian, with 2.6.24 kernel; this time we could compile the source files without any problems, and transferred the resulting binary files to our original system, where they were executed correctly.

We downloaded the *D-ITG-2.6.1d-WINbinaryIPv4.zip* and *D-ITG-2.6.1d-WINbinaryIPv6.zip* files for Windows. They contain the binary files ready for execution. No installation process is required.

Documentation can be found on the official website. It includes a manual (which can also be accessed from the tool with -*h* option) along with several examples of use.

3.2.2   Testing and Usability

D-ITG is able to measure one-way-delay, round-trip-time, packet loss rate, jitter and throughput.

---

[3]http://www.octave.org

This tool can work in two modes: single mode and script mode. The single mode allows users to generate a single flow of packets from a client (*ITGSend*) to the server (*ITGRecv*). The script mode enables *ITGSend* to simultaneously generate several flows from a single client. Each flow is managed by a single thread, and an additional thread acts as master and coordinates the other threads. To generate *n* flows, the script file (input file) must contain *n* lines, each of which is used to specify the characteristics of one flow.

Execution is started via console (no GUI is offered) using different options provided by the tool, which can be consulted with the -*h* option. Some configurable parameters include the TTL (Time to Live), inter-departure time, payload size and protocol type.

3.2.3   Supported Platforms and Protocols

D-ITG is available for Unix platforms as well as Windows platforms. IPv4 and IPv6 are both supported as network layer protocols; in Unix, to select one or the other users must specify the right type of IP address (IPv4 address or IPv6 address) with the –*a* option in *ITGSend*. In Windows there are two different binary files, one for IPv4 support and another one for IPv6 support.

By default, UDP is used as the transport layer protocol, but it is possible to use TCP and even ICMP. Upper layer protocols supported include Telnet, DNS, and RTP for VoIP applications.

3.3   NetStress

NetStress[4] is a simple benchmark tool used to measure network performance for both wired and wireless networks. It implements bulk data transfer using TCP. Network performance is reported in terms of throughput. Once again, the client-server model is used during the testing process.

3.3.1   Installation and Documentation

NetStress installation is very easy. Once the setup file was downloaded, we only had to execute it and follow the installation assistant's instructions.

On the official website there is a link to the *NetStress* help file that contains a brief description of the tool and its purpose, system requirements, and descriptions on how to run the server and the client. Additionally, it includes a section

---

[4]http://www.performancewifi.net/performance-wifi/main/netstress.htm

explaining how to interpret the results. This help can also be accessed within the application, in the *Help* menu choosing the Contents option.

3.3.2 Testing and Usability

NetStress only measures the maximum network throughput. It provides a GUI where users can select different options. Once started, a mode must be selected: server or client. The server listens for network packets, whereas the client transmits network packets. The tool reports the amount of bytes sent, bytes received, and the throughput, in text mode and graphically.

There is no way to specify the duration of the experiment or the amount of data to transfer, so users must close the application to stop the experiment.

3.3.3 Supported Platforms and Protocols

NetStress was developed for Windows platforms. It employs IPv4 at the network layer and TCP at the transport layer. It does not offer support for other protocols.

3.4 MGEN

The Multi-Generator[5] (*MGEN*) is an open source benchmark developed by the PROTocol Engineering Advanced Networking (PROTEAN) research group at the Naval Research Laboratory (NRL). *MGEN* provides the ability to perform IP network performance tests and measurements using UDP/IP traffic. It supports both, unicast and multicast traffic generation. It follows the client-server model, using the same program for both ends.

Currently, two different versions (3.X and 4.X) are available, which are not interoperable. Some versions of *MGEN* (such as 3.X) have a graphical user interface. *MGEN* version 4.X must be launched currently from the command-line.

3.4.1 Installation and Documentation

Two packages are available for Linux platforms: source code and a precompiled version. We installed the source files to guarantee system compatibility. We started to uncompress the tar ball and used the command *make –f Makefile.linux*, since there are several makefiles for Unix systems and one must be chosen according to

---

[5]http://cs.itd.nrl.navy.mil/work/mgen

the target system. For Windows platforms there is no need for installation, since the ZIP file includes an executable file ready to use.

The distribution contains an *HTML* file with extensive documentation. It explains the tool's usage and the different options. It also offers several examples on how to use the tool to evaluate performance.

3.4.2 Testing and Usability

Input files (configuration files) can be used to drive the generated loading patterns over the time. A command-line option is also available.

Input files can be used to specify the traffic patterns of unicast and/or multicast UDP/IP applications. Defined data flows can follow periodic (CBR), Poisson, and burst patterns. These data flows can be modified during the experiment, since the input file allows users to change a given data flow at specific times. Some fields of the IP header can be set. When a multicast flow is defined, users can specify the TTL (Time to Live) value. For both unicast and multicast flows, the ToS (Type of Service) value can also be specified. For IPv6 flows, a *Flow Label* value can be defined. For multicast flows, users can control when to join or leave the multicast group, indicating the IP address of the multicast group to join or leave and the time to do so.

Results are shown on standard output, or they can be redirected to a logfile for later analysis. These results are only the report of the packets exchanged; no additional information is given.

To obtain statistics, users must record the results in a logfile, which can be later used as an input for the *trpr*[6] (Trace Plot Real-time) program. *trpr* analyzes the output of *MGEN* and creates an output that is suitable for plotting. It also supports a range of functionalities for specific uses of the *gnuplot*[7] graphing program. Important results, such as the throughput, delivery latency, loss rate, message reordering, and multicast join/leave latency can be calculated from the information in the output logfile. However, it is left to users to do this calculation.

3.4.3 Supported Platforms and Protocols

The updated version of the *MGEN* toolset, *MGEN 4.0*, provides support for Win32 platforms in addition to a broad base of Unix-based platforms, including MacOS X. Several enhancements are planned, including support for TCP, since MGEN currently only supports UDP. IPv4 and IPv6 can be used as the network layer protocol.

---

[6]http://pf.itd.nrl.navy.mil/protools/trpr.html

[7]http://www.gnuplot.info

3.5   *LANforge*

Candela Technologies' *LANforge*[8] consists of two tools: *LANforge-FIRE* configuration, used for traffic generation that simulates the edge of the tested network; and *LANforge-ICE* feature set that is used to simulate the core of a network.

LANforge is a proprietary tool that can be purchased at its website. Candela Technologies offers several options of purchase. To obtain this tool we subscribed and created an account which enabled us to download a trial version of the tool.

3.5.1   Installation and Documentation

To install *LANforge* on a Linux platform, the *LANforgeServer-X.Y.Z_Linux-x64. tar.gz* file is needed (where *X.Y.Z* stands for the version, in our case it was version 5.0.9). Once uncompressed, we executed the *install.bash* file. A home directory for the tool is required. By default, */home/lanforge* is used, but users can change it by executing *install.bash –d <directory>*. Once executed *install.bash*, we had to run the *lfconfig* script (with the *–cwd* options) from the tool's home directory. Both scripts are self explanatory, and give users enough feedback. For the GUI (Graphic User Interface), we had to install the *LANforgeGUI_X.Y.Z_Linux.tar.bz2* package, and execute the *lfgui_install.bash* file.

For the installation on the Windows platform, we downloaded two files: *LANforge-Server-X.Y.Z-Installer.exe*   and   *LANforge-GUI-X.Y.Z-Installer.exe*. Then, we just followed the installation assistant's instructions. The installation process is easy and fast.

A complete documentation is available at the website, including a user's guide, an installation guide, and an upgrade guide for every package. The installation guide includes a step-by-step description of the installation process; as well as a troubleshooting section. Also, tutorials are available on the website. These include explicit examples on how to use the tool for different scenarios, along with detailed screenshots to describe the configuration and testing process.

3.5.2   Testing and Usability

LANforge is a tool that allows users to simulate networks and perform tests over them. It includes a GUI that makes the testing process easier. There are several tests that can be performed with this tool, including configuration of virtual routers (only for the Linux version). Some of the results that can be obtained with this tool are: bytes received and transmitted, packets received and transmitted, bits per second (bps) received and transmitted, collisions and errors; these results are shown in text and graphically, in real time.

---

[8]http://www.candelatech.com/lanforge_v3/datasheet.html

Users can configure many aspects of the test, including the number of packets to send, interface type of the endpoint to simulate (ISDN, T1, modem, etc), and even custom payloads can be defined. *LANforge* also incorporates a VoIP Call Generator which currently supports H.323 and SIP (Session Initiated Protocol). The voice payload is transmitted with RTP (Real-time Transport Protocol) which runs over UDP. RTCP (Real-time Transport Control Protocol) is used for latency and other accounting. Jitter buffers are used to smooth out network jitter inherent in RTP traffic.

WAN links can also be simulated and tested. Various characteristics can be added to the traffic flowing though them, including maximum bandwidth, latency, jitter, dropped packets, duplicated packets, bit and byte errors, etc. Results are reported as text and graphically, using a vertical bar graph.

The *LANforge* GUI has many tabs grouping different tests, and it can be confusing trying to use it because of the many options offered. However, there is a detailed description of each tab at the GUI User Guide that can be found at the website. This guide shows several examples and screenshots that clarify the process.

3.5.3 Supported Platforms and Protocols

LANforge is available for Linux, Windows, and Solaris platforms (the Solaris platform was not included in this research). Supported protocols include, but are not limited to, raw Ethernet, MPLS, IPv4, IPv6, ICMP, OSPF, BGP, UDP, TCP, FTP, HTTP, and HTTPS.

3.6 WLAN Traffic Visualizer

WLAN Traffic Visualizer[9] (*WlanTV* for short) provides measurements of traffic load and visualization of sequence of frames in IEEE 802.11 WLANs [4]. *WlanTV* is published under the GPL terms. It is developed in Java and both source and JAR distributions are available for download at the website.

3.6.1 Installation and Documentation

WlanTV requires *Wireshark*[10] (Windows) or *TShark* (Linux), and JRE 1.6 (Java Runtime Environment) or later to build and run the program. Only standard Java packets are needed. It uses *TShark* to parse the log files and to perform live captures.

---

[9]http://sourceforge.net/projects/wlantv

[10]http://www.wireshark.org

We installed *WlanTV* from the JAR distribution and used the command *java –jar wlantv.jar*, which opens a GUI from which users can run the experiments.

Documentation is not very extensive. The distribution includes a very small *README* file with a few instructions about system prerequisites and not many instructions about installation and execution. However, an example of capture is available for download at the website.

3.6.2   Testing and Usability

This tool does not use the client-server model. Only one computer is needed to conduct the experiments. It sniffs the traffic of an IEEE 802.11 network and reports the results of that capture. For the experiments, users can employ a file from a previous capture, or they can start a new live capture. In the later option, users must stop the live capture in order to observe the results; which include the frame count, byte count, capture duration, transmission rate during the capture, detailed information for each packet (protocols, length, addresses, etc.), and bandwidth distribution. Results are shown in text mode and graphically.

3.6.3   Supported Platforms and Protocols

WlanTV is available for Linux and Windows platforms. This tool reports statistics for 802.11 protocols.

3.7   TTCP

TTCP[11] (Test TCP) is a command-line sockets-based benchmark tool for measuring TCP and UDP performance between two systems. *TTCP* is designed around the client-server model. There is only one executable, *ttcp*, and users must specify which end will be transmitting (option -*t*) and which end will be receiving (option -*r*) the data during the experiment.

3.7.1   Installation and Documentation

To install *TTCP*, *gcc*[12] is needed to compile the file *ttcp.c* and generate the executable *ttcp*, which includes both the sender and the receiver. Documentation is offered via man pages, and also by invoking the tool without options.

---

[11]ftp://ftp.eenet.ee/pub/FreeBSD/distfiles/ttcp

[12]http://gcc.gnu.org

3.7.2 Testing and Usability

Users must start the receiver and then start the transmitter, both via command-line. The transmitting side sends a specified number of packets (option –n, 2048 by default) to the receiving side. At the end of the test, the two sides display the number of bytes transmitted and the time elapsed for the packets to travel from one end to the other.

Results are shown on standard output. These results include the throughput and an average of system calls. It is possible to change the units for the output, but they have to be set at the beginning of the test.

3.7.3 Supported Platforms and Protocols

TTCP is only supported on Unix platforms. However, some ports for Windows are available (e.g. *PCATTCP*). IPv4 is used as the network layer protocol. It allows the usage of both TCP and UDP as the transport layer protocol.

4   Comparative Analysis

Table 1 presents the main characteristics of the reviewed tools. We considered some interesting features such as: date of the last release, A/P, privileges, supported platforms, protocols, reported results, and clock synchronization.

The date of the last release is important since it can indicate how much time the project has been inactive. The A/P category indicates if the tool is Active (A) or Passive (P). Active tools will affect the normal traffic in the network, by injecting their own packets; while passive tools only capture the traffic that passes through the interface. The *Privileges* row shows information about the user permission required to install or run the application. We limited the study of supported platforms to Linux and Windows. For the protocols, we focused on network and transport layer protocols. For the *Reported results* we include the delay, jitter and throughput. In the row labeled *Sync required*, we indicate if clock synchronization is required between computers that run any of the processes involved in the tests. We used '–' to denote features that are not available, not supported, nor documented by some tools.

According to Table 1, some projects are more active than others. It seems that *MGEN* has been inactive for a while, since its last version was released more than 5 years ago, which may be a reason for its lack of support of IPv6. *TTCP* seems to be dead, however *NTTCP* (New Test TCP) and *ETTCP* (Enhanced Test TCP) are recent forks of *TTCP*. Another important issue is the cost of the tools. Most of the evaluated tools are free, except *LANforge*. *LANforge* price starts at $1999 (USD); this could be an impediment for some researchers or network administrators. It is also important to consider if the tool is active or passive. Among the studied tools,

Table 1 Tool's main features

Feature	Netperf	D-ITG	NetStress	MGEN	LANforge	WlanTV	TTCP
Evaluated version	2.4.4	2.6.1d	1.0.8350	4.2b6	5.0.9	1.3.0	–
Last release	10/2007	09/2008	09/2008	10/2005	04/2009	12/2008	–
Free	Yes	Yes	Yes	Yes	No	Yes	Yes
Open source	Yes	Yes	No	Yes	No	Yes	Yes
A/P	Active	Active	Active	Active	Active	Passive	Active
Privileges	User	User	User	User	User	Root	User
Supported platforms	Linux	Linux, Windows	Windows	Linux, Windows	Linux, Windows	Linux, Windows	Linux
Network protocols	IPv4, IPv6	IPv4, IPv6, ICMPv4, ICMPv6	IPv4	IPv4, IPv6	IPv4, IPv6, ICMPv4, ICMPv6	IPv4, IPv6	IPv4
Transport protocols	TCP, UDP	TCP, UDP	TCP	UDP	TCP, UDP	TCP, UDP	TCP, UDP
Reported results	Throughput	Delay, jitter and throughput	Throughput	Delay and throughput	Delay, jitter and throughput	Throughput	Throughput
User interface	Console	Console	GUI	Console	GUI	GUI	Console
Sync required	No	No	No	No	No	–	No

only *WlanTV* does not introduce additional overhead to the network by injecting traffic; however this also implies that this tool is not able to measure the throughput and only reports results of the transmission rate for the captured packets during the test.

We also noticed that *WlanTV* users must have root privileges for execution since the tool requires the network interface card to be set in promiscuous mode. Also, we noticed that most of the studied tools already have IPv6 support. Network throughput is the most common reported result; followed by delay and jitter.

In relation to the user interface, only a few tools have a GUI (*NetStress*, *LANforge*, and *WlanTV*). To ease the testing process, some tools (*D-ITG* and *MGEN*) use an input file that contains the parameters and the streams description for the tests, which is more flexible than the traditional command line arguments.

One last issue of great importance is the clock synchronization, especially when measuring delays. Manual clock synchronization is not recommended due to of its poor accuracy. One solution is to use NTP (Network Time Protocol); this requires additional knowledge and configuration.

5 Conclusions and Future Work

In this work, we presented the results of our analysis in which we evaluated and compared several current network benchmark tools. We studied seven tools: *Netperf, D-ITG, NetStress, MGEN, LANforge, WlanTV* and *TTCP*. The evaluation was categorized in: installation and documentation, testing and usability, and supported platforms and protocols. Table 1 summarizes the results of our study. We consider that none of the seven tools studied is complete, so in general, users must install several of these tools to perform basic performance tests. *Netperf* is a simple tool and users can easily learn how to use it; however, the results of the tests are limited to throughput. *D-ITG* is a more powerful tool and reports more results, such as jitter, delay, and throughput; however, the times reported by *D-ITG* for the delay are not realistic. In our experiments, we had delays that were over an hour. We believe it was not an installation problem, since there are some test results reported on the *D-ITG* manual with similar times. Furthermore, users must face some problems during installation in some Debian versions. *NetStress* is an easy tool to work with that offers a GUI; but has poor protocol support, and tests cannot be neither parameterized nor stopped. *MGEN* allows the parameterization of tests, and also offers a complete documentation; however, it is not that simple to use, because of the many options it offers. Moreover, it does not show results directly, and users must calculate them from the information of the exchanged packets. *LANforge* provides support for many different protocols and tests; it is easy to install and offers good documentation. *WlanTV* is a passive tool that offers relevant indicators of performance for traffic captured on WiFi networks. *TTCP* is an easy tool to work with that reports the throughput of a network; but offers just a few parameters to customize the experiment (buffer size and number) as well as a poor protocol support.

For future work, we plan to design and implement our own network benchmark tool that will incorporate the strengths of the evaluated tools, but eliminates their weaknesses, following the client-server model. Our tool will be compliant with RFCs 1242 [13], 2544 [14] and 5180 [15], where a set of recommendations are suggested to develop benchmarks. One of our main interests is to offer support for IPv6 protocol. Also, it will be distributed under the GNU General Public License, allowing users to modify the source code for any particular requirement.

Among the features we plan to include in our tool is the ability to measure the packet loss rate, throughput and RTT for different protocols (e.g. TCP, UDP, ICMP, and IP). Tests also may be parameterized, allowing users to define custom traffic generation that follow different models (random variables), including CBR, burst traffic, Poisson, Exponential and Pareto. Also, users should be able to specify values for some header fields, such as the TTL and ToS in IPv4, and the *Traffic Class*, *Hop Limit* and *Flow Label* in IPv6.

As shown in our analysis, most of the tools do not have specific support for WiFi. For IEEE 802.11, only end-to-end performance evaluation is offered. Users have no way to obtain performance result (bandwidth, delay, packet loss rate) between a mobile station and an access point (or wireless router). So we also plan to adapt the tool that we will develop for use in access points. Since the shipped firmware of access points (or wireless routers) does not allow users to customize them, several free Linux-based firmwares (*DD-WRT*[13], *OpenWrt*[14], *HyperWRT*[15], *Tomato*[16]) were developed by the Internet community. We evaluated them and we decided to use *DD-WRT* mega version 2.24, for its flexibility and because we were able to successfully install it on an *ASUS WL-500W* wireless router. A cross compiler suite (also called toolchain) is a set of tools that includes a compiler capable of creating executable code for a platform other than the one on which the compiler is run. Cross compiler tools are used to generate executables for embedded system or small devices upon which it is not feasible to do the compiling process, because of the limited resources (memory, disk space, computational power, etc). From the source codes of different GNU packages (*binutils*, *gcc*, g++, *glibc* and *uClibc*), we created the cross compiler suite that generates the appropriate executable code for our *ASUS WL-500W* wireless router (MIPS ELF based wireless router). *uClibc*[17] is a C library for developing applications for embedded Linux systems. It is much smaller than the GNU C Library (*glibc*), but nearly all applications supported by *glibc* also work perfectly with *uClibc* as well. To test our cross compiler suite, we developed several small applications for our wireless router and we observed that the code

---

[13]http://www.dd-wrt.com

[14]http://www.openwrt.org

[15]http://sourceforge.net/projects/hyperwrt

[16]http://www.polarcloud.com/tomato

[17]http://www.uclibc.org

generated by the cross compiler is much smaller when *uClibc* is used rather than when *glibc* is used. This result is important due to the limited capacity of flash memory of the wireless router, so we will use *uClibc* when possible, and *glibc* when using advanced library functions not supported by *uClibc*.

Acknowledgment We would like to thank Candela Technologies for the trial version of their tool (*LANforge*), which allowed us to include it in our study.

References

1. E. Gamess, K. Velásquez, IPv4 and IPv6 Forwarding Performance Evaluation on Different Operating Systems, in XXXIV Conferencia Latinoamericana de Informática (CLEI 2008), Sept 2008
2. S. Parker, C. Schmechel, *Some Testing Tools for TCP Implementors*. RFC 2398. Aug 1998
3. A. Botta, A. Pescapé, G. Ventre, On the Performance of Bandwidth Estimation Tools, in *Proceedings of the 2005 Systems Communications (ICW'05)*, Aug 2005
4. S. Ubik, A. Král, End-to-End Bandwidth Estimation Tools. CESNET Technical Report. 25/2003. Nov 2003
5. D. Gupta, D. Wu, P. Mohapatra, C. Chuah, Experimental Comparison of Bandwidth Estimation Tools for Wireless Mesh Networks. IEEE INFOCOM mini-conference, Apr 2009
6. J. Strauss, D. Katabi, F. Kaashoek, *A Measurement Study of Available Bandwidth Estimation Tools*. IMC2003. Oct 2003
7. F. Montesino, Assessing active bandwidth estimation tools in high performance networks. TERENA networking conference, June 2004
8. L. Mingzhe, M. Claypool, R. Kinicki, WBest: a bandwidth estimation tool for IEEE 802.11 wireless networks, in *Proceedings of 33rd IEEE Conference on Local Computer Networks (LCN)*, Oct 2008
9. V. Ribeiro, R. Riedi, R. Baraniuk, J. Navratil, L. Cottrell, pathChirp: efficient available bandwidth estimation for network paths, in *Proceedings of Passive and Active Measurements (PAM) Workshop*, Apr 2003
10. M. Jain, C. Dovrolis, Pathload: a measurement tool for end-to-end available bandwidth, in *Proceedings of Passive and Active Measurements (PAM) Workshop*, Mar 2002
11. R. Prasad, M. Murray, C. Dovrolis, K. Claffy, Bandwidth estimation: metrics, measurement techniques, and tools. IEEE Netw. **17**(6), Nov-Dec 2003, 27–35 (Apr 2004)
12. A. Botta, A. Dainotti, A. Pescapé, Multi-Protocol and Multi-Platform Traffic Generation and Measurement. INFOCOM 2007 Demo Session. May 2007
13. S. Bradner, *Benchmark Terminology for Network Interconnection Devices*. RFC 1242. July 1991
14. S. Bradner, J. McQuaid, *Benchmarking Methodology for Network Interconnect Devices*. RFC 2544. Mar 1999
15. C. Popoviciu, A. Hamza, G. Van de Velde, D. Dugatkin, *IPv6 Benchmarking Methodology for Network Interconnect Devices*. RFC 5180. May 2008

Chapter 37
Hybrid Stock Investment Strategy Decision Support System

Integration of Data Mining, Artificial Intelligence and Decision Support

Chee Yong Chai and Shakirah Mohd Taib

Abstract This chapter discusses the continuous effort to explore stock price and trend prediction from finance perspective as well as from the integration of three major research areas namely data mining, artificial intelligence and decision support. These areas have been explored to design a hybrid stock price prediction model with relevant techniques into the stock price analysis and prediction activities.

1 Introduction

Throughout the centuries and in various countries, many people had tried to predict the movement of share prices and beat the market but no one can really accurately predict the movement of a particular share prices for company listed in the stock exchange. No one knows the futures; hence no one knows what will happen to a company in the future where such uncertainty caused the movement of the share price unpredictable. The advances and great movements in computing give implications to finance and economics today. New area of research known as financial engineering or computational finance is highly mathematical and cross-disciplinary field which relies on finance, mathematics and programming. Thus it enables financial analyst to analyze data efficiently. Much work can be done using mathematical model in spreadsheet program but the more demanding work needs one to write a computer program. This chapter presents the study on designing a decision model for investment strategy by utilizing financial methods and computation techniques.

The movement of the share price is unpredictable. There was a study done to estimate and predict the movement of share prices by using statistics, math and

C.Y. Chai (✉)
Department of Engineering, Development Division, Petronas Carigali Sdn. Bhd., Tower 1, Petronas Twin Towers, 50088 Kuala Lumpur, Malaysia
e-mail: chai_cheeyong@petronas.com.my

S.-I. Ao et al. (eds.), *Machine Learning and Systems Engineering*,
Lecture Notes in Electrical Engineering 68,
DOI 10.1007/978-90-481-9419-3_37, © Springer Science+Business Media B.V. 2010

forecasting methods based on the historical share price data and this named as Technical Analysis (TA). According to Rockefeller [1], TA has been around for 100 years and before the existence of computers, it was done only by professionals who had access to real time price on the stock exchange.

There are tools and software created to automatically illustrates and generate various graphs and perform various complex calculations in the TA. These tools eliminate the haste of Financial Analyst in performing TA manually by paper and pen in the past. With computer capabilities, now investors also perform TA themselves with the help of appropriate software that help them in making their trading and investment decision.

Other than assisting and making the TA process more efficient and effective, IT field also brought a totally new perspective and thinking into stock analysis through the researches and studies on the implementation of AI and DM concepts in the stock analysis. For example, there were studies conducted on the combining the Candlestick analysis (one form of the TA) with the Genetic Algorithms in stock analysis [2]. In addition, Quah [3] stated that with the significant advancement in the field of Artificial Neural Network (ANN), the generalization ability of the ANN is able to create an effective and efficient tool for stock selection.

1.1 High Risk Investment

Investors face the highest risks compared to other form of financial investments such as bond, Treasury bill and fixed deposit when they invest in the stock market. Stock price fluctuates almost every second and most people see the share price movement as unpredictable or in other term "random walk". Some investors involve in this activity mostly based on speculation where their aim is more to obtaining capital gain rather than earning dividend as their investment return. Investing based on speculation sometimes caused many individual investors had their 'hand burn' where they lose their initial investment. Kamich [4] defined that stock market is a private or public market for the trading of company stocks at an agreed price. One of the main objectives of the company to be listed in the stock market is to raise capital to fund future expansion.

If the investors able to purchase the correct share when the price at its lowest range and it can be sold at a better price later, they will be able to earn big chunk of money from the stock market. Of course, this is not easy to be done as no one able to predict the future, as well as the movement of the share price accurately.

2   Finance Theories and Analysis in Stock Price Prediction

The attempt of analyzing the stock performance using financial theories can be traced back up to late 1800s, where the oldest approach to common stock selection and price prediction, TA is created and used [5]. Edward and Magee [6], stated that throughout the years of stock market study, two distinct schools of thought has arisen each with radically different methods of obtaining answers on what share to buy and when to buy and sell the share. These two different schools of thought in stock analysis are Fundamental Analysis (FA) and TA.

Jones [5] defined FA as a method of security valuation which involves analyzing basic financial variables such as sales, profit margin, depreciation, tax rate, sources of financing, asset utilization and other factors. Edward and Magee [6] notified that FA relies heavily on statistics and people who performing FA will be looking through the auditor's reports, the profit-and-loss statement, balance sheet, dividend records, and policies of the companies whose shares is under their observation. They will also analyze business activities and daily news to estimate the company's future business condition. The investors who use FA will purchase stocks that are viewed as underpriced by the analysts who believe that in the future, the share price will increase to the level that it should be as computed in FA.

Another theory, TA is defined as the study of how securities prices behave and how to exploit that information to make money while avoiding losses [1]. TA focuses on the share price to assess and evaluate the demand and the supply for the shares based on the market price itself and do not listen to chatter about securities. Jones [5] claimed that technical analyst believes that the market itself is its own best source of data. It is believed that all the investors' reactions towards all the information regarding the security already embedded in the share price.

According to Rockefeller [1], TA works because people constantly repeat behavior under similar circumstances and TA is a forecasting method where it uses past and current behavior (price) to predict future behavior (price). Concurrent to this behavior repetition concept, Edward and Magee [6] noted that share price move in trends and they tend to continue until something happens that will change the supply and demand balance. As a conclusion, TA charts and analyzes the historical share prices to reveal the pattern, formation and investor's behavior and interpret it to predict the possible future share price trend.

3 Data Mining (DM) and Artificial Intelligence (AI)

Generally, prediction of stock is a very difficult task as it behaves like 'random walk' process and the prediction might be run off due to some unexpected news that have direct impacts on the respective company. The obvious complexity of the problem paved the way for the importance of intelligent prediction paradigms [7]. One of the most popular researches in this field is on the usage and implementation of ANN in the stock market analysis. Kamruzzaman et al. [8] said that stock analysis has been one of the most important applications of neural networks in finance where numerous researches had been done on ANN in stock market index prediction, stock performance/selection prediction, and stock risk prediction. Quah [3] further explained that ANN generalization ability is able to infer the characteristics of performing stocks from the historical pattern. ANN is used to discover the non-linear relationship between financial variables and stock price which cannot be done by the computerized rule-based expert system. In stock prediction using ANN, the input will be historical data such financial variables and the output of it will be the performance of stock.

TA is about find the pattern in the historical data to predict the future price movement, therefore researchers found DM is a suitable technique to reveal the hidden pattern in the historical data and predict the price trend. Both Y.Y. Shi and Z.K. Shi [9] has conducted a study on clustering technique (one of DM techniques) in stock prediction. In their studies, they identified that clustering technique is one of the efficient methods for stock investment analysis and they also noticed the weakness of this technique in stock prediction where constructed cluster might not completely reflecting the characteristics of the stock. In the study, the idea of rules mining of stock market using the statistical pattern recognition is proposed in order to overcome the weakness in clustering techniques. From techno-index in TA, Shi et al. [9] developed an algorithm of fuzzy kern clustering and tested this idea on Shanghai and Shenzhen Stock Market historical data since 1997. The result of the study clearly indicates that statistic rules in essential trends of the stock market and implied rules can be recognized to a certain degree through the construction of proper clustering algorithm [9].

The authors' proposed model for stock price prediction is based on Decision Support System (DSS) model by Turban et al. [10]. AI and DM are included in one of the subsystem which is Knowledge based subsystem.

4   DSS Model for Stock Investment Strategy

The authors have designed DSS model and architecture with some minor modification for stock investment strategy. This model can be used to adapt AI and DM elements into the system and utilizing TA and FA methods. The flow of the system is shown in Fig. 1.

5   Architecture of Stock Investment Strategy Decision Support System

Based on the architecture that shown in Fig. 2, the flow of the system activities can be divided into two which reflect the two main processes of the system. The two main processes of the system are analyzing the raw financial data from TA concept (TA process) and analyzing the financial data from FA concept (AI process) with both processes have the same aim of predicting stock price and trends. The first flow of activities will be started from database where it provides raw financial data to DM component in knowledge-based subsystem, the output from DM component

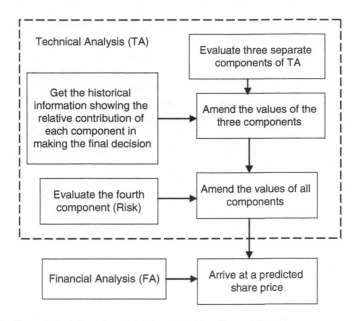

Fig. 1 The flow of Stock Investment Strategy Decision Support System

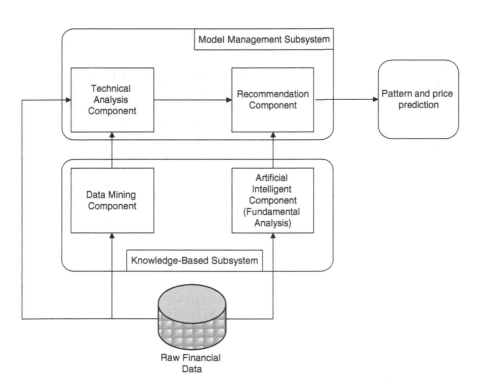

Fig. 2 The architecture of Stock Investment Strategy Decision Support System

together with raw financial data from database will be sent to TA component in model management subsystem for further processing and the output of TA component will be the input for Recommendation Component. The second flow of activities will be from raw financial data to AI component in knowledge-based subsystem. Finally the output from AI component will be sent to Recommendation Component in model management subsystem.

TA component focuses on illustrating historical stock price data into the form of graph as well as other TA indicators to interpret investors and market sentiment in the effort of predicting the stock price movement and trend. Among the most popular charting techniques that being used in TA are Line Graph, Open-High-Low-Close chart and Japanese Candlestick chart. The mechanism proposed here is the system that able to indentify and recognized pattern directly from the raw financial data using rules generated from the underlying reasoning and interpretation in various TA techniques. In achieving this, TA component of the system will use rule-based TA model that is constructed from underlying rules in various TA charting techniques.

5.1 DM Component

Rockefeller [1] highlighted that the same TA method would not work in the same stock analysis all the time. The same TA method that proven working on a particular stock does not necessary mean the same method will also work in other stocks. Therefore, instead of identifying and interpreting patterns from the raw financial based on the rules generated from one TA method only, the system will identify the patterns based on underlying rules in various TA methods. This will reduce the dependency on a single TA method in analyzing a particular stock as well as provide better reasoning and analysis capability in estimating future trend of the stock as we compare the same pattern using different rules from different techniques which give higher accuracy of the prediction. This system will use the association rules mining to determine the accuracy and effectiveness of TA methods in predicting price movement and trend for a particular stock. Therefore, the flow of activities and information in analyzing the stock in TA concept will start from database and end at recommendation component as illustrated in Fig. 3

The first phase of TA process will be the financial raw data being mined using the association rules mining. With association rules mining, all the underlying rules that reflecting the interpretation of TA in stock price and trend prediction will be mined by calculating their respective support and confidence values for a particular stock using the historical price of the stock. By using association rules mining, the system will be able to identify which association rules that actually work on that particular stock based on the historical data stored in the database based on the confidence and support. Those confidence and support value together with raw financial data will be the input for TA component.

Fig. 3 Activities and
Information Flow for TA
process

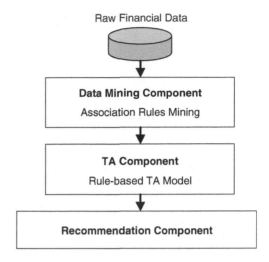

Raw Financial Data

In the rule-based TA model, the recent historical stock price movement and its related information will be analyzed by all association rules to identify pattern. If recent historical price information triggered more than one association rules, the output from the association rules will be taken into consideration based on weight that are determined based on the support and confidence level. If the triggered association rules have a confidence level that is more than 0.50 and it's the highest among other triggered association rules, that association rules will have the highest weight. On the other hand, if an association rule has the confidence level below than 0.50, it means that although the rule has been triggered in the past, the actual result is different from the expected result. The rule has wrongly predicted the result more than 50% based on historical data. This indicates that the association rule does not work well with the stock, hence this association rule will receive lower weight. Besides that, if an association rule is triggered has the support value of 0, it means the rule never being triggered before in the past and the system will assign a new confidence value of 0.5 for the associate rule because since the rule never triggered before, hence the rule will have a 50% chance that it will predict correctly. If all the triggered associate rules have confidence level below than 0.50 the system will notify the user that prediction might have higher risk to be incorrect and the system would still provide the stock trend prediction. If from all association rules, the predicted stock trend is a mixture of up and down, the system will compute the likelihood of the expected trend to be happening based on the expected trend that has the highest accumulated weight which this process will be done by the Inference Engine in the TA Component.

5.2   TA Component

There are six main sub-components included in TA component as shown in Fig. 4.

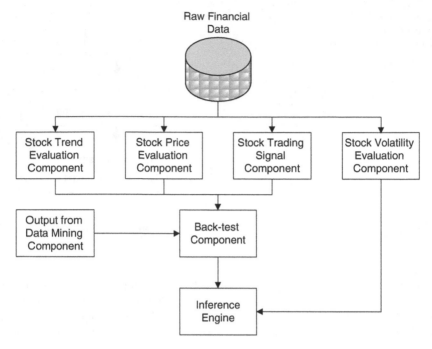

Fig. 4 Processes in TA Component

5.2.1 Stock Trend Evaluation Component

The first process would be in the Stock Trend Evaluation Component where the latest two weeks historical data is used to sun through the underlying rules derived from the Open-High-Low-Close chart and Japanese Candlestick chart. This process will detect the pattern, gauge the investor's sentiment on the stock and discover the current trend of a particular stock. This component has the most association rules compared to other components. The output for this process will be the trend of the particular stock in the near future which will be uptrend, downtrend, indecision or no trend. Example of association rules from this component illustrated in Fig. 5.

5.2.2 Stock Price Evaluation Component

This component will evaluate either the stock has been overvalued or undervalued using the Williams %R Indicator and Relative Strength Index. Williams %R indicator is developed with the aim to assess either the particular stock is being overbought or oversold. There will be two outputs from two indicators which the result might be contradicted with one to another. This contradiction issue will be sorted out in back-test component. The output for this process is the evaluation

> Rule 1:- If the "high" is higher than previous day's "high" for 3 consecutive day AND
> "low" is higher than previous day's "low" for 3 consecutive days AND
> "close" is higher than previous day's "close" for 3 consecutive days THEN
> the stock is in uptrend.

Fig. 5 Sample of association rule from Stock Trend Evaluation Component

Fig. 6 Sample of association rule from Stock Price Evaluation Component

> **Williams %R Indicator**
> Rule 1:-   If the today's Williams %R is above −20%
> THEN the stock is overvalued.
>
> **Relative Strength Index**
> Rule 1:-   If the today's RSI is above 70%
> THEN the stock is overvalued.

on the stock either it's overvalued, undervalued or reasonable. A sample of association rules used in this component is shown in Fig. 6

5.2.3   Stock Trading Signal Component

Another component is used to identify any buying or selling signal generated based on TA for that particular stock. There will be six TA methods being used to trace the trading signals which are Weighted Multiple moving Average, Momentum, Engulfing, Doji, Long Shadow and Gap. Unlike 'Stock Price Evaluation Component', association rules in 'Stock Trading Signal Component' are not being triggered all the time. There is high tendency that none of the association rules in this component is being triggered most of the time because trend reversal itself does not happen all the time and this component focuses on tracing trend reversal that leads to the generation of trading signal. The output of this component will be trading signal and forecasted stock trend. Trading signal is not generated everyday for a particular stock and the same goes to the forecasted stock trend. Some of the association rules from various indicators used in this component is illustrated in Fig. 7.

5.2.4   Stock Volatility Component

The forth component in the TA component is used to evaluate the risks and the noise level of a particular stock. Volatility is a measure of price variation which is directly linked to risk level and noise level. Stock with high volatility means trading is riskier but has more profit potential, while low volatility means less immediate risk. Therefore, this component serves to evaluate the stock's volatility level and provides a warning if the volatility level is high because high volatility will cause the system's forecast and prediction less accurate due to the high noise level in the stock movement.

Weighted Multiple Moving Average

Rule 1:- If 5-days Moving Average decline and moves towards 10-days Moving Average
         AND the trend is uptrend
         THEN the uptrend will continue.

10-days Momentum Indicator

Rule 1:- If today's momentum higher than yesterday's momentum
         AND yesterday's momentum is above 100
         AND last 2 day's momentum is below 100
         THEN buying signal is generated.

Fig. 7 Sample of association rule from Stock Trading Signal Component

There are two indicators embedded in this component which are Average True Range and Average Volume Indicator. The Average True Range is a useful measure of volatility. The increase in Average True Range signifies that highs and lows are getting farther apart which means volatility is rising and risk has increased. The Average Volume Indicator is basically an indicator that shows the average of the stock's volume over a period. Volume is an important confirming indicator for any forecasts especially trading signal where any trading signal generated without accompanied by signification increase in volume is normally a false trading signal. The output from this component will be the percentage of changes in the today's Average True Range as compared to the result from the previous day and the percentage of changes in volume as compared to 10-days average volume.

5.2.5 Back-Test Component

Back-test component is one of the main parts in the TA component. Output from the various components explained earlier might have contradicted forecasts that even occurred within a particular component itself. Back-test component will weight each output from all of the TA indicators based on the performance of those indicators for that particular stock in the past where indicators that performed the best in the past will be given the highest weight. This component receives an important input from the DM component which is the support and confidence values of each of the association rules used in the three components explained earlier based on the historical stock price. All the outputs from Stock Trend Evaluation Component, Stock Price Evaluation Component and Stock Trading Signal Component together with their respective weights are sent to Inference Engine.

5.2.6 Inference Engine

Inference Engine will produce the final decision on the various different forecasts and predictions generated by other respective components. Inference engine works based on all the forecasts and their respective weights. If there are contradict forecasts incurred, those forecasts with higher weight will be chosen as the decision. All of the outputs from four other components will be included because each of the indicators in all the components are complementary to each other. Verification of forecast done by indicators is very important to reduce the unprofitable trade resulted from false trading signal generated by a particular indicator. A final recommendation will be generated by Inference Engine that includes Trading Signal (Trading Signal is not generated everyday), Forecasted Stock Trend, Stock Price Evaluation and risks level. Provided together is the possibility level for the generated trading signal, forecasted stock trend and stock price evaluation which is derived from the confidence level from each indicators.

5.2.7   AI Component

TA techniques exclusively cannot recognize the non-linear relationship between financial variables and stock price. Therefore, AI component that particularly using ANN will be included in the system. This ANN is based on the concepts in FA. With the inputs of ANN which are extracted from raw financial data, the system would be able to discover the non-linear relationship between financial variables and stock price. Furthermore, by integrating FA concept into ANN, the system is able to compare and back-test the stock price and trend prediction done by TA to achieve higher prediction accuracy than only use TA or FA in ANN alone.

The inputs for the ANN will be from company and economic factors. Inputs from company factor will be Return on Equity, Historical P/E Ratio, Prospective P/E ratio, Cash flow yield, Payout Ratio, Operating Margin, Current Ratio and Debt to Capital at book. As for economic factor, the inputs will be Inflation, Interest rate, GDP Growth. Hence, the output from the ANN in the AI component will be the forecasted stock price in the next 4 month. The forecasted stock price will be in the format of percentage of increase or decrease such as less than 5%, between 5% and 10% or above 10% from the base stock price which is the price of the day the forecast is computed by ANN.

5.2.8   Recommendation Component

Both outputs from TA component and AI component are the inputs for the Recommendation component. Forecasted Stock Trend, Stock Price Evaluation, risks level and Trading Signal are the inputs obtained from TA component while 4-month target stock market price is the input obtained from AI component. The main reason that the output from AI does not undergo the back-test component like in TA component is because the ANN itself has been trained using historical financial data before producing the 4 month target stock price prediction. This component task is to assess whether the 4 month target stock price is achievable or not based on the predicted near future stock trend, evaluation of the current stock price as well as trading signals. Hence, it actually integrates both long term and short term forecast for the stock and present it to the user. The final outputs from this component will be the predicted near future stock trend to enable user to know the current and possible trend of the stock with the historical accuracy that the prediction accurate, stock price evaluation on either the stock price is reasonable, undervalued or overvalued and with the historical accuracy, risks level to alert users on the possibility that the system's prediction is wrong due to high level of volatility, trading signal (buy or sell signal) if there are trading signal generated for that stock and with historical accuracy and finally the 4 month targeted stock price.

The recommendation will be provided together with all the outputs given by various TA indicators and their historical accuracy in presenting the fine-grained information to the users for their use in making trading decision making concurrent with the role of the system as a Decision Support System. The system does not dictate, but recommend. Any final decision is based on the user's judgment and evaluation themselves.

6   Conclusion

A study on the FA and TA has been done in order to understand how stock price prediction works in Finance world as well as to create the TA Model and ANN based on FA. The further study on DSS model and architecture is done to ensure Stock Investment Strategy Decision Support System follows the actual basic framework and model of a DSS. The combination of AI and DM is applied in this hybrid model particularly in the knowledge-based subsystem to generate better analysis of the historical data. The output from the system will be the result analysis and recommendation to the user. The analysis will be generated by the TA component and AI component where it will be presented to the users to aid them in their trading decision making with the presentation of this fine-grained information. However, it will only present the information and outputs derived from various TA and FA techniques used in the system. The recommendation component will judge, evaluate and provide trading recommendation based on the same outputs that are presented to the user in the analysis.

References

1. B.Rockefeller, *TA for Dummies* (Wiley Publishing, New York, 2004)
2. P.Belford, *Candlestick Stock Analysis with Genetic Algorithms* (ACM Digital Library, 2006)
3. T.S.Quah, *Improving Returns on Stock Investment through Neural Network Selection* (Nanyang Technical University, Singapore, 2006)
4. B.M.Kamich, *How TA Works* (New York Institute of Finance, New York, 2003)
5. C.P.Jones, *Investments*, 14th edn. (John Wiley & Sons, 2007)
6. R.D.Edward, J. Magee, *TA of Stock Trends*, 8th edn. (St. Lucie Press, FL, 2001)

7. A.Abraham, B. Nath, P.K. Mahanti, *Hybrid Intelligent Systems for Stock Market Analysis* (Springer-Verlag Berlin Heidelberg, 2001)
8. J.Kamruzzaman, R.Begg, R. Sarker, *Artificial Neural Networks in Finance and Manufacturing* (Idea Group Publishing, Hershey, PA, 2006)
9. Y.Shi, Z. Shi, *Clustering Based Stocks Recognition* (ACM Digital Library, 2006)
10. E.Turban, J.E.Aronson, T.P.Liang, R. Sharda, *Decision Support System and Intelligence Systems* (Pearson Education, Upper Saddle River, NJ, 2007)

Chapter 38
Towards Performance Analysis of Ad hoc Multimedia Network

Kamal Sharma, Hemant Sharma, and Dr. A.K. Ramani

Abstract A primary challenge in mobile ad hoc network scenario is the dynamic differentiation of provided levels of Quality of Service (QoS) depending on client characteristics and current resource availability. In this context the paper focuses on characterizing the average end-to-end delay and maximum achievable per-node throughput for In-vehicle ad hoc multimedia network with stationary and mobile nodes. This work considers an approximation for the expected packet delay based on the assumption that the traffic at each link acts as an independent M/D/1 queue. The chapter establishes the network model, throughput model and delay model for ad hoc multimedia network. It further presents analysis and evaluation of the delay model based on an experimental multimedia communication scenario.

During the last years, the evolutionary development of in-car electronic systems has led to a significant increase of the number of connecting cables within a car. To reduce the amount of cabling and to simplify the networking of dedicated devices, currently appropriate wired bus systems are being considered. These systems are related with high costs and effort regarding the installation of cables and accessory components. Thus, wireless systems are a flexible and very advanced alterative to wired connections. However, the realization of a fully wireless car bus system is still far away. Though, cost-effective wireless subsystems which could extend or partly replace wired bus systems are already nowadays conceivable. A promising technology in this context is specified by the latest Bluetooth standard [1].

In the wireless networks, Bluetooth is one of the communication technologies emerging as the de facto standard for "last-meter" connectivity to the traditional fixed network infrastructure [2]. Bluetooth-enabled portable devices can interconnect to form a particular incarnation of Personal Area Networks (PAN) called piconet, which consists of one master and up to seven slaves. The master device has direct

K. Sharma (✉)
Institue of Computer Science and Information Technology, Devi Ahilya University, Khandwa Road, Indore, India
e-mail: kamal.sharma74@rediffmail.com

S.-I. Ao et al. (eds.), *Machine Learning and Systems Engineering,*
Lecture Notes in Electrical Engineering 68,
DOI 10.1007/978-90-481-9419-3_38, © Springer Science+Business Media B.V. 2010

visibility of all slaves in its piconet and can handle three types of communication services, which in Bluetooth terminology are called logical transports: unicast packet-oriented Asynchronous Connection-oriented Logical transports (ACL), broadcast packet-oriented Active Slave Broadcast (ASB), and circuit-oriented Synchronous Connection Oriented (SCO). Bluetooth logical transports have very different quality characteristics, e.g., in terms of reliability and maximum throughput. It is also vital to predict actual performance of Bluetooth systems so that it can be applied to the design of a system.

The use of Bluetooth has been restricted to (safety) uncritical applications in and around the car, e.g. communication and infotainment applications or applications regarding the exchange of pure car-specific information (e.g. control and status data of the body electronics). Inside a car, Bluetooth allows the wireless connection of a variety of car-embedded electronic devices such as control panels, small displays or headsets with other electronic in-car systems. But beyond replacing wires between in-car devices, the scope of the application of Bluetooth in automotive environments is to ensure wireless interoperability between mobile devices (e.g. PDA, notebook, mobile phone) and car-embedded devices (e.g. navigation system, car radio, car-embedded phone). The main principle behind is to allow devices to co-operate and share resources, and to control mobile devices by using the car embedded user interface, or the other way round, to access car-embedded systems by mobile devices. In this way, the functionality of an in-car electronics environment can be greatly extended, and beyond that, the personalization of the car and its services can be realized.

In addition, current implementations of the Bluetooth software stack do not allow applications to exploit the limited performance functions included in the specification in a portable way. The result is that the development of Bluetooth operations in multimedia ad hoc applications currently depends on specific implementation details of the target Bluetooth hardware/software platform. This relevantly complicates service design and implementation, limits the portability of developed applications, and calls for the adequate modeling of performance parameters corresponding to potential ad hoc applications and services.

This chapter discusses the model and analysis for common ad hoc network performance attributes for pervasive in-vehicle multimedia services for packet based data communication.

1 In-Vehicle Multimedia Network

An ad hoc multimedia network in a car and participating devices are shown in Fig. 1. The diagram presents an ad hoc network containing a smart phone, an infotainment system, and rear seat entertainment system. The devices are equipped with Bluetooth module and are capable of establishing a Bluetooth connection. A piconet can be established between Infotainment system and iPhone when the driver or any other occupant of the vehicle pairs the phone. Similarly a piconet could be established between *Infotainment System* and *Rear Seat Entertainment*

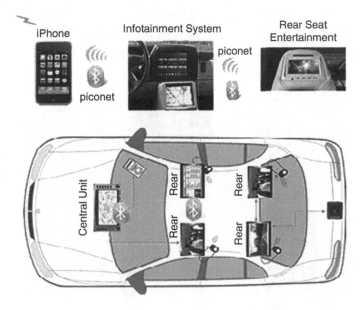

Fig. 1 Typical pervasive vehicle network with multimedia devices

system. The piconet shall enable sharing of iPhone contents or access to contents from internet, if appropriate application framework is available at the infotainment system.

The network contains a pure multimedia part, represented by Multimedia Bus, and a control and external communication part, represented by Networking for Automotive. Multimedia Bus represents the wired multimedia resources. External multimedia resources can form an ad hoc network via the communication interfaces of the vehicle network.

Bluetooth technology is based on a master-slave concept where the master device controls data transmissions through a polling procedure. The master is defined as the device that initiates the connection. A collection of slave devices associated with a single master device is referred to as a piconet.

The master dictates packet transmissions within a piconet according to a time-slot process. The channel is divided into time slots that are numbered according to an internal clock running on the master. A time division duplex (TDD) scheme is used where the master and slaves alternatively transmit packets, where even numbered time slots are reserved for master-slave transmissions, while odd numbered time slots are reserved for slave-master transmissions.

Figure 2 provides an overview of the multimedia PAN inside the car with the role of different nodes. The piconet here consists of a master, the infotainment device, and two slaves, the smart phone (iPhone) and rear-seat entertainment device. As soon as the infotainment system is up and running, its Bluetooth module is ready to pair with other available device. Pairing with iPhone or with rear-seat entertainment unit or with both establishes the network.

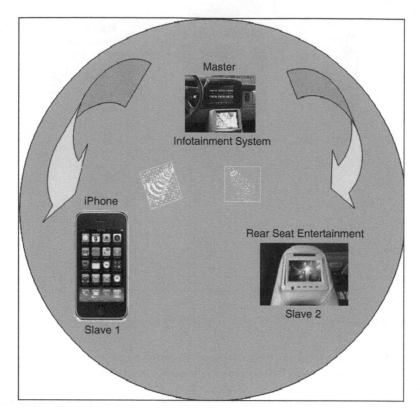

Fig. 2 In-car ad hoc multimedia PAN

1.1 System Architecture

The network organizes infotainment application contents into communication channels to facilitate identification and the communication of intended data streams. Therefore, every application subscribes to the channels that it is interested in, and his corresponding device node will try to retrieve any contents belonging to those channels. The framework follows the approach used in Internet-based podcasting protocols and structures channels into different media streams. To make efficient use of contacts with a small duration, the streams are further divided into data packets, transport-level data units of a size that can typically be downloaded in an individual node encounter. The data packets are further organized into protocol data units (PDU), the atomic transport unit of the network.

The system architecture for ad hoc multimedia network is illustrated in Fig. 3. The transport layer acts directly on top of the link-layer without any routing layer. To distribute contents among the communicating nodes, the framework does not rely on any explicit multi-hop routing scheme. Instead of explicitly routing the data

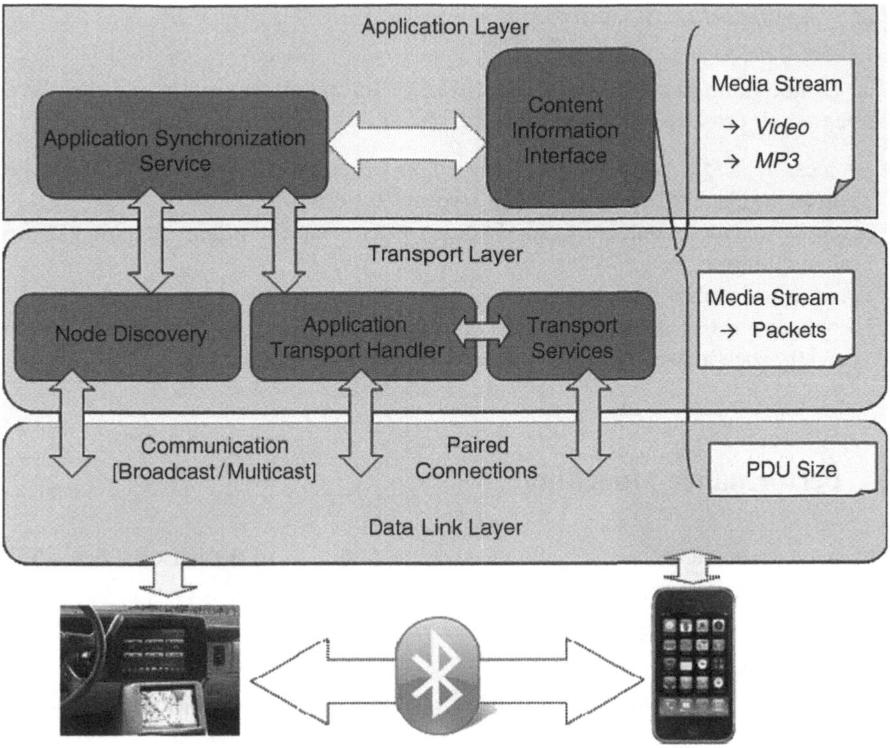

Fig. 3 Overview of ad hoc multimedia network

through specific nodes, it relies on a receiver-driven application-level dissemination model, where contents are routed implicitly as nodes retrieve contents that they request from neighboring nodes. It distinguishes between an application layer, a transport layer, and the data link layer. In a first level of aggregation, the applications organize their data contents in media stream channels.

Below the application layer, the transport layer organizes the data into data streams. Streams are smaller data units that should be able to communicate over short contacts. The use of smaller file blocks is also supported by the idea of integrating forward error correction, for instance the use of fountain codes, to speed up and secure the data transfer, specifically when packets are received unordered. The streams packets themselves are then again cut into smaller parts to optimize the interaction with the data link layer; i.e., the size is set to the PDU size of the data link layer.

The proposed system is designed to work on any MAC architecture, however, to be effective even in the presence of short contact durations, short setup times and high data rates are important for achieving high application communication throughput.

1.2   Application Scenarios

The ad hoc multimedia network consisting of Infotainment system and the slave devices shall provide following services:

- Access to audio contents from iPhone to Infotainment system that could be played and hearable on the vehicle's sound system
- Access to video contents from iPhone to rear-seat entertainment unit via the piconet master.
- Access to internet from rear-seat unit using iPhone via the piconet master
- Applications based on Information contents received by the iPhone to help safe driving, such as weather information or traffic information

2   Performance Modelling

The main parameters that can influence the performance of the network described in previous section are, basically,

- The number of application channels
- The offered traffic per application
- The application profile, defined by the objective parameters (thresholds) and weights

The performance indicators that should be considered for this network are: the packet delay, the packet error rate, the latency, and the throughput. Packet transmission and reception are synchronized by the system clock (625 per s slot interval). This paper focuses on throughput and delay performance in this point-to-multipoint communication environment. We try to simplify and abstract the system as much as possible while preserving the essence of the network which affects the actual performance of the system.

2.1   Network Model

The model considered here is that of a wireless ad hoc network with nodes assumed either fixed or mobile. The network consists of a normalized unit area torus containing n nodes [9]. For the case of fixed nodes, the position of node i is given by Xi. A node i is capable of transmitting at a given transmission rate of W bits/s to j if,

$$\left| X_k - X_j \right| \geq (1 + \Delta) \left| X_i - X_j \right|,$$

Where $\Delta > 0$, such that the node X_k will not impede X_i and X_j communication. This is called the *protocol model* [9].

For the case of mobile nodes, the position of node i at any time is now a function of time. A successful transmission between nodes i and j is governed again by above equation, where the positions of the nodes are time dependent. Time is slotted to simplify the analysis. Also, at each time step, a scheduler decides which nodes are sources, relays, or destinations, in such a manner that the association pair (source-destination) does not change with time.

The multimedia network presented in previous section resembles to multihop wireless ad hoc network, and therefore, can be modeled as a queuing network as shown in Fig. 4a. The stations of the queuing network correspond to the nodes of the

a

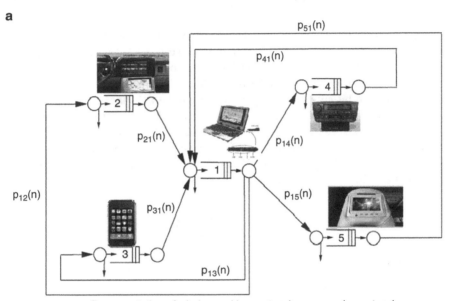

Representation of wireless ad hoc network as a queuing network

b

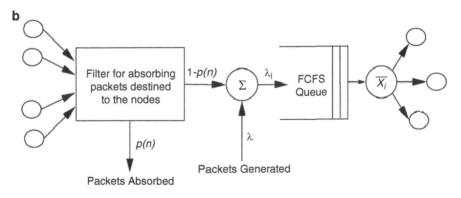

Representation of Node

Fig. 4 Network model for ad hoc multimedia network

wireless network. The forwarding probabilities in the queuing network, denoted by p_{ij}, correspond to the probability that a packet that is transmitted by node i enters the queue of node j. Figure 4b shows a representation of a node in the ad hoc network as a station in the queuing network.

The delay in a wireless network equals the sum of queuing and transmission delays at source and intermediate nodes. We will use the queuing network model, shown in figures above, in order to mathematically analyze the packet delay.

2.2   Packet Delay Model

In order to model the delay, we consider one to five ad hoc multimedia applications with ACL connections in the piconet. In this scenario, packet types are adaptively selected among DM 1, DM3 and DM5 according to the amount of data waiting for transmission in buffers. We consider only DM packets assuming that data packets are vulnerable to bit errors due to error-prone wireless channel characteristics. To ensure reliable data transmission, FEC (Forward Error Correction) and ARQ (Automatic Repeat Request) [1] are employed. We simulated simple wireless channel model, where BER (Bit Error Rate) is constant in a Simulation.

The evaluation of the probability of the queuing delay exceeding a threshold D_{th}, the delay threshold, starts from the expression of the delay

$$D = \sum_{i=0}^{Pq} \sum_{j=0}^{N_{active}(i)-1} Lij$$

where p_q denotes the number of packets in the queue of the desired application, $N_{active}(i)$ is the number of active applications during the transmission of packet i, and l_{ij} is the length of the packet transmitted by application j during the ith round.

The limiting effect to the delay, for very high values, is determined by the control imposed to the incoming traffic when the buffer occupancy gets close to its limits.

2.3   Throughput Model

Taking into account that the main purpose of this ad hoc network is to provide access to multimedia contents, the throughput model has to reflect the characteristics of the network traffic. Throughput performance is evaluated when there are one or more slave device applications, for which an ACL or SCO connection is established.

It is assumed that packets are always transmitted and received in the corresponding slots. In other words, the master device polls a slave device with an

ACL connection in a round-robin fashion and it transmits a packet to the polled slave. When the slave device is polled by the master, the slave sends a packet to the master in the consecutive T_X slot of the reception of the packet from the application at master. Various packet types can be used for ACL connections when there are no SCO connections in a piconet. Only DH1 or DM1 packets are assumed to be transmitted if at least one ACL connection is present in a piconet. In order to see the maximum possible performance, asymmetric T_X is considered, where different packet types are used for down (master to slave) and up (slave to master) stream traffic.

A buffer analysis can be conducted for the network throughput resorting to the classical queuing theory [2] since the arrival process is modelled by an Interrupted Poisson Process (IPP). The number of packets in the queue is well approximated by a geometric random variable with parameter $(1 - \rho)$, where ρ is the activity coefficient [3] given by the ratio between the arrival rate, given by the sum of the arrival rates of the N_{active} active applications in the piconet and the service rate μ:

$$\rho = \frac{(N_{active})\lambda}{\mu}$$

To take into account the effects of a finite buffer, a feedback is considered, blocking the incoming traffic whenever the buffer occupancy exceeds 90% of its capacity. In the simulations, the buffer capacity is considered at 256 Kb.

3   Performance Evaluation

The performance of network has been evaluated when the applications on piconet nodes and transferring audio or data packets. We present here the analysis based on data packets.

3.1   Simulation Setup

The number of applications in the modelled network is relatively small and they are assumed to transfer data traffic between the nodes in the piconet. For the data traffic, this combination is expected to result in burst traffic streams in the network. The voice traffic was simulated using a measured trace of PCM audio stream, but not considered here.

Transmitted packets may be lost due to bit errors and is modeled with a constant loss probability. All lost ACL packets are resent according to the Bluetooth ARQ scheme. In all the simulations presented herein, the packet loss probability was set to 10^{-3}.

3.2   Delay Analysis

Average packet delay, time between arrival of a new packet (data) into a buffer for transmission and reception of the packet at a receiver, is measured for each of traffic load as the number of ACL connections changes.

Figure 5 presents the packet latency as a function of the offered traffic per application for the network, where the number of applications is limited to 50, the objective thresholds are packet error $ERR_{th} = 10^{-3}$ and delay threshold $D_{th} = 10$ ms, and the average duration of the vertical handover procedure is assumed to be 100 ms.

The weights reflect the relative importance of each objective, according to the type of service. The objective values are used as threshold to determine a performance dissatisfaction value U on the basis of the weights assigned to the profile parameters.

For low traffic and a profile where packet error EER has higher priority, the delay can even exceed that of the abstract Bluetooth system; otherwise, the delay lies between the performance of the ad hoc network and abstract Bluetooth network.

3.3   Throughput Analysis

The throughput performance of the system is considered without SCO connections as the number of ACL connections increases. In terms of throughput, in Fig. 6, the

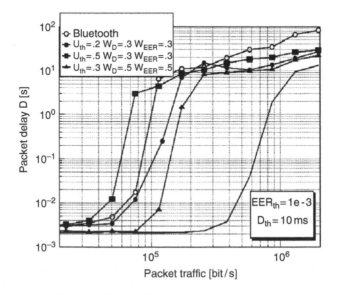

Fig. 5 Packet delay as function of buffered packet traffic

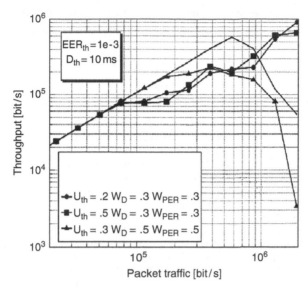

Fig. 6 Throughput as function of buffered packet traffic

available bandwidth is shown as a function of the offered traffic per application, comparing the three scenarios, with packet $EER_{th} = 10^{-3}$ and $D_{th} = 10$ ms. Again, the number of applications in the network is considered to be around 5 and other parameters have the same values as that for packet delay measurement.

Figure 6 below indicates an improvement in the throughput can be obtained by choosing a high weight for the packet error EER, thus forcing a control on the reliability of the transmission.

On the other hand, if the focus is on the delay, the performance in terms of available throughput drops to low values in correspondence to a high offered traffic load. Moreover, note that, in this case, a control on the packet error rate and delay is performed.

4 Summary

This chapter presented the network model, delay model and throughput model for ad hoc In-car multimedia network. We have modelled and evaluated the performance of the multimedia piconet in Bluetooth supporting point-to-multipoint communications. Our performance models provide a performance metric when ACL slave applications are supported. An approximation for the packet error probabilities has been used and the influence of the main parameters on the network performance has been studied. In order to guarantee low delay requirements, the number of application running simultaneously may need to be limited.

These results can serve as valuable inputs to the optimal design of Bluetooth based in-car multimedia applications or interfaces with special throughput and delay requirements.

References

1. Bluetooth 3.0 Specifications, www.bluetooth.com
2. L. Kleinrock, Queuing Systems (Wiley, 1975)
3. G. Lin, G. Noubir, R. Rajaraman, Mobility models for ad hoc network simulation, in Proceedings of the IEEE INFOCOM Conf., Mar 2004, pp. 454–463
4. Q. Zhang, W. Zhu, Y.-q. Zhang, End-to-end QoS for video delivery over wireless internet. Proc. IEEE 03 93(1), 123–134 (2005)
5. K.K. Sharma, H. Sharma, Dr. A.K. Ramani, mCAR: software framework architecture for in-vehicle pervasive multimedia services. Proc. IMECS I, 1100–1105 (2009)
6. K.K.[AU3] Sharma, H. Sharma, Dr. A.K. Ramani, Performance analysis of ad hoc multimedia services in automotive environment, in Proceedings of WCECS, 2009, pp. 218–222
7. X. Masip-Bruin, M. Yannuzzi, J. Domingo-Pascual, A. Fonte, M. Curado, E. Monteiro, F. Kuipers, P. Van Mieghem, S. Avallone, G. Ventre, P. Aranda-Guti´errez, M. Hollick, R. Steinmetz, L. Iannone, K. Salamatian, Research challenges in QoS routing. Comput. Commun. 29, 563–581 (2006)
8. X. Zhou, J. Wei, C.-Z. Xu, Quality-of-service differentiation on the internet: A taxonomy. J. Netw. Comput. Appl. 30(1), 354–383 (2007)
9. J. Li, C. Blake, D.S.D. Couto, H.I. Lee, R. Morris. Capacity of ad hoc wireless networks, in *MobiCom '01: Proceedings of the 7th annual international conference on Mobile computing and networking* (ACM Press, New York, 2001), pp. 61–69

Chapter 39
Towards the Performance Optimization of Public-key Algorithms Using Fuzzy Modular Arithematic and Addition Chain

Gopinath Ganapathy and K. Mani

Abstract In most of the public-key cryptosystems like RSA, ElGamal, etc.; modular exponentiation plays a vital role for performing encryption/decryption operations. In other public-key cryptosystems like ECC, scalar point multiplication, kP where k is an arbitrary integer in the range $1 < k < ord(P)$ and P is a point in the elliptic curve is the central operation. In cryptographic algorithms, exponent is always an integer and can be performed faster than the traditional square and multiply method by iteratively reducing the small gain may be made if the numbers of multiplications are organized properly. For that some integer can be represented in the form of sum of squares and based on the sum of squares larger exponent e can be reduced into smaller one. Then, the addition chain is used to minimize the number of multiplications in the smaller exponent to speed up the operations. Similarly, in the case of ECC to speed up kP fuzzy modular arithmetic is considered.

1 Introduction

Data security and cryptographic techniques are essential for safety relevant applications. Cryptography keeps the message communication secure so that eavesdroppers cannot decipher the transmitted message. It provides the various security services like confidentiality, integrity, authentication, and non-repudiation. There are several cryptographic algorithms available for both symmetric and public-key cryptosystem [1]. Slow running cryptographic algorithms cause customer dissatisfaction and inconvenience. On the other hand, fast running encryption algorithms

G. Ganapathy (✉)
Professor and Head, Department of Computer Science, University of Bharthidasan, Trichy, 620 024, India
e-mail: gganapathy@gmail.com

S.-I. Ao et al. (eds.), *Machine Learning and Systems Engineering*,
Lecture Notes in Electrical Engineering 68,
DOI 10.1007/978-90-481-9419-3_39, © Springer Science+Business Media B.V. 2010

lead high speed. To achieve this, fast exponentiation methods and high speed custom hardware devices are needed.

Public-key cryptosystems like RSA, ElGamal often involve raising large elements of some group fields to large powers. The performance and practicality of such cryptosystems is primarily determined by the implementation efficiency of the modular exponentiation. Fast exponentiation is becoming increasingly important with the widening use of encryption. In order to improve the time requirements of the encryption process, minimizing the number of modular multiplication is essential. For example, to compute $C = x^e$ mod n, based on paper-and-pencil method, it requires $(e - 1)$ modular multiplication of x. i.e. $x–x^2–x^3–x^{e-1}–x^e$. Several authors have proposed the fast modular exponentiation like left-to-right, right-to-left, multi-exponentiation by Interleaved exponentiation, sliding window and window NAF exponential methods etc. [2]. But in this paper, in order to reduce larger exponent into smaller an alternative approach based on sum of squares is thought of because an integer is represented as several forms of sum of squares Then, the addition chain is used to reduce the number of multiplications.

ECC was proposed in 1985 by Neal Koblitz and Victor Miller. It is an alternative to the established cryptosystems like RSA, ElGamal, and Rabin etc. It guarantees all the security services with the shorter keys. The use of shorter length implies less space for key storage, less arithmetic cost and time saving when keys are transmitted. These characteristics make ECC the best choice to provide security in wireless networks. ECC has increased as evidenced by its inclusion in standards by in credited standards organizations such as American National Standards Institute *(ANSI)*, Institute of Electrical and Electronic Engineers (*IEEE*), International Standards Organization (*ISO*), and National Institute of Standards and Technology (*NIST*). The security of ECC is based on the discrete logarithm problem over the points on an elliptic curve. But, it is more difficult problem than the prime number problem of RSA algorithm [3]. These two problems are closely related to the key length of cryptosystems. If the security problem is more difficult, then smaller key length can be used with sufficient security. This smaller key length makes ECC suitable for practical applications such as embedded systems and wireless applications [4].

A vast amount of research has been conducted on secure and efficient implementation since then. All focus on scalar point multiplication that is kP, where k is a scalar and P is a point on elliptic curve. Efficient hardware and software implementation of scalar point multiplication is the main research topic on ECC in recent years. To encrypt and decrypt the message using any public key cryptosystem like ECC, modular arithmetic plays a vital role and it is the key operation in modern cryptology. Even though, many algorithms have been developed for faster implementation, none of the work reveals the use of fuzzy modular arithmetic to speed up the encryption and decryption operations on ECC.

Fuzzy modular arithmetic was proposed by Wael Adi. The key idea used in fuzzy modular arithmetic is not to compute the result exactly as in the traditional modular arithmetic because the traditional modular arithmetic uses division operation for modulo reduction m. Thus, instead of the full reduction, a pseudo *fuzzy* randomized partial reduction is performed.

The rest of the paper is organized as follows. A brief introduction to sum of squares, addition chain, Fermat's Theorem and EC is provided in Section 2. A short description of fuzzy modular arithmetic is presented in Section 3. Section 4 describes the applications of sum of squares, and addition chain to reduce the number of multiplication in modular exponentiation. Section 5 presents the application of fuzzy modular arithmetic, and Fermat's theorem to speed up the operation of scalar multiplication in ECC. Finally, we draw our conclusion in Section 6.

2 Concept of Sum of Squares, Addition Chain, Elliptic Curve, and Fermat Theorem

In this section, we present the representation of some integers based on the sum of two, three, and four squares, the concept of addition chain which is used to find the minimal number of multiplications required for an exponent, the mathematical preliminaries of elliptic curve, addition of points in elliptic curve, and Fermat's theorem to find the modular inverse of an integer.

2.1 Sum of Squares

In this section, we explore integers that may (or may not) be expressed as a sum of two, three and four squares [5–7].

2.1.1 Sum of Two Squares

The Theorems 1 and 2, and Corollaries 1 and 2 are useful to represent an integer in the form of sum of two squares.

- Theorem 1 (*Fermat*)

  An odd prime p is expressible as a sum of two squares if and only if $p \equiv 1 (\mathrm{mod}\ 4)$

- Corollary 1

  Any prime p is of the form $4k + 1$ can be represented uniquely (aside from the order of the summands) as a sum of two squares.

- Theorem 2

  Let the positive integer n be written as $n = N^2 m$, where m is square-free. Then n can be represented as the sum of two squares if and only if m contains no prime factor of the form $4k + 3$.

- Corollary 2

A positive integer n is representable as a sum of two squares if and only if each of its prime factors of the form 4k + 3 occurs to an even power.

- Theorem 3

Any positive integer n can be represented as the difference of two squares if and only if n is not of the form 4k + 2.

- Example

$13 = 2^2 + 3^2$ (We do not consider $13 = 2^2 + 3^2$ as distinct from $13 = 3^2 + 2^2$). In this case, the total number of pairs (x, y) of integers such that $x^2 + y^2 = n$ is 8. For n = 13, we get the pairs (± 2, ± 3) and (± 3, ± 2). This result is due to Fermat.

2.1.2 Sum of Three Squares

- Theorem 4

The Diophantine equation $x_1{}^2 + x_2{}^2 + x_3{}^2 = n$ where n is a positive integer has integral solutions if and only if n is not of the form $4^k (8m + 7)$.

- Theorem 5

No positive integer of the form $4^n (8m + 7)$ can be represented as the sum of three squares.

- Example

The integer 14, 33, and 67 can be represented as sum of three squares.

$14 = 3^2 + 2^2 + 1^2$
$33 = 5^2 + 2^2 + 2^2$
$67 = 7^2 + 3^2 + 3^2$

2.1.3 Sum of Four Squares

- Theorem 6 (*Lagrange*)

Any positive integer n can be written as the sum of four squares, some of which may be zero.

- Theorem 7

Any prime p of the form 4m + 3 can be expressed as the sum of four squares.

- Lemma 1 (*Euler*)

If the integers m and n are each the some of four squares, then m $\times$ n is likewise so representable.

- Example

  To represent the integer 459 as sum of four squares, we have
  $$459 = 3^2.3.17 \text{ (Euler Identity)}$$
  $$= 3^2(1^2 + 1^2 + 1^2 + 0^2)(4^2 + 1^2 + 0^2 + 0^2)$$
  $$= 3^2[(4+1+0+0)^2 + (1-4+0-0)^2 + (0-0-4+0)^2 + (0+0-1-0)^2]$$
  $$= 3^2[5^2 + 3^2 + 4^2 + 1^2]$$
  $$459 = 15^2 + 9^2 + 12^2 + 3^2$$

2.1.4 Integers Which Cannot Be Represented as Sum of Squares

- If $n \equiv 3 \pmod 4$, then n cannot be written as sum of two squares. More generally, if n contains a prime factor $p \equiv 3 \pmod 4$ raised to an odd power, then n cannot be written as a sum of two squares.
- Example
  $78 = 2 \times 3 \times 13$ is such a number.

For the general case, when n is an arbitrary positive integer, a formula for the number of pairs (x, y) of integers such that $x^2 + y^2 = n$ was first found by Jacobi. It is expressed in terms of the number of divisors of n, thus: Let the number of divisors of n of the type 1 (mod 4) be $d_1(n)$, and let the number of divisors of n of the type 3 (mod 4) be $d_3(n)$. Then the number of pairs (x, y) of integers such that $x^2 + y^2 = n$ is equal to $4[d_1(n) - d_3(n)]$.

- Example:
  Take $n = 65$. The divisors of 65 are 1, 5, 13, 65. These are all of the form 1 (mod 4), so $d_1(n) = 4$, $d_3(n) = 0$, $4[d_1(n) - d_3(n)] = 16$. The integer pairs (x, y) for which $x^2 + y^2 = 65$ are $(\pm1, \pm8)$, $(\pm8, \pm1)$, $(\pm4, \pm7)$, $(\pm7, \pm4)$, and these are indeed 16 in number. Or take $n = 39$. The divisors of 39 are 1, 3, 13, 39, so that $d_1(n) = 2$, $d_3(n) = 2$, $d_1(n) - d_3(n) = 0$. And indeed there are no pairs (x, y) of integers for which $x^2 + y^2 = 39$.

2.2 Addition Chain

To compute x^e by multiplication, the most economical way is to reduce into addition, since the exponents are additive. Formally, an addition chain can be defined as follows.

2.2.1 Definition (Addition Chain)

An addition chain for e is a sequence of integers

$$1 = a_0, a_1, a_2, \ldots, a_n = e \tag{1}$$

with the property that

$$a_i = a_j + a_k, \text{ for some } k \le j < i \text{ for all } i = 1, 2, \ldots, r.$$

without loss of generality that an addition chain is "ascending"

$$1 = a_0, a_1, a_2, \ldots, a_n = e \tag{2}$$

For if any two a's are equal, one of them may be dropped; and arrange the sequence (1) into ascending order and remove terms $>e$ without destroying the addition chain property (2). The shortest length, r, for which there exists an addition chain for e is denoted by $l(r)$. Clearly, the problem of finding the shortest ones becomes more and more complicated as r grows larger [8].

2.3   Elliptic Curve

This section focus on the concept of elliptic curve, addition of two points in elliptic curve, and scalar point multiplication.

2.3.1   Definition (Elliptic Curve over GF(p))

Let p be a prime greater than 3 and let a and b be two integers such that $4a^3 + 27b^2 \ne 0$ (mod p). An elliptic curve E over the finite field Fp is the set of points $(x, y) \in Fp \times Fp$ satisfying the Weierstrass equation

$$E : y^2 = x^3 + ax + b \tag{3}$$

together with point at infinity O. The inverse of the point $P = (x_1, y_1)$ is $-P = (x_1, -y_1)$. The sum $P + Q$ of the points $P = (x_1, y_1)$ and $Q = P = (x_2, y_2)$ (where P, Q $\ne$ O and $P = \pm Q$) is the point $R = (x_3, y_3)$ where

$$\lambda = \frac{(y_2 - y_1)}{(x_2 - x_1)}, x_3 = \lambda^2 - x_1 - x_2, y_3 = (x_1 - x_3)\lambda - y_1. \tag{4}$$

For $P = Q$, the doubling formulae are

$$\lambda = \frac{3x_1^2 + a}{2y_1}, x_3 = \lambda^2 - 2x_1, y_3 = (x_1 - x_3)\lambda - y_1. \tag{5}$$

The point at infinity O plays a rule similar to that of the number 0 in normal addition. Thus, $P + Q = P$, and $P + (-P) = O$ for all points P. The point on E together with the operation of "addition" forms an abelian group [9]. Since the point

Table 1 Double-and-add algorithm

Algorithm 1 Elliptic Curve point Multiplication: Binary Method
Input: A point P an *l*-bit integer $k = \sum_{j=0}^{l-1} k_j 2^j, k_j \in \{0, 1\}$
Output: Q = [k]P
1. Q → P
2. for j from *l*–1 to 0 by −1 do:
3. Q → [2] Q
4. If $k_j = 1$ then
Q ← P + Q
5. end if
6. end for

or scalar multiplication is the main operation on ECC, the 'kP' multiplication is based on point doubling and point addition and it can be calculated by using double-and-add algorithm, which is shown in Table 1 [10].

2.3.2 Modular Multiplication Inversion

It is done according to Fermat's theorem, $a^{-1} = a^{p-2}$ mod p, if gcd $(a, p) = 1$. In this paper, we are interested the E defined over GF (p), p is a prime. Thus, Fermat's theorem is used to find multiplicative inverse modulo p. The multiplication inversion can be performed by modular exponentiation of a by $p - 2$ and it can be realized by using the square and multiply algorithm [11].

3 Fuzzy Modular Arithmetic

The concept of fuzzy modular arithmetic for cryptographic schemes was proposed by Wael Adi. As illustrated by Wael Adi [12, 13], in partial modulo reduction instead of performing division, the division is replaced by repeated subtraction. For that some random multiples of m is used. Since the operation used is not exact but involves some random multiples of m from the value to be reduced modulo m, and hence this approach is called fuzzy. The result of the process is not the rest of the division usually computed, but another value and it can be used for calculation. The numerical example of serial fuzzy modular multiplication is shown in Fig. 2.

The modulus m = 13, a multiple of m i.e., m = m·t = 13·2 = 26 is subtracted each time. The 2's complement of m is . . .1100110.

In Fig. 1, the conventional multiplication of two binary integers, say A = 35, and B = 33 with $m = 13$ is shown and the result is 11 (mod 13). Since, fuzzy modular multiplication mainly focuses on repeated subtraction instead of division for modulo reduction m, in this work $m = 2 \times 13 = 26$ is considered and the process is shown in Fig. 2. It is noted that subtraction is performed whenever the bit position in B is zero.

Fig. 1 Conventional
multiplication

$A = (35)_2 = \times$	100011	$35 \times$
$B = (33)_2 =$	100001	33
	100011	$1155 = 11 \pmod{13}$
	000000	
	000000	
	000000	
	000000	
	100011	
	10010000011	

Fig. 2 Numerical example
for fuzzy modular
multiplication

$A = (35)_2 =$	$100011 \times$	$35 \times$
$B = (33)_2 =$	100001	33
	000000100011	$+ 35$
	111111111100110	$- 2 \times 26$
	1111111111100110	$- 4 \times 26$
	1111111111100110	$- 8 \times 26$
	11111111100110	16×26
	00000100011	32×35
	00000000101110111	$375 = 11 \pmod{13}$

4 Applications of Sum of Squares, and Addition Chain in Reducing the Number of Multiplication in Modular Exponentiation

In many public-key cryptosystems such as RSA, the modular exponentiation x^e mod n, where x, e, and n are positive integer is a corner stone operation. It is performed using successive modular multiplications. Several authors have proposed the methods to reduce the total number of multiplication. Nadia Nedjah et al proposed an idea to obtain addition chains with a minimal number of multiplication and hence implementing efficiently the exponentiation operation using genetic algorithms. To find optimal addition chains a genetic algorithm approach was illustrated by Nareli Cruz-Cortes et al. [14]. But, in this work, we proposed an algorithm to reduce the higher power of an exponent into smaller one using the sum of squares. The proposed pesudocode is shown in Section 4.1

4.1 Pseudocode

The pseudocode illustrates the representation of an integer as sum of squares and the number multiplications required for an exponent.

1. Read the exponent e.
2. Check, whether e is prime or not.
   a. If e is a prime and $e \equiv 1 \mod 4$ then express e as sum of two squares.
   b. If e is a prime and $e \equiv 3 \mod 4$, then if e is of the form $4k + 3$ it cannot be expressed as sum of squares.
   c. If e is not a prime and one of the prime factor say p is $p \equiv (3 \mod 4)$, express the factor as sum of squares.
   d. If e is not a prime and if e is not of the form $4^k (8m + 7)$,based on Diophantine equation, express $e = x_1^2 + x_2^2 + x_3^2$ as sum of three squares.
   e. If e not a prime and of the form $4m + 3$, express e as the sum of four squares $e = x_1^2 + x_2^2 + x_3^2 + x_4^2$.

Once the exponent is expressed as the sum of squares, use the addition chain to find the minimal number of multiplication if it is possible.

4.2 Example

In order to understand the pseudocode, the example illustrated in various cases are useful.

- Case 1: When e is a prime and $e \equiv 1 \mod 4$

  Suppose e is 29, then $e \equiv 1 \mod 4$. Using sum of two squares, say

  $e = 5^2 + 2^2$; $x^{29} = x^{25} + {}^4 = (x^4)^7 x$
  we first evaluate $y = x^4$
  Then, we form $y1 = y^2 = (x^{\,4})^2$; $y_2 = y_1^2$
  $x^{29} = y * y_1 * y_2 * x$

  Thus, the whole process takes 7 multiplication, which is equal to shorter addition chain as illustrated by Nareli Cruz-Cortes et al.

- Case 2: when e is a prime and $e \equiv 3 \pmod 4$

  Suppose e is 11, and e is in the form of $4k + 3$, where $k = 1$. It cannot be expressed as sum of squares.

  Thus, we form $x^{11} = x(x^5)^2$
  We first evaluate $y = x(x^4) = x(x^2)^2$;
  Then, we form $x^{11} = y^2 * x$ which requires 5 multiplication

- Case 3: when e is not a prime and of the prime factor is of the form, $p \equiv 3 \pmod 4$

  Suppose e is 78. Then, the factors of 78 are 2, 3, and 13. One of the prime factor, say $p = 3$ is of the form $p \equiv 3 \pmod 4$. Thus, 78 cannot be expressed as a sum of two squares. In this case, we have to consider all factors of 78.

Table 2 Minimum number of multiplication for some exponent e

Exponent (e)	Shortest addition chain l(r)	No. multiplication
11	1–2–3–5–10– 11	5
15	1–2–3–5–10–15	5
27	1–2–3–6–12–15–27	6
29	1–2–3–5–7–14–28-29	7
78	1–2–3–6–12–15–27–39-78	8

Thus, $78 = 2 \times 3 \times 13$

$x^{78} = ((x^2)^3)^{13}$

We first evaluate $y_1 = x * x$

$$y_2 = y_1 * y_1 * y_1$$
$$x^{78} = y_2^{13} = y_2^{12} y_2$$
$$y_3 = y_2 * y_2$$
$$y_4 = y_3^6$$
$$x^{78} = y_4 * y_2$$

Thus, the exponent 78 requires 8 multiplications.

- Case 4: when e is expressed as sum of three squares

  Suppose $e = 27$, then $27 = 1^2 + 1^2 + 5^2$

  $x^{27} = x^2(x^5)^5$

  Thus, the exponent 27 requires 6 multiplication

- Case 5: when e is expressed as sum of four squares

  Suppose $e = 15$, then $15 = 1^2 + 1^2 + 2^2 + 3^2$
  $$x^{15} = x^2 \cdot x^4 \cdot x^9$$
  $$= x^2(x^4)^3 x$$
  $$= x\, x^2(x^4)^3$$
  Thus, the exponent 15 requires 5 multiplications.

  Similar calculations can be performed other exponents also. The addition chain for the exponents 11, 15, 27, 29, and 78 are shown in the Table 2.

5   Implementation of ECC Using Fuzzy Modular Arithmetic

The execution of ECC scheme is mostly dominated by the efficient implementation of finite field arithmetic and point multiplication kP. To compute kP, several methods such as binary method, binary NAF method, windowed NAF etc. have been proposed [15]. To implement these methods, efficient hardwired architecture and a platform that support the efficient computation of such methods are needed. Previous hardware work includes: the first ASIC implementation with Motorola M68008 microcomputer, reconfigurable finite-field multiplier, parallelized field

Table 3 Addition of two points in EC using fuzzy modular arithmetic and Fermat's theorem

Algorithm 2 EC point addition and doubling using fuzzy modular arithmetic	
Input: $p_1 = (x_1, y_1)$, $p_2 = (x_2, y_2)$ and m	*Input*: $p_1 = (x_1, y_1)$, a and m
Output: $p_1 + p_2 = p_3 = (x_3, y_3)$	*Output*: $2\, p_1 = p_3 = (x_3, y_3)$
1. $T_{\lambda_1} \leftarrow y_2 - y_1$	1. $T_{\lambda_1} \leftarrow 3x_1^2 + a$
2. $T_{\lambda_2} \leftarrow x_2 - x_1$	2. $T_{\lambda_2} \leftarrow 2\, y_1$
3. $T_{\lambda_{m2}} \leftarrow T_{\lambda_2}{}^{m-2}$ mod m	3. $T_{\lambda_{m2}} \leftarrow T_{\lambda_2}{}^{m-2}$ mod m
(Fermat's Theorem)	(Fermat's Theorem)
4. $T_{\lambda_{m2}}^{-1} \leftarrow T_{\lambda_{m2}} - z_1 m (z_1 m \leq T_{\lambda_{m2}})$	4. $T_{\lambda_{m2}}^{-1} \leftarrow T_{\lambda_{m2}} - z_1 m (z_1\, m \leq T_{\lambda_{m2}})$
5. $\lambda \leftarrow T_{\lambda_1} T_{\lambda_{m2}}^{-1}$	5. $\lambda \leftarrow T_{\lambda_1} T_{\lambda_{m2}}^{-1}$
6. $T_{x_3} \leftarrow \lambda^2 - x_1 - x_2$	6. $T_{x_3} \leftarrow \lambda^2 - 2\, x_1$
7. **If** $T_{x_3} > m$ **then**	7. **If** $T_{x_3} > m$ **then**
8. $x_3 \leftarrow T_{x_3} - z_2 m$ $(z_2 m \leq T_{x_3})$	8. $x_3 \leftarrow T_{x_3} - z_2 m$ $(z_2 m \leq T_{x_3})$
Else	**Else**
9. $x_3 \leftarrow T_{x_3}$	9. $x_3 \leftarrow T_{x_3}$
10. **end if**	10. **end if**
11. $T_{y_3} \leftarrow (x_1 - x_3)\, \lambda - y_1$	11. $T_{y_3} \leftarrow (x_1 - x_3)\, \lambda - y_1$
12. **If** $T_{y_3} > m$ **then**	12. **If** $T_{y_3} > m$ **then**
13. $y_3 \leftarrow T_{y_3} - z_3 m$	13. $y_3 \leftarrow T_{y_3} - z_3 m$
else	**else**
14. $y_3 \leftarrow T_{y_3}$	14. $y_3 \leftarrow T_{y_3}$
15. **end if**	15. **end if**

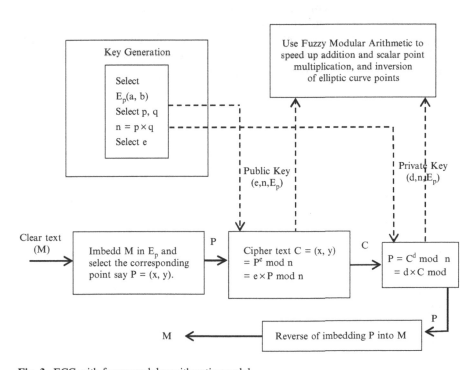

Fig. 3 ECC with fuzzy modular arithmetic model

multiplier etc. [16]. The main objective of this work is to perform modular operations without trail division because the hardware implementation of trial division is an expensive. Thus, this work emphasizes fuzzy modular arithmetic as illustrated by W. Adi no division is required and division is repeatedly replaced by subtraction and hence the hardware implementation is relatively easy. Moreover, to perform kP, repeated addition is used. To add two points of EC, in both formula modular multiplication inversions is involved and it takes considerable amount of time. To avoid it, Fermat's Theorem is used in this paper which is illustrated in Section 2.3.2. The algorithm which describes the EC point addition and doubling using Fuzzy Modular Arithmetic and Fermat's theorem is shown in Table 3 [17].

The proposed methodology to speed up the modular arithmetic is shown in Fig. 3.

6 Conclusion

A straight forward, cheap algorithm for reducing the number of multiplications normally performed in an exponentiation has been described and it can be used for any public-key cryptosystem. Further, to increase the speed of kP, which is the key operations in performing the encryption and decryption process of ECC, fuzzy modular arithmetic and Fermat's theorem are used. In fuzzy modular arithmetic to perform modular operation, repeated subtraction is used instead of division and hence hardware complexity is relatively reduced than division operation. To perform inversion operation with respect to modulus, Fermat's theorem is used which is further simplified by repeated multiplication.

References

1. A. Menezes, P. Van Oorschot, S. Vanstone, *Handbook of Applied Cryptography*, (CRC Press, Boca Raton, FL, 1997)
2. B. Moller, Improved techniques for fast exponentiation, *Information Security and Cryptology – ICISC 2002* (Springer-Verlag, LNCS 2587), p. 298–312
3. R. Cheung, W. Telle, P. Cheung, Customizable elliptic curve cryptosystems. IEEE TVLSI, **13**(a), 1048–1049 (Sept 2005)
4. D. Hankerson, L. Herandez, A. Menezes, *Software Implementation of Elliptic Curve Cryptography over Binary Fields* (CHES, LNCS 1965, Springer-Verlag, Berlin Heidelberg, 2000), pp. 1–24
5. J.A. Anderson, J.M. Bell, *Number Theory with Applications* (Prentice Hall, Englewood. Cliffs, NJ, 1997)
6. M.D. Burton, *Elementary Number Theory*, 6th edn. (Tata Mc-Graw-Hill, New Delhi, 2007)
7. N. Koblitz, *A Course in Number Theory and Cryptography*, 2nd edn. (Springer-Verlag, Berlin, 1994)
8. D. Knuth, *The Art of Computer Programming – Semi Numerical Algorithms*, vol. 2, 3rd edn. (Addison-Wesley, 1998)

9. N. Koblitz, Elliptic curve cryptosystems. Math. Comput. **48**(177), 203–209 (Nov 1982)
10. J. Lutz, A. Hasan, High performance FPGA based elliptic curve cryptographic co-processor. ITCC 04, international conference on information technology coding and computing, vol. 2, 2004, p. 486
11. I. Blake, G. Seroussi, N.P. Smart, *Elliptic Curves in Cryptography*, (Ser. London Math. Soc. Lecture Note Series, Cambridge University Press, New York, 1999)
12. W. Adi, Fuzzy modular arithmetic for cryptographic schemes with applications for mobile security, IEEE, Institute of Technology Technical University of Braunschweig, Braunschweig, 2000, p. 263–265
13. A. Hanoun, W. Adi, F. Mayer-Lindenberg, B. Soundan, Fuzzy modular multiplication architecture and low complexity IPR-protection for FPGA technology. IEEE, 2006, p. 325–326
14. N. Cruz-Cortes et al., *Finding Optimal Addition Chain Using a Genetic Algorithm Approach*, *LNCS*, vol. 3801 (Spring Berlin/Heidberg, 2005), pp. 208–215
15. N. Koblitz, Elliptic curve cryptosystems. Math. Comput. **48**(177), 203–209 (Nov 1982)
16. Y.-J. Choi, M.-S. Kim, H.-R. Lee, H.-W. Kim, Implementation and analysis of Elliptic Curve Cryptosystems over Polynomial basis and ONB. PWAEST, **10**, 130–134 (Dec 2005)
17. G. Ganapathy, K. Mani, *Maximization of Speed in Elliptic Curve Cryptography Using Fuzzy Modular Arithmetic over a Microcontroller based Environment*, Lecture Notes in Engineering and Computer Science WCECS (San Francisco, CA, 20–22 Oct 2009), pp. 328–332

Chapter 40
RBDT-1 Method: Combining Rules and Decision Tree Capabilities

Amany Abdelhalim and Issa Traore

Abstract Most of the methods that generate decision trees for a specific problem use examples of data instances in the decision tree generation process. This chapter proposes a method called *"RBDT-1"* - rule based decision tree - for learning a decision tree from a set of decision rules that cover the data instances rather than from the data instances themselves. *RBDT-1* method uses a set of declarative rules as an input for generating a decision tree. The method's goal is to create on-demand a short and accurate decision tree from a stable or dynamically changing set of rules. We conduct a comparative study of *RBDT-1* with existing decision tree methods based on different problems. The outcome of the study shows that in terms of tree complexity (number of nodes and leaves in the decision tree) *RBDT-1* compares favorably to *AQDT-1, AQDT-2* which are methods that create decision trees from rules. *RBDT-1* compares favorably also to *ID3* which is a famous method that generates decision trees from data examples. Experiments show that the classification accuracies of the different decision trees produced by the different methods under comparison are equal.

1 Introduction

Decision Tree is one of the most popular classification algorithms used in data mining and machine learning to create knowledge structures that guide the decision making process.

The most common methods for creating decision trees are those that create decision trees from a set of examples (data records). We refer to these methods as data-based decision tree methods.

A. Abdelhalim (✉)
Department of Electrical and Computer Engineering, University of Victoria, 3055 STN CSC, Victoria, B.C.,V8W 3P6, Canada
e-mail: amany@ece.uvic.ca

S.-I. Ao et al. (eds.), *Machine Learning and Systems Engineering*,
Lecture Notes in Electrical Engineering 68,
DOI 10.1007/978-90-481-9419-3_40, © Springer Science+Business Media B.V. 2010

On the other hand, to our knowledge there are only few approaches that create decision trees from rules proposed in the literature which we refer to as rule-based decision tree methods.

A decision tree can be an effective tool for guiding a decision process as long as no changes occur in the dataset used to create the decision tree. Thus, for the data-based decision tree methods once there is a significant change in the data, restructuring the decision tree becomes a desirable task. However, it is difficult to manipulate or restructure decision trees. This is because a decision tree is a procedural knowledge representation, which imposes an evaluation order on the attributes. In contrast, rule-based decision tree methods handle manipulations in the data through the rules induced from the data not the decision tree itself. A declarative representation, such as a set of decision rules is much easier to modify and adapt to different situations than a procedural one. This easiness is due to the absence of constraints on the order of evaluating the rules [1].

On the other hand, in order to be able to make a decision for some situation we need to decide the best order in which tests should be evaluated in those rules. In that case a decision structure (e.g. decision tree) will be created from the rules. So, rule-based decision tree methods combine the best of both worlds. On one hand they easily allow changes to the data (when needed) by modifying the rules rather than the decision tree itself. On the other hand they take advantage of the structure of the decision tree to organize the rules in a concise and efficient way required to take the best decision. So knowledge can be stored in a declarative rule form and then be transformed (on the fly) into a decision tree only when needed for a decision making situation [1].

In addition to that, generating a decision structure from decision rules can potentially be performed faster than generating it from training examples because the number of decision rules per decision class is usually much smaller than the number of training examples per class. Thus, this process could be done on demand without any noticeable delay [2, 3]. Data-based decision tree methods require examining the complete tree to extract information about any single classification. Otherwise, with rule-based decision tree methods, extracting information about any single classification can be done directly from the declarative rules themselves [4].

Although rule-based decision tree methods create decision trees from rules, they could be used also to create decision trees from examples by considering each example as a rule. Data-based decision tree methods create decision trees from data only. Thus, to generate a decision tree for problems where rules are provided, e.g. by an expert, and no data is available, rule-based decision tree methods are the only applicable solution.

This chapter presents a new rule-based decision tree method called *RBDT-1*. To generate a decision tree, the *RBDT-1* method uses in sequence three different criteria to determine the fit (best) attribute for each node of the tree, referred to as *the attribute effectiveness (AE)*, *the attribute autonomy (AA)*, and *the minimum value distribution (MVD)*. In this chapter, the *RBDT-1* method is compared to the *AQDT-1* and *AQDT-2* methods which are rule-based decision tree methods, along with *ID3* which is one of the most famous data-based decision tree methods.

2   Related Work

There are few published works on creating decision structures from declarative rules.

The *AQDT-1* method introduced in [1] is the first approach proposed in the literature to create a decision tree from decision rules. The *AQDT-1* method uses four criteria to select the fit attribute that will be placed at each node of the tree. Those criteria are the *cost*[1], the *disjointness*, the *dominance*, and the *extent*, which are applied in the same specified order in the method's default setting.

The *AQDT-2* method introduced in [4] is a variant of *AQDT-1*. *AQDT-2* uses five criteria in selecting the fit attribute for each node of the tree. Those criteria are the *cost*[1], *disjointness, information importance, value distribution*, and *dominance*, which are applied in the same specified order in the method's default setting. In both the *AQDT-1* & 2 methods, the order of each criterion expresses its level of importance in deciding which attribute will be selected for a node in the decision tree. Another point is that the calculation of the second criterion – the *information importance* – in *AQDT-2* method depends on the training examples as well as the rules, which contradicts the method's fundamental idea of being a rule-based decision tree method. *AQDT-2* requires both the examples and the rules to calculate the *information importance* at certain nodes where the first criterion – *Disjointness* – is not enough in choosing the fit attribute. *AQDT-2* being both dependent on the examples as well as the rules increases the running time of the algorithm remarkably in large datasets especially those with large number of attributes.

In contrast the *RBDT-1* method, proposed in this work, depends only on the rules induced from the examples, and does not require the presence of the examples themselves.

Akiba et al. [5] proposed a rule-based decision tree method for learning a single decision tree that approximates the classification decision of a majority voting classifier. In their proposed method, if-then rules are generated from each classifier (a *C4.5* based decision tree) and then a single decision tree is learned from these rules. Since the final learning result is represented as a single decision tree, problems of intelligibility and classification speed and storage consumption are improved. The procedure that they follow in selecting the best attribute at each node of the tree is based on the *C4.5* method which is a data-based decision tree method. The input to the proposed method requires both the real examples used to create the

---

[1]In the default setting, the *cost* equals 1 for all the attributes. Thus, the *disjointness* criterion is treated as the first criterion of the *AQDT-1* and *AQDT-2* methods in the decision tree building experiments throughout this paper.

classifiers (decision trees) and the rules extracted from the classifiers, which are used to create a set of training examples to be used in the method.

In [6], the authors proposed a method called Associative Classification Tree (ACT) for building a decision tree from association rules rather than from data. They proposed two splitting algorithms for choosing attributes in the *ACT* one based on the confidence gain criterion and one based on the entropy gain criterion. The attribute selection process at each node in both splitting algorithms relies on both the existence of rules and the data itself as well. Unlike our proposed method *RBDT-1, ACT* is not capable of building a decision tree from the rules in the absence of data, or from data (considering them as rules) in the absence of rules.

3   Rule Generation and Notations

In this section, we present the notations used to describe the rules used by our method. We also present the methods used to generate the rules that will serve as input to the rule-based decision tree methods in our experiments.

3.1   *Notations*

In order for our proposed method to be capable of generating a decision tree for a given dataset, it has to be presented with a set of rules that cover the dataset. The rules will be used as input to *RBDT-1* which will produce a decision tree as an output. The rules can either be provided up front, for instance, by an expert or can be generated algorithmically.

Let $a_1, ..., a_n$ denote the attributes characterizing the data under consideration, and let $D_1, ..., D_n$ denote the corresponding domains, respectively (i.e. D_i represents the set of values for attribute a_i). Let $c_1, ..., c_m$ represent the decision classes associated with the dataset.

3.2   *Rule Generation Method*

The *RBDT-1* takes rules as input and produces a decision tree as output. Thus, to illustrate our approach and compare it to existing similar approaches, we use in this chapter two options for generating a set of disjoint rules for each dataset. The first option is based on extracting rules from a decision tree generated by the *ID3* method, where we convert each branch – from the root to a leaf – of the decision tree to an if–then rule whose condition part is a pure conjunction. This scenario will ensure that we will have a collection of disjoint rules. We refer to these rules as *ID3-based* rules.

Using *ID3-based* rules will give us an opportunity to illustrate our method's capability of producing a smaller tree while using the same rules extracted from a decision tree generated by a data-based decision tree method without reducing the tree's accuracy.

The second rule generation option consists of using an *AQ-type* rule induction program. We use the *AQ19* program for creating logically disjoint rules which we refer to as *AQ-based* rules.

4   RBDT-1 Method

In this section, we describe the *RBDT-1* method by first outlining the format of the input rules used by the method and the attribute selection criteria of the method. Then we summarize the main steps of the underlying decision tree building process and present the technique used to prune the rules.

4.1   Attribute Selection Criteria

The *RBDT-1* method applies three criteria on the attributes to select the fittest attribute that will be assigned to each node of the decision tree. These criteria are *the Attribute Effectiveness, the Attribute Autonomy, and the Minimum Value Distribution.*

4.1.1   Attribute Effectiveness (AE)

AE is the first criterion to be examined for the attributes. It prefers an attribute which has the most influence in determining the decision classes. In other words, it prefers the attribute that has the least number of "don't care" values for the class decisions in the rules, as this indicates its high relevance for discriminating among rule sets of given decision classes. On the other hand, an attribute which is omitted from all the rules (i.e. has a "don't care" value) for a certain class decision does not contribute in producing that corresponding decision. Choosing attributes based on this criterion maximizes the chances of reaching leaf nodes faster which on its turn minimizes the branching process and leads to producing a smaller tree.

Using the notation provided above (see Section 3), let V_{ij} denote the set of values for attribute a_j involved in the rules in R_i, which denote the set of rules associated with decision class c_i, $1 \leq i \leq m$. Let DC denote the 'don't care' value, we calculate $C_{ij}(DC)$ as shown in (1):

$$C_{ij}(DC) = \begin{cases} 1 & \text{if } DC \in V_{ij} \\ 0 & \text{otherwise} \end{cases} \tag{1}$$

Given an attribute a_j, where $1 \leq j \leq n$, the corresponding attribute effectiveness is given in (2).

$$AE(a_j) = \frac{m - \sum_{i=1}^{m} C_{ij}(DC)}{m} \tag{2}$$

(where m is the total number of different classes in the set of rules).

The attribute with the highest AE is selected as the fit attribute.

4.1.2   Attribute Autonomy (AA)

AA is the second criterion to be examined for the attributes. This criterion is examined when the highest AE score is obtained by more than one attribute. This criterion prefers the attribute that will decrease the number of subsequent nodes required ahead in the branch before reaching a leaf node. Thus, it selects the attribute that is less dependent on the other attributes in deciding on the decision classes. We calculate the attribute autonomy for each attribute and the one with the highest score will be selected as the fit attribute.

For the sake of simplicity, let us assume that the set of attributes that achieved the highest AE score are $a_1, ..., a_s, 2 \leq s \leq n$. Let $v_{j1}, ..., v_{jp_j}$ denote the set of possible values for attribute a_j including the "don't care", and R_{ji} denote the rule subset consisting of the rules that have a_j appearing with the value v_{ji}, where $1 \leq j \leq s$ and $1 \leq i \leq p_j$. Note that R_{ji} will include the rules that have don't care values for a_j as well.

The AA criterion is computed in terms of the Attribute Disjointness Score (ADS), which was introduced by Imam and Michalski [1]. For each rule subset R_{ji}, let $MaxADS_{ji}$ denote the maximum ADS value and let ADS_List_{ji} denote a list that contains the ADS score for each attribute a_k, where $1 \leq k \leq s, k \neq j$.

According to [1], given an attribute a_j and two decision classes c_i and c_k (where $1 \leq i, k \leq m; 1 \leq j \leq s$), the degree of disjointness between the rule set for c_i and the rule set for c_j with respect to attribute a_j is defined as shown in (3):

$$ADS(A_j, C_i, C_k) = \begin{cases} 0 & \text{if } V_{ij} \subseteq V_{kj} \\ 1 & \text{if } V_{ij} \supseteq V_{kj} \\ 2 & \text{if } V_{ij} \cap V_{kj} \neq (\emptyset \text{ or } V_{ij} \text{ or } V_{kj}) \\ 3 & \text{if } V_{ij} \cap V_{kj} = \emptyset \end{cases} \tag{3}$$

The *Attribute Disjointness* of the attribute a_j, $ADS(a_j)$ score, is the summation of the degrees of class disjointness $ADS(a_j, c_i, c_k)$ given in (4):

$$ADS(a_j) = \sum_{i=1}^{m} \sum_{\substack{1 \leq k \leq s \\ i \neq k}} ADS(a_j, c_i, c_k) \tag{4}$$

Thus, the number of *ADS_List* that will be created for each attribute a_j as well as the number of *MaxADS* values that are calculated will be equal to p_j. The *MaxADS$_{ji}$* value as defined by Wojtusiak [8] is $3 \times m \times (m-1)$ where m is the total number of classes in R_{ji}. We introduce the *AA* as a new criterion for attribute a_j as given in (5):

$$AA(a_j) = \frac{1}{\sum\limits_{i=1}^{p_j} AA(a_j, i)} \tag{5}$$

where $AA(a_j, i)$ is defined as shown in (6):

$$AA(a_j, i) = \begin{cases} 0 & \text{if } MaxADS_{ji} = 0 \\ 1 & \text{if} \\ \quad \left((MaxADS_{ji} \neq 0) \wedge \left(\begin{array}{l} (s = 2) \vee \\ (\exists l : MaxADS_{ji} = ADS\_List_{ji}[l]) \end{array} \right) \right) \\ 1 + \left[(s-1) \times MaxADS_{ji} - \sum\limits_{l=1, l \neq j}^{s-1} ADS\_List_{ji}[l] \right] & \text{otherwise} \end{cases} \tag{6}$$

The *AA* for each of the attributes is calculated using the above formula and the attribute with the highest *AA* score is selected as the fit attribute. According to the above formula, $AA(a_j, i)$, equals zero when the class decision for the rule subset examined corresponds to one class, in that case $MaxADS = 0$, which indicates that a leaf node is reached (best case for a branch). $AA(a_j, i)$ equals 1 when s equals 2 or when one of the attributes in the *ADS_list* has an *ADS* score equal to *MaxADS* value (second best case). The second best case indicates that only one extra node will be required to reach a leaf node. Otherwise $AA(a_j, i)$ will be equal to 1 + (the difference between the *ADS* scores of the attributes in the *ADS_list* and the *MaxADS* value) which indicates that more than one node will be required until reaching a leaf node.

4.1.3   Minimum Value Distribution (MVD)

The *MVD* criterion is concerned with the number of values that an attribute has in the current rules. When the highest *AA* score is obtained by more than one attribute, this criterion selects the attribute with the minimum number of values in the current rules. *MVD* criterion minimizes the size of the tree because the fewer the number of values of the attributes the fewer the number of branches involved and consequently the smaller the tree will become [1]. For the sake of simplicity, let us assume that the set of attributes that achieved the highest *AA* score are $a_1, \ldots, a_q, 2 \leq q \leq s$. Given an attribute a_j (where $1 \leq j \leq q$), we compute corresponding *MVD* value as shown in (7).

$$MVD(A_j) = \left| \bigcup_{1 \leq i \leq m} V_{ij} \right| \tag{7}$$

(where $|X|$ denote the cardinality of set X).

When the lowest MVD score is obtained by more than one attribute, any of these attributes can be selected randomly as the fit attribute. In our experiments in case where more than two attributes have the lowest MVD score we take the first attribute.

4.2   Building the Decision Tree

In the decision tree building process, we select the fit attribute that will be assigned to each node from the current set of rules CR based on the attribute selection criteria outlined in the previous section. CR is a subset of the decision rules that satisfy the combination of attribute values assigned to the path from the root to the current node. CR will correspond to the whole set of rules at the root node.

From each node a number of branches are pulled out according to the total number of values available for the corresponding attribute in CR.

Each branch is associated with a reduced set of rules RR which is a subset of CR that satisfies the value of the corresponding attribute. If RR is empty, then a single node will be returned with the value of the most frequent class found in the whole set of rules. Otherwise, if all the rules in RR assigned to the branch belong to the same decision class, a leaf node will be created and assigned a value of that decision class. The process continues until each branch from the root node is terminated with a leaf node and no more further branching is required.

5   Illustration of the RBDT-1 Method

In this section the *RBDT-1* method is illustrated by a dataset named the weekend problem which is a publicly available dataset.

5.1   Illustration

The Weekend problem is a dataset that consists of ten data records obtained from [12]. For this example, we used the *AQ19* rule induction program to induce the rule set shown in Table 1 which will serve as the input to our proposed method *RBDT-1*. *AQ19* was used with the mode of generating disjoint rules and with producing a complete set of rules without truncating any of the rules. The corresponding

Table 1 The weekend rule set induced by AQ19

Rule#	Description
1:	*Cinema ← Parents-visiting="yes" & weather = "don't care" & Money ="rich"*
2:	*Tennis ← Parents-visiting=" "no" & weather ="sunny" & Money ="don't care"*
3:	*Shopping ← Parents-visiting="no" & weather ="windy" & Money ="rich"*
4:	*Cinema ← Parents-visiting="no" & weather ="windy" & Money ="poor"*
5:	*Stay-in ← Parents-visiting="no" & weather ="rainy" & Money ="poor"*

Fig. 1 The decision tree generated by RBDT-1 for the weekend problem

```
parents-visiting
  yes
              ->   cinema
  no
  weather
      windy
      money
          poor
                  ->   cinema
          rich
                      ->  shopping
      sunny
          ->    tennis
      rainy
          ->    stay-in
```

decision tree created by the proposed *RBDT-1* method for the weekend problem is shown in Fig. 1. It consists of three nodes and five leaves with 100% classification accuracy for the data.

Figure 2 depicts the tree created by *AQDT-1 & 2* and the *ID3* methods consisting of five nodes and seven leaves, which is bigger than the decision tree created by the *RBDT-1* method. Both trees have the same classification accuracy as *RBDT-1*. Due to space limitations, we are unable to include here all of our work. We refer the reader to our extended technical report on *RBDT-1* [13] for more details on the decision tree generation and discussion on tree sizes.

6   Experiments

In this section, we present an evaluation of our proposed method by comparing it with the *AQDT-1 & 2* and the *ID3* methods based on 16 public datasets. Our evaluation consisted of comparing the decision trees produced by the methods for each dataset in terms of tree complexity (number of nodes and leaves) and accuracy. Other than the weekend dataset, all the datasets were obtained from the UCI machine learning repository [14].

We conducted two experiments for comparing *RBDT-1* with *AQDT-1 & 2* and the *ID3*; the results of these experiments are summarized in Tables 2 and 3. The

Fig. 2 The decision tree generated by AQDT-1, AQDT-2 & ID3 for the weekend problem

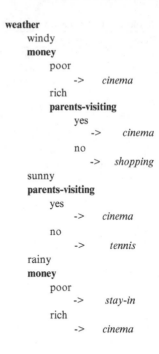

weather
  windy
  money
    poor
        ->   *cinema*
    rich
    parents-visiting
      yes
          ->   *cinema*
      no
          ->   *shopping*
  sunny
  parents-visiting
    yes
        ->   *cinema*
    no
        ->   *tennis*
  rainy
  money
    poor
        ->   *stay-in*
    rich
        ->   *cinema*

Table 2 Comparision of tree complexities of the RBDT-1, AQDT-1, AQDT-2 & ID3 methods using ID3-based rules

Dataset	Method	Dataset	Method
Weekend	*RBDT-1*	**MONK's 1**	*RBDT-1*
Lenses	*RBDT-1, ID3*	**MONK's 2**	*RBDT-1, AQDT-1 & 2*
Chess	=	**MONK's 3**	=
Car	*RBDT-1, ID3*	**Zoo**	=
Shuttle-L-C	=	**Nursery**	*RBDT-1*
Connect-4	*RBDT-1*	**Balance**	*RBDT-1, ID3*

Table 3 Comparison of tree complexities of the RBDT-1, AQDT-1, AQDT-2 & ID3 methods using AQ-based rules

Dataset	Method	Dataset	Method
Weekend	*RBDT-1*	**Monk's2**	*RBDT-1*
Lenses	*RBDT-1, ID3*	**Monk's3**	=
Zoo	*RBDT-1*	**Chess**	=
Car	*RBDT-1*	**Balance**	*RBDT-1, ID3*
Monk's1	*RBDT-1*	**Shuttle-L-C**	*RBDT-1, AQDT-1 & 2*

results in Table 2 are based on comparing the methods using *ID3-based* rules as input for the rule-based methods. The *ID3-based* rules were extracted from the decision tree generated by the *ID3* method from the whole set of examples of each dataset used. Thus, they cover the whole set of examples (100% coverage). The size of the extracted *ID3-based* rules is equal to the number of leaves in the *ID3* tree.

In Table 3, the experiment was conducted using *AQ-based* rules. We used *AQ-based* rules generated by the *AQ19* rule induction program with 100% correct recognition on the datasets used as input for the rule-based methods under comparison. No pruning is applied by any method in both comparisons presented in Tables 2 and 3.

In the following tables, the name of the method that produced a less complex tree appears in the table in the *method* column; the "=" symbol indicates that the same tree was obtained by all methods under comparison. All four methods run under the assumption that they will produce a complete and consistent decision tree yielding 100% correct recognition on the training examples. For 7 of the datasets listed in Table 2, *RBDT-1* produced a smaller tree than that produced by *AQDT-1 & 2* with an average of 274 nodes less while producing the same tree for the rest of the datasets. On the other hand, *RBDT-1* produced a smaller tree with 5 of these datasets compared to *ID3* with an average of 142.6 nodes less while producing the same tree for the rest of the datasets. The decision tree classification accuracies of all four methods were equal.

Based on the results of the comparison in Table 3, the *RBDT-1* method performed better than the *AQDT-1 & 2* and the *ID3* methods in most cases in terms of tree complexity by an average of 33.1 nodes less than *AQDT-1 & 2* and by an average of 88.5 nodes less than the *ID3* method. The decision tree classification accuracies of all four methods were equal.

7 Conclusions

The *RBDT-1* method proposed in this work allows generating a decision tree from a set of rules rather than from the whole set of examples. Following this methodology, knowledge can be stored in a declarative rule form and transformed into a decision structure when it is needed for decision making. Generating a decision structure from decision rules can potentially be performed much faster than by generating it from training examples.

In our experiments, our proposed method performed better than the other three methods under comparison in most cases in terms of tree complexity and achieves at least the same level of accuracy.

References

1. I. F. Imam, R.S. Michalski, Should decision trees be learned from examples of from decision rules? Source Lecture Notes in Computer Science, in *Proceedings of the 7th International Symposium on Methodologies*, vol. 689. (Springer, Berlin/Heidelberg, 1993), pp. 395–404
2. J.R. Quinlan, Discovering rules by induction from large collections of examples, in *Expert Systems in the Microelectronic Age*, ed. by D. Michie (Edinburgh University Press, Scotland, 1979), pp. 168–201

3. I. H. Witten, B.A. MacDonald, Using concept learning for knowledge acquisition, Int. J. Man Mach. Stud. 27(4), 349–370 (1988)
4. R.S. Michalski, I.F. Imam, Learning problem-oriented decision structures from decision rules: the AQDT-2 system, Lecture Notes in Artificial Intelligence, in *Proceedings of 8th International Symposium Methodologies for Intelligent Systems*, vol. 869 (Springer Verlag, Heidelberg, 1994), pp. 416–426
5. Y. Akiba, S. Kaneda, H. Almuallim, Turning majority voting classifiers into a single decision tree, in *Proceedings of the 10th IEEE International Conference on Tools with Artificial Intelligence*, 1998, pp. 224–230
6. Y. Chen, L.T. Hung, Using decision trees to summarize associative classification rules. Expert Syst. Appl. Pergamon Press, Inc. Publisher, 2009, 36(2), 2338–2351
7. R.S. Michalski, K. Kaufman, The aq19 system for machine learning and pattern discovery: a general description and user's guide, Reports of the Machine Learning and Inference Laboratory, MLI 01-2, George Mason University, Fairfax, VA, 2001
8. J. Wojtusiak, AQ21 user's guide. Reports of the Machine Learning and Inference Laboratory, MLI 04-5, George Mason University, 2004
9. R.S. Michalski, I.F. Imam, On learning decision structures. fundamenta informaticae **31**(1),: 49–64 (1997)
10. R.S. Michalski, I. Mozetic, J. Hong, N. Lavrac, The multi-purpose incremental learning system AQ15 and its testing application to three medical domains, in *Proceedings of AAAI-86*, Philadelphia, PA, 1986, pp. 1041–1045
11. F. Bergadano, S. Matwin, R.S. Michalski, J. Zhang, Learning two-tiered descriptions of flexible concepts: the POSEIDON system. Mach. Learning **8**(1), pp. 5–43 (1992)
12. A. Abdelhalim, I. Traore, The RBDT-1 method for rule-based decision tree generation, Technical report #ECE-09-1, ECE Department, University of Victoria, PO Box 3055, STN CSC, Victoria, BC, Canada, July 2009
13. A. Asuncion, D.J. Newman, UCI machine learning repository [http://www.ics.uci.edu/~mlearn/MLRepository.html]. University of California, School of Information and Computer Science, Irvine, CA, 2007

Chapter 41
Computational and Theoretical Concepts for Regulating Stem Cells Using Viral and Physical Methods

Aya Sedky Adly, Olfat A. Diab Kandil, M. Shaarawy Ibrahim, Mahmoud Sedky Adly, Ahmad Sedky Adly, and Afnan Sedky Adly

Abstract Regenerative medicine is the application of tissue, sciences, engineering, computations, related biological, and biochemical principles that restore the structure and function of damaged tissues and organs. This new field encompasses many novel approaches to treatment of disease and restoration of biological function. Scientists are one-step closer to create a gene therapy/stem cell combination to combat genetic diseases. This research may lead to not only curing the disease, but also repairing the damage left behind. However, the development of gene therapy vectors with sufficient targeting ability, efficiency, and safety must be achieved before gene therapy can be routinely used in human. Delivery systems based on both viruses and non-viral systems are being explored, characterized, and used for in vitro and in vivo gene delivery. Although advances in gene transfer technology have been made, an ideal system has not yet been constructed. The development of scalable computer systems constitutes one-step toward understanding dynamics and potential of this process. Therefore, the primary goal of this work is to develop a computer model that will support investigations of both viral and non-viral methods for gene transfer on regenerative tissues including genetically modified stem cells. Different simulation scenarios were implemented, and their results were encouraging compared to ex-vivo experiments, where, the error rate lies in the range of acceptable values in this domain of application.

1 Introduction

The transfer of genes into cells has played a decisive role in our knowledge of modern cell biology and associated biotechnology especially in the field of regenerative medicine. It is an exciting new area of medicine that has attracted much

A.S. Adly (✉)
Faculty of Computers and Information, Computer Science department, Helwan University, Cairo, Egypt
e-mail: ayasedky@helwan.edu.eg

S.-I. Ao et al. (eds.), *Machine Learning and Systems Engineering*,
Lecture Notes in Electrical Engineering 68,
DOI 10.1007/978-90-481-9419-3_41, © Springer Science+Business Media B.V. 2010

interest since the first submission of clinical trials. Preliminary results were very encouraging and prompted many investigators and researchers. However, the ability of stem cells to differentiate into specific cell types holds immense potential for therapeutic use in gene therapy. Realization of this potential depends on efficient and optimized protocols for genetic manipulation of stem cells. It is widely recognized that gain/loss of function approaches using gene therapy are essential and valuable for realizing and understanding specific functions in studies involving stem cells.

Integrating all these aspects into a successful therapy is an exceedingly complex process that requires expertise from many disciplines, including molecular and cell biology, genetics and virology, in addition to bioinformatics, bioprocess manufacturing capability and clinical laboratory infrastructure. Gene therapy delivers genetic material to the cell to generate a therapeutic effect by correcting an existing abnormality or providing cells with a new function. Researchers have successfully induced differentiated cells that can behave like pluripotent stem cells. These cells are artificially derived from a non-pluripotent cells, typically adult somatic cells, by inserting certain genes. Induced Pluripotent Stem Cells are believed to be identical to natural pluripotent stem cells in many respects, but the full extent of their relation is still being assessed [1–4]. This new field holds the realistic promise of regenerating damaged tissues, and empowers scientists to grow them in vitro. This is a completely new concept compared to the traditional medicine [5–7]. Whereas modeling of gene delivery systems for regulating stem cell lineage systems is critical for the development of an integrated attempt to develop ideas in a systematic manner, it has not been a research area that has received the desired mount of attention. Over the last few years or so, there has been a noticeable change in this respect, and there is now a growing awareness of the need to use models and computer simulation to understand its processes and behaviors.

2 Methods Used in Gene Therapy

Vectors are agents used to carry foreign genes into host cells that will produce the protein encoded by those genes. They contain an origin of replication, which enables the vector, together with the foreign DNA fragment inserted into it, to replicate. The development of efficient model for vector systems is crucial in supporting success of gene therapy [8–10]. The ideal vector for gene therapy would introduce a transcriptionally regulated gene into all organs or tissues relevant to the genetic disorder by a single safe application. However, it rapidly became obvious that there is no universally applicable ideal vector for gene transfer; each has its own advantages and disadvantages, see Table 1 [11–14]. In general, a vector used for gene delivery would have at least the following characteristics: (i) specificity for the targeted cells; (ii) resistance to metabolic degradation and/or attack by the immune system; (iii) safety, i.e., minimal side effects; and (iv) an ability to

Table 1 Commonly used viral and non-viral vectors

Retrovirus: Viral vector (can carry up to 8,000 base pairs)	
Target	Dividing cells (possible to target specific types by engineering proteins on virus surface)
Activation	Its RNA travels to the cell's nucleus to be converted to DNA before genes are activated
Integration	The DNA made from the virus' RNA template integrates into the host cell's genome in random locations, then duplicated when the cell divides
Side effects	There is the chance that it will disrupt another gene. Can cause an immune response, which can be reduced by removing proteins on the surface of virus that trigger it
Letivirus: Viral vector (can carry up to 8,000 base pairs)	
Target	Dividing cells and non-dividing cells (possible to target specific cell types by engineering proteins on the virus surface)
Activation	Via infectious particles. Infecting neighboring cells via direct contact with the host cells
Integration	Long-term stable gene expression
Side effects	Possible proviral insertional mutagenesis in target cells.
Adenovirus: Viral vector (can carry up to 7,500 base pairs)	
Target	Dividing cells and non-dividing cells (possible to target specific cell types by engineering proteins on the virus surface)
Activation	Its DNA travels to the cell's nucleus, where its genes are activated
Integration	Its DNA will not integrate, and after a week or two, the cell will discard it
Side effects	Can cause an immune response in the patient, which can be reduced by removing proteins on the surface of the virus that trigger the response
Adeno-associated virus: viral vector (can carry up to 5,000 base pairs)	
Target	Dividing and non-dividing cells but need assistance of a 'helper' virus to replicate inside cells (possible to target specific cell types by engineering proteins on the virus surface)
Activation	Its DNA travels to the cell's nucleus, where its genes are activated
Integration	Its DNA will integrate into the host cell genome. 95% of the time, it will integrate into a specific region on Chromosome 19, greatly reducing the chance other genes disruptions
Side effects	These viruses typically will not cause an immune response in the patient
Herpes simplex virus: viral vector (can carry up to 20,000 base pairs)	
Target	Cells of the nervous system
Activation	Its DNA travels to the cell's nucleus, where its genes are activated
Integration	It will not integrate, but can stay in the cell's nucleus for a long time, as a separate circular piece of DNA that replicates when the cell divides. It will not disrupt other genes
Side effects	Can cause an immune response in the patient, which can be reduced by removing proteins on the surface of the virus that trigger the response
Liposome: non-viral vector (there is no maximum length)	
Target	Not specific for any cell type (enter cells far less effectively than viruses)
Activation	The plasmid DNA is transported to nucleus, where its genes will be activated
Side effects	It will not generate an immune response, but some types can be toxic
Naked DNA: non-viral vector (there is no maximum length)	
Target	Not specific for any cell type (enter cells far less effectively than viruses)
Activation	Unless it is specifically engineered to do so, plasmid DNA will not integrate
Side effects	It will not generate an immune response, and it is generally not toxic

express, in an appropriately regulated fashion, the therapeutic gene for as long as required. In general terms, gene delivery methods can be sub-divided into two categories: (a) the use of biological vectors and (b) techniques employing either chemical or physical approaches [11, 13, 15].

Genes may be introduced into cells by transfection or transduction. Transfection utilizes non-viral methods as chemical or physical methods to introduce new genes into cells. Usually, small molecules, such as liposomes, are employed to facilitate the entry of DNA encoding the gene of interest into the cells. Brief electric shocks are additionally used to facilitate the process. These non-viral methods have a number of advantages [16, 17]: they are safer, cell independent, and there is no limitation to the size of the genes that can be delivered. In addition, gene transfection is a fundamental technology indispensable to the further research development of basic biology and medicine regarding stem cells. However, despite the fact that it is widely used to transfer genes into living cells, the steps that limit DNA transfer remain to be determined. On the other hand, transduction utilizes viral vectors for DNA transfer. Engineered viruses are very efficient and can be used to introduce almost any genetic information into cells. However, there are usually limitations in the size of the introduced gene. Additionally, some viruses only infect specific cells effectively. In most cases, a genetic material carried by viral vectors is stably integrated into the host cell genome [18, 19].

The development of efficient computer model for such systems is crucial in supporting the success of gene therapy. A computational approach towards understanding these processes therefore has two major objectives. The first objective is to provide a qualitative and quantitative explanatory model describing the inner workings of cell regulation. Computer simulations can help linking observations to hypotheses by reproducing the expected behavior from a theoretical model. The second objective is concerned with predictive models capable of extrapolating from a given state of the cell or its components to subsequent states.

3   Proposed Model

Models are used particularly in the natural sciences and engineering disciplines such as physics, biology, and electrical engineering and can be understood as a representation of the essential aspects of a certain system, and description of its behavior in order to presents its knowledge in usable form. This work applies the idea of intelligent agents where its structure is defined as a specific environment in which agents are situated. The Capabilities of agents include both ability to adapt and ability to learn.

Adaptation implies sensing the environment and reconfiguring in response. This can be achieved through the choice of alternative problem solving rules or algorithms, or through the discovery of problem solving strategies. Learning may proceed through observing actual data, and then it implies a capability of introspection and analysis of behavior and success. Alternatively, learning may proceed by example and generalization, and then it implies a capacity to abstract and generalize. We could group agents used in this work into five classes based on their degree of perceived intelligence and capability: Simple reflex agent, Model-based reflex agent, Goal-based agent, Utility-based agent, and learning agent Figs. 1 through 5 [20].

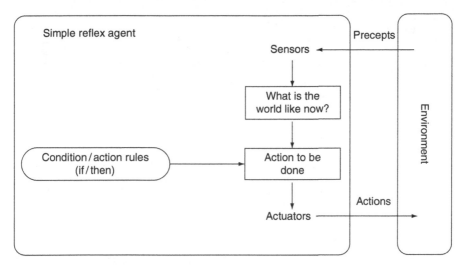

Fig. 1 Block diagram of a simple reflex agent. This type acts only on the basis of the current percept. The agent function is based on the condition-action rule

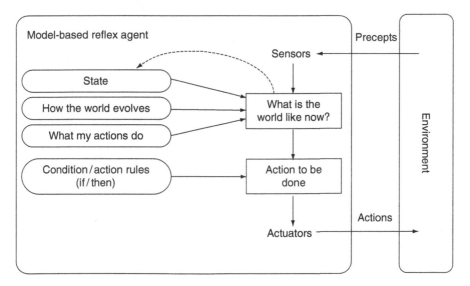

Fig. 2 Block diagram of a model-based reflex agent. This type can handle partially observable environments by keeping track of the current state of its world

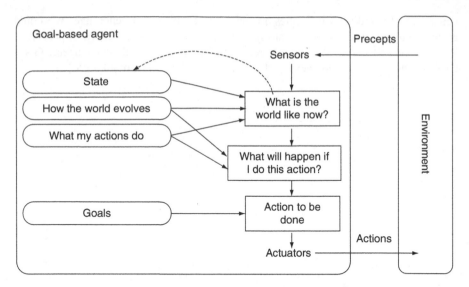

Fig. 3 Block diagram of a goal-based agent. These agents are model-based agents, which store information regarding desirable situations

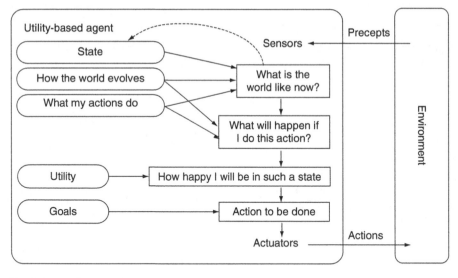

Fig. 4 Block diagram of a utility-based agent. It is possible to define a measure of how desirable a particular state is. This measure can be obtained through the use of a utility function which maps a state to a measure of the utility of the state

The motion for all agents is performed in a stepwise fashion over the space defined by a regular 2-dimensional grid. Within the grid each object or molecule occupies a square. The system development will be considered at discrete time

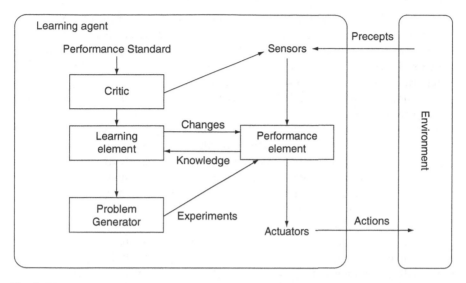

Fig. 5 Block diagram of a learning agent: Learning has an advantage that it allows the agents to initially operate in unknown environments and to be-come more competent than its initial knowledge alone might allow

steps $t_m = m * \Delta t$, $(m = 0, 1, 2, \ldots)$ with a fixed length Δt. In general, models are constructed to enable reasoning within an idealized logical framework about such processes and are an important component of scientific theories. Idealized here means that the model may make explicit assumptions that are known to be incomplete in some detail. Such assumptions may be justified on the grounds that they simplify the model while, at the same time, allowing the production of acceptably accurate solutions. We could summarize some of the essential points that were taken into consideration:

- Rule-based agents are the basic entities in the system whose:
  Behavior is non-deterministic.
  Interactions and responding abilities are governed by a set of rules that define their behavior.
  Rules will often contradict and there must be some mechanism for selecting over them.
  Not aware of the system macro-view although having the capability of responding to local environmental factors as well as external dynamics and signaling under various conditions.
  Not able to know the state and current behavior of every other agent.
- Only four common types of viral vectors and two of non-viral vectors are supported: Retroviruses, Adenoviruses, Adeno-associated viruses, Herpes simplex viruses, Liposomes, and Naked DNA.
- Each vector have six main properties that affect its behavior and uses: Type, Target, Max carrying length, Activation, Integration, Side effects.

- A virus has a glycoprotein spikes on its surface that helps it to enter or exit the cell. The H spikes help a virus enter a cell, and the N spikes help it leave.
- After virus transcription, viral proteins are made in the cytoplasm using the host cell's ribosomes. Some of these viral proteins move to the nucleus to aid in viral replication, others move to aid in virus assembly.
- Probability of infection is highly dependent on both viral propagation and concentration.
- Applying electric pulses to a cell, a process called electroporation, create transient pores in the plasma membrane that allow transport of traditionally nonpermeant molecules into cells via both diffusion and electrophoresis.
- In order to successfully deliver non-permeant molecules into cells they must be located within a critical distance during electroporation in order to reach the plasma membrane before the pores are closed.
- Delivering non-permeant molecules into cells using electric pulsation depends on several factors including the field magnitude, pulse duration, pulse interval, pore size, molecules concentration, cell size, cell type and its cellular response.
- Cells differentiation and regeneration are dynamic phenomena, which are highly related to time factors, interaction dependent and appear to be stochastic.
- Cell differentiation follows a hierarchical organization.
- Cells which have oscillatory chemical reactions within, could differentiate through interaction with other cells.
- Differentiated state of a cell is preserved by the cell division and transmitted to its offspring.
- During cell division, each offspring has two choices, either to preserve its parent state or to change it, and the probability of each choice seems to depend also on both its stability and differentiability.
- A kind of memory is formed through the transfer of initial conditions (e.g. chemical compositions) to offspring.
- Cells with high level of stemness could lead to several distinct types of tissues, these abilities decrease with lower levels.
- Stem cells normally reside in a specific micro-environmental niche. These cells can migrate from one crypt to another within their system.
- Maturity is the capability level of a cell for being able to accomplish tasks and functions and usually when it increases the stemness level decreases.
- The cell society as a whole is stabilized through cell-to-cell interactions e.g. when the number of one type of cells is decreased by external removal, the distribution is recovered by further differentiations, replications, renewal, and regeneration.
- Each cell has its own internal dynamics and genetic switching according to its state and condition.
- Interactions among cells are given by the diffusive transport of chemicals between each cell and its neighboring according to their environment.
- Within each cell, a network of biochemical reactions also includes reactions associated with genetic expressions, signaling, pathways, chemical concentrations, and so on.

- A cellular state is represented by both its internal and external dynamics.
- Cell volume is treated as a dynamical variable that increases as a result of transportation of chemicals from the environment, which leads to the cell division when it is larger than a given threshold.
- Here the environment doesn't mean external environment for individual organism, but is intended as interstitial environment of each cell.
- The concentrations of chemicals in the daughter cells are almost equal to the concentrations of the mother cell.
- The rates of chemicals transported into a cell are proportional to differences of chemicals concentrations between cell inside and outside.
- Some chemicals can penetrate the membrane and others cannot. The behavior of dynamics depends on their number and ability. This penetration ability should be different for different cell states.
- Each cell takes penetrable chemicals from the medium, while the reactions in the cell usually transform them to impenetrable chemicals.
- The nature of internal dynamics depends on the choice of the reaction network, and on the number of paths in the reaction matrix. When the number of paths is small, cellular dynamics may falls into a steady state without oscillation. On the other hand, when it is large, many chemicals generate each other. Then chemical concentrations take constant values, which are often almost equal. Any way it is not easy to estimate the number of paths in real biochemical data, although they may suggest the medium number of paths.
- Number of penetrating chemicals is another control parameter for capacity of oscillation or differentiation. When it is small, the rate of reaction networks, which show oscillatory dynamics, is small. On the other hand, when it is too large, it would be difficult to have oscillatory dynamics.
- The oscillatory dynamics is rather common as the number of auto-catalytic paths is increased.
- Tiny changes during division amplify the cell instability. When it exceeds some threshold, the differentiations start. Then the emergence of another cell type stabilizes the dynamics of each cell again.
- Each internal dynamics of a certain type of cells is gradually modified with the change of distribution of other cells.
- There is a higher level dynamics, which controls the rate of cell division and differentiation according to the number of each cell type. These dynamics should be stochastic, and robust against external perturbations.

4 Simulation Results

Stem cell lines vary in their characteristics and behavior not only because they are derived from genetically outbred populations, but also because they may undergo progressive adaptation upon long-term culture in vitro.

Hence, the task is to model the dynamic processes that drive and control the cellular attributes. Clearly, it is presently impossible to describe these processes in any reasonable detail. Therefore, it will be necessary to propose a simplified basic scheme of the cellular dynamics.

Therefore, different simulation scenarios were observed, each with different results, and different error rates lies in the range of acceptable values in this domain of application. We present the results for some of these simulations along with comparisons to experimental results.

To validate utility and performance we attempt to simulate and observe 16 fundamentally different but important cellular phenomena for which experimental measurements were available or for which results are well characterized. These included: (1) Intracellular Signaling; (2) Signaling Ligands; (3) Transcription Factors; (4) Cell Surface Receptors; (5) Cell Cycle; (6) Apoptosis; (7) DNA Repair; (8) Protein Synthesis; (9) Protein Folding; (10) Protein Processing; (11) RNA Binding; (12) Metabolism; (13) Transport; (14) Chromatin Regulators; (15) Cytoskeleton; (16) Adhesion. These processes span the range of most kinds of cellular behavior, from simple chemical kinetics to more complex metabolism and genetic circuits. For a number of these phenomena we investigate its utility in monitoring or simulating the effects of finite numbers and stochasticity on certain processes.

For observing non-viral transfection effect we attempt to simulate the process of electroporation of Chinese hamster ovary cells (CHO), which are spherical in suspension with a mean diameter of 13.5F1.4 Am, that isn't significantly different from control cells (P = 0.01) [21, 22]. After the aphidicolin (a reversible inhibitor of nuclear DNA replication, used for blocking the cell cycle at early S phase) treatment, cells synchronized in G_2/M phase significantly increased in size, their diameter reaching 17.1F1.55 Am. Indeed, in aphidicolin synchronized cells, synthesis of proteins and lipids has been shown to be still present while cell cycle was blocked. These syntheses led to an increase of cell volume, which is in agreement with previous experimental values investigated by Muriel, et al. [21], and Sukhorukov, et al. [22]. However, cells were pulsed under optimum conditions for gene transfer. Ten pulses, lasting 5 ms, were applied. Under those conditions, 63% of control cells were permeabilized at 0.8 kV/cm and 81% at 0.9 kV/cm (Fig. 6a). The associated fluorescence intensity, related to the number of molecules incorporated into the electropermeabilized cells (Fig. 6b), increased with an increase in the electric field intensity. The amount of electroloaded molecules in permeabilized cell increased with the electric field intensity, in agreement with previous experimental results [21, 23]. The percentage of permeabilized cells and the permeabilization efficiency for cells synchronized in G1 phase were slightly smaller than for control cells (Fig. 6a, b). Cells synchronized in G2/M phase exhibited a one and a half fold increase in the percentage of permeabilization at 0.8 kV/cm and a slight increase at 0.9 kV/cm (Fig. 6a). This synchronization effect was more pronounced when fluorescence intensities were compared. Electropermeabilized cells in G2/M phase exhibited a two-fold increase in fluorescence intensity at 0.8 and 0.9 kV/cm (Fig. 6b). However, this increase had to be associated with the fact that in G2/M phase, cells had 4N instead of 2N chromosomes and therefore two times more sites

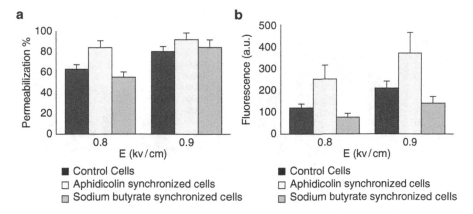

Fig. 6 Simulation results: (**a**) Percentage of permeabilized cells. Cells were submitted to 10 pulses of 5 ms duration in pulsing buffer having propidium iodide. Permeabilization was determined based on total number of cells. (**b**) Associated mean fluorescence level of the permeabilized cells. "Values represent means"

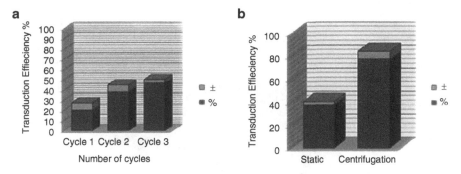

Fig. 7 (**a**) Effect of transduction cycle number on efficiency of transduction, error rate = 11.64%. (**b**) Static transduction versus centrifugational transduction, error rate = 5.22%

for PI interaction [21, 24]. These simulation values were in agreement with previous experimental values investigated by Muriel, et al. [21].

For observing viral transduction effect we attempt to simulate the process of self-renewal and differentiation of Human Fetal Mesenchymal Stem Cells (hfMSCs) after transduction with onco-retroviral and lentiviral vectors. Retroviral and lentiviral approaches offered the initial methodology that launched the field, and established the technological basis of nuclear reprogramming with rapid confirmation across integrating vector systems [25, 26]. However a static transduction of three samples of first trimester fetal blood-derived hfMSCs from 9^{+1}, 9^{+6}, and 10^{+6} weeks of gestation revealed a mean efficiency of 20.3% ± 7.2%, 38.6% ± 6.4%, and 47% ± 3.2% after 1, 2, and 3 cycles of transduction, respectively with an error rate = 11.64% (Fig. 6c, 7, 8 and 9). Simulating transduction under centrifugational forces

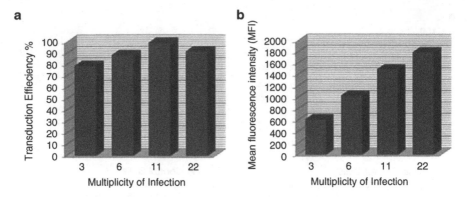

Fig. 8 (a) Percentage of lentiviral-transduced cells with varying MOI, error = 5.4%. (b) Mean fluorescence intensity with varying MOI after lentiviral transduction, error = 9.8%

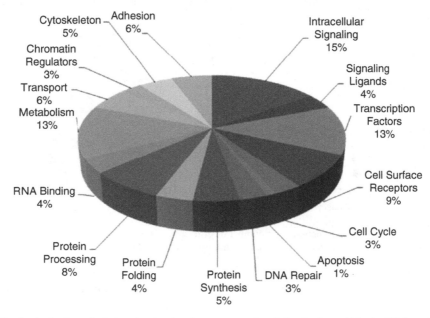

Fig. 9 A pie chart depicting the simulated average results of distribution within the HSC profile for gene products with known or putative functions

also demonstrated roughly a twofold increase in transduction efficiency (n = 3) from 37.8% ± 2.6% to 77.3% ± 6.8% after 2 cycles (paired t-test; p = 0.002) with an error rate = 5.22% (Fig. 6d). And using a lentiviral vector agent, transduction efficiencies were of 98.9% ± 1.1% having error rate = 5.4% with MFIs of 2,117 ± 633 were achieved with the same three samples and with error rate = 9.8% (Fig. 6e,f). This was achieved after a single cycle at MOI of 11. Raising viral titers was also detrimental to the hfMSCs, with massive cell death observed after MOI of >22.

These simulation values were in agreement with previous experimental values investigated by Chan, et al. [27].

Generally, for observing the basic cellular phenomena we attempt to simulate the genome wide transcriptional changes during early stages of hematopoietic differentiation (Fig. 6g) and the simulated values were in agreement with previous experimental values investigated by Ivanova, Dimos et al. [28].

References

1. A.L. David, D. Peebles, Gene therapy for the fetus: is there a future? Best Pract. Res. Clin. Obstet. Gynaecol. 22(1), 203–218 (2008)
2. M. Baker, (2007-12-06). Adult cells reprogrammed to pluripotency, without tumors. Nature Reports Stem Cells. Retrieved on 2007-12-11
3. R.S. Oliveri. Epigenetic dedifferentiation of somatic cells into pluripotency: cellular alchemy in the age of regenerative medicine? Regen. Med. 2(5), 795–816 (Sept 2007)
4. P. Collas, C.K. Taranger, Toward reprogramming cells to pluripotency. Ernst Schering Res Found Workshop. 2006;(60):47–67
5. J. Yu, J. Domen, D. Panchision, T.P. Zwaka, M.L. Rohrbaugh, et al. Regenerative medicine, NIH Report, 2006
6. A. Atala, Recent developments in tissue engineering and regenerative medicine. Curr. Opin. Pediatr. 18(2), 167–171, 2006
7. Y. Junying, J.A. Thomson, *Regenerative Medicine*, 2006
8. M. Colavito, M.A. Palladino, *Gene Therapy*, 1st edn. (Benjamin Cummings, Aug 2006) ISBN-10: 0805338195
9. D.V. Schaffer, W. Zhou, *Gene Therapy and Gene Delivery Systems* (Springer, Feb 2006) ISBN: 3540284044
10. A. Battler, J. Leor, *Stem Cell and Gene-Based Therapy: Frontiers in Regenerative Medicine* (Springer, June 2007) ISBN: 184628676X
11. J. Phillips, C. Gersbach, A. Garcia, Virus-based gene therapy strategies for bone regeneration. Biomaterials 28, 211–229 (2007)
12. Y. Yi, S.H. Hahm, KH. Lee, Retroviral gene therapy: safety issues and possible solutions. Curr. Gene Ther.5, 25–35 (2005)
13. L.F. Maxfield, C.D. Fraize, J.M. Coffin, Relationship between retroviral DNA-integration-site selection and host cell transcription. Proc. Natl. Acad. Sci. USA 102, 1436–41 (2005)
14. W. Weber, M. Fussenegger Pharmacologic transgene control systems for gene therapy. J Gene Med 8(5), 535–556 (2006)
15. S. Mehier-Humbert, R.H. Guy, Physical methods for gene transfer: improving the kinetics of gene delivery into cells. Adv. Drug Deliv. Rev. 57 733–753 (2005)
16. C. Capaccio, N.S. Stoykov, R. Sundararajan, D.A. Dean, Quantifying induced electric field strengths during gene transfer to the intact rat vasculature. J. Electrostat. 67 652–662, (2009)
17. J.P. Roberts, Gene therapy's fall and rise (again). Scientist 18(18), 22–24 (2004)
18. U. Bastolla, M. Porto, H.E. Roman, M. Vendruscolo, *Structural Approaches to Sequence Evolution: Molecules, Networks, Populations*, 1 edn. (Springer, Jan 2007) ISBN-10: 3540353054
19. A.J. Nair, *Introduction to Biotechnology and Genetic Engineering*. Firewall/isp (May 2008) ISBN: 8131803392
20. S.J. Russell, P. Norvig, Artificial Intelligence: A Modern Approach, 2nd edn. (Prentice Hall, Upper Saddle River, NJ, 2003) ISBN: 0-13-790395-2

21. M. Golzio, J. Teissie, M.-P. Rols, Cell synchronization effect on mammalian cell permeabilization and gene delivery by electric field. Biochimica et Biophysica Acta 1563(2002), 23–28, (2002)
22. V. Sukhorukov, C. Djuzenova, W. Arnold, U. Zimmermann, J. Membr. Biol. 142 77–92 (1994)
23. M.P. Rols, J. Teissie, Biophys. J. 58 1089–1098 (1990)
24. H.A. Crissman, J.A. Steinkamp, Eur. J. Histochem. 37 (1993) 129–138, 1993
25. T.J. Nelson, A. Terzic Induced pluripotent stem cells: reprogrammed without a trace. Regen. Med. (2009). doi:10.2217/rme.09.16
26. C.A. Sommer, M. Stadtfeld, G.J. Murphy et al. Induced pluripotent stem cell generation using a single lentiviral stem cell cassette. Stem Cells (2009). doi:10.1634/stemcells.2008-1075
27. J. Chan, K. O'Donoghue, J. de la Fuente et al. Human fetal mesenchymal stem cells as vehicles for gene delivery. Stem Cells 23, 93–102 (2005)
28. N.B. Ivanova, J.T. Dimos, C. Schaniel et al. A stem cell molecular signature (Sept 2002). doi:10.1126/science.1073823

Chapter 42
DFA, a Biomedical Checking Tool for the Heart Control System

Toru Yazawa, Yukio Shimoda, and Tomoo Katsuyama

Abstract We made our own detrended fluctuation analysis (DFA) program. We applied it to the cardio-vascular system for checking characteristics of the heartbeat of various individuals. Healthy subjects showed a normal scaling exponent, which is near 1.0. This is in agreement with the original report by Peng et al. published in 1990s. We found that the healthy exponents span from 0.9 to 1.19 in our temporary guideline based on our own experiments. In the present study, we investigated the persons who have an extra-systole heartbeat, so called as premature ventricular contractions (PVCs), and revealed that their arrhythmic heartbeat exhibited a low scaling exponent approaching to 0.7 with no exceptions. In addition, alternans, which is the heartbeats in period-2 rhythms, and which is also called as the harbinger of death, exhibited a low scaling exponent like the PVCs. We may conclude that if it would be possible to make a device that equips a DFA program, it might be useful to check the heart condition, and contribute not only in statistical physics but also in biomedical engineering and clinical practice fields; especially for health check. The device is applicable for people who are spending an ordinary life, before they get seriously heart sick.

1 Introduction

The healthy heart muscles can contract periodically and rhythmically, so the heart can supply the blood to the entire body through the cardio-vascular system. Since significant decrease or complete stop of the blood flow caused by heart's illness can immediately be lethal, the sick hearts are evidently facing death at every turn more than the healthy heart. Regardless of advancement of medical technology,

T. Yazawa (✉)
Department of Biological Science, Tokyo Metropolitan University, 1-1 Minami-Ohsawa, Hachioji, 192-0397 Tokyo, Japan
e-mail: yazawa-tohru@tmu.ac.jp

S.-I. Ao et al. (eds.), *Machine Learning and Systems Engineering,*
Lecture Notes in Electrical Engineering 68,
DOI 10.1007/978-90-481-9419-3_42, © Springer Science+Business Media B.V. 2010

cardio-vascular failure especially life threatening attacks are still taking place unpredictably in normally looking subjects. Gradually changing activity of cellular and tissue systems must be involved. Therefore, detection of "hidden" early symptom of normally looking subjects is one of the main goals. For understanding hidden information in the control system, nonlinear analysis has been proposed to be an applicable method since decades ago. Indeed, the DFA is a method that is potentially useful to diagnose a heart condition [1]. However, it has not been appeared as a practical tool for the use in a bio-medical engineering and clinical practice field. We have made our own DFA program (by K. Tanaka) [2]. We here demonstrate that our method can be useful to calculate the scaling exponent for persons, who have an extra-systolic heartbeat, for persons who have alternans-heartbeats, and for other types of heartbeats, including normal, healthy hearts.

2 Methods

2.1 Finger Blood-Pressure Pulse

Heartbeats have been recorded by a conventional electrophysiological method. For sensing the heartbeats, we performed it either electrically (EKGs) or mechanically (blood pressure pulses) in our series of studies. From human subjects we mostly used the finger pulse recording with a Piezo-crystal mechano-electric sensor, connected to a Power Lab System (AD Instruments, Australia). All data were registered at 1 kHz in rate of sampling.

2.2 DFA Methods

We made our own programs for measuring beat-to-beat intervals (by Tanaka) and for calculating the approximate scaling exponent. The DFA program (by Tanaka) can instantly calculate the scaling exponents if a time series data exist.

After recording EKG or finger pressure pulses, we first made the beat-to-beat interval time series. Preferable data length could be from some hundreds to about 2,000 heartbeats. From the time series we calculated the approximate scaling exponent. For preparing figures in this report, we presented (1) heartbeat-data by plotting heart rates against time, which is the representation of the original data used for the DFA, (2) he approximate scaling exponents calculated by the DFA, and (3) a graph with one or sometime three least mean square regression lines. From the graph, one can read an approximate slope to which the exponent corresponds directly. While beating hearts show a fluctuating pattern, we hardly predict the scaling exponent by eye-observation of such dynamic patterns visualized as the time series. Only a DFA program can do that. The scaling exponent reflects hidden information carried by quantitative measures in a dynamically changing interval of heartbeat and velocity of blood flow under the functional control of the autonomic nervous system. An EKG-chart data cannot tell us what the scaling exponent is. Mathematical procedures of DFA and explanation of biomedical meaning have been documented elsewhere [1, 3].

2.3   *Volunteers and Ethics*

Subjects were selected from colleagues of university laboratories including ourselves and companies, individuals of which visited to our exhibition place and asked us to record and test his/her heartbeats at the Innovation Japan 2006, 2007, 2008, and 2009, for example, NOMS Co. Ltd., Nihon-Koden Co. Ltd. and Toshiba Co. Ltd, across over 100 companies. All subjects were treated under the ethical control regulation of our University Committee.

3   Results

3.1   *Extra-Systole*

The subject, whose heart had an arrhythmia due to the ventricular extra-systole, exhibited an abnormal record such as shown in Fig. 1. This is so called as premature ventricular contractions (PVCs). We here show one example obtained from a female subject age 58. We have over 20 subjects who exhibited the ventricular extra-systole. All of them, and importantly with no exceptions, showed the same results; that is, a lower exponent.

In the ventricle extra-systole, ventricle muscle generally tried to generate its own rhythm before normal rhythmic excitation arrived to the ventricle. Therefore a

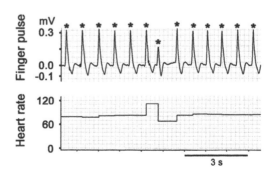

Fig. 1 Typical extra-systole. Recorded from a finger by a piezo-crystal pulse-recording device. We consider a small beat is one heartbeat (see asterisks). 58 years female

lower/smaller pulse, comparing to a normal pulse, was generated like a small pulse in Fig. 1. But this PVC is considered by physicians as a not so serious problem of the heart, if this happens less frequently. Thus, it is "benign" arrhythmia, though we cannot agree with the idea, because of DFA results as shown below. If it may happen a lot, like more than 10 times per 1 min, physicians may take great care of it finally.

As shown in Fig. 2, this person's extra-systole appeared less frequently, the count of which was 12 within the span of the record. The period length of this recording was for about 5,800 heartbeats, which corresponded to about 1.5 h. Therefore, this person's heart is "normal" in terms of a physician's guideline. However, we found out that our DFA's results was that the subject's heart shown in Figs.1 and 2 exhibited low exponents, about 0.7. Followings are brief description how we calculate the scaling exponents.

For the computation of the scaling exponents, at a final step, we need to calculate the slope in the graph of *Box-size* vs. *Variance* as shown in Fig. 3. Here we used *Box-size* across a range 100–1,000. Before this final step, there are procedures that are much complex. In the mid-way of the DFA, the least-mean square methods were involved which are linear-, quadratic-, cubic-, and bi-quadratic-fitting methods. Details of those "mid-way" procedures are not shown here. The graph Fig. 3 is only the final step. The scaling exponents, being made after a measurement of such a least-mean square fitting functions are indicated as inset. When we repeated linear-, quadratic-, cubic-, and bi-quadratic fitting, we are finally able to obtain a steady result of calculation. For example, there is no big difference between B and C

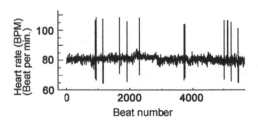

Fig. 2 Time series data of a long term recording. Twelve PVCs (premature ventricular contractions) are recorded. 58 years female shown in Fig. 1

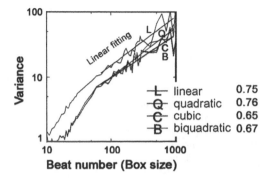

Fig. 3 An example DFA. The scaling exponent for this heart is 0.65 or 0.67. 58 years female

in Fig. 3. Consequently, we learn that the result B (and C) is the scaling exponent. From this calculation, an approximate scaling exponent was finalized for this subject. The figure indicates that the scaling exponent of this person is way down below from the normal value (i.e., 1.0). We determined that the scaling exponent for this heart is about 0.7.

This (0.7) is not normal. We were intrigued her heartbeats and she kindly allowed us to measure finger pulses multiple times over the course of 6 years. Figure 4 shows the result of our "automatic calculation" program. The data length was about 7,000 beats. The scaling exponents shown here is 0.91 and 0.65. The former value corresponds to 30–90 box-size and the latter to 70–240 box-size. In a short window size, i.e., 30–90, her heart looks fine but in long window size, 70–240, which is similar range to those shown in Fig. 3, we obtained a low scaling exponent, 0.65. Her body system is not perfect. Indeed, we never see the perfect 1/f rhythm in her data.

The 'skipping heartbeat' makes the subject feel very uncomfortable. She described that she was not feeling very well, even though the physician says it is okay. It might be not very good psychologically. Needless to say, ventricular tachycardia is a serious cause for a life threatening cardiac dysfunction. Importantly, this tachycardia is derived from randomly repetitive, very fast beatings, in other words, a series of extra-systoles. Although the causes of the extra-systole have

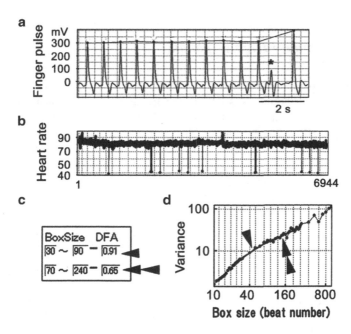

Fig. 4 An example of automatic computation of the DFA. The same subject shown in Figs. 1 through 3. (**a**) Finger-pulse test, "*" An asterisk shows PVC. (**b**) About 7,000 heartbeats. Sporadically appearing extra-systoles can be seen by downward swings. (**c**) DFA results at different box size. (**d**) Slopes corresponding to the scaling exponents in (**c**). 58 years female

not been fully understood physiologically by now, we confirmed by DFA that at least one reason of this person's extra-systole is due to the autonomic nervous system dysfunction, because, surprisingly enough, this person's extra-systole totally disappeared later. After 2 years, she nicely explained that her (58 years) environments had improved, in other words, the psychological stress of her life had been decreased during the years. As a result, we luckily discovered that the reason why the person's scaling exponent was pushed up as close as to 0.9 (data not shown).

3.2 Alternans with Low Exponent

We have examined DFA of an Alternans person (Fig. 5). The Alternans was documented in 1872 by Traube. However, the alternans did not get particular attention from doctors until recently, indeed before alternans was noticed as harbinger of death. We finally found that alternans heartbeat always exhibited a low scaling exponent when we measure heartbeats of animal models (crustacean animals such as crabs and lobsters, n = 13) and humans (n = 8, five Japanese and three Americans).

3.3 Extraordinary High Exponent

We have met a person who had a high scaling exponent, which was 1.5 at 30–60 box-size (Fig. 6). This man was tall but not fat, and apparently healthy. The time series look like as being normal (Fig. 9b). However, his heart rate at rest was relatively high (HR = 70 per min, Fig. 9b) comparing to that of normal man (HR = 60 per min, Fig. 9a). We wondered why. We asked him a possible explanation, and finally found out that he has two more brothers, who both had already died because

Fig. 5 Alternans shows very low scaling exponents. Heart rate trace clearly shows an abnormality of its rhythm. The scaling exponents were 0.5–0.7 at different box size. 58 years male. We have four human alternans subjects and many model animals too. All of them exhibited a low scaling exponent like this example

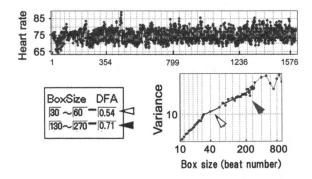

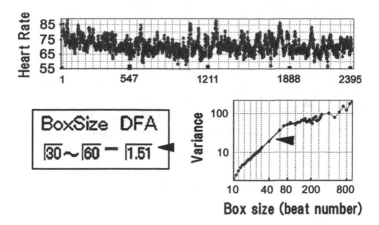

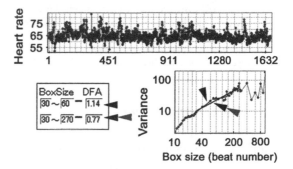

Fig. 6 A DFA example for a person who has a high scaling exponent. 59 years male

Fig. 7 A typical normal DFA for a middle aged subject. 56 years female

of detail-unknown heart problem. This high exponent might be some sign for a certain heart or heart-control system problems, although we have no idea for the reason. We consider that cardiac physicians and geneticists may identify the reason, why the family has a potential risk, if they study his family.

3.4 Normal Exponent

The DFA revealed that middle age people often show a normal exponent or a little lower exponent (Fig. 7). As shown in Fig. 7, the scaling exponents were unstable at a different box size. We would say that this person's heart could be normal, but having a stressful life, because the DFA showed roughly 0.77 when measuring the entire range of the box size. If a person is in perfect health, the person would have an exponent of 1.0 at entire range of the box size.

Fig. 8 EKG obtained at
hospital health check.
54 years male

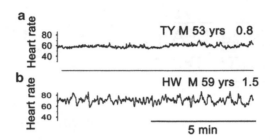

Fig. 9 Time series data of two persons. (a) Normal size healthy mail, age 53. This heartbeat looks normal. This person's EKG during this health check (Fig. 8) showed no problem with his heart, but his scaling exponent was 0.8, slightly below than normal (1.0). The detailed examination of his pulse record finally revealed that he had abnormal heartbeats. (b) A tall, close to 2 m high, healthy mail, age 59. This time the series again looked kind of normal. But this person had a scaling exponent of 1.5 (Fig. 6). Both (a) and (b) were recorded simultaneously at rest. Note that heart rate of (b) is greater than that of (a) in average, though (b) is greatly taller than (a). Generally, big body mass lowers heart rate but here not the case

3.5 DFA Is Beneficial

For the medical examination of EKG, usually five or six pulses has been recorded as shown in Fig. 8. In this record, the each wave form looks normal and also it shows that there are no Q-T elongation and no S-T elevation. However, the DFA revealed, that his heart exhibits a low scaling exponent, 0.8 (see Fig. 9a). So we performed a detailed examination of his pulse record for a period of 2 h. And finally we discovered that he had an abnormal heartbeat, such as an extra-systole and even tachycardia too, although those events occurred rarely (data not shown). The findings imply that his cardio-vascular control system must have some problems. Even though we cannot tell what the physiological mechanism is at the moment, DFA can find out the some physiological problem on cardio-vascular system.

4 Discussion

To perform DFA we normally need a pretty long term recording of heartbeats [1]. On the other hand, for an ordinal EKG done for a hospital health check, it would take a very short period of time. Indeed, a typical EKG needs only 6 beats or so, as shown in Fig. 8. Therefore, EKG recordings at medical health check unfortunately are useless for the DFA due to the short period length of recording.

Other important recording condition is the rate of digital sampling. Our DFA method needs the heartbeat recording at 1 kHz sampling rate, i.e., 1,000 dot per second. This notion was derived from our own previous research. According to our own cardio-vascular experiments on crabs and lobsters and much older great heritages of the crustacean heart research [4, 5, 11], we have had experienced that cardio-regulatory nerve impulses exhibits fluctuation of about 1 ms in time scale [6, 7]. Since the autonomic nerve activities directly affect the fluctuation of the rate of heartbeat [8], we must observe the fluctuation of the heartbeat in the time scale of 1 ms at a lowest limit. Therefore, we considered that the heartbeat fluctuation must be observed with a time scale in millisecond order, although the size of data increases dramatically comparing to the way of ordinal EKG recording (normally 4 ms). This notion which is derived from our crustacean studies must be applicable to the human heart too, because life on earth uses identical DNA-genetic control systems as the conventional way in the development of structure and function of the cardio-vascular system [9–11]. Unfortunately, the conventional Holter EKG recording system has not equipped 1 ms resolution in time scale (about 4–8 ms). So, we needed to record human heartbeat by our own machine. This is the reason why we preferred our own system to collect data instead of using the existing data base. Thus, we are always performing recording at a rate of 1 ms (1 kHz sampling) for the rate of data sampling.

Using the DFA, and using the principles of 1/f scaling in the dynamics of the cardio-vascular system [12], we have demonstrated the scaling exponents reflect the balance of the control system, that is, the autonomic nervous system. In other words, the DFA can determine whether the entire body system is functioning in sick or in healthy, simply by the scaling exponent. If we notice that the scaling exponents are lower, we must assume that the subjects are not normal and the subjects might be exhibiting abnormal "stressful" heartbeats.

Empirical evidences illustrated the dynamics of adaptive heart movement behavior. The dynamic behavior results from in a range of multi-elementally actions in the heart and nerve cells. Those are including molecules contributing for heart muscle contraction, molecules for regulating ionic flow of the myocardium, molecules for generating pace-making signals for contraction, molecules for neuro-muscular transmission, and so on. We expect that future research will identify parameters that determine the interaction between a heart function (e.g., generating rhythmically from inside the heart) and control processes (e.g., generating rhythmically from inside the brain). Moreover, we must consider the interaction between the heart and external factors in the environment (e.g., circadian rhythms) and biological, behavioral and medical factors (e.g., stress, fighting, and disease). However, we must pay attention to the fact that all of such elements are nonlinearly connected. That means that the best way to identify parameters is the way of complex dynamics analysis performing together with traditional physiological methods. Therefore, collaborating intimately together of biology and physics is the evocative solution in the future study.

We might conclude that the DFA has benefits. As a tool for diagnosing the condition of the cardio-vascular system, DFA can find out something more than those tests, which are not be able to be checked out by a simple traditional EKG.

Acknowledgment We are very grateful to all volunteers, for allowing us to test their heartbeats by finger pulse tests. This research was supported by the grants 2008-JST04-065, H20DG407 NOMS, and the grant TMU-San-Gaku-Ko-H20Project-D56I0106.

References

1. C-K. Peng, S. Havlin, HE. Stanley, AL. Goldberger, Quantification of scaling exponents and crossover phenomena in nonstationary heartbeat time series. Chaos **5**, 82–87 (1995)
2. T. Yazawa, K. Tanaka, T. Katsuyama, Neurodynamical control of the heart of healthy and dying crustacean animals. Proceedings CCCT05, vol. 1 (Austin, TX, USA, July 24–27 2005), pp. 367–372
3. T. Katsuyama, T. Yazawa, K. Kiyono, K. Tanaka, M. Otokawa, Scaling analysis of heart-interval fluctuation in the in-situ and in-vivo heart of spiny lobster. Panulirus japonicus, Bull. Univ. Housei Tama **18**, 97–108 (2003) (Japanese)
4. J.S. Alexandrowicz, The innervation of the heart of the crustacea. I. Decapoda, Quat. J. Microsc. Sci. **75**, 181–249 (1932)
5. D.M. Maynard, Circulation and heart function. *The Physiology of Crustacea*, vol. 1 (Academic, New York, 1961), pp. 161–226
6. T. Yazawa, K. Kuwasawa, Intrinsic and extrinsic neural and neurohymoral control of the decapod heart. *Experientia* **48**, 834–840 (1992)
7. T. Yazawa, T. Katsuyama, Spontaneous and repetitive cardiac slowdown in the freely moving spiny lobster, Panulirus japonicus. J. Comp. Physiol. A **87**, 817–824 (2001)
8. J.P. Saul, Beat-to-beat variations of heart rate reflect modulation of cardiac autonomic outflow. News Physiol. Sci. **5**, 32–27 (1990)
9. W.J. Gehring, *Master Control Genes in Development and Evolution: The Homeobox Story* (Yale University Press, New Haven, 1998)
10. I. Sabirzhanova, B. Sabirzhanov, J. Bjordahl, J. Brandt, P.Y. Jay, T.G. Clark, Activation of tolloid-like 1 gene expression by the cardiac specific homeobox gene Nkx2-5. *Dev. Growth Differ.* **51**, 403–410 (2009)
11. J.L. Wilkens, Evolution of the cardiovascular system in Crustacea. Am. Zool. **39**, 199–214 (1999)
12. M. Kobayashi, T. Musha, 1/f fluctuation of heartbeat period. IEEE Trans. Biomed. Eng. **29**, 456–457 (1982)

Chapter 43
Generalizations in Mathematical Epidemiology Using Computer Algebra and Intuitive Mechanized Reasoning

Davinson Castaño Cano

Abstract We are concerned by imminent future problems caused by biological dangers, here we think of a way to solve them. One of them is analyzing endemic models, for this we make a study supported by Computer Algebra Systems (CAS) and Mechanized Reasoning (MR). Also we show the advantages of the use of "CAS" and "MR" to obtain in that case, an epidemic threshold theorem. We prove a previously obtained theorem for S^nIR endemic model. Moreover using "CAS+MR" we obtain a new epidemic threshold theorem for the S^nI^mR epidemic model and for the staged progressive SI^mR model. Finally we discuss the relevance of the theorems and some future applications.

1 Introduction

At the moment, we are at the edge of a possible biological problem. Some people say that the nineteenth century was the century of chemistry, the twentieth was the century of physics, and they say that the twenty-first will be the century of biology. If we think, the advances in the biological field in the recent years have been incredible, and like the physics and its atomic bomb, with biology could create global epidemics diseases. Also the climate change could produce a new virus better than the existing virus, creating an atmosphere of panic. For these reasons and others, we think in a solution using mathematical models with computer algebra and mechanized reasoning. Specifically we consider the SIR (Susceptible-Infective-Removed) model, with differential susceptibility and multiple kinds of infected individuals. The objective is to derive three epidemic threshold

D.C. Cano
Logic and Computation Group, Engineering Physics Program, EAFIT University, Carrera 49 N° 7 Sur, 50, Medellín, Suramérica, Colombia
e-mail: dcasta12@eafit.edu.co

S.-I. Ao et al. (eds.), *Machine Learning and Systems Engineering*,   557
Lecture Notes in Electrical Engineering 68,
DOI 10.1007/978-90-481-9419-3_43, © Springer Science+Business Media B.V. 2010

theorems by using the algorithm MKNW given in [1] and a little bit of mechanized reasoning.

Briefly the MKNW runs on: Initially we have a system of ordinary non-lineal differential equations **S**, whose coefficients are polynomial. We start setting all derivates to zero for finding equilibrium; we solve the system finding the equilibrium point **T**. Then we compute the Jacobian **Jb** for the system **S** and replace **T** in **S**. We compute the eigenvalues for **Jb**; from the eigenvalues we obtain the stability conditions when each eigenvalue is less than zero. Finally we obtain the reproductive number for the system **S** in the particular cases. Using deductive reasoning we obtain some theorems based on the particular cases.

The MKNW algorithm is not sufficient to prove the threshold theorems that will be considered here and for this reason, it is necessary to use some form of mechanized reasoning, specifically some strategy of mechanized induction.

The threshold theorem that we probe in Section 2 was originally presented in [2] using only pen and paper and human intelligence. A first contribution of this paper is a mechanized derivation of such theorem using CAS.

The threshold theorem to be proved in Section 3 is original and some particular cases of this theorem were previously considered via CAS in [3, 4] and without CAS in [5].

The threshold theorem to be proved in Section 4 is original and similar models were before considered without CAS in [6].

2   CA And MR Applied to the S^NIR Epidemic Model

We introduce the system for the S^nIR epidemic model, which has n groups of susceptible individuals and which is described by the next system of Eq. [2]:

$$\frac{\mathrm{d}}{\mathrm{d}t} X_i(t) = \mu(p_i X_0 - X_i(t)) - \lambda_i X_i(t)$$

$$\frac{\mathrm{d}}{\mathrm{d}t} Y(t) = \sum_{k=1}^{n} \lambda_k X_k(t) - (\mu + \gamma + \delta)\, Y(t)$$

$$\frac{\mathrm{d}}{\mathrm{d}t} Z(t) = \gamma Y(t) - (\mu + \varepsilon)\, Z(t) \tag{1}$$

we define the rate of infection as:

$$\lambda_i = \alpha_i \beta \eta Y(t) \tag{2}$$

and we define p_i as follow:

$$\sum_{i=1}^{n} p_i = 1 \tag{3}$$

This is a system with $(n + 2)$ equations and each previous constant is defined like it is shown:

μ is the natural death rate.
γ is the rate at which infectives are removed or become immune.
δ is the disease-induced mortality rate for the infectives.
ε is the disease-induced mortality rate for removed individuals.
α_i is the susceptibility of susceptible individuals.
β is the infectious rate of infected individuals.
η is the average number of contacts per individual.

Each function or group is defined as follow:

$X_i(t)$ are the n groups of susceptible in the time equal t
$Y(t)$ is the group of infectives in the time equal t.
$Z(t)$ is the group of removed in the time equal t.

2.1 The Standard SIR Model

As a particular case we analyze the standard SIR model [7] which has just one group of susceptible and is described in the next equation system:

$$\frac{d}{dt}X_1(t) = \mu(p_1 X_0 - X_1(t)) - \lambda_1 X_1(t)$$

$$\frac{d}{dt}Y(t) = \lambda_1 X_1(t) - (\mu + \gamma + \delta)Y(t)$$

$$\frac{d}{dt}Z(t) = \gamma Y(t) - (\mu + \varepsilon)Z(t) \tag{4}$$

In the infection-free equilibrium there is no variation in time. So the derivates are canceled and the solution for the previous system, it's given by:

$$X_1 = p_1 X_0, Y = 0 \tag{5}$$

After, we generate the Jacobian matrix for the equations system:

$$\begin{bmatrix} -\mu - \alpha_1 \eta \beta Y & -\alpha_1 \eta X_1 \beta \\ -\alpha_1 \eta \beta Y & \alpha_1 \eta X_1 \beta - \mu - \gamma - \delta \end{bmatrix}, \tag{6}$$

we substitute the solution (5) in the Jacobian:

$$\begin{bmatrix} -\mu & -\alpha_1 \eta p_1 X_0 \beta \\ 0 & \alpha_1 \eta p_1 X_0 \beta - \mu - \gamma - \delta \end{bmatrix} \tag{7}$$

Now, we find the eigenvalues for the previous matrix:

$$-\mu, \alpha_1 \eta p_1 X_0 \beta - \mu - \gamma - \delta \tag{8}$$

and its corresponding stability condition is:

$$\alpha_1 \eta p_1 X_0 \beta - \mu - \gamma - \delta < 0 \tag{9}$$

this can be rewritten as:

$$\frac{\alpha_1 \eta p_1 X_0 \beta}{\mu + \gamma + \delta} < 1 \tag{10}$$

using the next expression:

$$R_0 < 1 \tag{11}$$

Finally we find the basic reproduction number which it represents the condition of equilibrium:

$$R_0 = \frac{\alpha_1 \eta p_1 X_0 \beta}{\mu + \gamma + \beta} \tag{12}$$

2.2   The S^2IR Model

As another particular case we analyze the S^2IR model where there are two groups of susceptible and the equations for this system are:

$$\frac{d}{dt}X_1(t) = \mu(p_1 X_0 - X_1(t)) - \lambda_1 X_1(t)$$

$$\frac{d}{dt}X_2(t) = \mu(p_2 X_0 - X_2(t)) - \lambda_2 X_2(t)$$

$$\frac{d}{dt}Y(t) = \lambda_1 X_1(t) + \lambda_2 X_2(t) - Y(t)\mu - \gamma Y(t) - Y(t)\delta$$

$$\frac{d}{dt}Z(t) = \gamma Y(t) - (\mu + \varepsilon)Z(t) \tag{13}$$

Initially we find the infection-free equilibrium solution for the previous system:

$$X_1 = p_1 X_0, X_2 = p_2 X_0, Y = 0 \tag{14}$$

Equally we generate the Jacobian matrix for the equations system and substituting the infection-free equilibrium point in the Jacobian:

$$
\begin{bmatrix}
-\mu & 0 & -\alpha_1\eta p_1 X_0\beta \\
0 & -\mu & -\alpha_2\eta p_2 X_0\beta \\
0 & 0 & \alpha_1\eta p_1 X_0\beta + \alpha_2\eta p_2 X_0\beta - \mu - \gamma - \delta
\end{bmatrix}
\tag{15}
$$

we find the eigenvalues for the previous Jacobian,

$$
-\mu, -\mu, \alpha_1\eta p_1 X_0\beta + \alpha_2\eta p_2 X_0\beta - \mu - \gamma - \delta
\tag{16}
$$

and the corresponding stability condition is:

$$
\alpha_1\eta p_1 X_0\beta + \alpha_2\eta p_2 X_0\beta - \mu - \gamma - \delta < 0
\tag{17}
$$

this can be rewritten as:

$$
\frac{\alpha_1\eta p_1 X_0\beta + \alpha_2\eta p_2 X_0\beta}{\mu + \gamma + \delta} < 1
\tag{18}
$$

also it can be written using (11) as:

$$
R_0 = \frac{\alpha_1\eta p_1 X_0\beta + \alpha_2\eta p_2 X_0\beta}{\mu + \gamma + \delta}
\tag{19}
$$

this is the basic reproductive number for S^2IR model.

2.3   The S^3IR, The S^4IR and S^5IR Models

Here we show the S^3IR, the S^4IR and S^5IR models where there are three, four and five groups of susceptibles, respectively. With these models we do the same process, and we only show the basic reproductive number.

$$
R_0 = \frac{\alpha_1\eta p_1 X_0\beta + \alpha_2\eta p_2 X_0\beta + \alpha_3\beta\eta p_3 X_0}{\mu + \gamma + \delta}
\tag{20}
$$

this is the basic reproductive number for S^3IR model.

$$
R_0 = \frac{\alpha_1\eta p_1 X_0\beta + \alpha_2\eta p_2 X_0\beta + \alpha_3\beta\eta p_3 X_0 + \alpha_4\beta\eta p_4 X_0}{\mu + \gamma + \delta}
\tag{21}
$$

this is the basic reproductive number for S^4IR model.

$$
R_0 = \frac{X_0\beta\eta(\alpha_1 p_1 + \alpha_2 p_2 + \alpha_3 p_3 + \alpha_4 p_4 + \alpha_5 p_5)}{\mu + \gamma + \delta}
\tag{22}
$$

and this is the basic reproductive number for S^5IR model.

2.4   The $S^n IR$ Model

Theorem. *For the equations system given by (1). The infection-free equilibrium is locally stable if $R_0 < 1$, and is unstable if $R_0 > 1$, where:*

$$R_0 = \frac{X_0 \beta \eta \left(\sum_{i=1}^{n} \alpha_i p_i \right)}{\mu + \gamma + \delta} \tag{23}$$

We can probe the theorem looking the inequalities corresponding to stability conditions for each system previously considered, we have listed in the Fig. 1. Using mechanized induction we obtain the general expression for the stability conditions for a system with $(n + 2)$ equations.

$$\frac{X_0 \beta \eta \left(\sum_{i=1}^{n} \alpha_i p_i \right)}{\mu + \gamma + \delta} < 1 \tag{24}$$

We can represent into a schematic diagram, the deductive reasoning using "MR".

Here we have an idea for the MR, it finds the similar components in each item and it has a viewer or a detector that find the sequential form for the dissimilar parts. It is just an idea, we believe this system have to be improved by the scientific community.

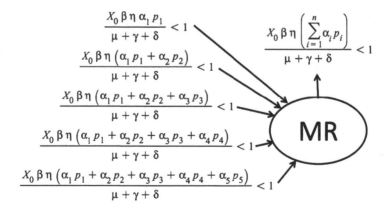

Fig. 1 Inductive mechanized reasoning for the $S^n IR$ Model

3 CA and MR Applied to the $S^N I^M R$ Epidemic Model

The $S^n I^m R$ epidemic model is made by a group of equations which has n groups of susceptible individuals and m groups of infected people, this system is illustrated in the next equations:

$$\frac{d}{dt}X_i(t) = \mu(p_i X_0 - X_i(t)) - \alpha_i \eta \left(\sum_{i=1}^{m} \beta_j Y_j(t) \right) X_i(t)$$

$$\frac{d}{dt}Y_j(t) = \eta \beta_j Y_j(t) \left(\sum_{i=1}^{n} \alpha_i X_i(t) \right) - (\mu + \gamma + \delta) Y_j(t)$$

$$\frac{d}{dt}Z(t) = \gamma Y_j(t) - (\mu + \varepsilon) Z(t) \tag{25}$$

p_i is defined in (3). This is a system with $(n + m + 1)$ equations and each constant were defined in the last model.

3.1 The SI^2R Model

Like in all cases, we analyze a particular case with one group of susceptibles and two groups of infectives. The following equations describe this case:

$$\frac{d}{dt}X_1(t) = \mu(p_1 X_0 - X_1(t)) - \alpha_1 \eta \left(\sum_{j=1}^{2} \beta_j Y_j(t) \right) X_1(t)$$

$$\frac{d}{dt}Y_1(t) = \eta \beta_1 Y_1(t) \left(\sum_{i=1}^{1} \beta_i X_i(t) \right) - (\mu + \gamma + \delta) Y_1(t)$$

$$\frac{d}{dt}Y_2(t) = \eta \beta_2 Y_2(t) \left(\sum_{i=1}^{1} \alpha_i X_i(t) \right) - (\mu + \gamma + \delta) Y_2(t)$$

$$\frac{d}{dt}Z(t) = \gamma Y_j(t) - (\mu + \varepsilon) Z(t) \tag{26}$$

Solving the system for the infection-free equilibrium, we find:

$$Y_1 = 0, Y_2 = 0, X_1 = \frac{\mu p_1 X_0}{\mu + \alpha_1 \eta \left(\sum_{j=1}^{2} \beta_j Y_j \right)} \tag{27}$$

We generate a Jacobian as in the others cases. After, we substitute the infection-free equilibrium point and we find the eigenvalues for the system,

$$-\mu, \alpha_1 \eta p_1 X_0 \beta_1 - \mu - \gamma - \delta, \alpha_1 \eta p_1 X_0 \beta_2 - \mu - \gamma - \delta \qquad (28)$$

The stability conditions shall satisfy,

$$\alpha_1 \eta p_1 X_0 \beta_2 - \mu - \gamma - \delta < 0$$
$$\alpha_1 \eta p_1 X_0 \beta_1 - \mu - \gamma - \delta < 0 \qquad (29)$$

Moreover the inequalities in (29) can be written like the others cases as:

$$R_{0,2} = \frac{\alpha_1 \eta p_1 X_0 \beta_2}{\mu + \gamma + \delta}$$
$$R_{0,1} = \frac{\alpha_1 \eta p_1 X_0 \beta_1}{\mu + \gamma + \delta} \qquad (30)$$

Here, we have the two basic reproductive numbers for SI^2R model.

3.2 The S^2I^2R Model

We analyze a particular case where we have two groups of susceptible and two groups of infective, in addition we solve the system for the infection-free equilibrium and we find:

$$Y_1 = 0, Y_2 = 0, X_1 = \frac{\mu p_1 X_0}{\mu + \alpha_1 \eta \left(\sum_{j=1}^{2} \beta_j Y_j \right)}, X_2 = \frac{\mu p_2 X_0}{\mu + \alpha_2 \eta \left(\sum_{j=1}^{2} \beta_j Y_j \right)} \qquad (31)$$

Also we generate a Jacobian; we substitute the infection-free equilibrium point and find the eigenvalues for this system:

$$-\mu, -\mu, \alpha_1 \eta p_1 X_0 \beta_1 + \alpha_2 \eta p_2 X_0 \beta_1 - \mu - \gamma - \delta, \alpha_1 \eta p_1 X_0 \beta_2$$
$$+\alpha_2 \eta p_2 X_0 \beta_2 - \mu - \gamma - \delta \qquad (32)$$

In like manner that we obtained the last reproductive numbers, we obtain the two basic reproductive numbers for S^2I^2R model:

$$R_{0,2} = \frac{\alpha_1 \eta p_1 X_0 \beta_2 + \alpha_2 \eta p_2 X_0 \beta_2}{\mu + \gamma + \delta}$$
$$R_{0,1} = \frac{\alpha_1 \eta p_1 X_0 \beta_1 + \alpha_2 \eta p_2 X_0 \beta_1}{\mu + \gamma + \delta} \qquad (33)$$

3.3   The $S^n I^m R$ Model

Theorem. *For the equations system given by (25). The infection-free equilibriums are locally stable if the reproductive numbers of infection $R_{0,j} < 1$, and is unstable if $R_{0,j} > 1$, with j from 1 to m, where:*

$$R_{0,j} = \frac{X_0 \eta \left(\sum\limits_{i=1}^{n} \alpha_i p_i \right) \beta_j}{\mu + \gamma + \delta} \qquad (34)$$

To prove the theorem, we used mechanized induction starting from the particular results previously obtained in (30) and (33). Finally, we find the general solution for the basic reproductive numbers for the $S^n I^m R$ model according with (Fig. 2).

4   CA and MR Applied to the Staged Progressive $SI^M R$ Epidemic Model

At this moment, we are concerned in the analysis of the staged progressive $SI^m R$ epidemic model. This has m groups of infected; in this case the infection is staged and progressive, which is described by next system:

$$\frac{d}{dt} X(t) = \mu(x_0 - X(t)) - \lambda X(t)$$

$$\frac{d}{dt} Y_1(t) = \lambda X(t) - (\mu + \gamma_1 + \delta_1) Y_1(t)$$

$$\frac{d}{dt} Y_j(t) = Y_{j-1} Y_{j-1}(t) - (\mu + \gamma_j + \delta_j) Y_j(t)$$

$$\frac{d}{dt} Z(t) = \gamma_m Y_m(t) - (\mu + \varepsilon) Z(t) \qquad (35)$$

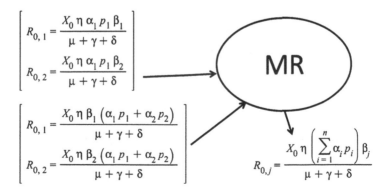

Fig. 2 Inductive mechanized reasoning for the $S^n I^m R$ Model

with the restriction for j is: $2 \leq j \leq m$. And where the rate of infection is:

$$\lambda = \alpha \left(\sum_{j=1}^{m} \beta_j Y_j(t) \right) \eta \qquad (36)$$

This is a system with $(m + 2)$ equations and all the constants were described before.

4.1   The Staged Progressive SI^2R Model

Now, we analyze a particular case with one group of susceptible and two groups of infectives. The following equations describe this case:

$$\frac{d}{dt}X(t) = \mu(x_0 - X(t)) - \lambda X(t)$$

$$\frac{d}{dt}Y_1(t) = \lambda X(t) - (\mu + \gamma_1 + \delta_1)Y_1(t)$$

$$\frac{d}{dt}Y_2(t) = \lambda_1 Y_1(t) - (\mu + \gamma_2 + \delta_2)Y_2(t)$$

$$\frac{d}{dt}Z(t) = \lambda_2 Y_2(t) - (\mu + \varepsilon)Z(t) \qquad (37)$$

Solving the system for the infection-free equilibrium, we find:

$$X = \frac{\mu x_0}{\mu + \alpha \eta \beta_1 Y_1 + \alpha \eta \beta_2 Y_2}, Y_1 = 0, Y_2 = 0 \qquad (38)$$

We generate a Jacobian coming off the equations system and substituting the infection-free equilibrium point:

$$\begin{vmatrix} -\mu & -\alpha \eta x_0 \beta_1 & -\alpha \eta x_0 \beta_2 \\ 0 & -\alpha \eta x_0 \beta_1 - \mu - \gamma_1 - \delta_1 & \alpha \eta x_0 \beta_2 \\ 0 & \gamma_1 & -\mu - \gamma_2 - \delta_2 \end{vmatrix} \qquad (39)$$

Now we obtain the characteristic polynomial:

$$(\lambda + \mu)(\lambda^2 + 2\lambda\mu + \lambda\gamma_2 + \lambda\delta_2 - \alpha\eta x_0\beta_1\lambda - \alpha\eta x_0\beta_1\mu$$
$$- \alpha\eta x_0\beta_1\gamma_2 - \alpha\eta x_0\beta_1\delta_2 + \mu^2 + \mu\gamma_2 + \mu\delta_2 + \gamma_1\lambda$$
$$+ \gamma_1\mu + \gamma_1\gamma_2 + \gamma_1\delta_2 + \delta_1\lambda + \delta_1\gamma_2 + \delta_1\delta_2$$
$$- \alpha\eta x_0\beta_2\gamma_1 \qquad (40)$$

From this polynomial, we obtain the basic reproduction number using the Routh-Hurwitz theorem:

$$R_0 = \frac{x_0 \alpha \eta (\beta_1 \delta_2 + \beta_1 \mu + \beta_1 \gamma_2 + \beta_2 \gamma_1)}{(\mu + \gamma_2 + \delta_2)(\mu + \gamma_1 + \delta_1)} \tag{41}$$

4.2 The Staged Progressive SI^3R Model

We analyze a particular case with three groups of infectives. We solve this system in the same way that we did in the Section 4.1. For this model the basic reproduction number that we find is:

$$\begin{aligned}
R_0 = (x_0 \alpha \eta (\beta_1 \delta_2 \mu &+ \beta_1 \delta_2 \lambda_3 + \beta_1 \mu \lambda_3 + \beta_1 \gamma_2 \mu + \beta_1 \gamma_2 \gamma_3 \\
&+ \beta_1 \gamma_2 \delta_3 + \beta_1 \mu^2 + \beta_1 \delta_2 \delta_3 + \gamma_1 \beta_2 \mu + \gamma_1 \beta_2 \gamma_3 + \gamma_1 \beta_2 \delta_3 \\
&+ \gamma_1 \beta_3 \gamma_2 + \beta_1 \mu \delta_3)) / ((\mu + \gamma_3 + \delta_3)(\mu + \gamma_2 + \delta_2)(\mu \\
&+ \gamma_1 + \delta_1))
\end{aligned} \tag{42}$$

4.3 Staged Progressive SI^mR Model

Theorem. *For the equations system given by (35). The infection-free equilibrium is locally stable if the reproductive numbers of infection $R_0 < 1$, and is unstable if $R_0 > 1$, with j from 1 to m, where:*

$$R_0 = \frac{x_0 \alpha \eta \left(\sum\limits_{j-1}^{m} \beta_j \prod\limits_{k=1}^{j-1} \gamma_k \prod\limits_{l=j+1}^{m} (\mu + \gamma_l + \delta_l) \right)}{\prod\limits_{l=1}^{m} (\mu + \gamma_l + \delta_l)} \tag{43}$$

To prove the theorem, we used mechanized induction starting from the particular results previously obtained. To conclude, we find the general solution for the basic reproductive numbers for the Staged Progressive SI^mR model according with (Fig. 3).

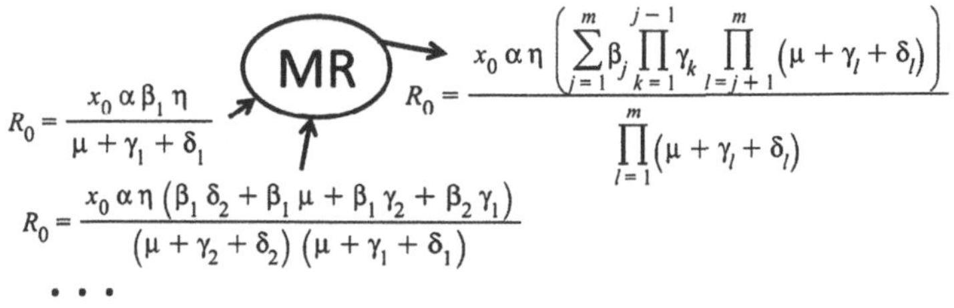

Fig. 3 Inductive mechanized reasoning for the staged progressive SI^mR Model

5 Conclusions

We finally obtain three theorems which can help us to demonstrate that CAS+MR are important tools for solving problems in every situation that mathematics could model. The theorems are useful to make strategies to fight against epidemic diseases in the future biological dangers.

Due to use CAS, in our case "Maple 11", we can proceed to solve the mathematical problem and we can obtain results very fast that without them could take us too much time.

The use of CAS+MR can help in teaching and learning the mathematics to engineering, whose don't have time and need to give quickly solutions. It can be implemented in engineer programs.

Through the MR we already found the general forms for the infection free equilibrium, we can see the importance of developing software it can do the mechanized reasoning automatically.

Acknowledgments The authors would like to thank Prof. Dr. Eng. Andrés Sicard and Prof. Mario Elkin Vélez for keeping company in this work in the Logic and Computation group, Prof. Eng. Maurio Arroyave for supporting this work from the engineering physics program. They are also very grateful for Prof. Dr. Eng. Félix Londoño for his help in the World Congress on Engineering and Computer Sciences and the Student Organization and Ms. Angela Echeverri for the economical assistance in the trip to the congress. This research was support by the Logic and Computation group at EAFIT University.

References

1. C.W. Brown, M.E. Kahoui, D. Novotni, A. Weber, Algorithmic methods for investigating equilibria in epidemic modeling. J. Symb. Comput. **41**, 1157–1173 (2006)
2. J.M. Hyman, J. Li, Differential susceptibility epidemic models. J. Math. Biol. **50**, 626–644 (2005)
3. J.A.M. Taborda, Epidemic thresholds via computer algebra, MSV 2008, pp. 178–181, http://dblp.uni-trier.de/rec/bibtex/conf/msv/Taborda08
4. D. Hincapié et al., Epidemic thresholds in SIR and SIIR models applying an algorithmic method, Lectures Notes Bioinformatics, **5354**, 119–130, 2008
5. S.B. Hsua, Y.H. Hsiehb, On the role of asymptomatic infection in transmission dyanamics of infectious diseases, Bull. Math. Biol. **70**, 134–155 (2008)
6. J.M. Hyman, J. Li, An intuitive formulation for the reproductive number for spread of diseases in heterogeneous populations. Math. Biosci. **167**, 65–86 (2000)
7. N.T.J. Bailey, *The Mathematical Theory of Epidemics*, 1957

Chapter 44
Review of Daily Physical Activity Monitoring System Based on Single Triaxial Accelerometer and Portable Data Measurement Unit

Mihee Lee, Jungchae Kim, Sun Ha Jee, and Sun Kook Yoo

Abstract Main objective of this pilot study was to present a method to convenient monitoring of detailed ambulatory movements in daily life, by use of a portable measurement device employing single tri-axial accelerometer. In addition, the purpose of this review article is to provide researchers with a guide to understanding some commonly-used accelerometers in physical activity assessment. Specially, we implemented a small-size wearable data storing system in real time that we used Micro SD-Memory card for convenient and long period habitual physical activity monitoring during daily life. Activity recognition on these features was performed using Fuzzy c means classification algorithm recognized standing, sitting, lying, walking and running with 99.5% accuracy. This study was pilot test for our developed system's feasibilities. Further application of the present technique may be helpful in the health promotion of both young and elderly.

1 Introduction

Over the past two decades, a striking increase in the number of people with the metabolic syndrome worldwide has taken place. This increase is associated with the global epidemic of obesity and diabetes. With the elevated risk not only of diabetes but also of cardiovascular disease from the metabolic syndrome, there is urgent need for strategies to prevent the emerging global epidemic [1, 2]. Although the metabolic syndrome appears to be more common in people who are genetically susceptible, acquired underlying risk factors-being overweight or obese, physical inactivity, and an atherogenic diet-commonly elicit clinical manifestations [3].

M. Lee (✉)
Graduate Programs of Biomedical Engineering, Yonsei University, 262 Seongsanno, Seodaemoon-Ku, Seoul 120-749, Korea
e-mail: leemihee76@yuhs.ac

S.-I. Ao et al. (eds.), *Machine Learning and Systems Engineering*,
Lecture Notes in Electrical Engineering 68,
DOI 10.1007/978-90-481-9419-3_44, © Springer Science+Business Media B.V. 2010

Current guidelines recommend practical, regular, and moderate regimens of physical activity (eq, 30 min moderate-intensity exercise daily) [4]. Regular and sustained physical activity will improve all risk factors of the metabolic syndrome. Sedentary activities in leisure time should be replaced by more active behavior such as brisk walking, jogging, swimming, biking, golfing, and team sports. Combination of weight loss and exercise to reduce the incidence of type 2 diabetes in patients with glucose intolerance should not be dismissed [5].

A variety of methods exist to quantify levels of habitual physical activity during daily life, including objective measures such as heart rate, one- and three-dimensional accelerometer, and pedometer, as well as subjective recall questionnaires like the International Physical Activity Questionnaire and physical activity logbooks [7, 8]. Yet all possess some important limitations. Heart rate monitors have been widely used to quantify physiological stress, but their efficacy at low intensities has been questioned due to the potential interference of environmental conditions and emotional stress [9]. A wide range of self-report activity questionnaires exist that are well suited to large surveillance studies but are limited due to their reliance on subjective recall. Pedometers are an inexpensive form of body motion sensor, yet many fail to measure slow walking speeds or upper body movements, and most are unable to log data to determine changes in exercise intensity [10]. The most common accelerometers used in human activity research measure accelerations either in a vertical plane (uni-axial), or in three planes (tri-axial), with excellent data-logging abilities [1, 10].

Main objective of this study was to present a method to convenient monitoring of detailed ambulatory movements in daily life, by use of a portable measurement device employing single tri-axial accelerometer. In addition, the purpose of this review article is to provide researchers with a guide to understanding some commonly-used accelerometers in physical activity assessment. The literature related to physical activity measurement is reviewed in this paper. For organizational purposes, the literature is presented under the following topics: (a) Measurement of Physical Activity, (b) Behavioral Observation, (c) Pedometers, and (d) Accelerometers.

1.1   Measurement of Physical Activity

Direct measurements of physical activity include behavioral observation and motion sensors (pedometers and accelerometers). An estimate of energy expenditure can be extrapolated from these direct measures of physical activity based on previously validated equations. The ability to examine health in relation to physical activity requires an accurate, precise and reproducible measurement of physical activity. Physical activity measurement tools should provide an objective measure of physical activity, be utilized under normal daily living conditions, have the capacity to record data over a prolonged period of time, and produce minimal discomfort to subjects [11]. Consideration should be given to the number of subjects to be monitored, and size and cost of the measurement tool, and the activity being monitored.

1.2   Behavioral Observation

Behavioral observation is a seldom utilized, time intensive technique that has primarily be applied when monitoring children and adolescents. This technique requires skilled observers who monitor individuals for a set period of time in an effort to code their behavior to describe overall physical activity patterns and estimate energy expenditure based on published MET values [12].

1.3   Pedometers

The concept of a pedometer was first introduced by Leonardo Da Vinci ~500 years ago. Pedometers are inexpensive tools that measure the number of steps taken by responding to the vertical acceleration of the trunk. Mechanical pedometers have a level arm is triggered when a step is taken and a ratchet within the pedometer that rotates to record the movement [13]. Electronic pedometers have a spring suspended pendulum arm that moves up and down thereby opening and closing an electrical circuit [7]. Output from the pedometer consists of number of steps taken, distance walked (with stride information), and estimated number of calories expended (with body weight considered) [13]. However, it is assumed that a person expends a constant amount of energy per step regardless of the speed (e.g. Yamax pedometer, 0.55 kcal $\cdot$ kg^{-1} $\cdot$ step^{-1}) [7].

Vertical acceleration of the hip during walking varies from 0.5 to 8 m $\cdot$ s^{-2}, however some pedometers do not register movement below 2.5 m $\cdot$ s^{-2} or distinguish vertical accelerations above a certain threshold, and therefore are inaccurate at slow or very fast walking speeds and during running [7, 14].

Limitations of the pedometer include insensitivity to: (1) static or sedentary activity, (2) isometric activity, (3) movement occurring with the arms, and (4) slow or fast walking velocities [7, 13, 14]. In addition, they lack internal clocks and data storage capability which makes an analysis of overall physical activity patterns difficult [7]. Pedometers are primarily utilized in health promotion and maintenance and may provide valuable feedback to a participant concerning their level of activity. Advantages of utilizing a pedometer include low cost, size, and compliance of subjects.

1.4   Accelerometers

Accelerometer is based on the assumption that limb movement and body acceleration are theoretically proportional to the muscular forces responsible for the accelerations, and therefore energy expenditure may be estimated by quantifying these accelerations [11, 13]. Interest in monitoring the acceleration of the body started in the 1950s when Broubha and Smith (1958) suggested an association between the integral of vertical acceleration versus time and energy expenditure [15].

Acceleration is the change in velocity over time, and the degree of acceleration detected provides an index of movement intensity. Accelerations increase from the head to the feet and are generally the greatest in the vertical direction. Running produces the greatest vertical direction accelerations (8.1–12 g) at the ankle and up to 5 g at the back [16]. Bouten et al. (1996) recommends that accelerometers should be able to measure accelerations up to ± 12 g for measurements occurring in daily living activities and exercise. Processing and filtering is required to identify and include accelerations that are outside the normal acceleration range of human movement.

Low pass filters remove high frequency signals that may occur as a result of external vibrations [17]. Factors which can influence accelerometer output include the acceleration due to body movement and external vibrations (loose straps, transportation) [18]. Most accelerometers are oriented to monitor the vertical plane such that linear acceleration with movement is not recorded [19].

1.4.1 Types of Accelerometer

Accelerometers can be classified based on the process by which acceleration is measured and recorded (piezoelectric versus piezoresistive) or the number of planes of movement it measures. Piezoresistive accelerometers utilize a spring element with strain sensitive gages connected via a Wheatstone bridge. The deflection of the spring element during acceleration causes a deformation of the gages, producing an electrical output in response to the change in resistance [13]. Piezoelectric accelerometers consist of a piezoelectric element and a seismic mass. When movement occurs, the sensor undergoes acceleration which causes deformation of the piezoelectric element (e.g. bending). Conformational changes in the element produce a voltage signal that is proportional to stress, or acceleration of the limb. When the body part to which the accelerometer is attached accelerates a charge is generated that is proportional to the force exerted by the subject. An acceleration-deceleration curve wave is created and the area under the wave is summed to yield the final count value [13, 19]. Ceramic transducers are more reliable than spring mechanisms, sensitive at very low frequencies, and require less power; however, aging of the ceramic may result in a loss of piezoelectric capabilities causing a decreased sensitivity to movement [20].

Accelerometers can be uni-axial (one plane of movement), tri-axial (three planes of movement) or "omni-directional", monitoring acceleration in every plane other than that perpendicular to the point of reference. The uni-axial accelerometer is most sensitive to bending in one direction and is typically utilized to monitor acceleration in the vertical plane. An omni-directional accelerometer is most sensitive in one direction (vertical), but has the ability to quantify deformations in other planes (horizontal and lateral). Tri-axial accelerometers quantify acceleration in the vertical, anterior-posterior and medial-lateral plane and provide a measure of counts in each plane as well as an overall vector magnitude [13, 19]. There appears little advantage to using the triaxial accelerometer versus a single axis when the monitored activity is primarily walking [21]. However, activity energy expenditure during sedentary activities, as determined by a tri-axial accelerometer, correlates better with measured energy expenditure when compared to a uni-axial accelerometer [22].

1.4.2   Placement of Monitoring Sensors

Accelerometers are worn while the participant performs different activities and the counts from the monitor are regressed against a criterion measure of energy expenditure (e.g. whole room calorimetry, indirect calorimetry, doubly labeled water, or heart rate). Output from the monitors can be utilized to assess frequency, duration and intensity of physical activity. Energy expenditure is predicted by converting monitor counts into a unit of energy expenditure (MET, kcal $\cdot$ min^{-1}, or kcal $\cdot$ kg $\cdot$ min^{-1}).

The characteristics of the population being studied and/or the specific activity may affect the accuracy of the motion sensors [23]. The output of the accelerometer will depend on the placement site (e.g. hip, wrist, ankle), the orientation relative to the subject, the posture of the subject, and the activity being performed. Accelerometer output is influenced by the placement of the monitors as body parts are differentially active depending upon the activity. Placing the accelerometer on the limbs while monitoring sedentary and daily living activities may produce more accurate results than hip placement as the limbs are likely to reflect activities which occur in these circumstances. The primary acceleration which occurs during walking or running is in the vertical direction, therefore placing the accelerometer at a site which reflects acceleration at the center of the body mass (e.g. hip) is likely to be more accurate than the limbs. The velocity of movement can also have an effect on the relationship between accelerometer counts and energy expenditure.

1.4.3   Limitations for Physical Activity Monitoring

The disadvantages of utilizing the accelerometer to estimate energy expenditure include subject compliance, altered activity patterns, and a cost of approximately $400 per monitor [13, 24]. One limitation of use is the inability to account for increased energy cost of walking uphill or cost of carrying a load. Accelerometers do not respond to activities where there is minimal movement at the body's center of gravity (e.g. rowing and cycling) and they are not responsive to isometric work as even thought the limbs are not moving, energy is being expended [25, 26].

However, Westerterp (1999) suggests that the effect of static exercise on the total level of physical activity is negligible. Furthermore, estimates of energy expenditure are derived from equations that were developed within specific populations performing primarily walking activities in the laboratory. Consequently, energy estimation equations may not be applicable to the daily living activities performed within the general population. The primary advantages of utilizing an accelerometer include the ability to respond to both frequency and intensity of movement, its relatively low cost, ability to store large amounts of data over extended periods of time, small size and minimal participant discomfort [27].

These findings indicate that differences between uniaxial, triaxial, and omnidirectional accelerometry devices are minimal and all three provide reasonable measures of physical activity during treadmill walking. Furthermore, the inclusion of stride length in accelerometry based energy expenditure prediction equations significantly improved energy expenditure estimates.

The purpose of this study was to present a method to convenient monitoring of detailed ambulatory movements in daily life, by use of a portable measurement device employing single tri-axial accelerometer. Specially, we implemented a small-size wearable data store system in real time that we used Micro SD-Memory card (Secure Digital memory card) for convenient and long period habitual physical activity monitoring during daily life. In this work, the performance of activity recognition algorithms under conditions akin to those found in real-world settings is assessed. Activity recognition results are based on acceleration data collected from single tri-axial accelerometer placed on subjects' waist under semi-naturalistic conditions for pilot test.

2 Material and Method

2.1 Portable Data Measurement Unit

Our long-term aim in developing an accelerometry monitoring system was to develop a practical system that could be used to monitor and assess physical activities in free-living subjects. Therefore, we developed a wearable device consisted of Micro SD-Memory card connector (AUTF-08WP01, AUSTONE Electronics, Inc., USA) with mini USB socket (5P, SMT type, SHIH HAN CO., LTD., Russia). Measured acceleration signal was stored on micro SD-Memory card or transmitted wirelessly using Zigbee-compatible 2.4G bandwidth for wireless communication, and CC2420 (Chipcon Co. Ltd., Norway) with a simple interface circuit around the chip. In addition, a ceramic chip antenna TI-AN048 (SMD type, Texas Instruments Co. USA) was applied for stable wireless transmission.

For this we used ADXL330 (Analog Devices, Inc., USA), an acceleration sensor that is composed of a single chip and can detect tri-axis acceleration information, and measured acceleration information of axis X, Y and Z according to the subject's posture and activity. Using the implemented system, we measured change in acceleration signal according to the change of activity pattern. Li-Ionic batteries, micro processor units and micro SD memory card, as shown in Fig. 1. This equipment was small (60 × 40 × 20 [mm]) and light enough to carry without and restriction. Sampling frequency was 100 Hz. Data was downloaded via USB, and processed offline by a PC. The equipment was designed to be attached on the waist (see Fig. 1).

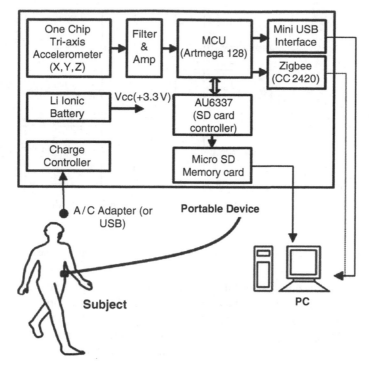

Fig. 1 Architecture of the wearable measurement device

Table 1 Participant characteristics

Case	Gender (M/F)	Age (year)	Height (cm)	Weight (kg)	BMI (kg $\bullet$ m^{-2})
A	M	26	182	76	22.9
B	M	28	179	85	26.5
C	M	28	177	69	22.0
D	F	24	167	55	19.7
E	F	33	160	55	21.5

2.2   Physical Activity Data Collection

The data for these experiments were gathered in an unsupervised pilot study in which healthy young (age 24–33) subjects performed a variety of activities in the three times on outdoor conditions. Characteristics of the participants are presented in Table 1.

We put the acceleration measuring sensor system on the left waist of the subjects, and measured change in acceleration signal according to change in ambulatory movement and physical activities. In the experiment performed in this research, the change of acceleration was measured while the subject was repeating postures such as standing, sitting, lying, walking and running. A representatives set of routine data is shown in Fig. 2.

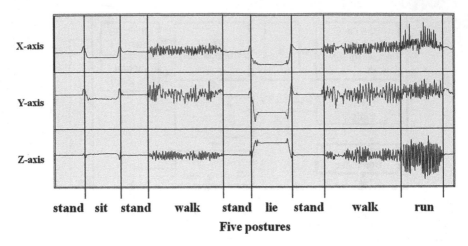

Fig. 2 Representative data from the daily routine for each of the three axes of the tri-axial device

While the movements and postures contained within the routine are by no means a complete set of all possible activities that a given person might perform, they do form a basic set of simple activities which form an underlying structure to a person's daily life, and are likely to provide a great deal of information in terms of the person's balance, gait and activity levels if they can be accurately identified.

2.3 Feature Extraction

Features were computed on 512 sample windows of acceleration data with 256 samples overlapping between consecutive windows. At a sampling frequency of 100 Hz, each window represents 5.2 s. Maximum acceleration, mean and standard deviation of acceleration channels were computed over sliding windows with 50% overlap has demonstrated success in past works. The 512 sample window size enabled fast computation of FFTs used for some of the features. The DC feature for normalization is the mean acceleration value of the signal over the window. Use of mean of maximum acceleration features has been shown to result in accurate recognition of certain postures and activities.

3 Results

This section describes the experiments and experimental results of the human posture recognition system. In the pilot test, a subject continuous posture change including standing, sitting, lying, walking and running. In the experiment, each posture was recognized third.

Table 2 Clustering results of different posture in a continuous motion	Parameters and real posture	Jaccard score	Purity	Efficiency
	Standing	0.99	0.99	1
	Sitting	1	1	1
	Lying	1	1	1
	Walking	0.99	1	0.99
	Running	1	1	1
	Average	0.98	0.99	0.99

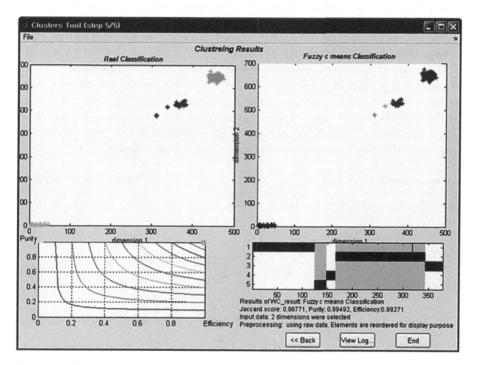

Fig. 3 Activity recognition result using Fuzzy c means classification algorithm

Mean and standard deviation of acceleration and correlation features were extracted from acceleration data. Activity recognition on these features was performed using Fuzzy c means classification algorithm recognized standing, sitting, lying, walking and running with 99.5% accuracy as shown in Table 2 and Fig. 3.

4   Discussion and Conclusion

This paper proposes an ambulatory movement's recognition system in daily life. A portable acceleration sensor module has been designed and implemented to measure human body motion. A small portable device utilizing single tri-axis accelerometer was developed, which detects features of ambulatory movements including vertical position shifts. The classification method based on Fuzzy c means classification algorithm, recognition accuracy of over 99% on a five activities (standing, sitting, lying, walking and running). These results are competitive with prior activity recognition results that only used laboratory data. However, several limitations are also observed for the system. Firstly, collected data was from younger (age 24–33) subjects. Secondly, single accelerometer of placed on body waist typically do not measure ascending and descending stairs walking.

Accelerometers are preferable to detect frequency and intensity of vibrational human motion [28]. Many studies have demonstrated the usefulness of accelerometer for the evaluation of physical activity, mostly focusing on the detection of level walking or active/rest discrimination [29–31].

This study was pilot test for our developed system's feasibilities. Further application of the present technique may be helpful in the health promotion of both young and elderly, and in the management of obese, diabetic, hyperlipidemic and cardiac patients. Efforts are being directed to make the device smaller and allow data collection for longer time periods. Implementation of real-time processing firmware and encapsulation of the hardware are our future studies.

Acknowledgments This study was supported by a grant of the Seoul R&BD Program, Republic of Korea (10526) and the Ministry of Knowledge Economy (MKE) and Korea Industrial Technology Foundation (KOTEF) through the Human Resource Training Project for Strategic Technology.

References

1. P. Zimmet, K.G. Alberti, J. Shaw, Global and societal implications of the diabetes epidemic. Nature **414**, 782–787 (2001)
2. S.M. Grundy, B. Hansen, S.C. Smith Jr, J.I. Cleeman, R.A. Kahn, Clinical management of metabolic syndrome: report of the American Heart Association/National Heart, Lung, and Blood Institute/American Diabetes Association conference on scientific issues related to management. Circulation **109**, 551–556 (2004)
3. R.H. Eckel, S.M. Grundy, P.Z. Zimmet, The metabolic syndrome. Lancet **365**, 1415–28
4. P.D. Thompson, D. Buchner, I.L. Pina, et al., Exercise and physical activity in the prevention and treatment of atherosclerotic cardiovascular disease: a statement from the Council on Clinical Cardiology (Subcommittee on Exercise, Rehabilitation, and Prevention) and the Council on Nutrition, Physical Activity, and Metabolism (Subcommittee on Physical Activity). Circulation **107**, 3109–3116 (2003)
5. J. Tuomilehto, J. Lindstrom, J.G. Eriksson, et al., Prevention of type 2 diabetes mellitus by changes in lifestyle among subjects with impaired glucose tolerance. N. Engl. J. Med. **344**, 1343–1350 (2001)
6. Y. Ohtaki, M. Susumago, A. Suzuki, et al., Automatic classification of ambulatory movements and evaluation of energy consumptions utilizing accelerometers and a barometer. Microsyst. Technol. **11**, 1034–1040 (2005)
7. D.R. Bassett Jr. Validity and reliability issues in objective monitoring of physical activity. Res. Q Exerc. Sport **71**, S30–S36 (2000)

8. C.L. Craig, A.L. Marshall, M. Sjostrom, A.E. Bauman, et al., International physical activity questionnaire: 12-country reliability and validity. Med. Sci. Sports Exerc. **35**, 1381–1395 (2003)

9. K.M. Allor, J.M. Pivarnik, Stability and convergent validity of three physical activity assessments. Med. Sci. Sports Exerc. **33**, 671–676 (2001)

10. M.J. LaMonte, B.E. Ainsworth, C. Tudor-Locke, Assessment of physical activity and energy expenditure, in *Obesity: Etiology, Assessment, Treatment and Prevention*, ed. by R.E. Andersen (Human Kinetics, Champaign, IL, 2003), pp. 111–117

11. C.V.C. Bouten, W.P.H.G. Verboeket-Van Venne, et al., Daily physical activity assessment: comparison between movement registration and doubly labeled water. J. Appl. Physiol. **81**, 1019–1026 (1996)

12. U.S. Department of Health & Human Services, *Physical Activity and Health: A Report of the Surgeon General* (U.S. Department of Health and Human Services, Centers for Disease Control and Prevention, National Center for Chronic Disease Prevention and Health Promotion, The President's Council on Physical Fitness and Sports, Atlanta, GA, 1996)

13. P.S. Freedson, K. Miller Objective monitoring of physical activity using motion sensors and heart rate. Res. Quart. Exerc. Sport **71**, 2129 (2000)

14. H.J. Montoye, H.C.G. Kemper, W.H.M. Saris, R.A. Washburn, *Measuring Physical Activity and Energy Expenditure* (Human Kinetics, Campaign, IL, 1996)

15. R.K. Dishman, R.A. Washburn, D.A. Schoeller, Measurement of physical activity. Quest. **53**, 295–309 (2001)

16. A. Bhattacharya, E.P. McCutcheon, E. Shvartz, J.E. Greenleaf, Body acceleration distribution and O2 uptake in humans during running and jumping. J. Appl. Physiol. **49**, 881–887 (1980)

17. G.J. Welk, *Physical Activity Assessments for Health-Related Research* (Human Kinetics, Champaign, IL, 2002)

18. C.V.C. Bouten, A.A.H.J. Sauren, M. Verduin, J.D. Janssen, Effects of placement and orientation of body-fixed accelerometers on the assessment of energy expenditure during walking. Med. Biol. Eng. Comput. **35**, 50–56 (1997)

19. K.Y. Chen, D.R. Bassett, The technology of accelerometry-based activity monitors: current and future. Med. Sci. Sport Exerc. **37**(11), S490–S500 (2005)

20. E.L. Melanson, P.S. Freedson Physical activity assessment: a review of methods. Crit. Rev. Food Sci. Nutr. **36**, 385–396 (1996)

21. T.G. Ayen, H.J. Montoye Estimation of energy expenditure with a simulated threedimensional accelerometer. J. Ambul. Monit. **1**(4), 293–301 (1988)

22. K.R. Westerterp, Physical activity assessment with accelerometers. Int. J. Obes. **23**(Suppl 3), S45–S49 (1999)

23. B.G. Steele, B. Belza, K. Cain, C. Warms, J. Coopersmith, J. Howard, Bodies in motion: monitoring daily activity and exercise with motion sensors in people with chronic pulmonary disease. J. Rehabil. Res. Dev. **40**(Suppl 2) 45–58 (2003)

24. M.J. Lamonte, B.E. Ainsworth, Quantifying energy expenditure and physical activity in the context of dose response. Med. Sci. Sports Exerc. **33**, S370–S378 (2001)

25. R.K. Dishman, R.A. Washburn, D.A. Schoeller, Measurement of physical activity. Quest. **53**, 295–309 (2001)

26. K.R. Westerterp, Physical activity assessment with accelerometers. Int. J. Obes. **23**(Suppl 3), S45–S49 (1999)

27. M.J. Mathie, A.C.F. Coster, N.H. Lovell, B.G. Celler, Accelerometry: providing an integrated, practical method for long-term, ambulatory monitoring of human movement. Physiol. Meas. **25**, R1–R20 (2004)

28. C.V.C. Bouten, K.T.M. Koekkoek, M. Verduin, R. Kodde, J.D. Janssen, A triaxial accelerometer and portable data processing unit for the assessment of daily physical activity. IEEE Trans. Biomed. Eng. **44**(3):136–147 (1997)

29. A.K. Nakahara, E.E. Sabelman, D.L. Jaffe, Development of a second generation wearable accelerometric motion analysis system. *Proceedings of the first joint EMBS/BMES conference*, 1999, p. 630
30. K. Aminian, P. Robert, E.E. Buchser, B. Rutschmann, D. Hayoz, M. Depairon, Physical activity monitoring based on accelerometry. Med. Biol. Eng. Comput. **37**, 304–308 (1999)
31. M.J. Mathie, N.H. Lovell, C.F. Coster, B.G. Celler, Determining activity using a triaxial accelerometer, in *Proceedings of the Second Joint EMBS/BMES Conference*, 2002, pp. 2481–2482

Chapter 45
A Study of the Protein Folding Problem by a Simulation Model

Omar Gaci

Abstract In this paper, we propose a simulation model to study the protein folding problem. We describe the main properties of proteins and describe the protein folding problem according to the existing approaches. Then, we propose to simulate the folding process when a protein is represented by an amino acid interaction network. This is a graph whose vertices are the proteins amino acids and whose edges are the interactions between them. We propose a genetic algorithm of reconstructing the graph of interactions between secondary structure elements which describe the structural motifs. The performance of our algorithms is validated experimentally.

1 Introduction

Proteins are biological macromolecules participating in the large majority of processes which govern organisms. The roles played by proteins are varied and complex. Certain proteins, called enzymes, act as catalysts and increase several orders of magnitude, with a remarkable specificity, the speed of multiple chemical reactions essential to the organism survival. Proteins are also used for storage and transport of small molecules or ions, control the passage of molecules through the cell membranes, etc. Hormones, which transmit information and allow the regulation of complex cellular processes, are also proteins.

Genome sequencing projects generate an ever increasing number of protein sequences. For example, the Human Genome Project has identified over 30,000 genes which may encode about 100,000 proteins. One of the first tasks when annotating a new genome is to assign functions to the proteins produced by the genes. To fully understand the biological functions of proteins, the knowledge of their structure is essential.

O. Gaci
Le Havre University, 25 rue Phillipe Lebon, 76600 Le Havre, France
e-mail: omar.gaci@gmail.com

S.-I. Ao et al. (eds.), *Machine Learning and Systems Engineering,*
Lecture Notes in Electrical Engineering 68,
DOI 10.1007/978-90-481-9419-3_45, © Springer Science+Business Media B.V. 2010

In their natural environment, proteins adopt a native compact three dimensional form. This process is called folding and is not fully understood. The process is a result of interactions between the protein's amino acids which form chemical bonds.

In this study, we propose to study the protein folding problem. We describe this biological process through the historical approaches to solve this problem. Then, we treat proteins as networks of interacting amino acid pairs [1]. In particular, we consider the subgraph induced by the set of amino acids participating in the secondary structure also called Secondary Structure Elements (SSE).We call this graph SSE interaction network (SSE-IN). We begin by recapitulating relative works about this kind of study model. Then, we present a genetic algorithm able to reconstruct the graph whose vertices represent the SSE and edges represent spatial interactions between them. In other words, this graph is another way to describe the motifs involved in the protein secondary structures.

2   The Protein Folding Problem

Several tens of thousands of protein sequences are encoded in the human genome. Λ protein is comparable to an amino acid chain which folds to adopt its tertiary structure. Thus, this 3D structure enables a protein to achieve its biological function. In vivo, each protein must quickly find its native structure, functional, among innumerable alternative conformations.

The protein 3D structure prediction is one of the most important problems of bioinformatics and remains however still irresolute in the majority of cases. The problem is summarized by the following question: being given a protein defined by its sequence of amino acids, which is its native structure? In other words, we want to determine the structure whose amino acids are correctly organized in three dimensions in order to this protein can achieve correctly its biological function.

Unfortunately, the exact answer is not always possible that is why the researchers have developed study models to provide a feasible solution for any unknown sequences. However, models to fold proteins bring back to NP-Hard optimization problems [2]. Those kinds of models consider a conformational space where the modeled protein tries to reach its minimum energy level which corresponds to its native structure.

Therefore, any algorithm of resolution seems improbable and ineffective; the fact is that in the absolute no study model is yet able to entirely define the general principles of the protein folding.

2.1   The Levinthal Paradox

The first observation of spontaneous and reversible folding in vitro was carried out by Anfinsen [3]. He deduced that the native structure of a protein corresponds to a conformation with a minimal free energy, at least under suitable environmental conditions. But if the protein folding is indeed under thermodynamic control, a judicious question is to know how a protein can find, in a reasonable time, its structure of lower energy among an astronomical number of possible conformations.

As example, a protein of 100 residues can adopt 2^{100} ($\approx 10^{30}$) distinct conformations when we suppose that only two possibilities are accessible to each residue. If the passage from a conformation to another is carried out in 10^{-13} s (which corresponds to time necessary for a rotation around a connection), this protein would need at least 10^{17} s, i.e. approximately three billion years, "to test" all possible conformations. The proteins however manage to find their native structures in a lapse of time which is about the millisecond at the second. The apparent incompatibility between these facts, raised initially by Levinthal [4], was quickly set up in paradox and made run enormously ink since.

Levinthal gives the solution of its paradox: proteins do not explore the integrality of their conformational space, and their folding needs to be "guided", for example, via the fast formation of certain interactions which would be determining for the continuation of the process.

2.2   Motivations

To be able to understand how a protein accomplishes its biological function, and to be able to act on the cellular processes in which the protein intervenes, it is essential to know its structure. Many protein native structures were determined experimentally – primarily by crystallography with X-rays or by Nuclear Magnetic Resonance (NMR) – and indexed in a database accessible to all, Protein Data Bank (PDB) [5].

However, the application of these experimental techniques consumes a considerable time [6, 7]. Indeed, the number of protein sequences known [8] is much more important than the number of solved structures [5], this gap continues to grow quickly.

The design of methods making it possible to predict the protein structure from its sequence is a problem whose stakes are major, and which fascine many of scientists for several decades. Various tracks were followed with an aim of solving this problem, elementary in theory but extremely complex in practice.

3   Approaches to Study the Protein Folding Problem

The existing models for the protein folding problem study depend, amongst other things, on the way that the native structure is supposed be reached. Either, a protein folds following a preferential folding path [9], or a protein folds by searching the native state among an energetic landscape organized as a funnel.

The first hypothesis implies the existence of preferential paths for the folding process. In the simplest case, the folding mechanism is comparable to a linear reaction. Thus, when the steps are enough specifics, only a local region of the conformational space will be explored. This concept is nowadays obsolete since we know the existence of parallel folding paths.

The second hypothesis defines the folding by the following way (Dill, 1997):

> a parallel flow process of an ensemble of chain molecules; folding is seen as more like trickle of water down mountainsides of complex shapes, and less like flow through a single gallery.

In other words, the folding can be described as a set of transitions between structures whose energies become weaker. It allows guiding the protein by a funnel effect toward the conformational state whose energy level is the minimum that is the native conformation. Then, the polypeptide chain explores only a fraction of the accessible states.

This last hypothesis is the one accepted in this chapter, the protein folding is a process by which a large number of conformations are accessible and which leads a sequence into its native structure with the lowest energy level (see Fig. 1).

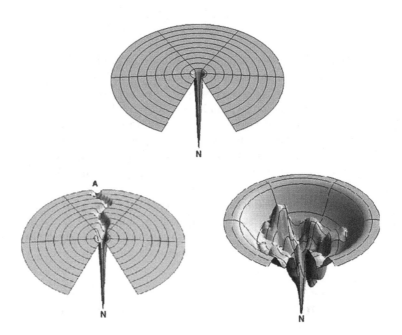

Fig. 1 Evolution in the description of folding paths in an energy landscape. Top, the protein folding according to the Anfinsen theory: a protein can adopt a large number of conformations. Bottom-left, a protein folds by following only one specific path in the conformational space. Bottom-right, from a denatured conformation, a protein searches its native structure whose energy level is minimum in a minimum time

3.1   Latest Approach

Many systems, both natural and artificial, can be represented by networks, that is, by sites or vertices bound by links. The study of these networks is interdisciplinary because they appear in scientific fields like physics, biology, computer science or information technology.

These studies are lead with the aim to explain how elements interact with each other inside the network and what the general laws which govern the observed network properties are.

From physics and computer science to biology and social sciences, researchers have found that a broad variety of systems can be represented as networks, and that there is much to be learned by studying these networks. Indeed, the studies of the Web [10], of social networks [11] or of metabolic networks [12] contribute to put in light common non-trivial properties of these networks which have a priori nothing in common. The ambition is to understand how the large networks are structured, how they evolve and what are the phenomena acting on their constitution and formation.

In [13], the authors propose to consider a protein as an interaction network whose vertices represent the amino acids and an edge describes a specific type of interaction (which is not the same according to the object of study). Thus, a protein, molecule composed of atoms becomes a set constituted by individuals (the amino acids), by interactions (to be defined according to the study) which evolves in a particular environment (describing the experimental conditions).

The vocabulary evolves but the aim remains the same, we want to better understand the protein folding process by the way of the modeling. The interaction network of a protein is initially the one built from the primary structure. The goal is to predict the graph of the tertiary structure through a discrete simulation process.

3.2   The Amino Acid Interaction Network

The 3D structure of a protein is represented by the coordinates of its atoms. This information is available in Protein Data Bank (PDB), which regroups all experimentally solved protein structures. Using the coordinates of two atoms, one can compute the distance between them. We define the distance between two amino acids as the distance between their C_α atoms. Considering the C_α atom as a "center" of the amino acid is an approximation, but it works well enough for our purposes. Let us denote by N the number of amino acids in the protein. A contact map matrix is a $N \times N$ 0-1 matrix, whose element (i, j) is one if there is a contact between amino acids i and j and zero otherwise. It provides useful information about the protein. For example, the secondary structure elements can be identified using this matrix. Indeed, α-helices spread along the main diagonal, while β-sheets appear as bands parallel or perpendicular to the main diagonal [14]. There are different ways to define the contact between two amino acids. Our notion is based on spatial proximity, so that the contact map can consider non-covalent interactions. We say that two amino acids are in contact if and only if the distance between them is below a given threshold. A commonly used threshold is 7 Å [1] and this is the value we use.

Consider a graph with N vertices (each vertex corresponds to an amino acid) and the contact map matrix as incidence matrix. It is called contact map graph. The contact map graph is an abstract description of the protein structure taking into account only the interactions between the amino acids.

First, we consider the graph induced by the entire set of amino acids participating in folded proteins. We call this graph the three dimensional structure elements interaction network (3DSE-IN), see Fig. 2.

As well, we consider the subgraph induced by the set of amino acids participating in SSE. We call this graph SSE interaction network (SSE-IN) (see Fig. 2).

In [15] the authors rely on amino acid interaction networks (more precisely they use SSE-IN) to study some of their properties, in particular concerning the role played by certain nodes or comparing the graph to general interaction networks models. Thus, thanks to this point of view the Protein Folding Problem can be tackle by the graph theory.

To manipulate a SSE-IN or a 3DSE-IN, we need a PDB file which is transformed by a parser we have developed. This parser generates a new file which is read by the GraphStream library [16] to display the SSE-IN in two or three dimensions.

4   Folding a Protein in a Topological Space by Bio-Inspired Methods

In this section, we treat proteins as amino acid interaction networks (see Fig. 2). We describe a bio-inspired method we use to fold amino acid interaction networks. In particular, we want to fold a SSE-IN to predict the motifs which describe the secondary structure.

4.1   Genetic Algorithms

The concept of genetic algorithms has been proposed by John Holland [17] to describe adaptive systems according to biological process.

The genetic algorithms are inspired from the concept of natural selection proposed by Charles Darwin. The vocabulary employed here is the one relative to the evolution theory and the genetic. We speak about individuals (potential solutions), populations, genes (which are the variables), chromosomes, parents, descendants, reproductions, etc.

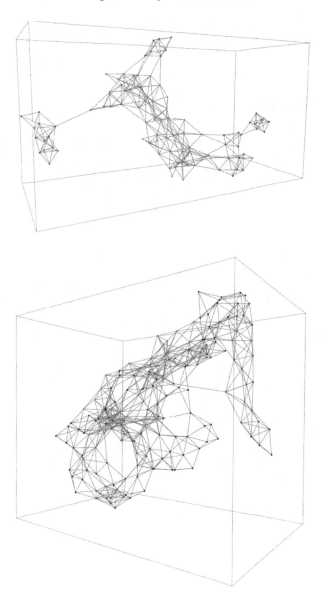

Fig. 2 Protein 1DTP SSE-IN (*top*) and the 1DTP 3DSE-IN (*bottom*). From a pdb file a parser we have developed produces a new file which corresponds to the SSE-IN graph displayed by the GraphStream library [16]

At the beginning, we conserve a population among which it exists a solution which is not yet optimal. Then, the genetic algorithm make evolves this population by an iterative process. Certain individuals reproduce themselves, others mute or disappear and only the well adapted individuals are supposed to survive. The genetic heritage between generations must help to produce individuals which are better and better adapted to correspond to the optimal solution.

4.2 Motif Prediction

In previous works [18], we have studied the protein SSE-IN. We have identified notably some of their properties like the degree distribution or also the way in which the amino acids interact. These works have allowed us to determine criteria discriminating the different structural families. We have established a parallel between structural families and topological metrics describing the protein SSE-IN.

Using these results, we have proposed a method to deduce the family of an unclassified protein based on the topological properties of its SSE-IN, see [19]. Thus, we consider a protein defined by its sequence in which the amino acids participating in the secondary structure are known. Then, we apply a method able to associate a family from which we rely to predict the fold shape of the protein. This work consists in associating the family which is the most compatible to the unknown sequence. The following step is to fold the unknown sequence SSE-IN relying on the family topological properties.

To fold a SSE-IN, we rely on the Levinthal hypothesis also called the kinetic hypothesis. Thus, the folding process is oriented and the proteins don't explore their entire conformational space. In this paper, we use the same approach: to fold a SSE-IN we limit the topological space by associating a structural family to a sequence [19]. Since the structural motifs which describe a structural family are limited, we propose a genetic algorithm (GA) to enumerate all possibilities.

In this section, we present a method based on a GA to predict the graph whose vertices represent the SSE and edges represent spatial interactions between two amino acids involved in two different SSE, further this graph is called Secondary Structure Interaction Network (SS-IN) (see Fig. 3).

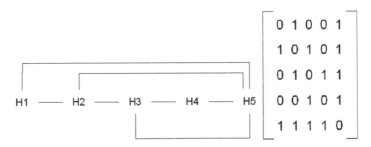

Fig. 3 2OUF SS-IN (*left*) and its associated incidence matrix (*right*). The vertices represent the different α-helices and an edge exists when two amino acids interact

4.3   Dataset

Thereafter, we use a dataset composed by proteins which have not fold families in the SCOP v1.73 classification and for which we have associated a family in [19].

4.4   Overall Description

The GA has to predict the adjacency matrix of an unknown sequence when it is represented by a chromosome. Then, the initial population is composed of proteins of the associated family with the same number of SSEs. During the genetic process, genetic operators are applied to create new individuals with new adjacency matrices. We want to predict the studied protein adjacency matrix when only its chromosome is known.

Here, we represent a protein by an array of alleles. Each allele represents a SSE notably considering its size that is the number of amino acids which compose it. The size is normalized contributing to produce genomes whose alleles describe a value between 0 and 100. Obviously, the position of an allele corresponds to the SSE position it represents in the sequence. In the same time, for each genome we associate its SS-IN incidence matrix.

The fitness function we use to evaluate the performance of a chromosome is the L_1 distance between this chromosome and the target sequence.

4.5   Genetic Operators

Our GA uses the common genetic operators and also a specific topological operator.

The crossover operator uses two parents to produce two children. It produces two new chromosomes and matrices. After generating two random cut positions, (one applied on chromosomes and another on matrices), we swap respectively the both chromosome parts and the both matrices parts. This operator can produce incidence matrices which are not compatible with the structural family, the topological operator solve this problem.

The mutation operator is used for a small fraction (about 1%) of the generated children. It modifies the chromosome and the associated matrix. For the chromosomes, we define two operators: the two position swapping and the one position mutation. Concerning the associated matrix, we define four operators: the row translation, the column translation, the two position swapping and the one position mutation.

These common operators may produce matrices which describe incoherent SS-IN compared to the associated sequence fold family. To eliminate the wrong cases we develop a topological operator.

The topological operator is used to exclude the incompatible children generated by our GA. The principle is the following; we have deduced a fold family for the sequence from which we extract an initial population of chromosomes. Thus, we compute the diameter, the characteristic path length and the mean degree to evaluate the average topological properties of the family for the particular SSE number. Then, after the GA generates a new individual by crossover or mutation, we compare the associated SS-IN matrix with the properties of the initial population by admitting an error rate up to 20%. If the new individual is not compatible, it is rejected.

4.6   Algorithm

Starting from an initial population of chromosomes from the associated family, the population evolves according to the genetic operators. When the global population fitness cannot increase between two generations, the process is stopped, see Algorithm 1.

Algorithm 1: Genetic algorithm for SS-IN adjacency matrix determination.

Data:
pop: Current chromosome population
parents: Set of parents
children: Set of children

begin
    pop ⟵ setInitialPopulation()
    while *fitness(pop) is increasing* **do**
        parents ⟵ parentExtraction(pop)
        children ⟵ parentCrossing(parents)
        children ⟵ childrenMutation(children)
        children ⟵ exclusionByTopology(children)
        pop ⟵ selection(pop, children)
end

The genetic process is the following: after the initial population is built, we extract a fraction of parents according to their fitness and we reproduce them to produce children. Then, we select the new generation by including the chromosomes which are not among the parents plus a fraction of parents plus a fraction of children. It remains to compute the new generation fitness.

When the algorithm stops, the final population is composed of individuals close to the target protein in terms of SSE length distribution because of the choice of our fitness function. As a side effect, their associated matrices are supposed to be close to the adjacency matrix of the studied protein that we want to predict.

In order to test the performance of our GA, we pick randomly three chromosomes from the final population and we compare their associated matrices to the sequence SS-IN adjacency matrix. To evaluate the difference between two matrices, we use an error rate defined as the number of wrong elements divided by the size of the matrix. The dataset we use is composed of 698 proteins belonging to the *All alpha* class and 413 proteins belonging to the *All beta* class. A structural family has been associated to this dataset in [19].

The average error rate for the *All alpha* class is 16.7% and for the *All beta* class it is 14.3%. The maximum error rate is 25%. As shown in Fig. 4, the error rate strongly depends on the initial population size. Indeed, when the initial population contains sufficient number of individuals, the genetic diversity ensures better SS-IN prediction. When we have sufficient number of sample proteins from the associated family, we expect more reliable results. Note for example that when

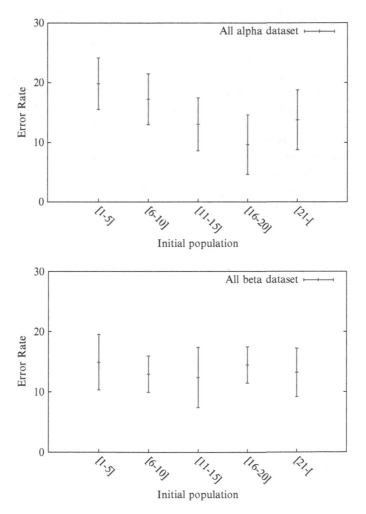

Fig. 4 Error rate as a function of the initial population size. When the initial population size is more than 10, the error rate becomes less than 15%

the initial population contains at least ten individuals, the error rate is always less than 15%.

5 Conclusions

In this paper, we present a simulation model to study the protein folding problem. We describe the reasons for which this biological process is a problem not yet solved.

We summarize relative works about how to fold an amino acid interaction networks. We need to limit the topological space so that the folding predictions become more accurate. We propose a genetic algorithm trying to construct the interaction network of SSEs (SS-IN). The GA starts with a population of real proteins from the predicted family. To complete the standard crossover and mutation operators, we introduce a topological operator which excludes the individuals incompatible with the fold family. The GA produces SS-IN with maximum error rate about 25% in the general case. The performance depends on the number of available sample proteins from the predicted family, when this number is greater than 10; the error rate is below 15%.

The characterization we propose constitutes a new approach to the protein folding problem. Here we propose to fold a protein SSE-IN relying on topological properties. We use these properties to guide a folding simulation in the topological pathway from unfolded to folded state.

References

1. A.R. Atilgan, P. Akan, C. Baysal. Small-world communication of residues and significance for protein dynamics. Biophys. J. **86**(1 Pt 1), 85–91 (2004)
2. K.A. Dill, S. Bromberg, K.Z. Yue, K.M. Fiebig, D.P. Yee, P.D. Thomas, H.S. Chan, Principles of protein folding: a perspective from simple exact models. Protein Sci. **4**(4), 561–602 (1995)
3. C.B. Anfinsen, Principles that govern the folding of protein chains. Science **181**, 223–230 (1973)
4. C. Levinthal. Are there pathways for protein folding? J. Chim. Phys. **65**, 44–45 (1968)
5. H.M. Berman, J. Westbrook, Z. Feng, G. Gilliland, T.N. Bhat, H. Weissig, P.E. Bourne, I.N. Shindyalov, The protein data bank. Nucleic Acids Res. **28**, 235–242 (2000)
6. M. Kataoka, Y. Goto, X-ray solution scattering studies of protein folding. Folding Des. **1**, 107–114 (1996)
7. K.W. Plaxco, C.M. Dobson, Time-relaxed biophysical methods in the study of protein folding. Curr. Opin. Struct. Biol. **6**, 630–636 (1996)
8. A. Bairoch, R. Apweiler, The swiss-prot protein sequence database and its supplement trembl. Nucleic Acids Res. **28**, 45–48 (2000)
9. R.L. Baldwin, Why is protein folding so fast? Proc. Natl. Acad. Sci. USA **93**, 2627–2628 (1996)
10. A. Broder, R. Kumar, F. Maghoul, P. Raghavan, S. Rajagopalan, R. Stata, A. Tomkins, J. Wiener. Graph structure in the Web. Comput. Netw. **33**(1–6), 309–320, (2000)

11. S. Wasserman, K. Faust, Social network analysis: methods and applications. *Structural Analysis in the Social Sciences*, vol. 8 (Cambridge University Press, Cambridge, 1994)
12. H. Jeong, B. Tombor, R. Albert, Z.N. Oltvai, A.-L. Barabàsi, The large-scale organization of metabolic networks. Nature **407**(6804), 651–654 (2000)
13. N.V. Dokholyan, L. Li, F. Ding, E.I. Shakhnovich, Topological determinants of protein folding. Proc. Natl. Acad. Sci. USA **99**(13), 8637–8641 (2002)
14. A. Ghosh, K.V. Brinda, S. Vishveshwara, Dynamics of lysozyme structure network: probing the process of unfolding. Biophys. J. **92**(7), 2523–2535, (2007)
15. U.K. Muppirala, Z. Li, A simple approach for protein structure discrimination based on the network pattern of conserved hydrophobic residues. Protein Eng. Des. Sel **19**(6), 265–275 (2006)
16. A. Dutot, F. Guinand, D. Olivier, Y. Pigné, GraphStream: A Tool for bridging the gap between Complex Systems and Dynamic Graphs, Proc of EPNACS: Emergent Properties in Natural and Artificial Complex Systems, Dresden, Germany, 137–143 (2007)
17. J.H. Holland, *Adaptation in Natural and Artificial System* (MIT Press, Cambridge, MA, 1992)
18. O. Gaci, Building a parallel between structural and topological properties. In Advances in Computational Biology (Springer, 2010)
19. O. Gaci, Building a topological inference exploiting qualitative criteria. Evol. Boinformatics (2010)

Chapter 46
Analysing Multiobjective Fitness Function with Finite State Automata

Nada M.A. Al Salami

Abstract This research analyses and discusses the use of Multiobjective fitness function to evolve Finite State Automata. Such automata can describe system's behavior mathematically in an efficient manner. However system's behavior must highly depend on its input-output specifications. Genetic Programming is used, and the fitness function is built to guide the evolutionary process in two different cases. First case: Single point fitness function is used where the only focus is on the correctness of the evolved automata. Second case: multiobjective fitness function is used since every real-world problem involves simultaneous optimization of several incommensurable and often competing objectives. Multiobjective optimization is defined as a problem of finding a Finite State Automata which satisfies: parsimony, efficiency, and correctness. It has been presented that for large and complex problems it is necessary to divide them into sub problem(s) and simultaneously breed both sub-program(s) and a calling program.

1 Introduction

The evolutionary multi-objective optimization (EMOO) is a popular and useful field of research and developing algorithms to solve many real-life multiobjective problems. Almost every real-world problem involves simultaneous optimization of

Nada M.A. Al Salami (1971) Assistance Professor in Management Information System Department, of Al Zaytoonah University, Amman, Jordan. She interested in the theory of computer science and evolutionary algorithms

N.M.A.A. Salami
Department of Management Information Systems, Faculty of Economic and Business, Al Zaytoonah University of Jordan, 130, Amman, 11733, Jordan
e-mail: nada.alsalami@yahoo.com

S.-I. Ao et al. (eds.), *Machine Learning and Systems Engineering*,
Lecture Notes in Electrical Engineering 68,
DOI 10.1007/978-90-481-9419-3_46, © Springer Science+Business Media B.V. 2010

several incommensurable and often competing objectives such as performance and cost. As a consequence, there is usually not a single optimum but rather a set of optimal trade-offs, which in turn contains the single-objective optimal. This makes clear that the optimization of multiple objectives adds a further level of complexity compared to the single-objective case. Multiopjective optimization defined as a problem of finding a vector of decision variables which satisfies constraints optimizes a vector function whose elements represent the objective functions. These function forms a mathematical description of performance criteria which are usually in conflict with each other. Consider a minimization problem; it tends to find a set of E for: $\mathbf{Min}_{E \in \emptyset} \, \mathbf{F}\,(\mathbf{E}), \mathbf{E} \subseteq \mathbf{R^n}$, where, $E = \{E_1, E_2, E_3, \ldots, E_n\}$ is an n-dimension vector having n decision variables or parameters and $\emptyset$ defines a feasible set of E. $F = \{f1, f_2, f_3, \ldots, f_m\}$ is an objective vector with m objective components to be minimized, which may be competing or noncommensurable to each other [1].

Evolutionary Computing is a research area within Computer Science, which draws inspiration from the process of natural evolution, application of evolutionary computing may includes: Bioinformatics, Numerical combinatorial Optimization, System Modeling and Identifications, Planning and Control, Engineering Design, Data Mining, Machine Learning, and Artificial Life. Evolutionary computing is needed for Developing automated problem solvers, where the most powerful natural problem solvers are human Brain, and evolutionary process. Designing the problem solvers based on human brain leads to the field of "neurocomputing". While the second one leads to evolutionary computing. The algorithms involved in Evolutionary computing are termed as Evolutionary algorithms (EA). Evolutionary algorithms (EAs) such as evolution strategies and genetic algorithms have become the method of choice for optimization problems that are too complex to be solved using deterministic techniques such as linear programming or gradient (Jacobian) methods. EAs require little knowledge about the problem being solved, and they are easy to implement, robust, and inherently parallel. To solve a certain optimization problem, it is enough to require that one is able to evaluate the objective (cost) function for a given set of input parameters. Because of their universality, ease of implementation, and fitness for parallel computing, EAs often take less time to find the optimal solution than gradient methods. However, most real-world problems involve simultaneous optimization of several often mutually concurrent objectives. Multi objective EAs are able to find optimal trade-offs in order to get a set of solutions that are optimal in an overall sense. In multi objective optimization, gradient based methods are often impossible to apply. Multiobjective EAs, however, can always be applied, and they inherit all of the favorable properties from their single objective relatives. As soon as there are many (possibly conflicting) objectives to be optimized simultaneously, there is no longer a single optimal solution but rather a whole set of possible solutions of equivalent quality. Consider, for example, the design of an automobile. Possible objectives could be: minimize cost, maximize speed, minimize fuel consumption and maximize luxury. These goals are clearly conflicting and, therefore, there is no single optimum to be found. Multiobjective EAs can yield a whole set of potential solutions – which are all optimal in some sense – and give the engineers the option to assess the trade-offs between different designs. One then could, for example, choose to create three different cars according to different marketing needs: a slow low-cost model which consumes least fuel, an intermediate solution, and a luxury sports car where speed is clearly the prime objective. Evolutionary algorithms are well suited to multi objective optimization problems as they are fundamentally based on biological processes which are inherently multiobjective [1–3]. After the first pioneering work on multi objective evolutionary optimization in the 1980s by Schatner [4, 5], several different algorithms have been proposed and successfully applied to various problems, for more details see [6, 7]. In this research, a new multiobjective evolutionary algorithm is analyzed and discussed; it is developed to deal with complex system. Since every real life problem are dynamic problem, thus their behaviors are much complex. Complex systems often include chaotic behavior [8] (the classic example of chaos theory is "the Butterfly effect") which is to say that the dynamics of these systems are nonlinear and difficult to predict over time, even while the systems themselves are deterministic machines following a strict sequence of cause and effect. Genetic programming [9–11] suffers from many serious problems that make it very difficult to predict chaotic behavior. Natural chaotic systems may be difficult to predict but they will still exhibit structure that is different than purely random systems [12] for more details see [13–15]. Such chaotic behavior of real live problem is reduced when the induction process is based on the meaning of the low-level primitives rather than their structure.

The meaning of a program P is specified by a set of function transformation from states to states, as given in [12, 16]; hence P effects a transformation on a state vector X, which consists of an association of the variable manipulated by the program and their values. A Program **P** is defined as 9-tuples, called Semantic Finite State Automata (**SFSA**): $\mathbf{P} = (\mathbf{x}, \mathbf{X}, \mathbf{T}, \mathbf{F}, \mathbf{Z}, \mathbf{I}, \mathbf{O}, \boldsymbol{\gamma}, \mathbf{X_{initial}})$, where: x is the set of system variables, X is the set of system states, $X = \{X_{initial}, - - - -, X_{final}\}$, T is the time scale: $T = [0, \infty)$, F is the set of primitive functions, Z is the state transition function, $Z = \{(f, X, t): (f, X, t) \in F \times X \times T, z(f, X, t) = (\bullet X, \bullet t)\}$, I is the set of inputs, O is the set of outputs, γ is the readout function, and $X_{initial}$ is the initial state of the system: $X_{initial} \in X$.

All sets involved in the definition of P are arbitrary, except T, and F. Time scale T must be some subset of the set $[0, \infty)$ of nonnegative integer numbers, while the set of primitive function **F** must be a subset of the set $C_L(F_L)$ of all computable functions in the language **L** and sufficient to generate the remainder functions. Two features characterize state transition function:

$$Z(-, -, t) = (X_{initial}, 1) \text{ if } t = 0 \tag{1}$$

$$Z(f, X, t) = Z(f, Z(f(t-1), X.t - 1)) \text{ if } t \neq 0 \tag{2}$$

The concepts of reusable parameterized sub-systems is implemented by restricting the transition functions of the main system, so that it have the ability to call and pass parameters to one or more such sub-systems. Suppose we have sub-system •P, and main-system P, then they are defined by the following 9-tuples:

$P (x, X, T, F, Z, 1, 0, X_{initial}, \gamma)$

${}^\bullet P ({}^\bullet x {}^\bullet X {}^\bullet T {}^\bullet F {}^\bullet Z {}^\bullet I, {}^\bullet O, {}^\bullet X_{initial}, {}^\bullet \gamma)$

where

${}^\bullet x \subseteq x$, ${}^\bullet X_{initial} \in X$, then there exit ${}^* f \in F$, $z \in Z^\bullet f$, $\in F$, and ${}^\bullet z \in {}^\bullet Z$, and h is a function defined over ${}^\bullet Z$ with value in ${}^\bullet X$ is defined as follows:

$$h = {}^\bullet z \left({}^* f, {}^\bullet X_{initial}, 1 \right) = X_{h,t_i} \tag{3}$$

$$z({}^* f, X, t) = z(h, X, t) = X_{h,t} \tag{4}$$

$*f$ is a special function we call it sub-SFSA function to distinguish it from other primitive functions in the set F. Also, we call the sub-system ${}^\bullet P$, sub-SFSA, to distinguish it from the main SFSA. Formally, a system ${}^\bullet P$ is a sub-system of a system P, iff: ${}^\bullet x \subseteq x$, ${}^\bullet T \subseteq T$, ${}^\bullet I \subseteq I$, ${}^\bullet O \subseteq O$, ${}^\bullet \gamma$ must be the restriction of γ to ${}^\bullet O$, and ${}^\bullet F \subseteq N$, where N is the set of restrictions of F to ${}^\bullet T$. If $({}^\bullet f, {}^\bullet X, {}^\bullet t)$ is an element of ${}^\bullet F \times {}^\bullet X \times {}^\bullet T$, then there exists $f \in F$, such that the restriction of f to ${}^\bullet T$ is ${}^\bullet f$, and ${}^\bullet z ({}^\bullet f, {}^\bullet X, {}^\bullet t)$ is $z (f, X, t)$.

2 Evolutionary Algorithm

On the basis of the theoretical approach sketch in the above section, the evolutionary algorithm is defined as 7-tuples: $(IOS, S, F, a_1, T_{max}, \beta, \upsilon)$, as given in [12, 16].

2.1 Input-Output Specification (IOS)

IOS is establishing the input-output boundaries of the system. It describes the inputs that the system is designed to handle and the outputs that the system is designed to produce. An IOS is a 6-tuples: $IOS = (T, I, O, Ti, TO, \eta)$. Where T, is the time scale of IOS, I is the set of inputs, O is a set of outputs, Ti is a set of input trajectories defined over T, with values in I, TO, is a set of output trajectories defined over T, with values in O, and η is a function defined over Ti whose values are subset of TO; that is, η matches with each given input trajectories the set of all output trajectories that might, or could be, or eligible to be produced by some systems as output, experiencing the given input trajectory. A system P satisfies IOS if there is a state X of P, and some subset U not empty of the time scale T of P, such that for every input trajectory g in Ti, there is an output trajectory h in TO matched with g by η such that the output trajectory generated by P, started in the state X is:

$$\gamma(Z(f(g), X, t) = \eta(h(t)), \text{ for every } t \in U \tag{5}$$

2.2   Syntax Term (S)

Refers to the written form of a program as far as possible independent from its meaning. In particular it concerns the legality or well-formedness of a program relative to a set of grammatical rules, and parsing algorithms for the discovery of the grammatical structure of such well-formed programs. S is a set of rules governing the construction of allowed or legal system forms.

2.3   Primitive Function (F)

Each function must be coupled with its effect on both the state vector X, and the time scale T of the system. Some primitive functions may serve as primitive building blocks for more complex functions or even sub-systems.

2.4   Learning Parameter (α_1)

Is a positive real number specifying the minimum accepted degree of matching between an IOS, and the real observed behavior of the system over the time scale, T_x, of IOS.

2.5   Complexity Parameters (T_{max}, β)

T_{max} and β parameters are merits of system complexity: size and time, respectively. It is important to note that there is a fundamental difference between a time scale T and an execution time of a system. T represents system size, it defines points within the overall system, whereas, β, is the time required by the machine to complete system execution, hence it is high sensitive to the machine type.

2.6   System Proof Plan (v)

Prove process should be a part of the statement of system induction problem especially when the IOS is imprecise or inadequate to generate an accurate system. We say P is correct iff it computes a certain function f from $X_{initial} \in X$ properly, that is if for each $X_i \in X$, $P(X_i)$ is defined, i.e. P does not loop for X_i, and is equal to $f(X_i)$. Broadly speaking, there have been two main approaches to the problem of developing methods for making programs more reliable: Systematized testing, and Mathematical proof.

Our works use systematized testing approach as a proof plane. The usual method for verifying that a program is correct by testing is by choosing a finite sample of states X_1, X_2, ..., X_n and running P on each of them to verify that: $P(X_1) = f(X_1)$, $P(X_2) = f(X_2)$, ..., $P(X_n) = f(X_n)$. Formally, if testing approach is used for system verification, a system proof is denoted $v = (\alpha 2, d)$, where $\alpha 2$ is a positive real parameter defining the maximum accepted error from testing process. $\alpha 2$ focus on the degree of generality, so that $\alpha 1$, and $\alpha 2$, parameters suggest a fundamental tradeoff between training and generality. On the other hand, d represents a set of test cases pairs (O_i, K_i), where K_i is a sequence of initial state $X_{initial}$ and input I_i.

In addition to using the idea of sub-system functions, i.e., sub-FSA, for complex software it is better to divide the process of system induction into sub-system(s) and main-system inductions. Sub-system induction must be accomplished with several objectives; first one is that a suitable solution to sub-problem must determine a solution to the next higher level problem. Second is to ensure that the figure of merit of the sub-system have relationships to the figure of merit of the top-level problem. The third objective is to ensure specific functional relationships of sub-system proof plans to the system proof plan of the top-level problems.

3   Evolutionary Process

3.1   Single Objective Evolutionary Process

The basic idea for single objective EAs is to imitate the natural process of biological evolution. The problem to be solved is therefore described using a SFSA with certain number of *parameters* (design variables). One then creates a group of $(n > 0)$ different parameter vectors and considers it as a *population of individuals*. The quantity is called the *population size*. The quality of a certain SFSA (i.e. an *individual* in the population) is expressed in terms of a scalar valued *fitness function* (objective function). Depending on whether one wants to minimize or maximize the objective function, individuals (i.e. parameter vectors) with lower or greater fitness are considered better, respectively. The algorithm then proceeds to choose the best individuals out of the population to become the parents of the next generation (natural *selection*, survival of the fittest). In this work we always try to minimize the objective function, i.e. search to find individual with lower fitness value. In this case, learning parameter $\alpha 1$ must be selected to the minimum accepted degree of matching between an IOS, and the real observed behavior of the SFSA over the time scale, T_X. Learning parameter $\alpha 1$ is used ultimately to guide the search process.

Within the context of the suggested mathematical approaches, to automatically generate a system means search to find an appropriate SFSA satisfying IOS efficiently, with regard to learning and complexity parameters. Then, proof plan υ must be applied to that SFSA for further assessing its performance and correctness. If that SFSA behaves well with υ, it may be considered as a solution or approximate solution to the problem, else, some or all terms in the statement of the problem of system induction must be modified, and the process is repeated until a good SFSA is found, or no further revisions can be made. The search space in Evolutionary algorithm is the space of all possible computer programs described as 9-tuples SFSA. Within the context of the suggested mathematical approaches, to automatically generate a system means search to find an appropriate SFSA satisfying IOS efficiently, with regard to learning and complexity parameters. Then, proof plan υ must be applied to that SFSA for further assessing its performance and correctness. If that SFSA behaves well with υ, it may be considered as a solution or approximate solution to the problem, else, some or all terms in the statement of the problem of system induction must be modified, and the process is repeated until a good SFSA is found, or no further revisions can be made. The search space in Evolutionary algorithm is the space of all possible computer programs described as 9-tuples SFSA. The following equation was used to express the quality of a SFSA (i), where $0 < i <= n$:

$$'fitness\,(i) = \left(\alpha - \sum_{j=0}^{Tx} |\eta T(j) - \eta R(j)| \right) \tag{6}$$

Therefore, to generate a new population individuals with smaller fitness value are chosen and regenerate them to the next population. In addition to reproduction genetic operation, mutation and crossover genetic operations are also used to generate new individuals from the old individuals with smaller fitness value. When the first iteration is completed, then the algorithm loops back to fitness evaluation phase.

3.2   Multi Objective Evolutionary Process

Two major problems must be addressed when an evolutionary algorithm is applied to multiobjective optimization. First: how to accomplish fitness assignment and selection, respectively, in order to guide the search towards the Pareto-optimal set. Second: how to maintain a diverse population in order to prevent premature convergence and achieve a well distributed trade-off front. Now, the objective function should be satisfying the following characteristic [1, 2]:

Complete: so that all pertinent aspects of the decision problem are presented
Operational: in that they can be used in a meaningful manner
Decomposable: if desegregation of objective functions is required
Nonredundant: so that no aspect of the decision problem is considered twice
Minimal: such that there is no other set of objective functions capable of representing the problem with a smaller number of elements

Since there are many objectives to be optimized simultaneously, there is no longer a single optimal solution but rather a whole set of possible solutions of equivalent quality. Multi objective EAs can yield a whole set of potential solutions which are all optimal in some sense, and give the engineers the option to assess the trade-offs between different designs. One then could, for example, choose to create three different SFSA according to different needs. Evolutionary algorithms are well suited to multi objective optimization problems as they are fundamentally based on biological processes which are inherently multi objective. Multiobjective fitness measure is adopted to incorporate a combination of three objectives that are:

Correctness: how long the SFSA satisfy IOS as described above.
Parsimony: the length of the evolved SFSA must be smaller as possible i.e. smallness T. It must contain nonredundant or useless transformations.
Efficiency: although the time required by any machine to complete system execution is high depended on machine type, however the evolved SFSA must has smallness β, as possible. Otherwise the complexity of that SFSA is affected.

Accordingly, multiobjective fitness is computed by the following equation:

$$'\text{fitness}(i) = \delta 1 \left(\alpha - \sum_{j=0}^{Tx} |\eta(T(j) - \eta(R(j))| \right) + \delta 2(Tmax - Ti) + \delta 3$$
$$\times (\beta - \beta i) \tag{7}$$

Where: $\delta 1$, $\delta 2$, and $\delta 3$ are weight parameters: $(\delta 2, \delta 3) \langle \delta 1 \rangle = 0$, βi is the run time of individual i, Ti is the time scale of the individual i, and Rj is the actual calculated input trajectory of individual j. Since the Main objective is to satisfy the IOS, the first weight parameter in Eq. (7), i.e. $\delta 1$ is always greater than $\delta 2$ and $\delta 3$, and grater than or equal to 1. These values are selected by expert when the algorithm is running many times. In addition the values of $\delta 2$ and $\delta 3$ are selected with respect to different factors concerning: hardware type, required complexity, problem nature, and other. In all cases they must be small real numbers greater than or equal to zero. Table 1, gives different setting for these weight parameters, when the value is equal to 0 it means this objective is not considered in the evolutionary process. Also if it equal to 1 it means the effect of this objective is not either decreased or decreased. It is clear that the first setting for the parameters represent single point fitness function, as

Table 1 Weight parameters with different setting

Objective type	Objective degree	$\delta 1$	$\delta 2$	$\delta 3$
IOS Correctness	High	1	0	0
Parsimony	High	5	2	0
Efficiency	High	5	0	2
Parsimony	Moderate	5	2	1
Efficiency	Moderate	5	1	2
Parsimony	Low	4	0.5	1
Efficiency	Low	6	2	0.2

given by Eq. (6). In other words, single-objective optimization can be considered a special case of multiobjective optimization (and not vice versa).

To give rise to the fitness variation in the overall population from one generation to the next, the fitness of each individual is computed F:

$$''fitness\ (i) = \frac{'fitness\ (i)}{\sum_{J=0}^{M} 'fitness\ (j)} \tag{8}$$

Where: M: is the population size. Eq. (8) is also adjusted so that the adjusted fitness lies between 0 and 1:

$$fitness\ (i) = \frac{1}{''fitness\ (i)} \tag{9}$$

Three types of points are defined in each individual: transition, function, and function arguments. When structure-preserving crossover is performed, any point type anywhere in the first selected individuals may be chosen as the crossover point of the first parent. The crossover point of the second parent must be chosen only from among points of this type. The restriction in the choice of the second crossover points ensures the syntactic validity of the offspring.

4 Result and Discussion

4.1 Input-Output Specification

Unfortunately, when we deal with complex systems and real live problem, strong feedback (positive as well as negative) and many interactions exist: i.e. chaotic behavior. Thus, we need to find a way to control chaos, understand, and predict what may happen long term. In these cases input and output specifications are self organized, which mean that trajectory data are collected and enhanced over time, when genetic generation process runs again and again. With high trajectory information converge to the solution in best than these populations with little trajectory information. Although trajectory data are changed over time, but by experiment, it still sensitive to initial configuration (sensitivity to the initial conditions). There is a fundamental difference between a crossovers occurring in a sub-SFSA versus one occurring in the main-SFSA. Since the later usually contains multiple references to the sub-SFSA(s), a crossover occurring in the sub-SFSA is usually leveraged in the sense that it simultaneously affects the main-SFSA in several places. In contrast, a crossover occurring in the main-SFSA provides no such leverage. In addition, because the population is architecturally diverse, parents selected to participate in the crossover operation will usually possess different numbers of sub-SFSA(s). The proposed architecture-altering operations are: *Creating sub-SFSA, Deleting sub-SFSA, Adding Variables, and Deleting Variables.*

4.2   Performance

If every run was successes to obtain a solution, the computational effort required to get the solution depends primarily on four factors: population size M, number of generation g (g must be less than or equal to the maximum number of generation G), the amount of processing required for fitness measure over all fitness cases (we assume that the processing time to measure the fitness of an individual is its run time), and the amount of processing required for test phase e. If success occurs on the same generation of every run, then the computational effort E would be computed as follows:

$$E = M \cdot g \cdot e \tag{10}$$

Since the value of **e** is too small with respect to other factors, we shall not consider it. However, in most cases, success occurs on different generations in different runs, then the computational effort E would be computed as follows:

$$E = M \cdot \text{gavr} \cdot \beta \tag{11}$$

where gavr is the average number of executed generations. Since we use a probabilistic algorithm, thus the computational effort is computed in this way: first determining the number of independent runs R needed to yield a success with a certain probability, second, multiply R by the amount of processing required for each run. The number of independent runs R required to satisfy the success predicate by generation i with a probability z which depends on both z and P(M, i), where z is the probability of satisfying the success predicate by generation i at least once in R runs defined by:

$$Z = 1 - [1 - P(M, i)]^R \tag{12}$$

P(M, i) is the cumulative probability of success for all the generations between generation 0 and generation i. P(M, i) is computed after experimentally obtaining an estimate for the instantaneous probability Y(M, i) that a particular run with a population size M yields, for the first time, on a specified generation i, an individual is satisfying the success predicate for the problem. This experimental measurement of Y(M, i) usually requires a substantial number of runs. After taking logarithms, we find:

$$R = \left| \frac{\log(1 - z)}{\log(1 - \rho(M, i))} \right| \tag{13}$$

The computational effort E, is the minimal value of the total number of individuals that must be processed to yield a solution for the problem with z probability (ex: z = 99%):

$$E = M \cdot (^*g + 1) \cdot \beta \cdot R \tag{14}$$

Where *g is the first generation. The computational effort ratio, R_E, is the ratio of the computational effort without sub-SFSA to the computational effort with sub-SFSA:

$$R_E = \frac{E\ without\ sub - SFSA}{E\ with\ sub - SFSA} \qquad (15)$$

The fitness ratio, $R_{fitness}$, is the ratio of the average fitness without sub-SFSA to the average fitness, with sub-SFSA, for a problem.

$$R_{fitness} = \frac{fitness\ average\ without\ sub - SFSA}{fitness\ average\ with\ sub - SFSA} \qquad (16)$$

5 Conclusion

There are many objectives to be optimized simultaneously, thus there is no longer a single optimal solution but rather a whole set of possible solutions of equivalent quality. Multi objective EAs can yield a whole set of potential solutions which are all optimal in some sense, and give the engineers the option to assess the trade-offs between different designs. The discussed evolutionary algorithm can be changed to produce programs in different implementation languages without significantly affecting existing problem specification, leading to an increase in the system productivity. Multiobjective fitness measure is adopted to incorporate a combination of three objectives: correctness, parsimony, and efficiency. Trajectory data of the system is self-organization to reflect the chaotic behavior in real live applications. Convergences time is highly sensitive to the initial input-output specification of the problem. Programs with fewer sub-programs tend to disappear because they accrue fitness from generation to generation, more slowly than those programs with more sub-programs. Sub-systems can be reused to solve multiple problems. They provide rational way to reduce software cost and increase software quality. Although complexity parameter provide a way to bound the search space, their effect in the induction process is relatively less than the effect of learning and generalization parameters. System analyst must select a value for $\alpha 2$, so that missing or incorrect specifications of some input-output relationships don't lead to un-convergence situation.

References

1. K.C. Tan, E.F. Khor, T.H. Lee, *Muliobjective Evolutionary Algorithms and Applications* (Springer-Verlag, London limited, 2005
2. E. Zitzler, K. Deb, L. Thiele, Comparison of multiobjective evolutionary algorithm: empirical result. Evol. Comput. **8**(2), 173–195 (2000), (Massachustts Institute of Technology)

3. E. Zitzler, Evolutionary algorithm for multiobjective optimization, *Evolutionary Methods for Design, Optimization and Control* (CIMNE, Barcelona, Spain, 200)

4. J.D. Schaffner, Multiple Objective Optimization with Vector Evaluated Genetic Algorithms. Unpublished Ph.D. Thesis, Vanderbilt University, Nashville, TN, 1984

5. J.D. Schaffner, Multiple objective optimization with vector evaluated genetic algorithm, in *Proceeding of an International Conference on Genetic Algorithms and their Applications, sponsored by Texas Instruments and the U.S. Navy Center for Applied Research in Artifficl intelligence (NCARAI)* pp. 93–100, 1985

6. C.M. Fonseca, P.J. Fleming, Am overview of evolutionary algorithms in multiobjective optimization. Evol. Comput. **3**(1), 1–16 (1995)

7. C.A.C. Coello, A comprehensive survey of evolutionary-based multiobjective optimization. Knowl. Inf. Syst. **1**(3), 269–308, 1999

8. L. Smith, *Chaos: A very Short Introduction* (Oxford University Press, UK, 2007

9. J.R. Koza, *Genetic Programming: on the Programming of Computer by Means of Natural Selection* (MIT Press, Cambridge, MA, 2004)

10. D.E. Golberg, *Genetic Algorithm in Search, Optimization, and Machine Learning* (Addison-Wesley, Boston, MA, 1989)

11. M. Mitchell, *An Introduction to Genetic Algorithm* (MIT Press, Cambridge, MA, 1996)

12. N.A. Salami, Evolutionary algorithm definition. AJEAS **2**(4), 789–795 (2009)

13. J.P. Koza, Two ways of discovering the size and shape of a computer program to solve a problem, pp. 287–294,1998

14. A. Kent, J.G. Williams, C.M. Hall, Genetic programming. *Encyclopedia of Computer Science and Technology* (Marcel Dekker, New York, 1998), pp. 29–43

15. R. Poli, W.B. Langdon, N.F. McPhee, J.R. Koza, Genetic Programming: An Introductory Tutorial and a Survey of Techniques and pplications, Technical Report CES-475 ISSN: 1744-8050. essex.ac.uk/dces/research/publications/. . ./2007/ces475.pdf. Oct 2007

16. N.A. Salami, Genetic system generation, in *Proceeding of the WCECS 2009*, vol. I. (San Francisco USA, 20–22 Oct 2009), pp. 23–27 ISBN:978-988-17012-6-8

Index